FUNDAMENTALS OF
SOIL BEHAVIOR

SERIES IN SOIL ENGINEERING

Edited by

T. William Lambe
Robert V. Whitman
Professors of Civil Engineering
Massachusetts Institute of Technology

The aim of this series is to present the modern concepts of soil engineering, which is the science and technology of soils and their application to problems in civil engineering. The word "soil" is interpreted broadly to include all earth materials whose properties and behavior influence civil engineering construction.

Soil engineering is founded upon many basic disciplines: mechanics and dynamics; physical geology and engineering geology; clay mineralogy and colloidal chemistry; and mechanics of granular systems and fluid mechanics. Principles from these basic disciplines are backed by experimental evidence from laboratory and field investigations and from observations on actual structures. Judgment derived from experience and engineering economics are central to soil engineering.

The books in this series are intended primarily for use in university courses, at both the undergraduate and graduate levels. The editors also expect that all of the books will serve as valuable reference material for practicing engineers.

T. William Lambe and Robert V. Whitman

Fundamentals of Soil Behavior

James K. Mitchell

University of California, Berkeley

1976

John Wiley & Sons, Inc.

New York   London   Sydney   Toronto

Library of Congress Cataloging in Publication Data:

Mitchell, James Kenneth, 1930-
  Fundamentals of soil behavior.

  (Series in soil engineering)
  Bibliography: p.
  1. Soil mechanics. I. Title.

TA710.M577   624'.1513   75-28096
ISBN 0-471-61168-9

Printed in the United States of America

10 9 8 7 6 5 4 3

Board of Advisors, Engineering

PREFACE

In many cases a knowledge of *how* a soil behaves is sufficient for solution of a geotechnical problem, and most previous books on soil mechanics emphasize this aspect of behavior. There are cases, however, where an understanding of *why* certain soils behave as they do can be important if a proper solution is to be obtained. A basic understanding of the *why* of soil behavior serves the geotechnical engineer in much the same manner as conventional engineering courses in properties of materials or materials science serve the structural and mechanical engineer.

This book has as its purpose the development of an understanding of the factors determining and controlling the engineering properties of soils, with emphasis on the *why* aspect of soil behavior. To meet this objective it is necessary to examine the composition of soils in terms of their mineralogy and pore fluid and the interactions of these phases with each other and the surrounding environment. The particulate nature of soils is given special attention in the development of considerations of soil fabric and structure and then in connection with volume change properties, strength and deformation behavior, and conduction phenomena.

This book is an outgrowth of a graduate course started in 1959 and modified continuously since to reflect the steady increases in the state of knowledge. Although the book is aimed primarily at graduate students and researchers in geotechnical engineering, it contains material of interest to students of geology and soil science and also should be a useful reference for practicing engineers faced with unusual problems.

More material is included than can be conveniently covered in a one-quarter course, thus some selectivity in choice of topics may be in order depending on the specific emphases desired. A number of the topics can be supplemented effectively by means of laboratory demonstrations and exercises. These are particularly helpful in connection with analysis of soil composition, properties of the clay minerals, colloidal phenomena in clays, fabric–property relationships, and coupled flows.

I am indebted to a large number of people who helped directly and indirectly in the preparation of this book. Professor T. William Lambe provided the original stimulation that led to my specialization in the general subject area. Professor H. Bolton Seed was responsible for the expansion of soil mechanics studies at Berkeley into basic areas of soil behavior. His insights and helpful discussions have been invaluable over the years.

Many colleagues provided constructive critical reviews of different parts of the book: Professor Roy E. Olson who reviewed most of the manuscript; Professor Tor L. Brekke, Chapters 3 and 4; Professor Lawrence Waldron, Chapter 4; Dr. Issaac Barshad, Chapter 5; Dr. R. Torrence Martin, Chapters 5, 6, and 8; Professor Charles A. Moore, Chapters 7 and 10; Professor Norbert R. Morgenstern, Chapters 8, 11, and 12; Professors Richard G. Campanella, James M. Duncan, and Charles C. Ladd, Chapter 14; and Professor Donald H. Gray and Dr. Harold W. Olsen, Chapter 15. Dr. James L. Sherard reviewed sections dealing with dispersive clays. Their assistance is greatly appreciated.

Finally, I owe the deepest gratitude to my wife, not only for her patience, understanding, and encouragement during preparation of the work, but also for her services as typist, editor, and proofreader.

James K. Mitchell

Berkeley, California
June 1975

PART II   Soil Behavior

FUNDAMENTALS OF
SOIL BEHAVIOR

CHAPTER 1

Introduction

1.1 CIVIL ENGINEERING, GEOTECHNICAL ENGINEERING, AND SOIL BEHAVIOR, PERSPECTIVE

Civil engineering embraces the analysis, design, and construction of a diversity of structures and systems. All of these facilities, for example, buildings, dams, tunnels, highways, airfields, bridges, are built on, in, or with soil or rock. Thus, the behavior of the soil and rock at the location of any project and the interactions of the earth materials during and after construction of the facility have a major influence on the success, economy, and safety of the work.

To deal properly with earth materials in any case requires knowledge, understanding, and appreciation of geology, materials science and testing, and mechanics. The subdiscipline within civil engineering that is concerned with all of these factors is geotechnical engineering. Geotechnical engineering is the broadest of the civil engineering specialities because it relates to all other areas of civil engineering. The successful practice of geotechnical engineering requires integration of knowledge from several fields.

Students of civil engineering are often quite surprised (and sometimes confused) by their first course in soil mechanics. After a series of courses in statics and mechanics of materials, wherein problems are quite clear-cut and well-defined, each having its own unambiguous solution, they are suddenly confronted by a situation where this is no longer the case. A first course in soil mechanics may not, for half to two-thirds of the class time, be mechanics at all. The reason for this is simple. Analyses and designs in geotechnical engineering, which may in the end be done using mechanics and elaborate computational methods, are useless if the boundary conditions and material properties are improperly defined.

Acquisition of the data needed for analysis and design on, in, and with soils and rocks may be a far more difficult and uncertain problem than is the case when dealing with other materials of construction and structures. There are at least three important reasons for this.

1. *No clearly defined boundaries.* An embankment resting on a soil foundation is shown in Fig. 1.1*a*, and a cantilever beam fixed at one end is shown in Fig. 1.1*b*. The free body of the cantilever beam, Fig. 1.1*c*, is readily analyzed for reactions, shears, moments, and deflections using standard methods of structural analysis. But what are the boundary conditions and what is the free body for analysis of the embankment foundation?

2. *Variable and unknown material properties.* The properties of most construction materials (e.g., steel, plastics, concrete, aluminum, and wood) are ordinarily known within rather narrow limits and can often be specified to meet certain needs. Although this may sometimes be the case in construction using earth and rock fills, at least part of every geotechnical problem involves interactions with the naturally occurring soil and rock. The engineer is then confronted with the problems of determining what materials are there and characterizing their properties. In most cases, more than one stratum is present, and conditions are nonhomogeneous and anisotropic.

3. *Stress-dependent material properties.* Soils, and also some rocks, have mechanical properties that depend on the stress history and present stress. This is because the volume change, stress–strain, and strength properties depend on stresses trans-

1

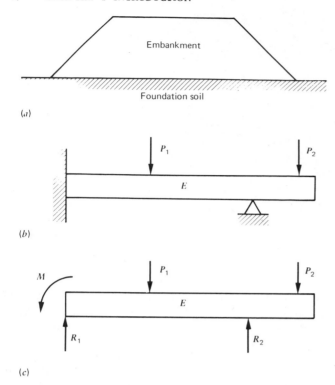

Fig. 1.1 The problem of boundary conditions in geotechnical problems. (a) Embankment of soil foundation. (b) Cantilever beam. (c) Free body for analysis of cantilever beam.

mitted between particles and particle groups. These stresses are, for the most part, generated by body forces and boundary stresses and not by internal forces of cohesion, as is the case with many other materials. Because of this stress dependency, any given geotechnical problem may involve not one or two, but an almost infinite number of different materials.

Add to the above three factors the fact that soil and rock properties may change with time and may be susceptible to influences from changes in environment (e.g., temperature, pressure, chemistry, wetting, and drying), and one might conclude that successful application of mechanics to earth materials is an almost hopeless proposition. It has been amply demonstrated, of course, that such is not the case; in fact, it is for these very reasons that geotechnical engineering offers such a great challenge for imaginative and creative work.

Modern theories of soil mechanics and the computational capabilities provided by high-speed digital computers make possible the analysis of a great diversity of static and dynamic problems of stress-deforma-

tion behavior and stability, as well as the transient and steady-state flow of fluids through the ground. It is probably safe to say that at the present time (1975) our ability to analyze and compute exceeds our ability to understand, measure, and characterize the properties of soils and rocks for use in the analyses. Fortunately, intensive efforts are being made on many fronts to narrow these areas of uncertainty.

The objective of this book is to develop a background and understanding of the engineering properties of soils, the factors controlling their magnitude, and the influences of environment and time. It is perhaps easier to state what this book is not, rather than what it is. It is not a book on soil and rock mechanics; it is not a book on soil exploration and testing; it is not a book that instructs in analysis and design; and it is not a book on geotechnical engineering practice. Excellent books dealing with each of these important areas are available. It is a book on the behavior of soils as engineering materials. It is intended for students, researchers, and practicing engineers and engineering geologists who have a more in-depth interest in the nature and behavior of soils than is provided by classical and conventional treatments of soil mechanics and soil engineering.

1.2 SCOPE AND ORGANIZATION

Following are a few examples of the types of questions that this book addresses:

What are soils composed of? Why?
How are soil properties related to geological history?
How do engineering properties relate to composition?
What is a clay?
Why are clays plastic, whereas sands are not?
What are friction and cohesion in soils?
Why are some soils expansive while others are not?
What is "effective" stress and why does Terzaghi's simple effective stress equation account so well for the volume change and strength behavior of most soils?
Why do soils exhibit creep and stress relaxation?
Why does failure sometimes occur under a sustained stress less than the normal strength?
Why are soil properties sensitive to disturbance?
How do environmental factors influence properties?
What are the influences of changes in these factors?

What are some practical consequences of ion exchange?

How are particle arrangements and soil properties related?

Why does a direct current electrical field cause water flow (electro-osmosis) in soils?

Why is the residual strength of many clays less than the peak strength?

What causes frost heave?

Of what value in engineering practice is knowledge of the mineralogical composition of a soil?

The development of answers to these and related questions requires application of underlying principles from chemistry, geology, materials science, and physics. Principles from these disciplines are introduced as necessary to develop background for the phenomenon or property under study. It is assumed that the reader has some basic knowledge of applied mechanics and soil mechanics, as well as a general familiarity with the engineering properties of soils and their determination.

The vagaries of soil are such, and the complexities of interactions are so great, that correct, unambiguous explanations for all phenomena discussed are not yet available. Some points made may be in dispute, and some discussions may appear incomplete. Hopefully, these uncertainties can be reduced in time.

The book is divided into two main parts. Part I, "The Nature of Soil," begins with a consideration of the structure of solid matter (Chapter 2). Soil mineralogy is treated in some detail (Chapter 3), followed by discussion of soil formation, sedimentary processes, and some relationships between geological history and geotechnical properties and problems (Chapter 4). Methods for determining soil composition are summarized in Chapter 5. Because water may comprise more than half the volume of a soil mass, water properties, interactions between water and soil particles, and the properties of adsorbed water are treated (Chapter 6). The behavior of the clay–water–electrolyte system is then analyzed from a colloid chemical standpoint (Chapter 7). Particle arrangements in soils and methods for their determination are described (Chapter 8) to complete Part I.

Part II draws on the developments of Part I to provide explanations for many facets of soil behavior of interest in geotechnical engineering. Relationships between composition and engineering properties are considered first (Chapter 9)), followed by an analysis of interparticle forces and effective, intergranular, and total stresses in soils (Chapter 10). Soil stucture and its stability are examined (Chapter 11), and relationships between fabric, structure, and properties are discussed in some detail (Chapter 12). The final three chapters deal with the three types of mechanical behavior of interest in most engineering problems: volume change (Chapter 13), strength and deformation (Chapter 14), and conduction phenomena, that is, flows through soils (Chapter 15).

Some of the derivations and equations may not be suitable for direct quantitative analysis of specific engineering problems. Nonetheless, they can be useful to gain insights into the relative importance of different factors controlling behavior. Some topics, for example, determination of fabric, fabric–property interrelationships, temperature effects on soil behavior, electro-osmosis, are covered in considerable detail, because comprehensive treatments are not available elsewhere without reference to several publications.

Several references for further study are listed at the end of each chapter, in addition to the detailed bibliography at the end of the book.

PART I

The Nature of Soil

CHAPTER 2

Bonding, Crystal Structure, and Surface Characteristics

2.1 INTRODUCTION

The material that we commonly refer to as soil* is composed usually of solid particles, liquid, and gas, and may range from very soft, highly organic deposits through less compressible clays and sands to soft rock. The particles comprising the solid phase may encompass a wide range of shapes from nearly spherical, bulky grains to thin, flat plates or long, slender needles. They may vary in size from large boulders to minute particles visible only with the aid of the electron microscope. Some organic material and non-crystalline inorganic components are found in most natural fine-grained soils. Soils may contain virtually any element found in the earth's crust; however, by far the most abundant elements are oxygen, silicon, hydrogen, and aluminum. Atoms of these elements are organized into various crystallographic forms along with lesser amounts of other elements to form the common soil-forming minerals.

The tremendous range in solid particle sizes which can be encountered in natural soil materials is illustrated in Fig. 2.1. The scale across the top of Fig. 2.1 shows the correspondence between the meter and two other metric length designations, micrometers (10^{-6} m) and Ångstrom units (10^{-10} m). Micrometers and Ångstrom units are used extensively throughout this book. Particles smaller than about 200 mesh sieve size (74 μm), which is the boundary between sand and silt sizes, cannot be seen by the naked eye. Thus, silt and clay particles can be observed only with the aid of the optical and electron microscope.

The liquid phase of most soil systems is composed of water containing various types and amounts of dis-

solved electrolytes. The gas phase, in partially saturated soils, is usually air, although organic gases may be present in zones of high biological activity. The mechanical properties of soils derive directly from the interactions of these phases with each other and with applied potentials, for example, stress, hydraulic head, electrical potential, and temperature difference. Because of these interactions, we cannot understand soil behavior in terms of the solid particles alone. However, the structure of these particles tells us something about their surface characteristics and interactions with adjacent phases.

Interatomic and intermolecular bonding forces hold matter together. Unbalanced forces exist at phase boundaries. The nature and magnitude of all these forces must be examined in order to understand the structure, size and shape of soil particles, the formation of soil minerals and the physico-chemical phenomena that influence engineering properties.

In this chapter some aspects of atomic and intermolecular forces, crystal structure, structure stability, and characteristics of surfaces which are pertinent to the understanding of soil behavior are discussed. References to detailed treatises are indicated in "Suggestions for Additional Study."

2.2 ATOMIC STRUCTURE

Current concepts of atomic structure and interatomic bonding forces are based on principles of quantum mechanics. According to quantum theory an electron can have only certain values of energy. It cannot have any intermediate values. Electronic energy can jump to a higher level by the absorption of radiant energy or drop to a lower level by emission of radiant energy. These changes in energy level are termed quantum jumps. No more than two electrons

---

* A variety of definitions has been proposed for soil. None will be attempted here, since some general familiarity with the material is assumed.

7

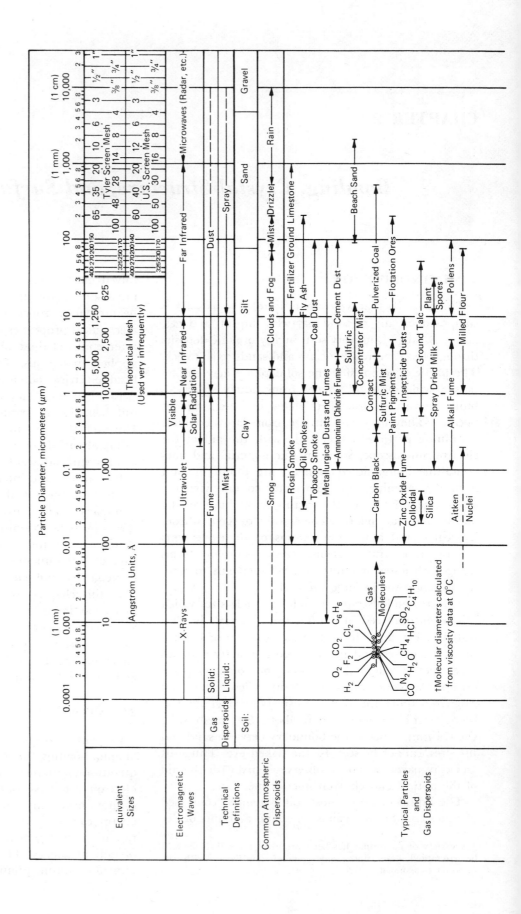

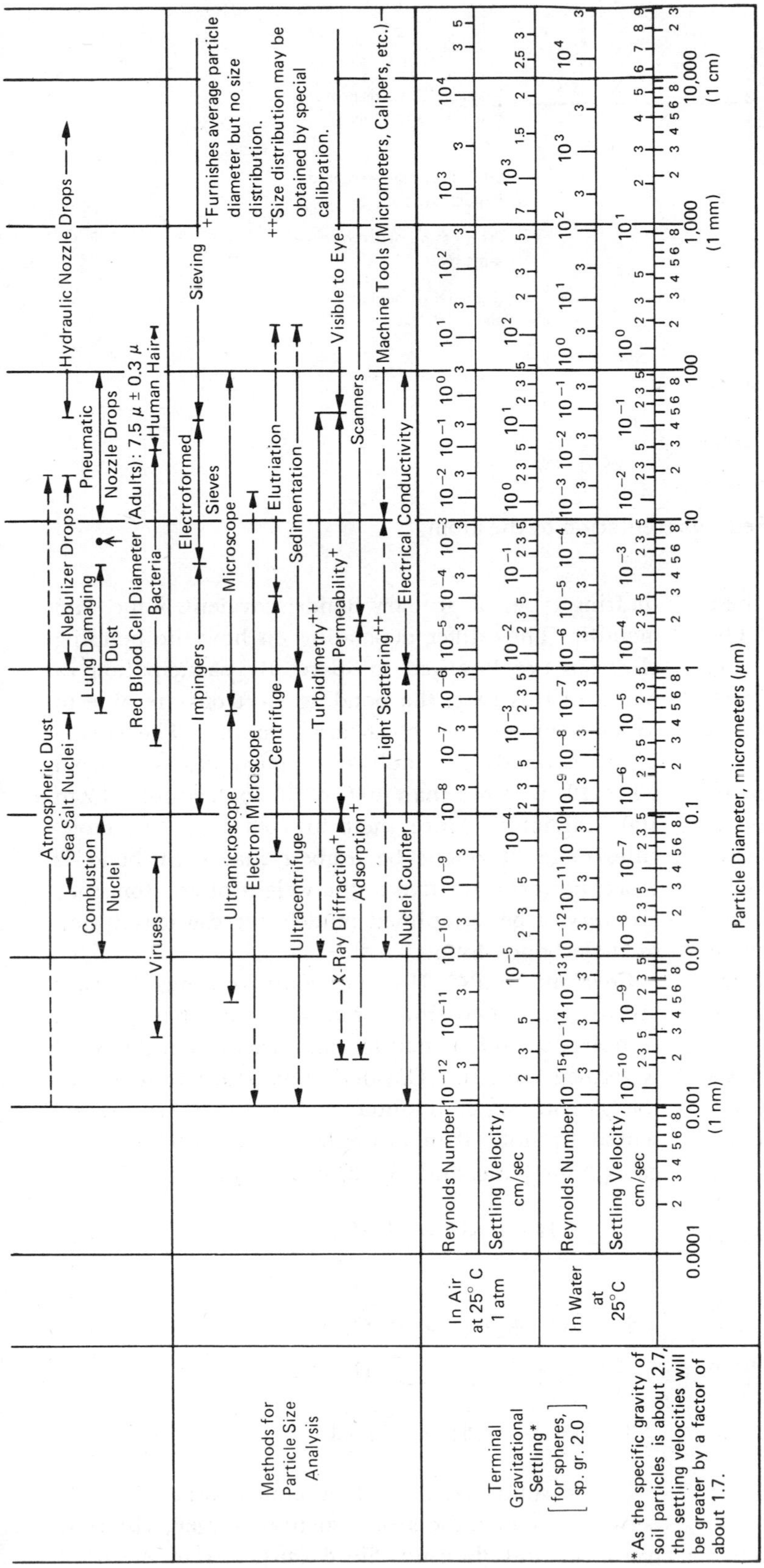

Fig. 2.1 Characteristics of particles and particle dispersoids (Adapted from Stanford Research Institute Journal, Third Quarter, 1961).

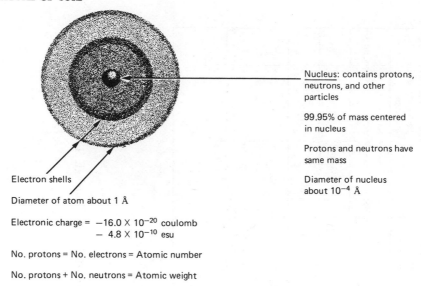

Nucleus: contains protons, neutrons, and other particles

99.95% of mass centered in nucleus

Protons and neutrons have same mass

Diameter of nucleus about 10^{-4} Å

Electron shells

Diameter of atom about 1 Å

Electronic charge $= -16.0 \times 10^{-20}$ coulomb
-4.8×10^{-10} esu

No. protons = No. electrons = Atomic number

No. protons + No. neutrons = Atomic weight

Fig. 2.2 Simplified representation of the atom.

can have the same energy level, and the spin of these two electrons must be in opposite directions. The combined effects of electronic energy quantization and the limitation on the number of electrons at each energy level are responsible for different bonding characteristics for different elements.

An atom may be represented in simplified form by a small nucleus surrounded by diffuse concentric "clouds" of electrons (Fig. 2.2). The results of quantum theory indiciate the maximum number of electrons that may be located in each diffuse shell surrounding the nucleus. The number and arrangement of the electrons in the outer-most shell are of prime importance in the development of different types of interatomic bonding and crystal structure.

Interatomic bonding develops when electrons associated with adjacent atoms can interact in such a way as to lower their energy levels. If the energy reduction of the bonding electrons is great, then a strong or primary bond develops. The way in which the bonding electrons are localized in space determines whether or not the bonds are directional in character. The strength and directionality of interatomic bonds together with the relative sizes of the bonded atoms determines the type of crystal structure assumed by a given composition.

2.3 INTERATOMIC BONDING

Primary bonds

In the formation of primary bonds, only the outer shell or valence electrons participate. There are three limiting types of primary bonds: covalent, ionic, and metallic. They differ depending on how the bonding electrons are localized in space. By participation in bonds of this type, the bonding electrons are able to lower their energies by a large amount. The energy of these bonds per mole of bonded atoms is from 60×10^3 to more than 400×10^3 joules/mole (15 to 100 kcal/mole). Since there are 6.023×10^{23} molecules/mole, it might be argued that each bond is weak; however, relative to the weight of an atom they are very large. Simplified models for the description of these bonds follow.

Covalent bonds. If one or more bonding electrons are shared by two atomic nuclei so that they serve to complete the outer shell for each atom, then the bond is termed *covalent*. Covalent bonds are common in the gaseous state. If outer shell electrons are represented by dots, then examples are for (1) hydrogen gas, (2) methane, and (3) chlorine gas:

1. $\text{H} \cdot \; + \; \cdot \text{H} \; = \; \text{H} : \text{H}$

2. $\cdot \overset{\cdot}{\underset{\cdot}{\text{C}}} \cdot \; + \; 4\text{H} \cdot \; = \; \text{H} : \overset{\overset{\text{H}}{..}}{\underset{\underset{\text{H}}{..}}{\text{C}}} : \text{H}$

3. $: \overset{..}{\text{Cl}} \cdot \; + \; \cdot \overset{..}{\text{Cl}} : \; = \; : \overset{..}{\underset{..}{\text{Cl}}} : \overset{..}{\underset{..}{\text{Cl}}} :$

In the solid state, covalent bonds form primarily between nonmetallic atoms such as oxygen, chlorine, nitrogen, and fluorine. Since only certain electrons participate in the bonding, covalent bonds are directional. Thus, atoms bonded covalently pack in such a way that fixed bond angles are developed.

Ionic bonding. Ionic bonds result from the electrostatic attraction between positive and negative ions formed from free atoms through gain or loss of electrons. *Cations* (positively charged atoms that are attracted by the *cathode* in an electric field) are formed from atoms by giving up one or more loosely held electrons having a high energy level and lying outside a completed electron shell. Metals, alkalies (e.g., sodium, potassium), and alkaline earths (e.g., calcium, magnesium) form cations. *Anions* (negatively charged atoms that are attracted to the *anode*) are those atoms requiring only a few electrons to complete their outer shell. Because outer shells of ions are complete, structures cannot be formed by sharing as is the case of the covalent bond. Since they are electrically charged, however, strong electrostatic attractions (and repulsions) may develop.

The ionic bond is nondirectional. Each cation attracts all neighboring anions and vice versa. In sodium chloride, one of the best examples of ionic bonding, a sodium ion attracts as many chlorine atoms as will fit around it. Geometric considerations and electrical neutrality determine the actual arrangement of ionically bonded atoms.

Since ionic bonding causes a separation between centers of positive and negative charge in a molecule, the molecule will tend to orient in an electric field forming a *dipole*. The strength of this dipole is expressed in terms of the *dipole moment μ*. If two electrical charges, $\pm \delta e$, where e is the electronic charge, are separated by a distance d then

$$\mu = d \cdot \delta e \qquad (2.1)$$

Covalently bonded atoms may also produce dipolar molecules.

Metallic bonding. Metallic bonds are nondirectional and can exist only among a large group of atoms. Metals are characterized by loosely held valence electrons that are more or less free to travel through the solid material. Positive metal ions are held together by a diffuse cloud of electrons. It is this cloud of electrons that makes metals such good conductors. The metallic bond is of little importance in the study of soils.

Bonding in soil minerals

Purely ionic and purely covalent bounds are limiting conditions that are the exception rather than the rule in most materials. A combination of the two types is usual in nonmetallic solids. Silicate minerals are the most abundant constituents of most soils. The interatomic bond in silica (SiO_2) is about half covalent and half ionic.

2.4 SECONDARY BONDS

In addition to the primary bonds described previously, other types of weaker, secondary bonds exist between units of matter. Although these bonds are weak relative to ionic and covalent bonds, they are often strong enough to determine the final arrangements of atoms in solids. They may also be important sources of attraction between very small particles and between liquids and solid particles.

The hydrogen bond

Attraction exists between oppositely charged ends of two permanent dipoles. If hydrogen is the positive end of the dipole, then the resulting bond is termed a hydrogen bond. Hydrogen bonds are formed only with strongly electronegative atoms such as oxygen and fluorine, because these atoms produce the strongest dipoles. When the electron is detached from a hydrogen atom, as for example when it combines to form water, only a proton remains. Since the electrons shared between the oxygen and hydrogen atoms in the water molecule tend to spend most of their time between the atoms, the oxygen atoms act as the negative end of a dipole and the hydrogen proton as the positive end. The positive end attaches to the negative end of another water molecule and bonds them together.

The strength of hydrogen bonds is considerably greater than for other types of secondary bonds because of the small size of the hydrogen ion. Hydrogen bonds play an important role in determining some of the clay mineral characteristics and in the interactions between soil particles and water.

van der Waals bonds

Permanent dipole bonds such as the hydrogen bond are directional bonds. Fluctuating dipole bonds, commonly termed van der Waals bonds may also arise, because at any one time there may be more electrons on one side of an atomic nucleus than the other, thus giving rise to weak instantaneous dipoles whose oppositely charged ends attract each other.

Although individual van der Waals bonds are weak, typically an order of magnitude weaker than a hydrogen bond, they are nondirectional and additive between atoms. Consequently, they decrease less rapidly with distance than do primary valence and hydrogen

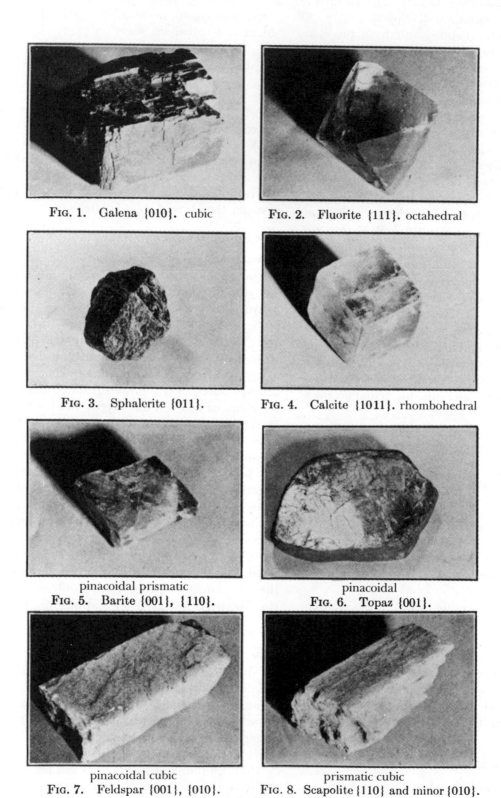

FIG. 1. Galena {010}. cubic   FIG. 2. Fluorite {111}. octahedral

FIG. 3. Sphalerite {011}.   FIG. 4. Calcite {1011}. rhombohedral

pinacoidal prismatic
FIG. 5. Barite {001}, {110}.

pinacoidal
FIG. 6. Topaz {001}.

pinacoidal cubic
FIG. 7. Feldspar {001}, {010}.

prismatic cubic
FIG. 8. Scapolite {110} and minor {010}.

Fig. 2.3   Examples of some common crystals {hkl} are cleavage plane
indicies (*Dana's Manual of Mineralogy,* 16th ed., p. 93, Plate V,
Wiley, New York).

bonds when large groups of atoms are considered. They are strong enough to determine the final arrangements of groups of atoms in some solids (e.g., many polymers), and they may be a major source of cohesion in fine-grained soils. van der Waals forces are described in more detail in Chapter 7.

Additional forces

A number of additional forces may be involved in the interactions between both soil particles and soil and water. These forces include both attractions and repulsions, and may be short or long range. These forces arise primarily from electric field effects surrounding clay particles and the presence of ions and water adjacent to clay surfaces. Discussion of these forces is deferred until after description of clay mineral structures and properties.

2.5 CRYSTALS AND THEIR PROPERTIES

By far the greatest proportions of the solid phase in soils consists of particles composed of mineral crystals. A crystal may be defined as a homogeneous body bounded by smooth plane surfaces that are the external expression of an orderly internal atomic arrangement. Examples of some common crystals are shown in Fig. 2.3.

Crystal formation

Crystals may be formed in the following ways:

1. *From solution.* Ions combine as they separate from solution and gradually build up a solid of definite structure and shape. Precipitation of sodium chloride is an example of this type of crystal formation.
2. *By fusion.* Crystals are formed directly from a liquid as a result of cooling. Solidification of molten rock magma to igneous rock minerals and water to ice are examples.
3. *From vapor.* Although not of particular importance in the formation of soil minerals, crystals can be formed directly from cooling vapors. Examples of crystals formed in this way include snowflakes and flowers of sulfur.

Characteristics of crystals

All crystals have certain characteristics that may be used to distinguish different classes or groups of minerals. Variations in these characteristics give rise to different properties.

1. *Structure.* The atoms in a crystal are arranged in a definite orderly manner to form a three-dimensional network termed a *lattice*. Positions within the lattice where atoms or atomic groups are located are termed lattice points. Each lattice point has surroundings identical with every other point. Only 14 different arrangements of lattice points in space are possible. These are termed the Bravais space lattices and are illustrated in Fig. 2.4.

   The smallest subdivision of a crystal that will still possess the characteristic properties of the crystal is termed the *unit cell*; that is, the unit cell is the basic repeating unit of the space lattice. Figure 2.5 shows the crystal structure of halite (NaCl), with alternating sodium and chloride ions packed in a simple cubic array. This unit cell requires four sodium and four chloride ions to establish the basic repeating three-dimensional network.
2. *Cleavage and outward form.* The angles between corresponding faces on crystals of the same substance are constant. The property of breaking along smooth plane surfaces is termed *cleavage*. Study of the relationship between crystal structure and cleavage shows that cleavage planes generally pass between planes of the most closely packed atoms. This is because the center-to-center distance between atoms located in opposing close-packed planes is greater than along other planes passing through the crystal; therefore, the strength along the interatomic plane is less than in other directions.
3. *Optical properties.* Because of the specific atomic arrangements within crystals, they are able to both refract and polarize light. Use of these properties is made in the identification and classification of different crystalline materials.
4. *X-Ray and electron diffraction.* The orderly atomic arrangements in crystals causes them to behave with respect to X-ray and electron beams in much the same way as does a diffraction grating with respect to visible light. Because different crystals give different diffraction patterns, X-ray and electron diffraction are powerful tools for the study of crystals, particularly for materials composed of very fine particles such as clays, which cannot be studied optically.
5. *Symmetry.* There are 32 distinct crystal classes based on symmetry considerations involving the arrangement and orientation of crystal faces. These 32 classes may be grouped into six crystal

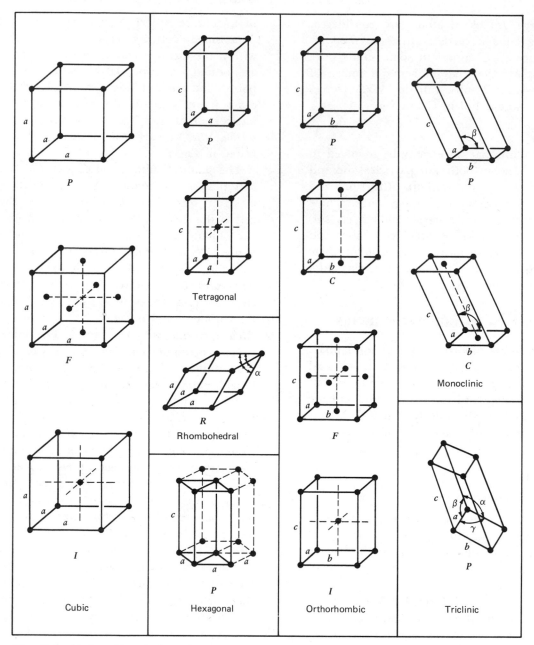

Fig. 2.4 Unit cells of the 14 Bravais space lattices. The capital letters refer to the type of cell—*P:* primitive cell; *C:* cell with a lattice point in the center of two parallel faces; *F:* cell with a lattice point in the center of each face; *I:* cell with a lattice point in the center of the interior; *R:* rhombohedral primitive cell. All points indicated are lattice points. There is no general agreement on the unit cell to use for the hexagonal Bravais lattice; some prefer the *P* cell shown with solid lines, and others prefer the *C* cell shown in dashed lines.

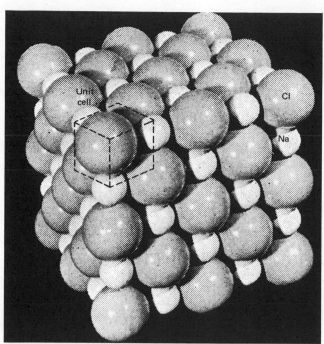

Fig. 2.5 Crystal structure of halite illustrating the unit cell.

systems with the classes within each system bearing close relationships to each other. These six systems are identified in Fig. 2.6.

Crystallographic axes parallel to the intersection edges of prominent crystal faces are established for each of the six crystal systems. In most crystals, these axes will also be symmetry axes or normals to symmetry planes. In five of the six systems, the crystals are referred to three crystallographic axes. In the sixth (the hexagonal system), four axes are used. The axes are denoted by a, b, c (a_1, a_2, a_3, and c in the hexagonal system) and the angles between the axes by α, β, and γ. Characteristics of the six crystal systems are shown in Fig. 2.6.

Isometic or cubic system. Crystals are referred to three mutually perpendicular axes of equal length. Examples are galena, halite, magnetite, and pyrite.

Hexagonal system. Crystals are referred to four axes; three equal horizontal axes intersect at 120°, and the fourth is of different length and perpendicular to the plane of the other three. Examples are quartz, calcite, brucite, and beryl.

Tetragonal system. Crystals are referred to three mutually perpendicular axes. The two horizontal axes are of equal length, but the vertical axis is of different length. Zircon is an example of a mineral in the tetragonal system.

Orthorhombic system. Crystals are referred to three mutually perpendicular axes, all of different length. Examples include sulfur, anhydrite, barite, diaspore, and topaz.

Monoclinic system. Crystals are referred to three unequal axes, two of which are inclined to each other at an oblique angle, with the third perpendicular to the other two. Examples of monoclinic crystals are orthoclase feldspar, gypsum, muscovite, biotite, gibbsite, and chlorite.

Triclinic system. Crystals are referred to three unequal axes all intersecting at oblique angles. Examples are plagioclase feldspar, kaolinite, albite, microcline, and turquoise.

2.6 CRYSTAL NOTATION

To describe planes and directions in a crystal, crystallographers have adopted a system of indices known as the Miller indices. In this system, all lengths are expressed in terms of unit cell lengths. Any plane through a crystal may be expressed as intercepts, in terms of unit cell lengths, on the three or four crystallographic axes for the system in which the crystal falls. The reciprocals of these intercepts are then used to index the plane. Reciprocals are used because the intercepts often are fractions less than one, and also to be able to account for planes parallel to an axis (infinite intercept).

An example serves to illustrate the determination and meaning of Miller indices. Consider the mineral muscovite, a member of the monoclinic system with unit cell dimensions $a = 5.2$ Å, $b = 9.0$ Å, $c = 20.0$ Å, $\beta = 95°\ 30'$. Both the composition and crystal structure of muscovite are similar to that of some of the important clay minerals.

The muscovite unit cell dimensions and intercepts are shown in Fig. 2.7a. To designate any plane through a muscovite crystal, the intercepts of the plane are first determined in terms of unit cell lengths. Suppose plane mnp in Fig. 2.7a is of interest. The intercepts of this plane are $a = 1$, $b = 1$, and $c = 1$. The Miller indices of this plane are found by taking the reciprocals of the intercepts and clearing of fractions. Thus,

Reciprocals are $\dfrac{1}{1}, \dfrac{1}{1}, \dfrac{1}{1}$

Miller indices are (111)

Note that the indices are always enclosed within parentheses and indicated in the order $a\ b\ c$ without

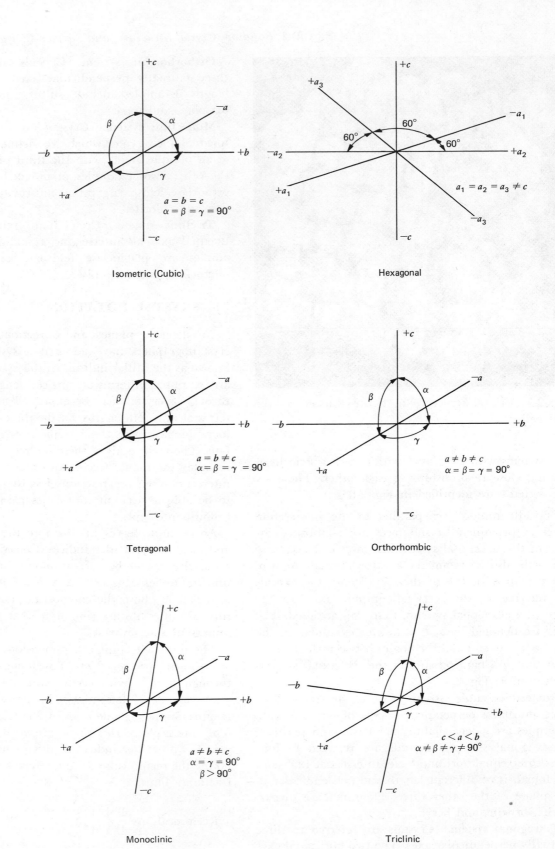

Fig. 2.6 The six crystal systems.

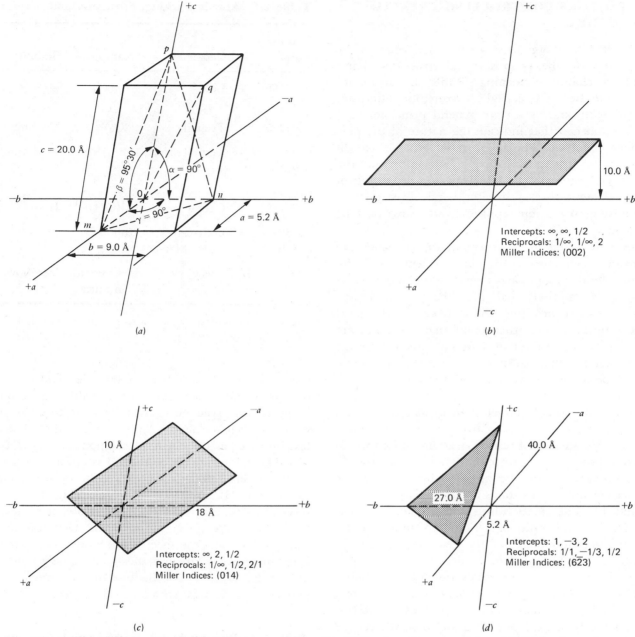

Fig. 2.7   Miller indices. (*a*) Unit cell of muscovite. (*b*) (002) plane for muscovite. (*c*) (014) plane for muscovite. (*d*) (6$\bar{2}$3) plane for muscovite.

commas. Parentheses are always used to indicate crystallographic planes; whereas, brackets are used to indictate directions; for example, [111] designates line *oq* in Fig. 2.7*a*.

Additional examples of Miller indices for planes through the muscovite crystal are shown in Figs. 2.7*b*, 2.7*c*, and 2.7*d*. A plane cutting a negative axis is designated by placing a bar over the index pertaining to the negative intercept (Fig. 2.7*d*). The general in-

dex (*hkl*) is used to refer to any plane that cuts all three axes. Similarly (*h*00) designates a plane cutting only the *a* axis, (*h0l*) designates a plane parallel to the *b* axis and so on. For crystals referred to the hexagonal system, the Miller index contains four numbers.

Of particular interest in the study of soils are the (001) planes for the clay minerals. These planes are characteristic for the different clays and can be used as a basis for their identification.

2.7 FACTORS CONTROLLING CRYSTAL STRUCTURE

The fact that organized atomic arrangements develop to form the various crystal structures is not simply a chance happening. While all details of atomic bonding and crystal structure formation are not presently known, certain general principles have been established that indicate the nature of the phenomena involved. The most stable arrangement of atoms in a structure is that which minimizes the energy per unit volume. This involves preserving of electrical neutrality, satisfying bond directionality, minimizing strong ion repulsions, and close packing of atoms.

Covalent bonds and permanent dipole bonds are directional. Of these, only the covalent bonds influence local atomic arrangements significantly. If bonding is nondirectional, then the relative atomic sizes have a controlling influence on packing. If bonds are directional, then bond angles and size are both important. If the bonds are nondirectional, then the closest possible packing will maximize the number of bonds per unit volume and minimize the bonding energy.

Anions are usually larger than cations as a result of the transfer of electrons from the cations to the anions. The number of nearest neighbor anions which a cation possesses in a structure is termed the coordination number (N) or ligancy. Possible values of coordination number in solids are 1 (trivial), 2, 3, 4, 6, 8, and 12. The relationships between atomic sizes, expressed as the ratio of cationic to anionic radii, coordination number, and the geometry formed by the anions are indicated in Table 2.1.

Considering that most solids do not have bonds that are completely nondirectional and that second nearest neighbors may influence packing as well as nearest neighbors, the predicted and observed coordinations are in quite good agreement for many materials. The valence of the cation divided by the number of coordinated anions provides a very approximate indication of the relative bond strength, which, in turn, is related to the structural stability of the unit. Table 2.2 lists some of the structural units commonly found in soil minerals and their relative bond strengths.

The basic coordination polyhedra are seldom electrically neutral. In crystals formed by ionic bonded polyhedra, packing is such as to maintain electrical neutrality and to minimize strong repulsions between ions with like charges. In such cases, the valence of

Table 2.1   Atomic Packing, Structure, and Stability

Radius Ratios [a]	N [b]	Geometry	Example	Stability
0–0.155	2	Line	—	—
0.155–0.225	3	Triangle	$(CO_3)^{2-}$	Very high
0.225–0.414	4	Tetrahedron	$(SiO_4)^{4-}$	Moderately high
0.414–0.732	6	Octahedron	$[Al(OH)_6]^{3-}$	High
0.732–1.0	8	Body-centered cube	Iron	Low
1.0	12	Sheet	K–O bond in mica	Very low

[a] Range over which stable coordination is expected.
[b] Coordination number.

the central cation equals the total valence of the coordinated anions, and the unit is really a molecule. Units of this type must then be held together by weaker, secondary bonds. An example is the mineral brucite that has the chemical composition $Mg(OH)_2$. The Mg^{2+} ions are in octahedral coordination with six (OH) ions forming a sheet structure in such a way that each $(OH)^-$ is shared by 3 Mg^{2+} (Fig. 2.8). In a sheet containing N Mg^{2+} ions, therefore, there must be $6N/3 = 2N$ $(OH)^-$ ions. Thus, electrical neutrality results, and the sheet is in reality a large molecule. Successive octahedral sheets are loosely bonded by van der Waals forces. Because of this, brucite has perfect basal cleavage parallel to the sheets.

Table 2.2   Relative Stabilities of Some Soil Mineral Structural Units

Structural Unit	Approximate Relative Bond Strength (Valence/N)
Silicon Tetrahedron, $(SiO_4)^{4-}$	$4/4 = 1$
Aluminum Tetrahedron, $[Al(OH)_4]^{1-}$	$3/4$
Aluminum Octahedron, $[Al(OH)_6]^{3-}$	$3/6 = 1/2$
Magnesium Octahedron, $[Mg(OH)_6]^{4-}$	$2/6 = 1/3$
$K-O_{12}^{-23}$	$1/12$

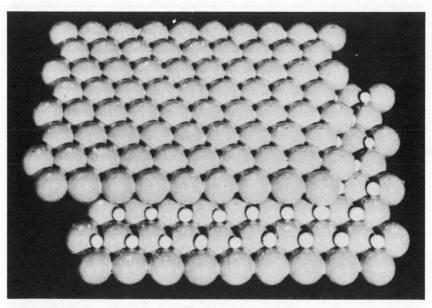

Fig. 2.8   Magnesium in octahedral coordination with hydroxyl (brucite)
(hydroxyl layers staggered to show Mg atoms) (Marshall, 1964).

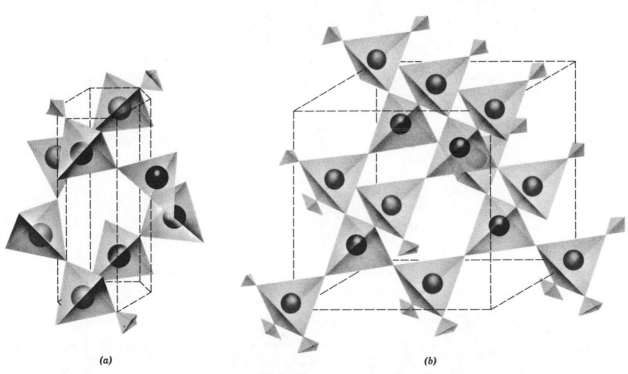

(a)   *(b)*

Fig. 2.9   Two of the crystal structures of SiO_2 showing the arrangements of the Si–O tetrahedra. *(a)*
The crystal structure of β-quartz in which two spiral chains wind around a hexagonal prism. The silicon
atoms in one chain are shaded darker than in the other chain to distinguish between the two. *(b)* The
crystal structure of cristobalite in which the tetrahedra are arranged similar to the carbon atoms in
diamond. The silicon atoms that are on FCC lattice points in the unit cell are shaded somewhat lighter
than those lying within the cell (Moffat et al. 1965).

Combination of Tetrahedra	Diagrammatic Representation of Structure	Si-O Group and Negative Charge	Oxygen to Silicon Ratio	Example
Island Independent	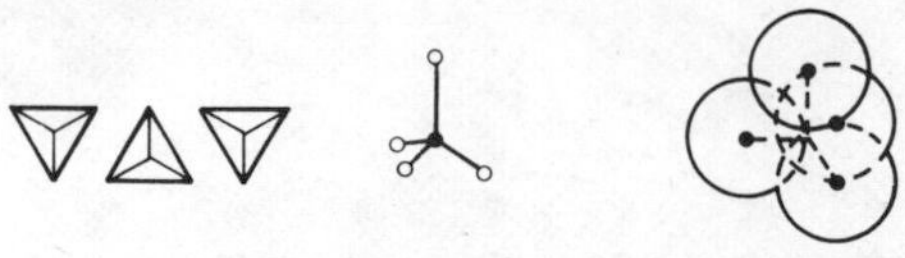	$(SiO_4)^{4-}$	4 : 1	Olivines $(Mg, Fe)_2SiO_4$
Double	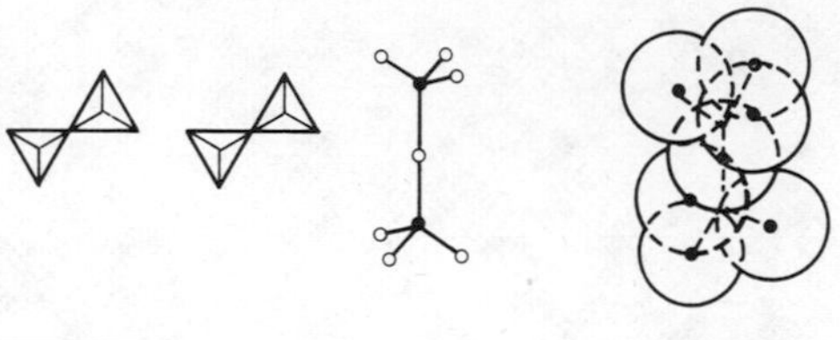	$(Si_2O_7)^{6-}$	7 : 2	Amermanite $Ca_2Mg_2Si_2O_7$
Rings	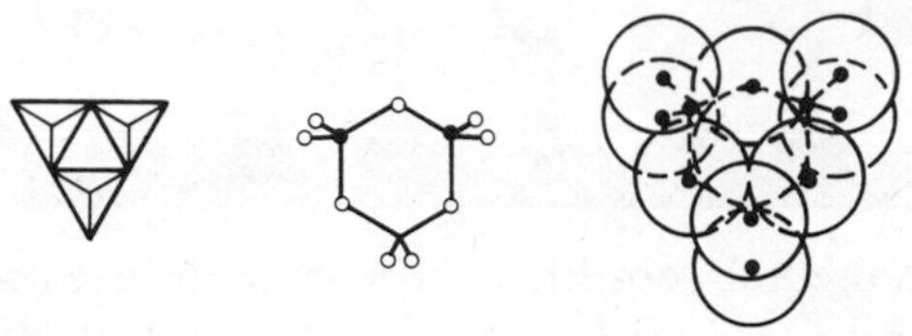	$(Si_3O_9)^{6-}$	3 : 1	Benitoite $BaTiSi_3O_9$
	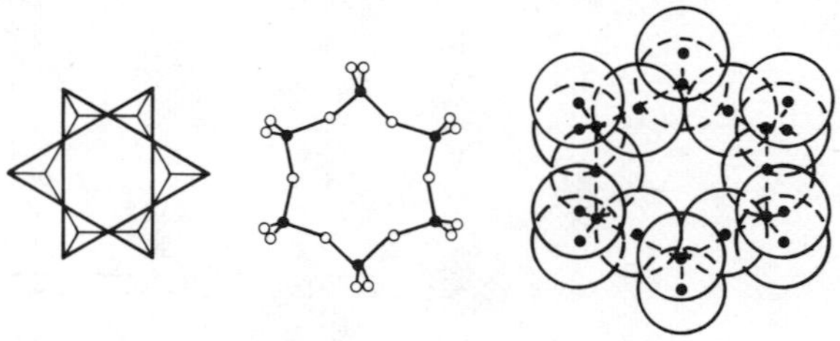	$(Si_6O_{18})^{12-}$		Beryl $Be_3Al_2Si_6O_{18}$
Chains	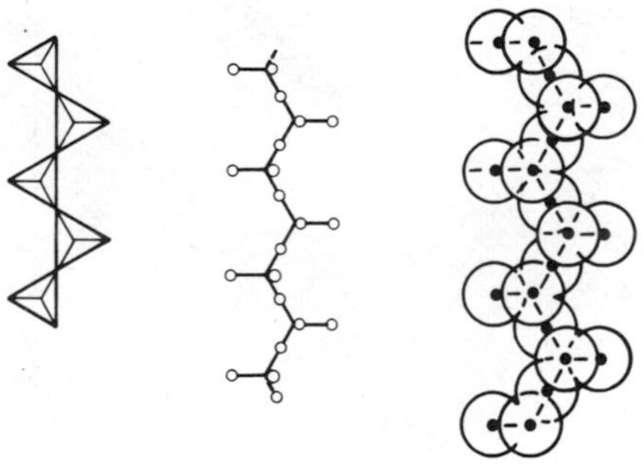	$(SiO_3)_n^{2-}$	3 : 1	Pyroxenes
Bands	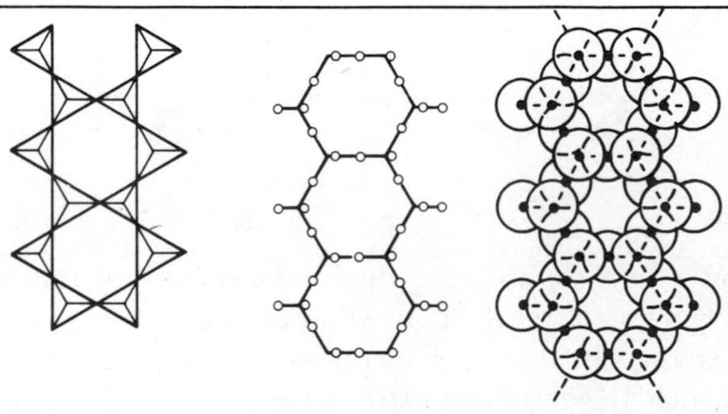	$(Si_4O_{11})_n^{6-}$	11 : 4	Amphiboles

Fig. 2.10   Crystal chemistry of silicates (After Gillott, 1968).

Combination of Tetrahedra	Diagrammatic Representation of Structure	Si-O Group and Negative Charge	Oxygen to Silicon Ratio	Example
Sheets	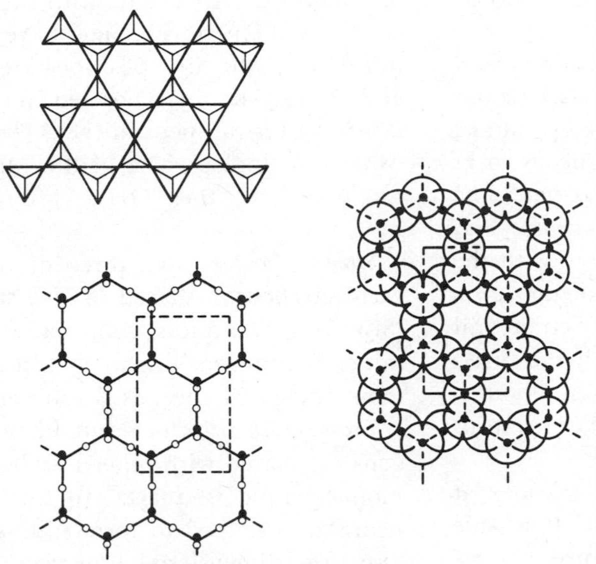	$(Si_4O_{10})_n^{4-}$	5 : 2	Micas
Frameworks		$(SiO_2)_n^0$	2 : 1	Quartz SiO_2
				Also feldspars, for example, orthoclase, $KAlSi_3O_8$

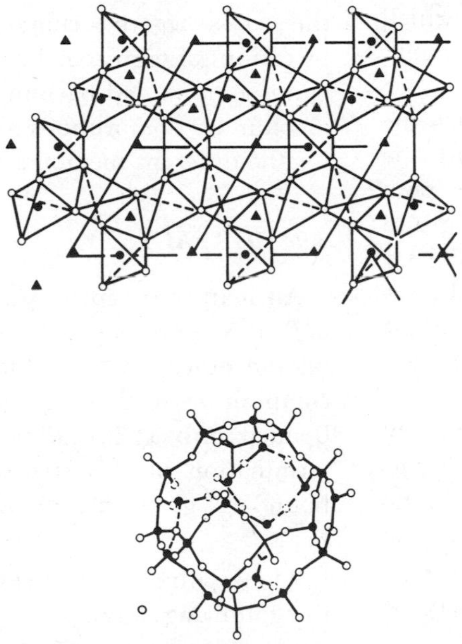

Figure 2.10 (*continued*).

Since cations concentrate their charge in a smaller volume than do anions, the repulsion between cations is greater than between anions. With cations located at the centers of coordination polyhedra, the cationic repulsions are minimized. If the cations have a low valence, the anion polyhedra will pack as closely as possible to minimize the energy per unit volume. On the other hand, if the cations are highly charged, the units will develop a variety of other packings in response to the repulsions generated between adjacent cations.

2.8 SILICATE CRYSTALS

Small cations give rise to structures with coordination numbers of three and four (Table 2.1). These cations are often highly charged leading to strong repulsions between adjacent triangles or tetrahedra. These structures will, therefore, share only corners and possibly edges, but never faces, since to do so would bring cations too close together.

The silica ion is one that fits in this category and one that is very abundant in earth materials (about 25 percent by weight but only 0.8 percent by volume because of its small size). Almost half of igneous rocks by weight and 91.8 percent by volume is oxygen. The radius of the silicon ion is 0.39 Å. Thus, silicon and oxygen combine in tetrahedral coordination, with the silicon occupying the space at the center of the tetrahedron formed by four oxygens (Fig. 2.9). Tetrahedral coordination satisfies requirements of both the directionality of the bonds (the silicon–oxygen bond is about half covalent and half ionic) and the radius ratio.

Because the silica tetrahedra are joined only at their corners, and in some instances not at all, a wide range of crystal structures is possible; hence, a large variety of silicate minerals results. The different silicates are classified according to the manner in which the $(SiO_4)^{-4}$ tetrahedra associate with one another, as shown in Fig. 2.10.

Island (independent) silicates are those in which the tetrahedra are not joined to each other. Instead, the four excess oxygen electrons are bonded to other positive ions in the crystal structure. In the olivine group the minerals have the composition $R_2^{2+} \cdot SiO_4^{4-}$. Garnets contain cations of different valences and coordination numbers, $R_3^{2+} \cdot R_2^{3+} (SiO_4)_3$. In zircon the negative charge on the SiO_4 group is all taken by a single Zr^{4+}.

Ring and chain silicates are formed when corners of tetrahedra are shared. The formula for these structures contains $(SiO_3)^{2-}$. The pyroxene group of minerals falls into this class. Enstatite, $MgSiO_3$, is a simple member of this group. Some of the positions normally occupied by Si^{4+} in single chain structures may be filled by Al^{3+}.

Substitution of ions of one kind by ions of another type, with the same or different valence, but with retention of the same crystal structure—is termed *isomorphous substitution*. The term "substitution" is misleading in that it implies a replacement whereby a cation in the structure is replaced at some time by a cation of another type. In reality, the replaced cations were never there, and the mineral formed with its present proportions of different cations in the structure.

Double chains of indefinite length may form having $(Si_4O_{11})^{6-}$ as part of the structure. The amphiboles fall into this group (Fig. 2.10). The simplest member of the group is tremolite, with the formula $(OH)_2Ca_2$-$Mg_5Si_8O_{22}$. The hornblendes have this basic structure, but some of the Si^{4+} positions are filled by Al^{3+}; Na and K can be incorporated into the structure to satisfy electrical neutrality; Al^{3+}, Fe^{3+}, Fe^{2+}, and Mn^{2+} can replace part of the Mg^{2+} in sixfold coordination, and the $(OH)^-$ group can be replaced by F^-.

Sheet silicates have three of the four oxygens of each tetrahedron shared to give structures containing $(Si_2O_5)^{2-}$. The micas, chlorites, and many of the clay minerals contain silica in a sheet structure. The common hexagonal sheet is illustrated in Fig. 2.10.

Framework silicates result when all four of the oxygens are shared with other tetrahedra. The most common example is quartz. In quartz, the silica tetrahedra are grouped to form spirals. The feldspars are also three-dimensional framework structures. Part of the silicon positions are occupied by aluminum, and the excess negative charge, thus created, is balanced by cations of high coordination such as potassium, calcium, sodium, and barium. Variations in the type and amount of isomorphous substitution are responsible for the different members of the feldspar family.

2.9   SURFACES

All matter, except possibly gas, is limited in extent and must terminate at a surface or phase boundary, on the other side of which is matter of a different composition or state. In solid substances, atoms are bonded into a three-dimensional structure, and the termination of this structure at a surface or phase boundary gives rise to unsatisfied attractive force fields.

These unbalanced forces may be satisfied in any of the following ways:

1. Attraction and adsorption of molecules from the adjacent phase. This may take the form of a simple physical adsorption, or it may involve chemical interaction between the surface material and the adsorbed substances. An example of the latter is the oxidation film formed on clean surfaces of metals when they are exposed to air.
2. Cohesion with the surface of another mass of the same substance.

3. Solid-state adjustments of structure beneath the surface.

Each unsatisfied bond at a surface or phase boundary is of significant magnitude relative to the mass of atoms and molecules. Their actual magnitude of 10^{-11} Newton or less, however, is infinitesimal compared to the mass of a piece of gravel or a grain of sand. On the other hand, consider the effect of reducing particle size. A cube 10 mm on an edge has a surface area of 6.0×10^{-4} m². If it is cut in half in the three directions, eight cubes result, with each 5 mm on an edge. The area now is 12.0×10^{-4} m². If the cubes are subdivided further to $1/\mu$m on an edge, the surface becomes 6.0 m² for the same 1000 mm³ of material. Thus, as a substance is subdivided into smaller and smaller units, the proportion of surface area to mass becomes larger and larger. For a given particle shape the ratio of surface to volume is inversely proportional to some effective particle diameter.

Experience shows that when particle size is reduced to 1 or 2 μm or less for many materials, then surface forces begin to exert a distinct influence on the behavior. Study of the behavior of particles of this size involves considerations of colloidal and surface chemistry. Most clay particles fall within the colloidal range in terms of both their size and the importance of surface forces to their behavior. In fact, due to the platy morphology of many of the clay minerals and to the fact that most of the clays have a residual negative electrostatic charge resulting from isomorphous substitutions in the crystal structure, surface and colloidal forces may exert a controlling influence.

Montmorillonite, one of the clay minerals, when in a dispersed state, may break down into platy particles only one unit cell in thickness (10 Å) and exhibit a specific surface of 800 m²/g. If all particles contained in about 10 g of this clay could be spread out side by side, they would cover a football field.

2.10 PRACTICAL IMPLICATIONS

A knowledge of bonding, crystal structure, and surface characteristics is essential to the understanding of the size, shape, and durability of soil particles and to the interactions of soil particles with liquids and gases. The fact that interatomic bonds in soil particles are of the strong primary valence type, whereas usual interparticle bonds are of the secondary valence or hydrogen type, means that particles are strong relative to groups of particles. Thus, in most cases, soil masses behave as assemblages of particles wherein deformation processes are dominated by displacements between particles and not by deformations of the particles themselves. In the case of sands and gravels under high stresses, however, there may be grain crushing as well.

The type of bonding between unit layers of some of the clay minerals, coupled with the adsorption properties of the particle surfaces greatly influences soil swelling.

The optical and X-ray diffraction properties of mineral crystals provide a basis for the identification of soil minerals. The structural stability of the different minerals controls their resistance to weathering, and hence accounts in part for the relative abundance of different minerals in different soils.

SUGGESTIONS FOR FURTHER STUDY

Adamson, A. W. (1960), *The Physical Chemistry of Surfaces*, Interscience, New York.

Brophy, J. H., Rose, R. M., and Wulff, J. (1964), *The Structure and Properties of Materials*, Vol. II, *Thermodynamics of Structure*, Wiley, New York.

Buerger, M. J. (1956), *Elementary Crystallography*, Wiley, New York.

Daniels, F., and Alberty, R. A. (1961), *Physical Chemistry*, 2d ed., Wiley, New York.

Hurlbut, C. S. (1957), *Dana's Manual of Mineralogy*, 16th ed., Wiley, New York.

Marshall, C. E. (1964), *The Physical Chemistry and Mineralogy of Soils*, Vol. 1, Soil Materials, Wiley, New York.

Moffatt, W. G., Pearsall, G. W., and Wulff, J. (1965), *The Structure and Properties of Materials*, Vol. 1, *Structure*, Wiley, New York.

Pauling, L. (1960), *The Nature of the Chemical Bond*, Cornell University Press, Ithaca, N. Y.

CHAPTER 3

Soil Mineralogy

3.1 INTRODUCTION

Although it is not possible at the present time to express the engineering properties of a soil quantitatively in terms of composition, a knowledge of soil mineralogy is essential to a fundamental understanding of soil behavior. Mineralogy is the primary factor controlling the size, shape, and physical and chemical properties of soil particles.

The solid phase of a given soil may contain various amounts of crystalline clay and nonclay minerals, noncrystalline clay material, organic matter, and precipitated salts. The inorganic, crystalline minerals comprise by far the greatest proportion of the solid phase in most soils encountered in engineering problems, and the amount of nonclay mineral is usually considerably greater than the proportion of clay minerals present. However, the clay may influence behavior to an extent much greater than in simple proportion to the amount present.

The term "clay" is sometimes ambiguous. When used as a particle size term, it refers to all constituents of a soil smaller than some given size, usually 2 μm. As a mineral term it refers to specific minerals termed the "clay minerals". Clays are small crystalline particles of one or more members of a small group of minerals. They are primarily hydrous aluminum silicates, with magnesium or iron occupying all or part of the aluminum positions in some minerals, and with alkalis (e.g., sodium, potassium) or alkaline earth (e.g., calcium, magnesium) also present as essential constituents in some of them (Grim, 1962, 1968). Not all clay mineral particles are finer than 2 μm (or whatever boundary may be chosen to define a clay size) nor are all nonclay minerals coarser than 2 μm. Thus, the amounts of clay size and clay mineral in any soil may not be the same. To avoid confusion, it is best to use *clay size* when referring to compositions in terms of particle size and *clay mineral content* or simply *clay content* when speaking of mineral compositions.

The physical characteristics of cohesionless, nonclay soils are determined mainly by particle size, shape, surface texture, and size distribution. The mineral composition is of importance primarily as it influences hardness, cleavage, and resistance to chemical attack that determine these characteristics. By and large, however, the nonclay particles may be treated as relatively inert materials whose interactions are predominantly physical in nature. Convincing evidence in support of this has been obtained recently through study of the properties of lunar soils. Soils on the Moon have a silty, fine sand gradation, but their composition is totally different than that of terrestrial soils. The engineering properties of lunar soils and terrestrial fine sands are remarkably similar, however.

The clay minerals occur in particles of such small size that physico-chemical interactions with each other and with the water–electrolyte phase of a soil may be great. The clay phase may exert an influence on physical properties far exceeding its relative abundance in the soil.

The small size of clay particles precluded study of details of their structure prior to development of X-ray diffraction methods. Since then the study of clay minerals has been very extensive, aided in addition by the development of such techniques as electron microscopy, differential thermal analysis, and the electron probe. A large number of different clay minerals have been identified on the basis of composition. On the basis of structure, however, these minerals fall into a relatively small number of groups, and, furthermore, only a limited number of the different minerals are found with a significant abundance in the soils commonly encountered in engineering practice.

3.2 NONCLAY MINERALS IN SOILS

The gravel and sand fractions as well as the bulk of the silt fraction of a soil are composed of nonclay minerals. As soils are the products of the breakdown of pre-existing rocks and soils, they represent products of weathering. It would be expected, therefore, that the predominant mineral constituents of any soil would be those that are one or more of the following:

1. Very abundant in the source material.
2. Highly resistant to weathering, abrasion, and impact.
3. Weathering products.

The clay minerals are generally derived from weathering of pre-existing materials. The predominant nonclays, however, are usually rock fragments or mineral grains of the common rock-forming minerals. In igneous rocks, which represent the original source material for many soils, the most prevalent minerals are the feldspars (about 60 percent) and the pyroxenes and amphiboles (about 17 percent). Quartz accounts for about 12 percent of these rocks, micas for 4 percent, and other minerals for about 8 percent.

In most soils, however, the most abundant nonclay mineral by far is quartz, with small amounts of feldspar and mica sometimes present. The pyroxenes and amphiboles are seldom found to any significant extent. Carbonate minerals, mainly calcite and dolomite, are also found in some soils and occur as bulky particles, shells, precipitates, or in solution. Carbonates may dominate the composition of some deep sea sediments. Iron and aluminum oxides are abundant in residual soils of tropical regions.

Quartz is composed of silica tetrahedra grouped in such a way as to form spirals, with all tetrahedral oxygens bonded to silicon. As discussed in Chapter 2, the tetrahedral structure has a high stability. In addition, the spiral grouping of tetrahedra produces a structure without cleavage planes, quartz is already an oxide, there are no weakly bonded ions in the structure, and the mineral has a high hardness. These factors account for the persistence of quartz in the nonclay fraction of soils.

In the case of the feldspars, Marshall (1964) states: "The predominance of the feldspars among minerals of the lithosphere* makes their disappearance by

* The solid part of the earth.

weathering quantitatively the most important chemical reaction since solidification of the earth's crust." The feldspars are silicate minerals with a three-dimensional framework structure wherein part of the silicon is replaced by aluminum. The excess negative charge resulting from this replacement is balanced by cations such as potassium, calcium, sodium, strontium, and barium. As these cations are relatively large, their coordination number is also large. This results in the formation of an open structure and low bond strengths between units. As a consequence, there are cleavage planes, the hardness is only moderate, and feldspars are relatively easily broken down, which accounts for their lack of abundance in soils as compared to their abundance in igneous rocks.

Mica has a sheet structure composed of tetrahedral and octahedral units. Sheets are stacked one on the other and bonded primarily by potassium ions in 12-fold coordination that provide an electrostatic bond of moderate strength. In comparison with the bonds within layers, however, this bond is weak, thus accounting for the perfect basal cleavage of the micas. As a result of the platy morphology of mica plates, sand and silts containing only a few percent mica

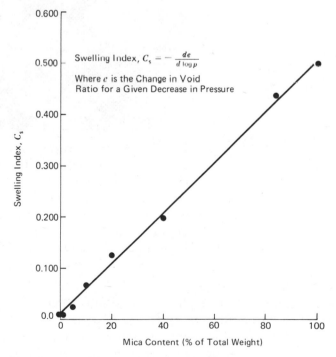

Fig. 3.1 Swelling index as a function of mica content for coarse-grained mixtures (Data from Terzaghi, 1931).

Fig. 3.2 Photomicrographs of sand and silt particles from several soils (a) Ottawa standard sand (Courtesy K. L. Lee). (b) Monterey sand (Courtesy K. L. Lee). (c) Sacramento River sand. (d) Eliot sand (Courtesy K. L. Lee). (e) Lunar soil mineral grains (Photo Courtesy Johnson Space Center). Squares in background area are 1 × 1 mm. (f) Recrystallized breccia particles from lunar soil (Photo Courtesy Johnson Space Center). Squares in background grid are 1 × 1 mm.

may exhibit both high compressibility and large swelling during unloading, as may be seen in Fig. 3.1 for coarse-grained mixtures of sand and mica.

The crystal structures and compositions of the amphiboles, pyroxene, and olivine are such that they are rapidly broken down by weathering; hence, they are absent from most soils.

Examples of silt and sand particles from different natural soils are shown in Fig. 3.2. Such particles can be classified in terms of angularity or roundness. Figure 3.3 shows one system that has been used for this purpose. Elongated and platy particles in soils can give rise to preferred orientations, which, in turn, can result in anisotropic mechanical properties.

Figure 3.2 (*continued*).

Another characteristic that influences mechanical behavior is the surface texture.

3.3 STRUCTURAL UNITS OF THE LAYER SILICATES

The clay minerals that are commonly found in soils belong to the larger mineral family termed *phyllosilicates,* which also contains other layer silicates such as serpentine, pyrophyllite, talc, mica, and chlorite that are themselves considered clay minerals by some when they occur in the clay-size fraction of a soil. The clay minerals usually occur in small particle sizes, and they ordinarily have unit cells with a residual negative charge. Most clay minerals exhibit plasticity when mixed with limited amounts of water, and they have relatively high resistance to weathering.

The structures of the common layer silicates can be considered in terms of two simple structural units. The different clay mineral groups are characterized by the stacking arrangements of *sheets** (sometimes chains) of these units and the manner in which two successive two- or three-sheet *layers* are held together.

* In conformity with the recommendations of the Nomenclature Committee of the Clay Minerals Society (Bailey et al., 1971), the following terms are used: a *plane* of atoms, a *sheet* of basic structural units; and a *layer* of unit cells composed of two, three, or four sheets.

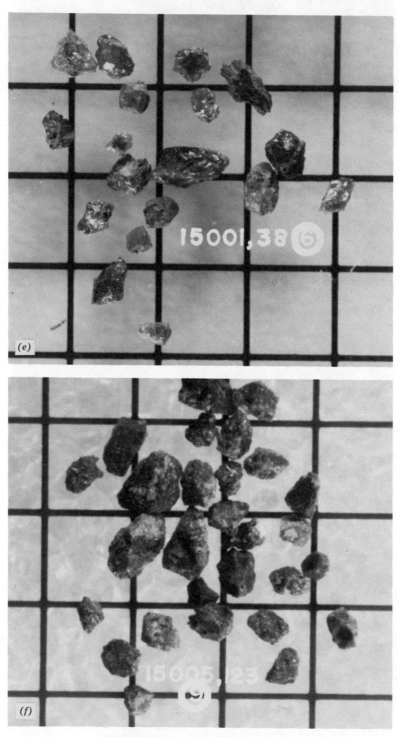

Figure 3.2 (*continued*).

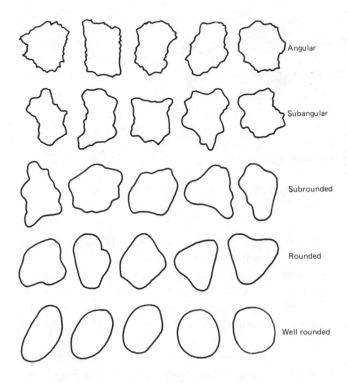

Fig. 3.3 Classification of shapes for sand and silt size particles. (Roundness of mineral particles as seen in silhouette; Müller, 1967).

Differences among minerals within clay mineral groups arise chiefly from differences in the type and amount of isomorphous substitution within the crystal structure. Because the possible substitutions are nearly endless in number, and because the development of crystal structure may range from very poor to nearly perfect, it is not surprising that the study of clay minerals appears imposing to the beginner, and the classification and naming of clay minerals provides an endless challenge to the expert. However, a knowledge of the structural and compositional characteristics of each group without detailed study of the subtleties of each specific mineral is adequate for many engineering purposes.

The two basic units in clay mineral structures are the *silica tetrahedron,* with a silicon ion tetrahedrally coordinated with four oxygens, and the *aluminum or magnesium octahedron,* wherein an aluminum or magnesium ion is octahedrally coordinated with six oxygens or hydroxyls. The make-up of these basic units is shown in Figs. 3.4 and 3.5.

Silica sheet

In most clay mineral structures the silica tetrahedra associate in a sheet structure. Three of the four oxygens of each tetrahedron are shared to form a hexagonal net, as shown in Fig. 3.6. The bases of the tetrahedra are all in the same plane and the tips all point in the same direction. The structure can repeat indefinitely and has the composition $(Si_4O_{10})^{4-}$. Electrical neutrality can be obtained by replacement of four oxygens by hydroxyls or by union with a sheet of different composition that is positively charged. The oxygen-to-oxygen distance is 2.55 Å, the space available for the silicon ion is 0.55 Å, and the thickness of the sheet in clay mineral structures is 4.63 Å (Grim, 1968).

Silica chains

In some of the less common clay minerals, silica tetrahedra are arranged in bands (double chains) of composition $(Si_4O_{11})^{6-}$. Electrical neutrality is achieved and the bands are bound together by means of aluminum and/or magnesium ions. A diagrammatic sketch of this structure is shown in Fig. 2.10. Minerals in this group resemble the amphiboles in structure.

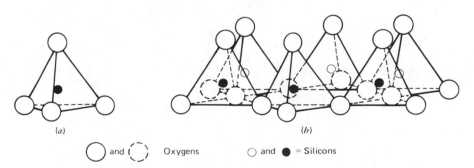

(a) *(b)*

◯ and (◌) Oxygens   ◯ and ● = Silicons

Fig. 3.4 Silica tetrahedron and silica tetrahedra arranged in a hexagonal network.

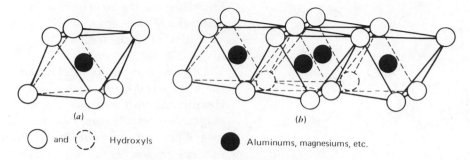

(a)   *(b)*

◯ and ⌇ Hydroxyls   ● Aluminums, magnesiums, etc.

Fig. 3.5   Octahedral unit and sheet structure of octahedral units.

Octahedral sheet

This sheet structure is composed of magnesium or aluminum coordinated octahedrally with oxygens or hydroxyls. In some cases, other cations are present in place of Al^{3+} and Mg^{2+}, such as Fe^{2+}, Fe^{3+}, Mn^{2+}, Ti^{4+}, Ni^{2+}, Cr^{3+}, and Li^{+}. Figure 3.5*b* is a schematic diagram of such a sheet structure. The oxygen-to-oxygen distance is 2.60 Å, the (OH)–(OH) distance is 2.94 Å, and the space available for the octahedrally coordinated cation is 0.61 Å. The thickness of the sheet is 5.05 Å in clays (Grim, 1968).

If the cation is trivalent, then only two-thirds the possible cationic spaces are normally filled, and the structure is termed *dioctahedral*. In the case of aluminum, the composition is $Al_2(OH)_6$ and, by itself, gives the mineral gibbsite. When found in clay mineral structures, an aluminum octahedral sheet is often referred to as a *gibbsite* sheet.

If the octahedrally coordinated cation is divalent, then all possible cation sites normally are filled, and the structure is *trioctahedral*. In the case of magnesium, the composition is $Mg_3(OH)_6$ giving the min-

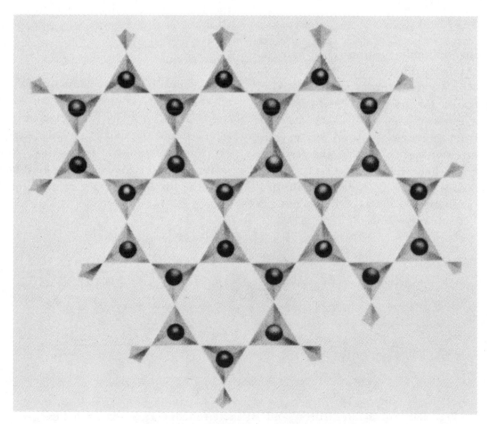

Fig. 3.6   Silica sheet in plan view.

eral brucite. In clay mineral structures, a sheet of magnesium octahedra is termed a *brucite* sheet.

In the following description of clay mineral structures schematic representations are used for the different structural units:

Silica sheet ⟋⟍ or ⟍⟋ (tips up)   (tips down)

Octahedral sheet ▭ (Various cations in octahedral coordination)

Gibbsite sheet [G] (Octahedral sheet cations are mainly aluminum)

Brucite sheet [B] (Octahedral sheet cations are mainly magnesium)

Water layers are found in some structures and may be represented by OOOOOOOO for a single layer, ○○○○○○○○ / ○○○○○○○○ for a double layer, and so on. Atoms of a specific type, for example, potassium are represented thus: ⓚ.

The diagrams are indicative of the clay mineral layer structure only. They do not indicate the correct width-to-length ratios for the minerals. The structures shown are idealized; in actual minerals, irregular substitutions and interlayering are common. Furthermore, the naturally occurring minerals are not necessarily formed by direct assembly of the basic units described above. The "building block" approach is useful, however, for the development of conceptual models.

3.4 CLASSIFICATION OF CLAY MINERALS

A complete classification system for the layer silicates that make up the various clay minerals should probably take into account three criteria (Warshaw and Roy, 1961). These are:

1. The height of the unit cell or "thickness of layer".
2. Composition, whether dioctahedral or trioctahedral, and ionic content of layer.
3. Stacking sequence of layers and degree of orderliness of stacking.

Grouping clay minerals according to crystal structure and stacking sequence of layers is convenient, since the members of the same group have somewhat similar engineering properties. The classification scheme for layer silicates recommended by the Nomenclature Committee of the Clay Minerals Society (Bailey *et al.* 1971) was developed on this basis, and is shown in Table 3.1.

The basic structure for each mineral group has been included schematically in Table 3.1. All the minerals have unit cells consisting of two, three, or four sheets. The two-sheet minerals are made up of a silica sheet and an octahedral sheet. The unit cell of the three-sheet minerals is composed of either a dioctahedral or trioctahedral sheet sandwiched between two silica sheets. Unit cells may be stacked closely together or water layers may intervene. The four-sheet structure of chlorite is composed of a 2:1 layer plus an interlayer hydroxide sheet.

Micalike clay minerals (illite, hydrous mica) are very common in soils encountered in practice, and hence their importance is much greater than implied by the table. In some soils inorganic clay-like material is found which has no detectable specific crystal structure and, thus, cannot be classified into one of the groups listed in Table 3.1. Such material is termed *allophane*.

The 2:1 mineral groups differ from each other mainly in terms of the type and amount of interlayer "glue". For example, the smectites are characterized by loosely held cations between layers, the micas by firmly fixed potassium ions between layers, and vermiculite by one or two layers of water and cations. The chlorite group may be viewed as an end member of a sequence leading to 2:1 layers bonded by an organized hydroxide sheet. The charge per formula unit is variable both within and between groups (Table 3.1), and this reflects the fact that the range of compositions is great as a result of varying amounts of *isomorphous substitution*. Because of this the boundaries between groups are somewhat arbitrary.

Isomorphous substitution

The concept of isomorphous substitution was introduced in Section 2.8 in connection with description of some of the silicate crystals. Isomorphous substitution is an important factor in the structure and behavior of the clay minerals. In an ideal gibbsite sheet, only two-thirds of the octahedral spaces are filled, and all of the cations are aluminum. In an ideal brucite sheet, all the octahedral spaces are filled by magnesium. In an ideal silica sheet, all the tetrahedral spaces are filled by silicons. In the naturally occurring clay minerals, however, some of the tetrahedral and octahedral positions are occupied by cations other than those in the ideal structure. Common examples are aluminum in place of some silicon, magnesium instead of aluminum, and ferrous iron (Fe^{2+}) for magnesium. This occupation of an octahedral or tetrahedral position by a cation other than the one

Table 3.1 Proposed Classification Scheme for Phyllosilicates Related to Clay Minerals

Type	Group (x = charge per formula unit)	Subgroup	Species[a]	Basic Structure
1:1	Kaolinite–serpentine $x \sim 0$	Kaolinites Serpentines	Kaolinite, halloysite Chrysotile, lizardite, anti-gorite	
2:1	Pyrophyllite–talc $x \sim 0$	Pyrophyllites Talcs	Pyrophyllite Talc	
	Smectite– or Montmorillonite– Saponite $x \sim 0.25–0.6$	Dioctahedral Smectites or Montmorillonites Trioctahedral Smectites or Saponites	Montmorillonite, beidellite, nontronite Saponite, hectorite, sauconite	
2:1	Vermiculite $x \sim 0.6–0.9$	Dioctahedral vermiculites Trioctahedral vermiculites	Dioctahedral vermiculite Trioctahedral vermiculite	
	Mica[b] $x \sim 1$	Dioctahedral micas Trioctahedral micas	Muscovite, paragonite Biotite, phlogopite	
	Brittle mica $x \sim 2$	Dioctahedral brittle micas Trioctahedral brittle micas	Margarite Clintonite	
2:1:1	Chlorite x variable	Dioctahedral chlorites Di,trioctahedral chlorites Trioctahedral chlorites	Donbassite Cookeite, sudoite Pennite, clinochlore, prochlorite	

[a] Only a few examples are given.
[b] The status of *illite* (or *hydromica*), *sericite*, and so on, are left open because it is not clear whether or at what level they would enter the table; many materials so indicated may be interstratified.
Bailey *et al.* (1971).

normally found, without change in crystal structure, is isomorphous substitution. The tetrahedral and octahedral cation distributions develop during initial formation of the mineral, not by later replacement.

3.5 INTERSHEET AND INTERLAYER BONDING IN THE CLAY MINERALS

The stacking of silica, gibbsite, and brucite sheets to form the basic clay mineral layer is such that a single plane of atoms is common to both the tetrahedral and octahedral sheets. Thus, bonding between these sheets is of the primary valence type and very strong. On the other hand, the bonds holding the unit cell layers together may be of several types, and they may be sufficiently weak that the physical and chemical behavior of the clay may be influenced by the response of these bonds to changes in environmental conditions.

Isomorphous substitution in all the clay minerals, with the possible exception of the kaolinites, gives clay particles a net negative charge. To preserve electrical neutrality cations are attracted and held on the surfaces and the edges, and in some clays, between the unit cells. These cations are termed "exchangeable cations" because in most instances cations of one type may be replaced by cations of another type. The quantity of exchangeable cations required to balance the charge deficiency of a clay is termed the "cation exchange capacity" (cec) and is usually expressed as milliequivalents* per 100 grams of dry clay.

* Equivalent weight = combining weight of an element = (atomic weight/valence).
Number of milliequivalents = 1000× (weight of element/atomic weight) × valence.
The number of ions in an equivalent = (Avogadro's number/valence).
Avogadro's number = 6.02×10^{23}. An equivalent contains 6.02×10^{23} electron charges or 96,500 coulombs, which is 1 Faraday.

Five types of interlayer bonding in the layer silicates are possible (Marshall, 1964).

1. Neutral parallel layers are held by van der Waals bonds. Bonding is relatively weak; however, relatively stable crystals of appreciable thickness may form as evidenced by the nonclay minerals pyrophyllite and talc, although cleavage parallel to the layers in these minerals is easy.

2. In minerals such as kaolinite, brucite, and gibbsite there are opposing layers of oxygen and hydroxyl or hydroxyl and hydroxyl. This gives rise to hydrogen bonding as well as van der Waals bonding between layers, which provides a fairly strong bond that will not separate in the presence of water but can still provide good cleavage.

3. Neutral silicate layers may be separated by layers of highly polar water molecules giving rise to hydrogen bonding.

4. Cations required to provide electrical neutrality may take up positions that control interlayer bonding characteristics. In the micas, some of the silicon is replaced by aluminum in the silica sheets. The resulting charge deficiency is balanced in part by potassium ions between mineral layers. Since the size of the potassium ion is such that it fits into the holes formed by the bases of the silica tetrahedra in the silica sheet (Fig. 3.4), it provides a strong bond between mica layers. In the chlorites, the charge deficiencies resulting from substitutions in the octahedral sheet in the 2:1 sandwich are balanced by a charge excess on the one-sheet layer interleaved between the three-sheet layers. This provides a strongly bonded structure that, while exhibiting cleavage, will not lead to separation in the presence of water or other polar liquids.

5. When the surface charge density is moderate, as in the case of the smectites and vermiculites, the silicate layers may readily take up polar molecules, and balancing cations may hydrate, resulting in a separation of layers and expansion. The strength of the interlayer bond is low and a sensitive function of charge distribution, hydration energy of the cation, geometry of cation in relation to the silicate surface, surface ion configuration, and structure of the polar molecule.

The smectite and vermiculite particles swell; whereas, particles of the nonclay minerals pyrophyllite and talc, which have comparable structures, do not. There are two possible explanations for this and direct evidence in support of either is difficult to obtain (van Olphen, 1963).

1. The interlayer cations in the smectites hydrate, and the hydration energy overcomes the attractive forces between unit layers. There are no interlayer cations in pyrophyllite, hence, no swelling.

2. Water does not hydrate the cations but is adsorbed on oxygen surfaces by hydrogen bonds. There is no swelling in pyrophyllites and talc because the surface hydration energy is too small to overcome the van der Waals forces between layers, which are greater in these minerals because of a smaller interlayer distance.

3.6 THE 1:1 MINERALS

Structure

The kaolinite–serpentine minerals are composed of alternating silica and octahedral sheets as shown schematically in Fig. 3.7. The tips of the silica tetrahedra and one of the planes of atoms in the octahedral sheet are common. The tips of the tetrahedra all point in the same direction, toward the center of the unit cell. In the plane of atoms common to both sheets, two-thirds of the atoms are oxygens and are shared by both silicon and the octahedral cations. The remaining atoms in this plane are (OH) and are located so that each is directly below the hole in the hexagonal net formed by the bases of the silica tetrahedra. If the octahedral layer is brucite, then a mineral of the serpentine subgroup results; whereas, dioctahedral gibbsite layers lead to clay minerals in the kaolinite subgroup. Trioctahedral 1:1 minerals are relatively rare, usually occur mixed with kaolinite or illite, and are hard to identify. A diagrammatic sketch of the kaolinite structure is shown in Fig. 3.8. The structural formula is $(OH)_8Si_4Al_4O_{10}$, and the charge distribution is indicated in Fig. 3.9.

Mineral particles of the kaolinite subgroup consist of the basic layers stacked in the c direction. The bonding between successive layers is both by van der Waals forces and hydrogen bonds. This bonding is of sufficient strength that there is no interlayer swelling.

Because of slight differences in the oxygen-to-oxygen distances in the tetrahedral and octahedral layers there is some distortion of the ideal hexagonal tetrahedral network. The upward directed Si–O bond is

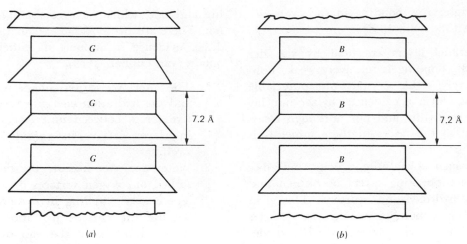

(a)   (b)

Fig. 3.7   Schematic diagrams of the structures of kaolinite and serpentine (a) Kaolinite. (b) Serpentine.

slightly tilted to fit the overlying octahedron. Apparently, because of this, the mineral kaolinite, which is the most important member of the subgroup and a common soil mineral, is triclinic instead of monoclinic. The unit cell dimensions are $a = 5.16$ Å, $b = 8.94$ Å, $c = 7.37$ Å, $\alpha = 91.8°$, $\beta = 104.5°$, and $\gamma = 90°$.

Variations between members of the kaolinite subgroup consist of the way layers are stacked above each other and possibly in the position of aluminum ions within the available sites in the octahedral sheet. Dickite and nacrite are rarely found. The dickite unit cell is made up of two unit layers, and the nacrite unit cell contains six. Both appear to form as a result of hydrothermal processes. Dickite is fairly common as a secondary clay in the pores of sandstone and in coal beds.

Halloysite

The mineral halloysite is a particularly interesting member of the kaolinite subgroup. Two distinct forms of this mineral exist as shown in Fig. 3.10; one a nonhydrated form having the same structural composition as kaolinite $((OH)_8Si_4Al_4O_{10})$ and the other a hydrated form consisting of unit kaolinite layers separated from each other by a single layer of water molecules having the composition $(OH)_8Si_4Al_4O_{10} \cdot 4H_2O$. The basal spacing in the c direction, $d_{(001)}$ for nonhydrated halloysite is about 7.2 Å, as for kaolinite. Because of the interleaved water layer, $d_{(001)}$ for hydrated halloysite is about 10.1 Å. The difference between these values, 2.9 Å, is the approximate thickness of a single layer of water molecules. A partially

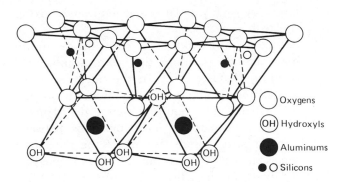

Fig. 3.8   Diagrammatic sketch of the kaolinite structure.

hydrated form with basal spacing in the range of 7.4 to 7.9 Å may also occur.

Hydrated halloysite, which has also been termed halloysite $(4H_2O)$, can dehydrate irreversibly to halloysite $(2H_2O)$ sometimes known as metahalloysite.

Isomorphous substitution and exchange capacity

Controversy exists as to whether or not any isomorphous substitution exists within the structure of the kaolinites. Nonetheless, values of cation exchange capacity for kaolinite in the range of 3 to 15 meq/100 g and from 5 to 40 meq/100 g for halloysite have been measured. Thus, clear evidence exists that the particles possess a net negative charge. Possible sources of this charge deficiency are:

1. Substitution of Al^{3+} for Si^{4+} in the silica sheet or a divalent ion for Al^{3+} in the octahedral sheet. Replacement of only one Si in every 400

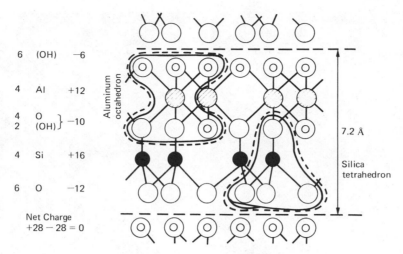

6	(OH)	−6
4	Al	+12
4 2	O (OH) }	−10
4	Si	+16
6	O	−12

Net Charge
+28 − 28 = 0

Fig. 3.9 Charge distribution in kaolinite.

would be adequate to account for the exchange capacity of many kaolinites, so proof of substitution using analytical methods is not easily obtained.

2. The hydrogen of exposed hydroxyls may be replaced by an exchangeable cation. This mechanism may be doubtful, however, since the hydrogen would probably not be replaceable under the conditions of most exchange reactions (Grim, 1968).

3. Broken bonds around particle edges give rise to unsatisfied charges that are balanced by adsorbed cations.

Considerable evidence exists that kaolinite particles are actually charged positively on their edges when in a low (acid) pH environment, but negatively charged in a high (basic) pH environment. Low exchange capacities are measured under low pH conditions and high exchange capacities are obtained for determinations at high pH. This supports broken bonds as a partial source of exchange capacity. That a positive cation exchange capacity is measured under low pH conditions when edges are positively charged indicates that some isomorphous substitution must exist as well.

Because interlayer separation does not occur in kaolinite, balancing cations must adsorb on the exterior surfaces of the particles.

Morphology and surface area

Well-crystallized particles of kaolinite (Fig. 3.11), nacrite, and dickite occur as well-formed six-sided plates. The lateral dimensions of these plates may range from about 0.1 to 4 μm, and their thickness may be about 0.05 to 2 μm. Stacks of kaolinite layers up to 4000 μm in thickness have been observed, although they are not common. Poorly crystallized kaolinite generally occurs as less distinct hexagonal plates, and the particle size is usually smaller than for the well-crystallized varieties.

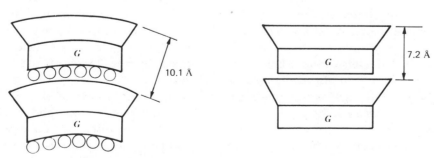

Fig. 3.10 Schematic diagrams of the structure of halloysite. (*a*) Hydrated halloysite. (*b*) Non-hydrated halloysite.

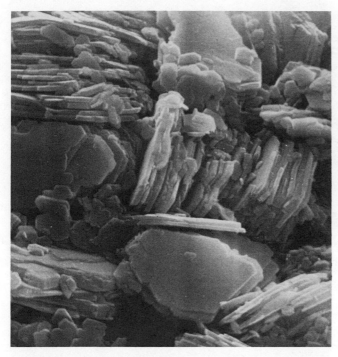

Fig. 3.11 Electron photomicrograph of well-crystallized kaolinite from St. Austell, Cornwall, England, Picture width is 17 μm (Tovey, 1971).

Fig. 3.12 Electron photomicrograph of halloysite from Bedford, Indiana. Picture width is 2 μm (Tovey, 1971).

The morphology of halloysite is among the most interesting of any of the clay minerals. The hydrated form of this mineral occurs as cylindrical tubes of overlapping sheets of the kaolinite type (Fig. 3.12). The c-axis at any point nearly coincides with the tube radius. The formation of tubes has been explained (Bates, Hildebrand, and Swineford, 1950) on the basis of a misfit in the b direction of the silica and gibbsite sheets. The b dimension of kaolinite is 8.93 Å. In gibbsite, however, it is only 8.62 Å. This implies that the (OH) spacings in kaolinite are somewhat stretched in order to obtain the proper fit with the silica sheet. Evidently in hydrated halloysite, the reduced interlayer bond, caused by the intervening layer of water molecules, enables the (OH) layer to revert to 8.62 Å resulting in a curvature of the unit with the hydroxyls on the inside and the bases of the silica tetrahedra on the outside. The outside diameters of the tubular particles range from about 0.05 to 0.20 μm with a median value of 0.07 μm. The wall thickness is about 0.02 μm (Bates, Hildebrand, and Swineford, 1950). The tubes may range in length from a fraction to several micrometers. Electron microscope studies have shown that drying of hydrated halloysite may result in a splitting or unrolling of the tubes.

The specific surface area of kaolinite is of the order of 10 to 20 m²/g of dry clay; whereas, that of hydrated halloysite is in the range of 35 to 70 m²/g.

3.7 THE SMECTITE MINERALS

Structure

The minerals of the smectite group have a prototype structure similar to that of pyrophyllite, consisting of an octahedral sheet sandwiched between two silica sheets, as shown schematically in Fig. 3.13 and diagrammatically in three dimensions in Fig. 3.14. All the tips of the tetrahedra point toward the center of the unit cell. The oxygens forming the tips of the tetrahedra are common to the octahedral sheet as well. The remaining anions in the octahedral sheet that fall directly above and below the hexagonal holes formed by the bases of the silica tetrahedra are hydroxyls.

The layers formed in this way are continuous in the a and b directions and stacked one above the other in the c direction. Bonding between successive layers is by van der Waals forces and by cations that may be present to balance charge deficiencies in the structure. These bonds are weak and easily separated

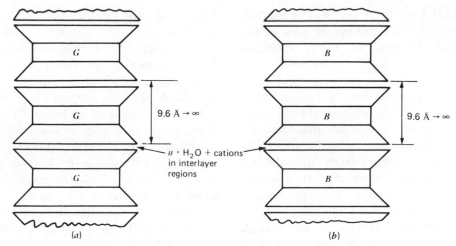

Fig. 3.13 Schematic diagrams of the structures of the smectite minerals (a) Montmorillonites. (b) Saponites.

by cleavage or adsorption of water or other polar liquids. The basal spacing in the c direction, $d_{(001)}$, is variable, ranging from about 9.6 Å to complete separation.

The theoretical composition in the absence of lattice substitutions is $(OH)_4Si_8Al_4O_{20} \cdot n$(interlayer) H_2O. The structural configuration and the corresponding charge distribution are shown in Fig. 3.15. The structure shown is electrically neutral overall, and it is the same as that of the nonclay mineral pyrophyllite.

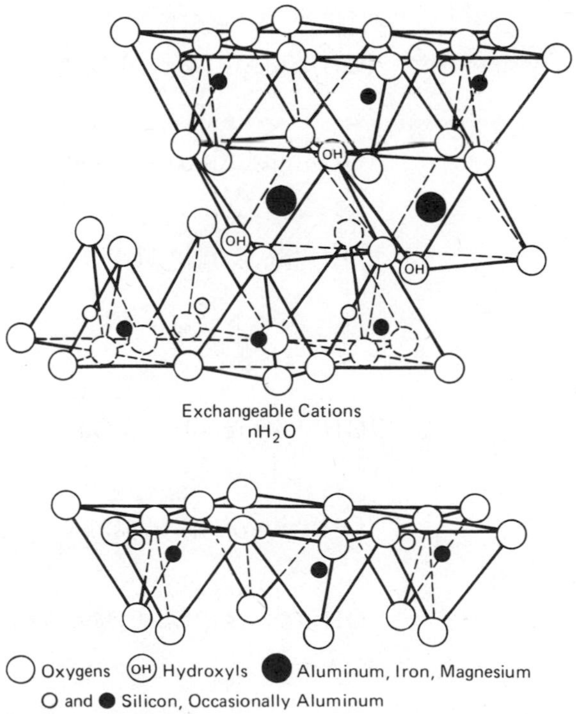

Isomorphous substitution in the smectite minerals

The factor that sets the minerals of the smectite group apart from pyrophyllite–talc as a clay mineral class is extensive substitution for aluminum and silicon within the lattice by other cations. Aluminum in the octahedral sheet may be replaced by magnesium, iron, zinc, nickel, lithium, or other cations.

Exchangeable Cations
nH_2O

○ Oxygens ⊙ Hydroxyls ● Aluminum, Iron, Magnesium
○ and ● Silicon, Occasionally Aluminum

Fig. 3.14 Diagrammatic sketch of the montmorillonite structure.

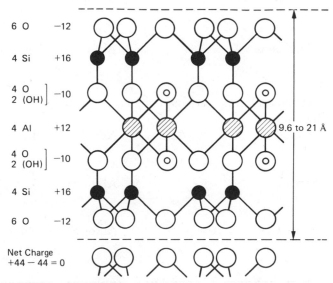

6 O	−12
4 Si	+16
4 O 2 (OH)	−10
4 Al	+12
4 O 2 (OH)	−10
4 Si	+16
6 O	−12

9.6 to 21 Å

Net Charge
+44 − 44 = 0

Fig. 3.15 Charge distribution in Pyrophyllite (type structure for montmorillonite).

Aluminum may replace up to 15 per cent of the silicon ions in the tetrahedral sheet. Possibly some of the silicon positions can be occupied by phosphorous (Grim, 1968).

Substitutions for aluminum in the octahedral sheet may be one-for-one or three-for-two (since aluminum occupies only two-thirds the possible octahedral sites) in any combination from a few to complete replacement. The only restriction appears to be that the resulting structure is either almost exactly dioctahedral (montmorillonite subgroup) or exactly trioctahedral (saponite subgroup). The charge deficiency resulting from these substitutions is in the range of 0.5 to 1.2, and it is usually close to 0.66 per unit cell. A charge deficiency of this amount is readily obtained by replacement of every sixth aluminum by a magnesium. Montmorillonite, the most common mineral of the group, has this composition. The charge deficiency resulting from these substitutions is balanced by exchangeable cations that take up positions between the unit cell layers and on the surfaces of particles.

Some minerals of the smectite group and their structural formulas are listed in Table 3.2. An arrow indicates the source of the charge deficiency, assumed to be 0.66 in each case, and the amount of exchangeable sodium required to balance the structure. More than one formula may be found in the literature for several of the minerals, differing in the relative amounts of substitution in different parts of the structure. This reflects both the great variety of compositions that may exist within the same basic crystal structure and the difficulties of identification and classification. Thus, the formulas given in Table 3.2 cannot be taken as absolute, but as indicative of the general character of the mineral.

Because of the large amount of unbalanced substitution in the smectite minerals, they exhibit high cation exchange capacities, generally in the range of 80 to 150 meq/100 g.

Table 3.2 Some Minerals of the Smectite Group

Mineral	Tetrahedral Sheet Substitutions	Octahedral Sheet Substitutions	Formula/Unit Cell[a]
Dioctahedral, Smectites or Montmorillonites			
Montmorillonite	None	1 Mg^{2+} for every sixth Al^{3+}	$(OH)_4Si_8(Al_{3.34}Mg_{0.66}) O_{20}$ $\downarrow$ $Na_{0.66}$
Beidellite	Al for Si	None	$(OH)_4(Si_{6.34}Al_{1.66}) Al_{4.34}O_{20}$ $\downarrow$ $Na_{0.66}$
Nontronite	Al for Si	Fe^{3+} for Al	$(OH)_4(Si_{7.34}Al_{0.66}) Fe_4{}^{3+}O_{20}$ $\downarrow$ $Na_{0.66}$
Trioctahedral, Smectites or Saponites			
Hectorite	None	Li for Mg	$(OH)_4Si_8(Mg_{5.34}Li_{0.66}) O_{20}$ $\downarrow$ $Na_{0.66}$
Saponite	Al for Si	Fe^{3+} for Mg	$(OH)_4Si_{7.34}Al_{0.66}) Mg_6O_{20}$ $\downarrow$ $Na_{0.66}$
Sauconite	Al for Si	Zn for Mg	$(OH)_4(Si_{8-y}Al_y)(Zn_{6-x}Mg_x) O_{20}$ $\downarrow$ $Na_{0.66}$

[a] Two formula units are needed to give one unit cell.

After Ross and Hendricks (1945); Marshall (1964); and Warshaw and Roy (1961).

Morphology and surface area

Because of the very small sizes in which particles of the smectite minerals are usually found and their tendency to break down into sheets of unit cell thickness when dispersed in water, clear electron micrographs are not easily obtained. Montmorillonite usually occurs as equidimensional flakes so thin as to appear more like films as shown in Fig. 3.16. Particles may range in thickness from 10 Å (unit cell) upwards to about 1/100 of the width. The long axis of the plates may be up to several micrometers in length; however, it is usually less than 1 or 2 μm.

In cases where there is a large amount of substitution of iron and/or magnesium for aluminum, for example, nontronite, particles have a lath or needle-like shape. It has been postulated (Grim, 1962) that these shapes result from the fact that the Mg^{2+} ion and the Fe^{3+} ion are somewhat large for the octahedral lattice, and, hence, the structure is subjected to a directional strain.

The specific surface of the smectites is extremely large. The primary surface, that is, surface due to particle surfaces exclusive of interlayer zones, is generally in the range of 50 to 120 m^2/g. The secondary specific surface that may be exposed by expanding the lattice so that polar fluids can penetrate between layers may range from 700 to 840 m^2/g.

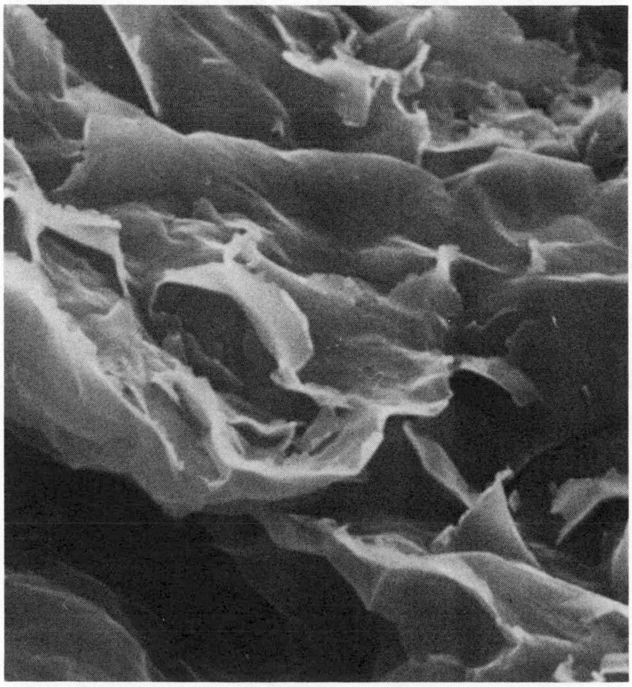

Bentonite

A very highly plastic, swelling clay material known as bentonite is widely used for a variety of purposes ranging from drilling muds for soil borings to clarification of beer and wine. The bentonite familiar to most geotechnical engineers is a highly colloidal, expansive clay that is an alteration product of volcanic ash and has a liquid limit of 500 percent or more. It is widely used as a backfill during the construction of slurry trench walls, as a grout material, as a sealant for piezometer installations, and for other special applications (Boyes, 1972).

When encountered naturally, or as a seam in rock formations or as a major constituent of soft shales, bentonite may be a continuing source of slope stability problems. Slide problems at Portugese Bend along the Pacific Ocean in southern California and in the Bearpaw shale in Saskatchewan can both be attributed in part to the presence of bentonite. Stability problems in underground construction may be caused by the presence of montmorillonite in joints and faults (Brekke and Selmer-Olsen, 1965).

3.8 THE MICALIKE CLAY MINERALS

Perhaps the most commonly occurring clay mineral found in the soils encountered in engineering practice has a structure similar to that of muscovite mica, and is termed "illite" or "hydrous mica." Although illite was not listed specifically in Table 3.1, it can be conveniently considered on the same level as kaolinite and montmorillonite. Vermiculite also is often found as a clay phase constituent of soils. Although it is classed as a separate group in Table 3.1, its structure is related to that of biotite mica.

Structure

The basic structural unit for the muscovite micas (white micas), shown schematically in Fig. 3.17a, is the three-layer silica–gibbsite–silica sandwich that forms pyrophyllite. The tips of all the tetrahedra in each silica sheet point toward the center and are common with octahedral sheet ions.

Muscovite differs from pyrophyllite in that about one-quarter of the silicon positions are filled by aluminum, and the resultant charge deficiency is balanced by potassium ions between the layers. The layers are continuous in the a and b directions and stacked in

Fig. 3.16 Electron photomicrograph of montmorillonite (bentonite) from Clay Spur, Wyoming. Picture width is 7.5 μm (Tovey, 1971).

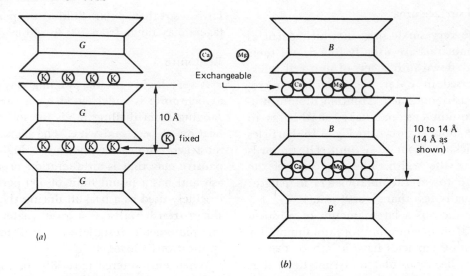

Fig. 3.17   Schematic diagram of the structures of muscovite, illite, and vermiculite. (a) Muscovite and illite. (b) Vermiculite.

the c direction. The radius of the potassium ion, 1.33 Å, is such that it fits snugly in the 1.32 Å radius hexagonal hole formed by the bases of the silica tetrahedra, where it is in 12-fold coordination with the six oxygens of each layer.

A diagrammatic three-dimensional sketch of the muscovite structure is shown in Fig. 3.18. The struc-

tural configuration and charge distribution are shown in Fig. 3.19, where it may be noted that the unit cell is electrically neutral and has the formula $(OH)_4K_2(Si_6Al_2)Al_4O_{20}$. Muscovite is the dioctahedral end member of the micas and contains only Al^{3+} in the octahedral layer. Phlogopite (brown mica) is the trioctahedral end member, having the octahedral positions filled entirely by magnesium. Its formula is $(OH)_4K_2(Si_6Al_2)Mg_6O_{20}$. The biotites (black micas) are trioctahedral, with the octahedral positions filled

Oxygens,   (OH) Hydroxyls,   ● Aluminum,   Potassium

○ and ● Silicons (One-Fourth Replaced by Aluminums)

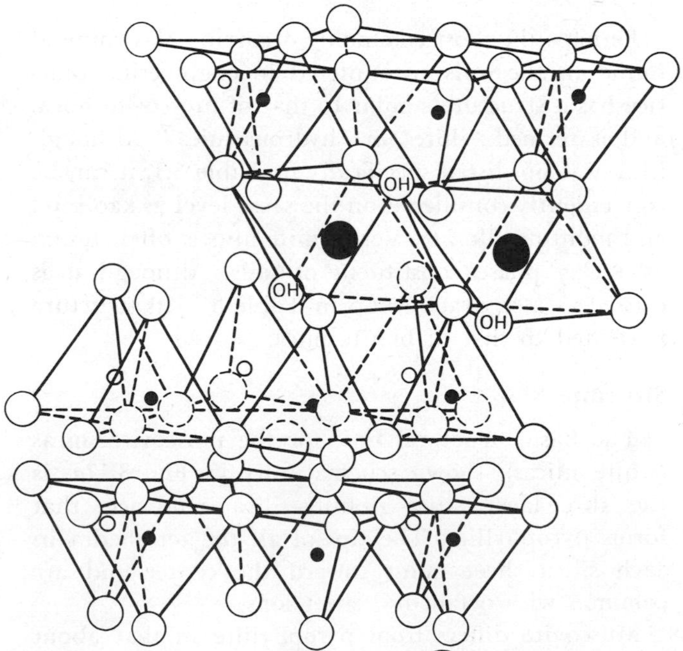

Fig. 3.18   Diagrammatic sketch of the structure of muscovite.

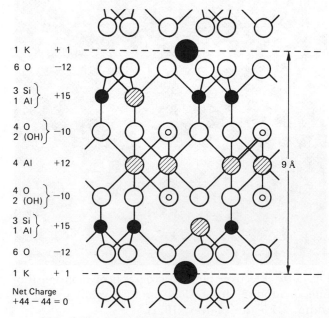

1 K	+1	
6 O	−12	
3 Si / 1 Al	+15	
4 O / 2 (OH)	−10	
4 Al	+12	
4 O / 2 (OH)	−10	
3 Si / 1 Al	+15	
6 O	−12	
1 K	+1	

Net Charge
+44 − 44 = 0

Fig. 3.19   Charge distribution in muscovite.

mostly by magnesium and iron. It has the general formula $(OH)_4K_2(Si_6Al_2)(MgFe)_6O_{20}$. The relative proportions of magnesium and iron may vary widely.

Illites differ from mica in the following ways (Grim, 1968):

1. Fewer of the Si^{4+} positions are occupied by Al^{3+} in illite.
2. There is some randomness in stacking of layers in illite.
3. There is relatively less potassium in illite. Well organized illite contains 9 to 10 percent K_2O (Weaver and Pollard, 1973).
4. The size of illite particles occurring naturally is very small.

Some illite may contain magnesium and iron in the octahedral sheet as well as aluminum (Marshall, 1964). Iron-rich illite, usually occurring as earthy green pellets, is termed glauconite.

The vermiculite structure consists of a regular interstratification of biotite mica layers and double molecular layers of water, as shown schematically in Fig. 3.17b. Actually, the thickness of the water layer between biotite units depends largely on the cation present in this region that balances charge deficiencies in the biotite-like layers. With magnesium or calcium present (the usual case in nature), there are two water layers, giving a basal spacing of 14 Å. A general formula for vermiculite is

$$(OH)_4(MgCa)_x(Si_{8-x}Al_x)\ (Mg \cdot Fe)_6O_{20}yH_2O$$

$$x \sim 1 \text{ to } 1.4, y \sim 8$$

Isomorphous substitution and exchange capacity

Isomorphous substitution in both the illite and vermiculite structure is extensive. The charge deficiency in illite is 1.3 to 1.5 per unit cell. In illite the charge deficiency is located primarily in the silica sheets and is balanced partly by the nonexchangeable potassium ions between layers. Thus, the cation exchange capacity of illite is less than that of montmorillonite, amounting to 10 to 40 meq/100 g. Values of exchange capacity greater than 10 to 15 meq/100 g may be indicative of some expanding layers (Weaver and Pollard, 1973). In the absence of the fixed potassiums, the exchange capacity would be about 150 meq/100 g. The interlayer bonding by potassium is sufficiently strong that the basal spacing of illite remains fixed at 10 Å in the presence of polar liquids.

In vermiculite, the charge deficiency is of the order of 1 to 1.4 per unit cell. Since the interlayer cations are exchangeable, the exchange capacity for this mineral is high, amounting to 100 to 150 meq/100 g. The basal spacing $d_{(001)}$ of the vermiculites is influenced by both the type of cation and dehydration. With potassium or ammonium in the exchange positions, the basal spacing is only 10.5 to 11 Å. Lithium gives 12.2 Å. The interlayer water can be driven off by heating to temperatures higher than 100°C. This dehydration is accompanied by a reduction in basal spacing to about 10 Å. The mineral quickly rehydrates and expands again to 14 Å when exposed to moist air at room temperature.

Morphology and surface area

Illites usually occur as very small, flaky particles mixed with other clay and nonclay materials. Illite deposits of high purity have not been located, as has been the case for kaolinite and montmorillonite. The flaky illite particles may have a hexagonal outline if well crystallized. The long axis dimension ranges from 0.1 μm or less to several micrometers, and the plate thickness may be as small as 30 Å. An electron photomicrograph of illite is shown in Fig. 3.20.

Vermiculite may occur in nature as large crystalline masses having a sheet structure somewhat similar in appearance to mica. In soils, vermiculite occurs as

Fig. 3.20 Electron photomicrograph of illite from Morris, Illinois. Picture width is 7.5 μm (Tovey, 1971).

small particles mixed with other clay minerals. Specific data concerning shape and size of these particles are not available.

Values of specific surface in the range of about 65 to 100 m²/g have been reported for illite. The primary surface area of the vermiculites is about 40 to 80 m²/g and the secondary (interlayer) surface may be as high as 870 m²/g.

3.9   THE CHLORITE MINERALS

Structure

The chlorite structure consists of alternating micalike and brucitelike layers as shown schematically in Fig. 3.21. The structure is similar to that of vermiculite, except that an organized octahedral sheet replaces the double water layer between mica layers. As with the kaolinites, smectites, and illites, the layers are continuous in the a and b directions and stacked in the c direction with basal cleavage. The basal spacing is fixed at 14 Å.

Isomorphous substitution

The central sheet of the mica layer is trioctahedral with magnesium as the predominate cation. There is often partial replacement of Mg^{2+} by Al^{3+}, Fe^{2+}, and Fe^{3+}. The silica sheets are unbalanced by substitution of Al^{3+} for Mg^{2+} in the brucite layer. The various members of the chlorite group differ in the kind and amounts of substitution and in the stacking of successive layers. The cation exchange capacity of the chlorites is in the range of 10 to 40 meq/100 g.

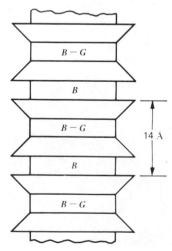

Fig. 3.21   Schematic diagram of the structure of chlorite.

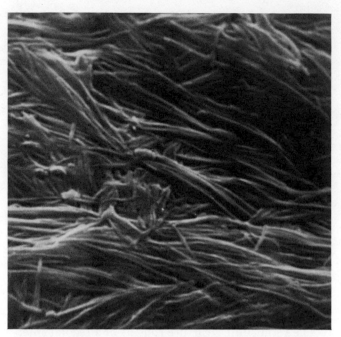

Fig. 3.22   Electron Photomicrograph of Attapulgite from Attapulgis, Georgia. Picture width is 4.7 μm (Tovey, 1971).

Morphology

Chlorite minerals are found as microscopic grains of platy morphology and poorly defined crystal edges in altered igneous and metamorphic rocks and soils derived therefrom. In soils, chlorites always appear to occur in mixtures with other clay minerals. Little specific information is available concerning particle size.

3.10   CHAIN STRUCTURE CLAY MINERALS

As indicated previously, some clay minerals are formed from bands (double chains) of silica tetrahedra. The minerals attapulgite, sepiolite, and palygorskite differ primarily in the replacements within the structure. These minerals have lathlike shapes with particle diameters of 50 to 100 Å and lengths of 4 to 5 μm. These minerals are not commonly encountered in soils of engineering interest. Figure 3.22 shows an electron photomicrograph of bundles of attapulgite particles.

3.11   SUMMARY OF CLAY MINERAL CHARACTERISTICS

Table 3.3 presents a summary of the important structural, compositional, and morphological characteristics of the important clay minerals. Data concerning the structural characteristics of tetrahedral and octahedral sheet structures are included, as well as values of ionic radii for the elements common in clays.

3.12 MIXED LAYER CLAYS

More than one type of clay mineral is usually found in most soils. Because of the great similarity in crystal structure among the different minerals, it is common, and not at all surprising, that interstratification of two or more different layer types often occurs within a single particle.

Interstratification may be regular, with a definite repetition of the different layers in sequence, or it may be random. According to Weaver and Pollard (1973), randomly interstratified minerals are second only to illite in abundance. The most abundant mixed-layer material is composed of expanded, water bearing layers and contracted nonwater bearing layers. Montmorillonite–illite is most common, although chlorite–vermiculite and chlorite–montmorillonite are often encountered. *Rectorite* is a regular interstratified clay with high charge, micalike layers with fixed interlayer cations alternating in a regular manner with low charge, montmorillonitelike layers with exchangeable cations capable of hydration.

3.13 NONCRYSTALLINE CLAY MATERIALS

Allophane

Noncrystalline silicate clay materials are generally termed allophane. These materials are noncrystalline, in the sense that they are amorphous to X rays, because there is insufficient long range order of the octahedral and tetrahedral units to produce sharp diffraction effects, although in some cases crude order may exist leading to diffraction bands. The allophanes have no definite composition or shape and may exhibit a wide range of physical properties. Some noncrystalline clay material is probably found in nearly all fine-grained soils. It may be particularly common in some soils formed from volcanic ash.

Oxides

There are probably no soils on earth that do not contain some amount of colloidal oxides and hydrous oxides (Marshall, 1964). The details of their occurrence and influence on the physical properties of a soil have not been studied much. The oxides and hydroxides of aluminum, iron, and silicon are of the greatest interest since they are the ones most frequently encountered. These materials may occur as gels or precipitates and coat mineral particles, or they may cement particles together. They may occur also as distinct crystalline units, for example, gibbsite, boehmite, hematite, and magnetite. Limonite and bauxite are sometimes found and represent amorphous mixtures of iron and aluminum hydroxides, respectively.

3.14 ORIGIN OF CLAY MINERALS

Clay minerals may be formed by one or more of the processes listed below (Keller, 1964):

1. Crystallization from solutions.
2. Weathering of silicate minerals and rocks.
3. Diagenesis, reconstitution, and ion exchange.
4. Hydrothermal alterations of minerals and rocks.
5. Laboratory synthesis.

Clay minerals are commonly found in the filling material (gouge) in joints, shears, and faults in rock, and often as an alteration product in the rock immediately adjacent to these discontinuities. Hydrothermal alterations may also lead to the formation of clay veins in rocks and zones of clay around hot springs and geysers. Clays formed in this way are not important constituents of soils; however, their presence in joints, shears, and faults is of great importance in the stability of underground openings and other rock structures.

The bulk of the clay minerals found in soils are formed by processes (1), (2), and (3). Further consideration of the nature of these processes is given in Chapter 4. In some cases, a silicate mineral, especially feldspar, may be replaced by a clay mineral within a granular rock structure as a result of various alteration processes. This may be of particular importance in the formation of decomposed granite, a broken-down rock material that is a frequent source of problems in foundation and earth work construction. The replacement mainly of feldspar by kaolinite is common.

3.15 PRACTICAL IMPLICATIONS

Mineralogy controls the sizes, shapes, and surface characteristics of the particles in a soil. These features, along with interactions with the fluid phase, determine plasticity, swelling, compression, strength,

Table 3.3 Summary of Clay

Structural

1. Silica Tetrahedron: Si atom at center. Tetrahedron units form hexagonal network = $Si_4O_6(OH)_4$
2. Gibbsite Sheet: Aluminum in octahedral coordination. Two-thirds of possible positions filled. $Al_2(OH)_6$—O–O = 2.60 Å.
3. Brucite Sheet: Magnesium in octahedral coordination. All possible positions filled. $Mg_3(OH)_6$—O–O = 2.60 Å

Type	Sub-Group and Schematic Structure	Mineral	Complete Formula/Unit Cell[a]	Octahedral Layer Cations	Tetrahedral Layer Cations	Structure Isomorphous Substitution	Interlayer Bond
	Allophane	Allophanes	Amorphous	—	—		
1:1	Kaolinite	Kaolinite	$(OH)_8Si_4Al_4O_{10}$	Al_4	Si_4	Little	O–OH Hydrogen Strong
		Dickite	$(OH)_8Si_4Al_4O_{10}$	Al_4	Si_4	Little	O–OH Hydrogen Strong
		Nacrite	$(OH)_8Si_4Al_4O_{10}$	Al_4	Si_4	Little	O–OH Hydrogen Strong
		Halloysite (dehydrated)	$(OH)_8Si_4Al_4O_{10}$	Al_4	Si_4	Little	O–OH Hydrogen Strong
		Halloysite (hydrated)	$(OH)_8Si_4Al_4O_{10}.4H_2O$	Al_4	Si_4	Little	O–OH Hydrogen Strong
2:1	Montmorillonite $(OH)_4Si_8Al_4O_{20}.nH_2O$ (Theoretical Unsubstituted)	Montmorillonite	$(OH)_4Si_8(Al_{3.34}.Mg_{.66})O_{20}.nH_2O$ ↓* $Na_{.66}$	$Al_{3.34}Mg_{.66}$	Si_8	Mg for Al, Net charge always = 0.66-/unit cell	O–O Very weak expanding lattice
		Beidellite	$(OH)_4(Si_{7.34}.Al_{.66})(Al_4)O_{20}.nH_2O$ ↓ $Na_{.66}$	Al_4	$Si_{7.34}Al_{.66}$	Al for Si, Net charge always = 0.66-/for unit cell	O–O Very weak expanding lattice
		Nontronite	$(OH)_4(Si_{7.34}.Al_{.66})Fe_4^{3+}O_{20}.nH_2O$ ↓ $Na_{.66}$	Fe_4	$Si_{7.34}Al_{.66}$	Fe for Al, Al for Si, Net charge always = 0.66-/for unit cell	O–O Very weak expanding lattice
	Saponite	Hectorite	$(OH)_4Si_8(Mg_{5.34}.Li_{.66})O_{20}.nH_2O$ ↓ $Na_{.66}$	$Mg_{5.34}Li_{.66}$	Si_8	Mg, Li for Al, Net charge always = 0.66-/unit cell	O–O Very weak expanding lattice
		Saponite	$(OH)_4(Si_{7.34}.Al_{.66})Mg_6O_{20}.nH_2O$ ↓ $Na_{.66}$	Mg, Fe^{3+}	$Si_{7.34}Al_{.66}$	Mg for Al, Al for Si, Net charge always = 0.66-/for unit cell	O–O Very weak expanding lattice
		Sauconite	$(Si_{6.94}Al_{1.06})Al_{.44}Fe_{.34}Mg_{.36}Zn_{4.80}O_{20}(OH)_4$ $.nH_2O$ ↓ $Na_{.66}$	$Al_{.44}Fe_{.34}Mg_{.36}Zn_{4.80}$	$Si_{6.94}Al_{1.06}$	Zn for Al	O–O Very weak expanding lattice
	Hydrous Mica (Illite)	Illites	$(K,H_2O)_2(Si)_8(Al,Mg,Fe)_{4.6}O_{20}(OH)_4$	$(Al,Mg,Fe)_{4-6}$	$(Al,Si)_8$	Some Si always replaced by Al. Balanced by K between layers.	K ions: strong
	Vermiculite	Vermiculite	$(OH)_4(Mg.Ca)_x(Si_{8-x}.Al_x)(Mg.Fe)_6O_{20}.yH_2O$ x = 1 to 1.4, y = 8	$(Mg,Fe)_6$	$(Si,Al)_8$	Al for Si net charge of 1 to 1.4/unit cell	Weak
2:1:1	Chlorite	Chlorite (Several varieties known)	$(OH)_4(SiAl)_8(Mg.Fe)_6O_{20}$ (2:1 layer) $(MgAl)_6(OH)_{12}$ Interlayer	$(Mg,Fe)_6$ (2:1 layer) $(Mg,Al)_6$ Interlayer	$(Si,Al)_8$	Al for Si in 2:1 layer Al for Mg in Interlayer	
Chain Structure		Sepiolite	$Si_4O_{11}(Mg.H_2)_3H_2O_2(H_2O)$			Fe or Al for Mg	
		Attapulgite	$(OH_2)_4-(OH)_2Mg_5Si_8O_{20}.4H_2O$			Some for Al for Si	Weak = chains linked by 0

[a] Arrows indicate source of charge deficiency. Equivalent Na listed as balancing cation. Two formula units (Table 3.2) are required per unit cell.
[b] Electron Microscope Data.

Mineral Characteristics

Units
All bases in same plane. O–O = 2.55 Å—Space for Si = 0.55 Å—Thicknesss 4.93 Å. C–C height = 2.1 Å.
OH–OH = 2.94 Å. Space for ion = 0.61 Å. Thickness of unit = 5.05 Å. Dioctahedral.
OH–OH = 2.94 Å. Space for ion = 0.61 Å. Thickness of unit = 5.05 Å. Trioctahedral.

Structure—Continued			Shape	Size[b]	Cation Exchange Cap.(meq/100 gm)	Specific Gravity	Specific Surface m²/gm.	Occurrence in Soils of Engineering Interest
Crystal Structure	Basal Spacing							
			Irregular, somewhat rounded	.05–1 μ				Common
Triclinic a = 5.14, b = 8.93, c = 7.37 α = 91.6°, β = 104.8°, γ = 89.9°	7.2 Å		6-sided flakes	0.1–4 μ × .05–2 μ } single to 3000 × 4000(stacks)	3–15	2.60–2.68	10–20	Very Common
Monoclinic a = 5.15, b = 8.95, c = 14.42 β = 96°48'	14.4 Å	Unit cell contains 2 unit layers	6-sided flakes	0.07–300 × 2.5–1000 μ	1–30			Rare
Almost Orthorhombic a = 5.15, b = 8.96, c = 43 β = 90°20'	43 Å	Unit cell contains 6 unit layers	Rounded flakes	1 μ × .025–.15 μ				Rare
a = 5.14 in O-Plane a = 5.06 in OH-Plane b = 8.93 in O-Plane	7.2 Å	Random stacking of unit cells	Tubes	.07 μ O.D. .04 μ I.D. 1 μ long.	5–10	2.55–2.56		Occasional
b = 8.62 in OH-Plane ∴ layers curve	10.1 Å	Water layer between unit cells	Tubes		5–40	2.0 –2.2	35–70	Occasional
	9.6 Å—Complete separation	Dioctahedral	Flakes (Equidimensional)	>10 Å × up to 10 μ	80–150	2.35–2.7	50–120 Primary 700–840 Secondary	Very Common
	9.6 Å—Complete separation	Dioctahedral						Rare
	9.6 Å—Complete separation	Dioctahedral	Laths	Breadth = 1/5 length to several μ × unit cell	110–150	2.2 –2.7		Rare
	9.6 Å—Complete separation	Trioctahedral		To 1 μ × unit cell breadth = 0.02 – 0.1 μ	17.5			Rare
		Trioctahedral	Similar to Mont.	Similar to Mont.	70–90	2.24–2.30		Rare
		Trioctahedral	Broad Laths	50 Å Thick				Rare
	10 Å	Both Dioctrahedral and Trioctahedral	Flakes	.003–.1 μ × up to 10 μ	10–40	2.6 –3.0	65–100	Very Common
a = 5.34, b = 9.20 c = 28.91, β = 93°15'	10.5–14	Alternating Mica and double H₂O layers	Similar ro Illite		100–150		40–80 Primary 870 Secondary	Fairly Common
Monoclinic (Mainly) a = 5.3, b = 9.3 c = 28.52, β = 97°8'	14 Å		Similar to Illite	1 μ	10–40	2.6 –2.96		Common
Monoclinic a = 2 × 11.6, b = 2 × 7.86 c = 5.33 a₀ Sin β = 12.9 b₀ = 18 c₀ = 5.2		Chain	Flakes or Fibers		20–30	2.08		Rare
		Double Silica Chains	Laths	Max, 4–5 μ × 50–100 Å Width = 2t	20–30			Occasional

References
Grim, R. E. (1968) *Clay Mineralogy*, 2d edition, McGraw-Hill, New York.
Brown, G. (editor) (1961) *The X-ray identification and Crystal Structure of Clay Materials*, Mineralogical Society (Clay Minerals Group), London.

Table 3.3 *Continued*

	IONIC RADII[a]						
ION	Nonhydrated Radius Angstroms	Hydrated Radius Angstroms	ION	Nonhydrated Radius Angstroms	Hydrated Radius Angstroms	ION	Nonhydrated Radius Angstroms
Li^{+1}	0.68–0.78	7.3–10.3	Ca^{+2}	1.06–1.17	9.6	O^{-2}	1.32
Na^{+1}	0.98	5.6– 7.90	Sr^{+2}	1.27–1.34	9.6	OH^{-1}	1.33
K^{+}	1.33	3.8– 5.32	Ba^{+2}	1.43–1.49	8.8	Mn	0.93
NH_4^{+1}	1.43	5.37	Al^{+3}	0.45–0.79		Fe^{+++}	0.67
Rb^{+1}	1.49	3.6– 5.09	La	1.22–1.30		Fe^{++}	0.82
Cs^{+1}	1.65	3.6– 5.05	Cl^{-1}	1.81			
Mg^{+2}	0.78–0.89	10.8	Si^{+4}	0.31–0.39			

[a] Some deviation from listed radii may be expected in specific crystal structures.

and hydraulic conductivity behavior. Thus, mineralogy can be considered fundamental to the understanding of geotechnical properties, even though mineralogical determinations are not made for many geotechnical investigations. Instead, other characteristics that reflect both composition and engineering properties, such as Atterberg limits and grain size distribution, are determined.

Mineralogy is related to soil properties in much the same way as the composition and structure of cement and aggregates are to concrete, or as the composition and crystal structure of a steel are to its strength and deformability. In the case of all three of these engineering materials—soil, concrete, and steel—the mechanical properties can be measured directly, but the properties cannot be explained without consideration of mineralogy.

SUGGESTIONS FOR FURTHER STUDY

Bohor, B. F., and Hughes, R. E. (1971), "Scanning Electron Microscopy of Clays and Clay Minerals", *Clays and Clay Minerals*, Vol. 19, no. 1. pp. 49–54.

Brown, G., Ed. (1961), *X-Ray Identification and Crystal Structure of the Clay Minerals,* 2nd ed., Mineralogical Society of London.

Clays and Clay Minerals (Journal of the Clay Minerals Society), published six times annually by the Pergamon Press, New York.

Grim, R. E. (1962), *Applied Clay Mineralogy*, McGraw-Hill, New York.

Grim, R. E. (1968), *Clay Mineralogy*, 2nd ed., McGraw-Hill, New York.

Hurlbut, C. S. (1957), *Dana's Manual of Mineralogy*, 16th ed., Wiley, New York.

Marshall, C. E. (1964), *The Physical Chemistry and Mineralogy of Soils*, Vol. 1—*Soil Materials, Wiley,* New York.

Millot, G. (1970), *Geology of Clays*, Chapman and Hall, London.

Rich, C. I., and Kunze, G. W. (Ed.) (1964), *Soil Clay Mineralogy*, A Symposium, University of North Carolina Press, Chapel Hill.

Weaver, C. E., and Pollard, L. D. (1973), *The Chemistry of Clay Minerals*, Developments in Sedimentology 15, Elsevier, Amsterdam, 213 pp.

CHAPTER 4

Soil Formation and Soil Deposits

4.1 INTRODUCTION

The variety of soil materials encountered in engineering problems is almost limitless, ranging from hard, dense, large pieces of rock through gravel, sand, silt, and clay to organic deposits of soft compressible peat. To compound the complexity, all of these materials may occur over a range of densities and water contents. At any given site, a number of different soil types may be present, and the composition may vary over intervals of as little as a few inches.

It is not surprising, therefore, that a considerable portion of the geotechnical engineer's effort is devoted to the identification of soils and the evaluation of the appropriate properties for use in a particular analysis. What is surprising, perhaps, is that the application of principles of mechanics to such a diverse material as soil meets with as much success as it does.

To understand and appreciate the characteristics of any soil deposit requires an understanding of what the material is and how it came to be in the state that it is in. Of greatest importance are considerations of soil and rock weathering, the erosion and transportation of soils, depositional processes, and postdepositional changes in sediments. Some of the important aspects of these processes are presented in this chapter. It should be noted that each has been the subject of numerous papers and books, and the amount of available information is enormous. Thus, it is possible only to summarize the subject and to encourage the reader to consult the references for more details.

4.2 THE GEOLOGIC CYCLE

Throughout geologic time the surface of the earth has been acted upon by four basic processes which

proceed in a never ending cycle as indicated in Fig. 4.1. *Denudation* includes all of those processes that act to wear down land masses. Of primary importance are *weathering*, which refers to all of the destructive mechanical and chemical processes which break down existing rock and soil masses in situ and *transportation* of weathering products by various agents from one region to another—generally from high areas to low. *Deposition* refers to the accumulation of sediments previously transported from some other area. *Sediment formation* pertains to the processes by which accumulated sediments are densified, altered in composition, converted into rock, and so on. *Crustal movement* involves both gradual rising of unloaded areas and slow subsidence of depositional basins (epirogenetic movements) and abrupt movements (tectonic movements) such as associated with faulting. Figure 4.2 presents a schematic diagram of the interrelationship of these processes.

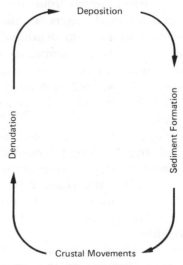

Fig. 4.1 The geologic cycle.

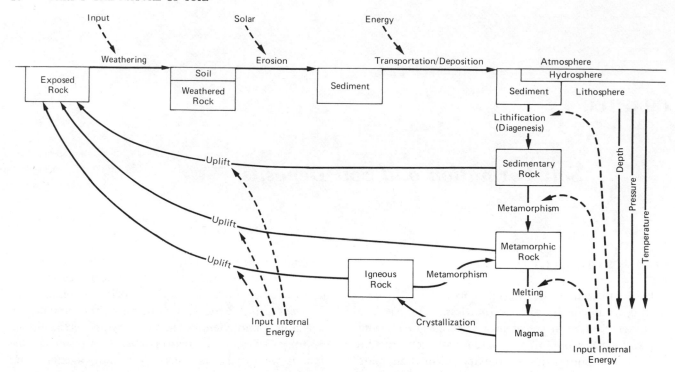

Fig. 4.2   The rock cycle.

In nature, more than one process acts simultaneously. For example, weathering and erosion take place at the surface during mountain building (crustal movements); deposition, sediment formation, and regional subsidence may be simultaneous. This accounts in part for the wide variety of conditions that may be observed in any area.

4.3   THE EARTH'S CRUST

Soils may be encountered from the surface to depths of several hundred meters. In the zone commonly encountered by man, to depths of 2 km, the rocks are 75 percent secondary (sedimentary and metamorphic) and 25 percent igneous. In many cases, distinction between soil and rock may be difficult, since differences between hard soil and soft rock are not precisely defined.

From depths of 2 to 15 km, the rocks are about 95 percent igneous and 5 percent secondary. Granitic rocks predominate in the continental parts of the earth's crust. The elemental composition of the top 16 km of the earth is shown in Fig. 4.3. Oxygen is the only anion that has an abundance of more than 1 percent by weight.

A temperature gradient of approximately 1°C/30 m exists between the bottom of the earth's crust at

1200°C and the surface.* The rate of cooling of igneous rock material as molten magma moves from the interior of the earth toward the surface has a significant influence on the characteristics of the resulting rock. As molten material cools, the thermal energy of the constituent atoms is reduced, their motions slow down, and they attempt to take up positions of minimum free energy. If cooling is slow, then large mineral crystals will form. The specific crystal structures that develop depend on the elemental composition.

The more rapid the cooling, the smaller the crystals that form because of the reduced time for atoms to attain minimum energy configurations. Cooling may be so rapid, as in a volcanic eruption, that no crystalline structure develops before solidification, and an amorphous material, for example, obsidian (volcanic glass), is formed.

4.4   ROCK AND MINERAL STABILITY

Crystal size and structure have an important influence on the resistance of different rocks to weathering.

* In some localized areas, usually within regions of recent crustal movement; for example, faulting, volcanism, the gradient may exceed 20°C/100 m. Such regions are of current interest because of their potential as sources of geothermal energy.

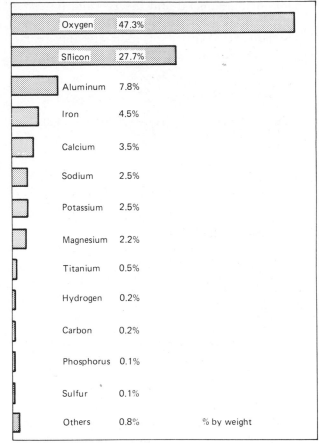

Fig. 4.3 Composition of the solid crust of Earth to a depth of 10 mi (15 km) (Clarke, 1920).

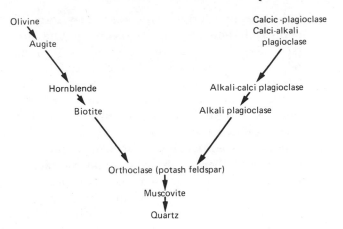

This reaction series parallels closely various weathering stability series.

4.5 WEATHERING

The weathering of soils and rocks is a destructive process whereby debris of various sizes, compositions, and shapes is formed.* The resulting new compositions are usually more stable than the old and involve a decrease in the internal energy of the materials. As erosion moves the ground surface downward, pressures and temperatures in the rocks are decreased, and they possess an internal energy in excess of that for equilibrium in the new environment. This, in conjunction with exposure to the atmosphere, water, and the various chemical and biological agents, results in processes of alteration.

A great variety of physical, chemical, and biological processes act to break down rock masses. Physical processes reduce particle size, increase surface area, and increase bulk volume. Chemical and biological processes may cause complete changes in both physical and chemical properties.

Physical processes of weathering

Physical weathering processes are generally considered to be those that cause in situ breakdown without chemical change. Five physical processes are important.

1. *Unloading.* When the effective confining pressure is lessened by uplift, erosion, or changes in

Factors controlling the stability of different crystal structures are considered in Chapter 2. The greatest electrochemical stability of a crystal is reached at its crystallization temperature. As temperature is reduced below the crystallization temperature, the structural stability decreases. For example, olivine is formed by crystallization of igneous rock magma at high temperature, and as a result, it is one of the most unstable igneous rock forming minerals. Quartz, on the other hand, does not assume its final crystal structure until the temperature drops to 573°C (inversion temperature from α to β quartz). Quartz is the most abundant nonclay mineral in soils, although it comprises but 12 percent of igneous rocks.

Bowen's reaction series indicates the crystallization sequence of the silicate minerals as temperature decreases from 1200°C. As magma cools, the minerals may form and remain, or they may react progressively to form the minerals below. The Bowen reaction series is as follows:

* A general definition of weathering (Reiche, 1945; Keller, 1957) is: ". . . the response of materials within the lithosphere to conditions at or near its contact with the atmosphere, the hydrosphere and perhaps still more importantly, the biosphere." *Biosphere* is the entire space occupied by living organisms; *hydrosphere* is the aqueous envelope of the earth; and *lithophere* is the solid part of the earth.

fluid pressures, cracks and joints may form to depths of thousands of feet below ground surface. *Exfoliation* is the spalling or peeling off of surface layers of rocks. Exfoliation frequently occurs during rock excavation and tunneling and has given rise to the term "popping rock" to describe the sudden spalling off of rock slabs as a result of stress release.

2. *Thermal expansion and contraction.* The importance of thermal expansion and contraction is considered to range from the development of planes of weakness from strains already present (Reiche, 1945) to the complete fracture of rocks. Repeated frost and insolation (daytime heating) may be the chief agents of mechanical weathering in some desert areas.

3. *Crystal growth, including frost action.* The crystallization pressures of salts, and especially the pressure associated with the freezing of water in saturated rocks, may be major factors responsible for a significant amount of disintegration. Typically, many talus deposits have been formed by frost action.

4. *Colloid plucking.* The shrinkage of colloidal materials on drying may exert a tensile stress on surfaces with which they are in contact.* Colloidal films on rock surfaces may be able to remove mineral flakes in this way.

5. *Organic activity.* The growth of plant roots in existing fractures in rocks is an important weathering process. The activities of worms, rodents, and man may cause considerable mixing within the zone of weathering.

Physical weathering processes are generally the forerunners of chemical weathering, and their main contributions are to loosen rock masses, reduce particle sizes, and increase the available surface for chemical attack.

Chemical processes of weathering

Practically all chemical weathering processes depend on the presence of water. Hydration, that is, the surface adsorption of water, is the forerunner of all the more complex chemical reactions, many of which proceed simultaneously. Some of the important chemical processes are listed below.

1. *Hydrolysis* is probably the most important chemical process, and is the reaction between

---

* To appreciate this phenomenon, smear a film of clay paste, prepared using a highly plastic clay, on the back of the hand and let it dry.

---

ions of the mineral and the H^+ and $(OH)^-$ of water. The small size of the H^+ ion enables it to enter the lattice of minerals and replace existing cations.

For example,

Orthoclase feldspar:

$$\boxed{K} + H^+OH^- \rightarrow \boxed{H} + K^+OH^- \quad \text{(alkaline)}$$

Anorthite:

$$\boxed{Ca} + 2H^+OH^- \rightarrow \boxed{\begin{array}{c}H\\H\end{array}} + Ca(OH)_2 \quad \text{(basic)}$$

A general expression for hydrolysis of a silicate mineral is (Reiche, 1945).

$$MSiAlO_{11} + H^+OH^- \rightleftharpoons M^+OH^- + [Si(OH)_{o-4}]_n + [Al(OH)_6]_n^o$$

or

$$\rightleftharpoons Al(OH)_3 + (M,H) AlSiAl^tO_n$$

where n refers to unspecified atomic ratios, and o and t refer to octahedral and tetrahedral coordinations, M indicates metal cations. (M,H) $AlSiAl^tO_n$ may represent clay minerals, zeolites, or "silicate wreckage." In the next step, the hydrogenated surface layers become unstable, and tetrahedra and octahedra peel off (Jenney, 1941). This is followed by the formation of ordered but variable chains and networks of $Si(OH)_4$, $Al(OH)_3$, KOH, and water.

Hydrolysis will not continue in the presence of static water. Continued driving of the reaction to the right requires removal of soluble material by leaching, complexing, adsorption, or precipitation, as well as the continued introduction of H^+ ions.

The pH of the system is important, because it influences the amount of available H^+, the solubility of SiO_2 and Al_2O_3, and the type of clay mineral that may form. Figure 4.4 shows the solubility of silica and alumina as a function of pH.

2. *Chelation* involves the complexing and removal of metal ions. It is effective in driving hydrolysis reactions. Ring structured organic compounds derived from humus are able to hold metal ions within the rings by covalent bonding.

3. *Cation exchange* is important in chemical weathering in at least three ways:

a. It may cause replacement of hydrogen on hydrogen bearing colloids. This reduces the

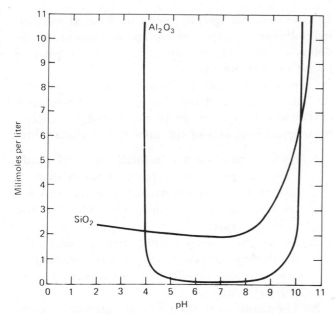

Fig. 4.4 Solubility of alumina and amorphous silica in water (Keller, 1964).

ability of colloids to bring H+ to unweathered surfaces.

  b. The ions held by Al_2O_3 and SiO_2 colloids influence the types of clay mineral that form.

  c. The physical properties of the system, such as the permeability, may depend on the adsorbed ion concentrations and types.

4. *Oxidation* is the loss of electrons by cations, and *reduction* is the gain of electrons. Both reactions are important in chemical weathering. Most of the important oxidation processes depend on dissolved oxygen in the water. The oxidation of pyrite is typical of many oxidation processes during weathering (Keller, 1957):

$$2FeS_2 + 2H_2O + 7O_2 \rightarrow 2FeSO_4 + 2H_2SO_4$$
$$FeSO_4 + 2H_2O \rightarrow Fe(OH)_2 + H_2SO_4$$
(hydrolysis)

Oxidation of $Fe(OH)_2$ gives

$$4Fe(OH)_2 + O_2 + 2H_2O \rightarrow 4Fe(OH)_3$$
$$2Fe(OH)_3 \rightarrow Fe_2O_3 \cdot nH_2O \qquad \text{(limonite)}$$

The H_2SO_4 formed in these reactions serves to rejuvenate the processes. It may also drive the hydrolysis of silicates and weather limestone to produce gypsum and carbonic acid.

Reduction reactions, which are of considerable importance relative to the influences of bacterial action and plants on weathering, store energy that may be used in later stages of weathering.

5. *Carbonation* is the combination of carbonate or bicarbonate ions with earth materials. Atmospheric CO_2 is the source of the ions. The carbonation of dolomitic limestone proceeds as follows:

$$CaMg(CO_3)_2 + 2CO_2 + 2H_2O \rightarrow Ca(HCO_3)_2 + Mg(HCO_3)_2$$

Weathering products

The products of weathering, several of which will generally coexist at any time include:

1. Unaltered minerals (either highly resistant or freshly exposed).
2. Newly formed more stable minerals having the same structure as the original mineral.
3. Newly formed minerals having a form similar to the original, but a changed internal structure.
4. Products of disrupted minerals, either at or transported from the site. Such materials might include:

  a. Colloidal gels of Al_2O_3 and SiO_2.
  b. Clay minerals.
  c. Zeolites.
  d. Cations and anions in solution.
  e. Mineral precipitates.

5. Unused guest reactants.

The relationship between minerals and different weathering stages is indicated in Table 4.1. Similarity between the order of representative minerals for the different weathering stages and Bowen's reaction series presented earlier is evident.

The interesting contrast in compositions between terrestrial and lunar soils can be accounted for largely in terms of differences in chemical weathering. On Earth the soils are composed mainly of quartz and clay minerals, because minerals of lower stability, for example, feldspar, olivine, hornblende, glasses, are rapidly weathered. On the Moon, however, the absence of water and free oxygen prevent chemical weathering. Hence, lunar soils are made up mainly of fragmented parent rock and rapidly crystallized glasses. Mineral fragments in lunar soils include plagioclase feldspar, pyroxene, ilmenite, olivine, and potassium feldspar. Quartz is extremely rare because it is not abundant in the source rocks. The nature of lunar soils is discussed by Carrier, Mitchell, and Mahmood (1973).

Table 4.1   Representative Minerals and Soils Associated with Weathering Stages

Weathering Stage	Representative Minerals	Typical Soil Groups
	Early weathering stages	
1	Gypsum (also halite, sodium nitrate)	Soils dominated by these minerals in the fine silt and clay fractions are the youthful soils all over the world, but mainly soils of the desert regions where limited water keeps chemical weathering to a minimum.
2	Calcite (also dolomite apatite)	
3	Olivine-hornblende (also pyroxenes)	
4	Biotite (also glauconite, nontronite)	
5	Albite (also anorthite microcline, orthoclase)	
	Intermediate weathering states	
6	Quartz	Soils dominated by these minerals in the fine silt and clay fractions are mainly those of temperate regions developed under grass or trees. Includes the major soils of the wheat and corn belts of the world.
7	Muscovite (also illite)	
8	2:1 layer silicates (including vermiculite, expanded hydrous mica)	
	Montmorillonite	
	Advanced weathering stages	
10	Kaolinite	Many intensely weathered soils of the warm and humid equatorial regions have clay fractions dominated by these minerals. They are frequently characterized by their infertility.
11	Gibbsite	
12	Hematite (also geothite, limonite)	
13	Anatase (also rutile, zircon)	

Effects of climate, topography, parent material, time, and biotic factors

Climate and parent material control the rate at which weathering can proceed. Topography, apart from its influence on climate, will determine primarily the rate of erosion, and thus control the depth of soil accumulation and time available for weathering prior to removal of material from the site. In areas of steep topography, rapid mechanical weathering followed by rapid downslope movement of the debris results in formation of talus slopes (piles of coarse rock fragments).

Climate determines the amount of water present, the temperature, and the character of the vegetative cover, which in turn affect the biologic complex. Some general aspects of the influence of climate are:

1. For a given amount of rainfall, chemical weathering proceeds more rapidly in warm than in cool climates. At normal temperatures reaction rates approximately double for each 10°C rise in temperature.
2. At a given temperature, and provided the weathering zone is freely drained, weathering proceeds more rapidly in a wet climate than in a dry climate.
3. The depth to the water table influences weathering through its effect on the depth to which air is available in gas or solution form and on the type of biotic activity. The depth to the water table at present may differ from that at an earlier time, so a soil may exhibit characteristics derived at an earlier time under different conditions.
4. Type of rainfall is important in that short, intense rains erode and run off; whereas, light intensity, long duration rains soak in and aid in leaching.

During the early stages of weathering and soil formation the nature of the parent material is much more important than after intense weathering for long periods of time. Although even the oldest soils may retain some characteristics derived directly from the original (parent) material, climate ultimately becomes a more dominant factor in residual soil formation than does parent material.

Of the igneous rock forming minerals, only quartz, and to a lesser extent muscovite, have sufficient chemical durability to persist over long periods of weathering.

Biological factors of importance include the influence of vegetation on erosion rate and the cycling of elements between plants and soils. Organic compounds can intensify chemical weathering. Examples include solution of minerals by carbonic acid, biological oxidation and reduction of iron and sulfur compounds, and chelation of iron compounds.

Quartz is most abundant in coarse granular rocks, such as granite, granodiorite, and gneiss, and it occurs in grains in the millimeter size range. Consequently, the granitic rocks are the main source of sand.

The time needed to weather different materials varies greatly. The more unconsolidated and permeable the parent material, and the warmer and more humid the climate, the shorter the time needed to achieve some given amount of soil formation. The rates of weathering and soil development decrease with increasing time.

The time for soil formation from hard rock parent materials may be very great; however, young soils can develop in less than 100 years from loessial, glacial, and volcanic parent materials (Millar, Turk, and Foth, 1965).

4.6 CLAY GENESIS BY WEATHERING

The behavior of nonclay colloids, for example, silica and alumina, during crystallization plays an important role in determining what specific clay minerals form from weathering processes. Certain general principles concerning this behavior can be stated:*

1. Alkaline earths (Ca^{2+}, Mg^{2+}) tend to flocculate silica.
2. Alkalis (K^+, Na^+, Li^+) tend to disperse silica.
3. A low pH tends to flocculate.
4. A high electrolyte content tends to flocculate.
5. Aluminous suspensions are more easily flocculated than siliceous suspensions.
6. Dispersed phases are more easily removed by groundwater than flocculated phases.

Kaolinite minerals

Since the kaolinite minerals are of a 1:1 silica: alumina structure, the formation of kaolinite is favored when alumina is abundant and silica is scarce. This is the case when the electrolyte content is low, pH is low, and ions tending to flocculate silica (Mg, Ca, Fe) are removed by leaching. Most kaolinite is formed, primarily from the feldspars and micas, by acid (pH 7) leaching of acid (SiO_2 rich) rocks (granitic rocks). Halloysite forms as a result of leaching of feldspar by H_2SO_4, which is formed by alteration of pyrite (Weaver and Pollard, 1973). Climatic conditions favoring kaolinite clays are those where precipitation is relatively high to insure leaching of cations and oxidation of iron. Good drainage is required for removal of these materials.

* The basis for these statements may be found in Chapter 7.

Smectite minerals

Smectites form when silica is abundant, as is the case where both silica and alumina are flocculated. Conditions favoring this are high pH, high electrolyte content, and the presence of more Mg^{2+} and Ca^{2+} than Na^+ and K^+. Rocks high in alkaline earths, such as the basic and intermediate igneous rocks, volcanic ash, and their derivatives containing ferromagnesian minerals and calcic plagioclase, are usual parent materials. Climatic conditions where evaporation exceeds precipitation favor formation of smectite, for example, arid and semiarid areas, where there is poor leaching.

Illite (hydrous mica) and vermiculite

The hydrous mica minerals form under conditions similar to those favoring formation of montmorillonite. Potassium is, in addition, essential, so acid igneous or metamorphic rocks and their derivatives are usual parent rocks. Alteration of muscovite to illite and biotite to vermiculite during weathering is also a significant source of these minerals. Interstratifications of vermiculite with mica and chlorite are common. The high stability of illite is responsible for its abundance and persistence in soils and sediments.

Chlorite minerals

Chlorites may form by alteration of smectite through introduction of sufficient Mg^{2+} to cause a crystalline brucite layer to replace the interlayer water. Biotite from igneous and metamorphic rocks may alter to form trioctahedral chlorites and mixed-layer chlorite–vermiculite. Chlorites also occur in low to medium grade metamorphic rocks as a rock-forming mineral and in many soils derived from such rocks.

The above conditions are greatly simplified, and there may be numerous ramifications, alterations, and qualifications to these processes. One clay type may alter to another by cation exchange and weathering under new conditions. Entire structures may change; for example, from 2:1 to 1:1 so that montmorillonite may form where magnesium rich rock weathers under humid, moderately drained conditions, but as leaching continues it may alter to kaolinite. Kaolinite cannot form in the presence of significant concentrations of calcium, and there is evidence that where kaolinite is transported to the sea it alters slowly to chlorite.

Relative proportions of potassium and magnesium determine how much montmorillonite and illite form.

Some montmorillonites tend to alter to illite in a marine environment due to the high K+ concentration. Mixed-layer clays often form during weathering or after burial by partial leaching of K or Mg(OH)$_2$ from between illite or chlorite layers and by incomplete adsorption of K or Mg(OH)$_2$ on montmorillonite or vermiculite.

The foregoing considerations are but a brief survey of factors controlling clay mineral formation. Detailed treatments of this complex subject are given by Keller (1957, 1964) and Weaver and Pollard (1973).

4.7 SOIL PROFILES AND THEIR DEVELOPMENT

Provided erosion does not rapidly remove materials from the site, *in situ* weathering processes tend to develop a sequence of *horizons* within the zone of alteration. These horizons may grade abruptly from one to the next or be hard to distinguish. Their thickness may range from a few millimeters to a few meters. The horizons may show differences in any or all of the following:

1. Degree of breakdown of parent material.
2. Organic material.
3. Kind and amount of secondary minerals.
4. pH.
5. Particle size distribution.

All the horizons considered together including the parent material form the *soil profile*.* That part of the profile above the parent material is termed the *solum*. *Eluviation* is the movement of soil material from one place to another within the soil, in solution or in suspension, when there is an excess of rainfall over evaporation. *Eluvial* horizons have lost material; whereas, *illuvial* horizons have gained material.

Major horizons are designated by the capital letters O, A, B, C, and R. Subdivisions of these horizons are denoted by number. The horizons and subdivisions are defined as follows (U. S. Department of Agriculture, 1951, 1962):

1. O—Organic horizons:
   1. Formed or forming above mineral part.
   2. Dominated by fresh or partly decomposed organic material.
   3. Contains more than 30 percent organic matter if mineral fraction contains more than 50

percent clay, or more than 20 percent organic if mineral fraction has no clay.
2. A—Mineral horizons consisting of:
   1. Humified organic matter accumulation at or adjacent to the surface. It is not recognizable as organic matter and usually imparts darker color to the horizon.
   2. Horizons leached of clay, iron, or aluminum with concentration of quartz or other resistant minerals of sand or silt size.
   3. Horizons dominated by 1 or 2 but transitional to an underlying B or C.
3. B—Horizons with one or more of the following:
   1. Illuvial concentration of clay, iron, aluminum or humus.
   2. Residual concentration of iron or aluminum oxides or clays, formed by means other than solution and removal of carbonates or more soluble salts.
   3. Coatings of oxides adequate to give conspicuously darker, stronger or redder colors than over- or underlying horizons.
   4. An alteration that obliterates original rock structure, that forms silicate clays, liberates oxides, or both, and that forms granular, blocky, or prismatic structure if textures are such that volume changes accompany changes in moisture.
4. C—A layer, excluding bedrock, either like or unlike the material from which the solum is presumed to have formed, only slightly affected by soil forming processes, and lacking properties diagnostic of A or B horizons.
5. R—Underlying consolidated bedrock.

Figure 4.5 shows a generalized profile of a soil developed in a humid climate with moderate temperature. Figure 4.6 traces the development of a typical soil profile in a temperate climate and under tall-grass vegetation as a function of time.

4.8 SURFICIAL SOILS

Agricultural soil maps may be available in areas where engineering data are lacking, and they may be useful for preliminary assessment of the soils. Surficial soils are of particular importance in highway engineering and land development.

A genetic classification, based on the idea that each soil has a definite morphology related to a particular set of soil forming factors, was suggested in the late 19th century. This system has been modified and ex-

---

* Profiles developed in this way should not be confused with soil profiles resulting from successive deposition of different soil types in alluvial, lake, or marine environments.

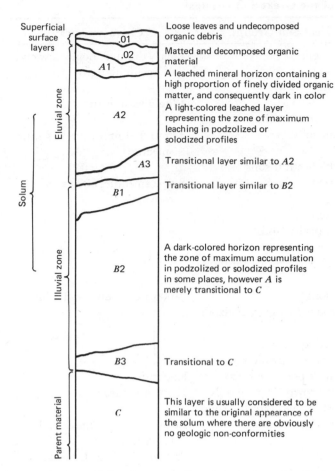

Fig. 4.5 A generalized profile of a timbered soil developed in a humid climate with moderate temperature (Millar, Turk, and Foth, 1965).

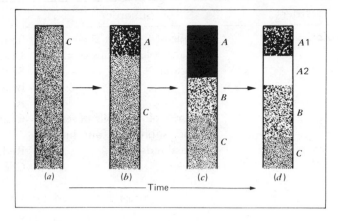

Fig. 4.6 A summary of the stages in the development of soils in central United States under tall grass vegetation (Millar, Turk, and Foth, 1965). (*a*) Parent material—original material before soil development begins. (*b*) Young soil (regosal)—thin solum, organic-matter accumulation in *A* horizon from which carbonates have been leached. Minimal weathering and eluviation. (*c*) Mature soil (brunizem)—organic-matter content is at a maximum. Has moderate clay accumulation in the *B* horizon, and the solum is acid. Stage of maximum productivity for corn. (*d*) Old soil (planosol)—very acid in reaction, severely weathered, and has less organic matter than mature stage. Clay accumulation in *B* horizon has formed a clay pan. An *A*2 horizon exists.

panded several times since. The most general unit is the *order* and the most specific is the *type* (or *phase* of a type). Soil *series* are given names, usually that of a county, town, city or river near the place where the soil was first identified. The soil *type* (series name plus texture class) is the mapping unit in soil surveys published by the U.S. Department of Agriculture (U.S.D.A.). A listing of the soil series of the United States, Puerto Rico, and the Virgin Islands was published by the Soil Conservation Service of the U.S.D.A. in 1972.

Orders. All soils can be placed in one of three orders:

1. *Zonal soils* are soils that are characterized by the dominating influence of climate. If climatic conditions are reasonably uniform and continuous and erosion is not too rapid, then soils from similar climates become alike regardless of parent material. Soils of the great soil groups of the world are zonal soils.

2. *Intrazonal soils* are associated with zonal soils but reflect the influence of some local condition such as poor drainage, alkali salts, or some other unique characteristic.

3. *Azonal soils* are soils without profile development. There is little or no alteration of the parent material because of their youth or environmental setting. Most alluvial, marine, aeolian, and glacial deposits are azonal soils.

Soil orders can be subdivided into suborders on the basis of the factors most significant in determining their characteristics or on the basis of the characteristics themselves. The suborders can be divided into the *great soil groups* of the world. Relationships between orders, suborders, and great soil groups are given in Table 4.2.

In 1965, the U.S.D.A. adopted a new soil order classification that is based more on properties than genetic factors. Table 4.3 lists the new orders and their approximate equivalents from the old system. A detailed treatise on soil classification by this system

Table 4.2 A Classification of the Great Soil Groups[a]

Order	Suborder	Great Soil Groups	Characteristics[b]
Zonal Soils	1. Soils of the cold zone	Tundra soils	Poorly drained, boggy low clay
	2. Light-colored soils of arid regions	Desert soils	Salt accumulations
		Sierozem	
		Brown soils	Illite and smectite
		Reddish-brown soils	
	3. Dark-colored soils of semi-arid, subhumid, and humid grasslands	Chestnut soils	375 to 560 mm rain, Ca salts, smectite
		Reddish chestnut soils	
		Chernozem soils	Black, smectite, 500 to 650 mm rain
		Prairie soils	
		Reddish prairie soils	
	4. Soils of the forest-grassland transition	Degraded chernozem	
		Noncalcic brown or shantung brown soils	
	5. Light-colored podzolized soils of the timbered regions	Podzol soils	>600 mm rain, acid, illite
		Gray wooded or gray podzolic soils[c]	
		Brown podzolic soils	
		Gray-brown podzolic soils	
		Red-yellow podzolic soils[c]	Kaolinite
	6. Lateritic soils of forested warm-temperature and tropical regions	Reddish-brown lateritic soils	High in oxides
		Yellowish-brown lateritic soils	
		Laterite soils[c]	Little clay, kaolinitic friable to hard rock
Intrazonal Soils	1. Halomorphic (saline and alkali) soils of imperfectly drained arid regions and littoral deposits	Sononchak or Saline soils	Salts of Na, Mg, Ca near surface, nonsaline, alkali
		Solonetz soils	
		Soloth soils	
	2. Hydromorphic soils of marshes, swamps, seep areas, and flats	Humic-gley soils[c] (includes Wiesenboden)	
		Alpine meadow soils	
		Bog soils	
		Half-bog soils	
		Low humic-gley[c] soils	
		Planosols	
		Ground water podzol soils	
		Ground water laterite soils	
	3. Calcimorphis soils	Brown forest soils (braunerde)	
		Rendzina soils	
		Grumosol (vertisol)[b]	Deep, clayey, climate with wet and dry seasons, high in smectite, expansive
Azonal Soils		Lithosols	Rock fragments, sand dunes, loess, glacial drift very important in geotechnical engineering
		Regosols (includes dry sands)	
		Alluvial soils	

[a] Prepared by James Thorp and Guy D. Smith, published in *Soil Science*, Vol. 67, No. 2, p. 118. Used through courtesy of Williams and Wilkins Co.   [b] Added by the author.   [c] New or recently modified great soil groups.

Table 4.3 New Soil Orders and Approximate Equivalents in Great Soil Groups[a]

Order	Formative Syllable	Derivation	Meaning	Approximate Equivalents
1. Entisol	ent	Coined syllable	Recent soil	Azonal soils and some low humic-gley soils
2. Vertisol	ert	L. verto, turn	Inverted soil	Gromusols
3. Inceptisol	ept	L. inceptum, beginning	Inception or young soil	Ando, sol brun acide, some brown forest, low humic-gley, and humic-gley soils
4. Aridisol	id	L. aridus, dry	Arid soil	Desert, reddish-desert, sierozem, solonchak, some brown and reddish-brown soils and associated solonetz
5. Mollisol	oll	L. mollis, soft	Soft soil	Chestnut, chernozem, brunizem (prairie), rendzinas, some brown, brown forest, and associated solonetz and humic-gley soils
6. Spodosol	od	Gk. spodos, wood ash	Ashy (podzol) soil	Podzols, brown podzolic soils, and ground-water podzols
7. Alfisol	alf	Coined syllable	Pedalfer (Al–Fe) soil	Gray-brown podzolic, gray wooded, noncalcic brown, degraded chernozem, and associated planosols and half-bog soils
8. Utisol	ult	L. ultimus, last	Ultimate (of leaching)	Red-yellow podzolic, reddish-brown lateritic (of U.S.), and associated planosols and half-bog soils
9. Oxisol	ox	F. oxide, oxide	Oxide soils	Laterite soils, latosols
10. Histosol	ist	G. histos, tissue	Tissue (organic) soils	Bog soils

[a] From The New Classification, E. Joseph Larsen, *Soil Conservation*, Vol. 30, No. 5, December 1964.

and mapping for soil surveys has been prepared by the Soil Survey staff (1974).

4.9 GREAT SOIL GROUPS

The great soil groups of the world and climate are related diagrammatically as shown in Fig. 4.7. Figure 4.8 shows the distribution of these residual soils throughout the world. In Fig. 4.7 the term "pedocal" refers to soils of drier climates (aridic soils) and means soils that tend to accumulate calcium carbonate. Soils of wetter climates (humid soils) accumulate alumina and ferric iron and are termed "pedalfers."

Important characteristics of the soils shown in Fig. 4.7 are as follows (Miller, Turk, and Foth, 1965):

Tundra is a dark grey peaty accumulation over grey mottled mineral horizons. The soil is poorly drained and boggy. Clay mineral content is low. Permafrost is frequently present in the substratum.

Podzols (spodosols in the new U.S.D.A. Classification System) are found south of the tundras in areas where rainfall exceeds 600 mm/yr, and summers are short and cool. Podzols are characterized by moderate humus accumulation (02), a thin Al horizon and a strongly eluviated A2 horizon. The B horizon is dark

brown to reddish-brown and often cemented by organic compounds and iron oxides. The texture of all horizons except 0 is often sandy. The soils are acid, have a low cation exchange capacity, and illites dominate the clay fraction.

Gray wooded (pedsolic) soils (alfisol) are found in subhumid to semiarid cool climates. They have a moderately thin O1, a thin organic mineral horizon (A1) over lighter, bleached A2. The B horizon is more clayey and blocky and grades into lighter colored, more friable B3 and C horizons.

Gray-brown podzolic soils (alfisol) are found south of the podzol region and east of the prairies in northeastern United States and southeastern Canada, and in the humid, temperate areas of western Europe and eastern Asia where rainfall averages 750 to 1300 mm annually. These soils are characterized by a thin A1 (50 to 150 mm) and a well-developed grey to yellowish A2. The B horizon is grey to reddish-brown, darker, and of finer texture than either the A or C horizons. They are acid soils, and kaolinite is the dominant clay mineral.

Red-yellow podzolic soils (utisol) are found in areas of high temperature and high rain (1000 to 1500 mm). Leaching is great and mineral decay is rapid. Surface

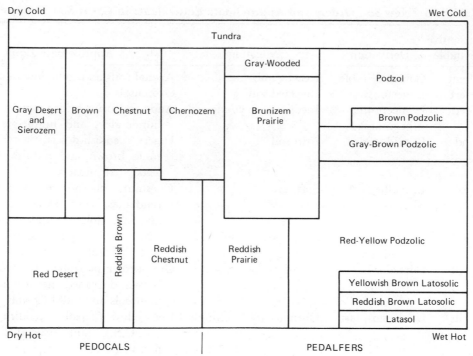

Fig. 4.7 A diagrammatic visualization of the relation between climatic conditions and the occurrence of the great soil groups (Millar, Turk, and Foth, 1965).

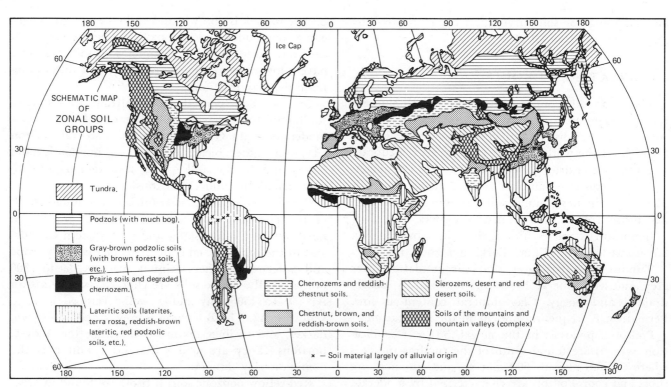

Fig. 4.8 A generalized map showing the approximate distribution of the zonal soil groups throughout the world (Kellogg, 1941).

accumulation of organic matter is small, and the leached A horizon is deep. A relatively thick B horizon may be brightly colored (red and yellow) as a result of oxidation and hydration of iron. The B horizon has more than twice the clay content of the A. The cation exchange capacity is low in all horizons, and the clay fraction is composed mainly of kaolinite, illite and quartz.

Latosol (oxisol) is an iron oxide and aluminum oxide rich, highly weathered clayey material that changes irreversibly to concretions, hardpans, or crusts when dehydrated. Clay minerals are rapidly broken down and removed. What little remains is usually kaolinitic. The process of *laterization* involves leaching of SiO_2 and deposition of Fe_2O_3 and Al_2O_3. A *laterite* is a soil whose ratio of SiO_2 to Al_2O_3 is less than 1.33; whereas a *lateritic soil* has a ratio between 1.33 and 2.00 (Bawa, 1957). Deposits of these soils may be up to 30 m or more in depth. They may range in texture from friable soils to hard rock. They may lack resistance to abrasion and lose their granular characteristic when worked, becoming soft, clayey, and impervious.

Brunizem or prairie soils (mollisol) are confined almost entirely to the corn belt area of the United States. They have a dark Al surface horizon greater than 150 mm thick. The organic content decreases with depth, and horizon boundaries are indistinct.

Chernozems (mollisol) form in cool areas having annual rainfall of 500 to 650 mm. Calcium salts accumulate at depths of 600 to 650 mm. The overlying material is dark in color and gives rise to the name Chernozem, which is Russian for black earth. Smectite predominates in the clay fraction over illite.

Chestnut soils (mollisol) concentrate calcium salts at depths of 350 to 600 mm and form where annual rainfall is 400 to 550 mm. Montmorillonite is the dominant clay mineral.

Brown soils (mollisol or aridosol) are found west of the chestnut soils in North America. $CaCO_3$ is found at 300 to 400 mm depth or less. Illite and montmorillonite are common. Local accumulations of sepiolite, palygorskite and attapulgite are found.

Desert soils (aridosol) are characterized by surface accumulations of salts from upward movement of water and usually consist of several centimeters of soil over a calcareous parent material. These soils may be alkaline.

Intrazonal soils

Three suborders of intrazonal soils are indicated in Table 4.2: halomorphic, hydromorphic, and calci-morphic. Some of the characteristics of these soils are as follows (Millar, Turk, and Foth, 1965):

1. **Halomorphic soils.** Saline and alkali soils of imperfectly drained arid regions and littoral deposits are classified as follows:
   a. **Solonchak.** This soil is saline, mildly alkaline, containing a high content of soluble salts of Ca, Mg, and Na near the surface. Illite and montmorillonite are found.
   b. **Solonetz.** Most of the soluble salt is removed by leaching. They have a high pH and sodium content; Illite and montmorillonite are found.
   c. **Soloth.** These soils are leached and have acid upper horizons and negligible exchangeable sodium.

2. **Hydromorphic soils.** These are soils of marshes, swamps, and flats.
   a. **Humic-gley (inceptosol).** These mineral soils are formed in poorly drained areas and possessing a sticky, compact, grey, or olive-grey B or C horizon. The A horizon may contain 5 to 10 percent organic matter.
   b. **Bog (histosol).** These soils are organic soils whose characteristics depend largely on the nature of the vegetation from which they form.
   c. **Planosol.** A horizon over a strongly differentiated B horizon. The B horizon is compacted, cemented or high in clay and acid. Common in upland flats of older landscapes.

3. **Calcimorphic soils.** These soils are formed from parent material with high calcium content.
   a. **Rendzina (mollisol).** The parent material has more than 40 percent $CaCO_3$. Dark Al horizon overlies a B horizon that has developed only structure and grades rapidly into parent material.
   b. **Grumosol (vertisol).** These soils are deep and clayey and are also known as black cotton, black earth, blackland soils, associated with a climate having very dry and very wet seasons. The texture of all horizons is clayey, and the montmorillonite content is high; there are no eluvial or illuvial horizons, and the soils exhibit expansive characteristics.

An additional type of residual soil not listed in Table 4.2 is *saprolite*. Saprolites retain the fabric and structure of the parent rock in terms of mineral concentrations and grain orientations.

Azonal soils

Azonal soils are soils without profile development.

1. *Lithosols (entisol)* have no clearly expressed morphology and consist of masses of rock fragments from imperfectly weathered, consolidated rocks. They are found primarily on steeply sloping land.
2. *Regosols (entisol)* consist of deep, soft mineral deposits (unconsolidated rock) without clearly expressed soil characteristics, such as sand dunes, loess, and glacial drift.
3. *Alluvial Soils* may range from clay to gravel deposits.

Azonal soils are of the greatest interest to the geotechnical engineer. The engineer encounters these soils more than any other, because large construction activities tend to concentrate in areas where they predominate such as in river valleys and in areas bounded by water. The majority of large urban areas are situated in such regions. To understand the characteristics of the azonal soils requires consideration of transportational, depositional, and postdepositional sedimentary processes.

4.10 SEDIMENT EROSION, TRANSPORT, AND DEPOSITION

Transportation

Streams, ocean currents, waves, wind, groundwater, gravity, and glaciers are continually at work transporting soils and rock debris away from the zone of weathering. Each may cause marked physical changes in the sediment it carries. Although detailed treatment of erosion, transportation, and deposition processes is outside the scope of this book,* a brief outline of the principles controlling these processes and their effects on the transported soil is helpful in understanding the properties of the transported material.

The suspension movement of all sediments carried by wind and water is controlled by the settling velocity of the particles and the laws of fluid motion. The settling velocity of small particles under laminar flow

---

* The mechanics of sediment transport are treated in detail in the Manual of Sedimentation Engineering in preparation by a Task Committee of the Hydraulics Division of the American Society of Civil Engineers. Several sections of this report have been published in the *Journal of the Hydraulics Division*, A.S.C.E., beginning in 1963. Section G. "Fundamentals of Sediment Transportation" appeared as Proc Paper 8591, Dec. 1971, pp. 1979–2022.

is proportional to the square of the particle diameter. For larger particles and turbulent flow conditions the settling velocity is proportional to the square root of the particle diameter. Particles stay in suspension once they have been set in motion as long as the turbulence of the stream is greater than the settling velocity.

Not all sediment is moved in suspension. The largest particles are carried by *traction,* which consists of rolling and dragging along the boundary between the transporting agent and the ground surface. Particles intermediate in size between the suspended load and the traction load may be carried by *saltation* movement, wherein they move in a series of leaps and bounds. Soluble materials are carried in solution and may later be precipitated due to some changed conditions. Thus, the concentration of sediment is not constant throughout the depth of the transporting agent, but is much greater near the stream bed than near the top. While fine particles may be fairly evenly distributed from top to bottom, coarser particles are distributed mainly within short distances from the bottom, as shown in Fig. 4.9, which pertains to a river following a straight course.

For sediment to be transported, it must first be picked up by the eroding agent, and greater average velocities may be necessary to erode than to transport particles. Particles are eroded when the drag and lift of the fluid exceeds the gravitational, cohesive, and frictional forces acting to hold them in place. The stream velocity required to erode does not decrease indefinitely with decreasing particle size, because small particles remain within the boundary layer adjacent to the stream bed where the actual velocity is much less than the average. Thus, transportation depends mainly on average velocity and boundary layer thickness relative to the projection of the particle. Relationships between particle size and average stream velocities for erosion and transportation by wind and water are shown in Fig. 4.10. The straight lines separate transportation and deposition; whereas, velocities must be above the curved lines to permit erosion.

Splash erosion is caused by the impact of rain on unprotected soil surfaces. The detachment and removal of thin sheets (sheet erosion) of soil may result.

The various transporting agents are compared in Table 4.4. The relative effect indicated in the last column of this table denotes the importance of the agent on a geological scale as regards the overall amount of sediment moved, with 1 representing the greatest amount.

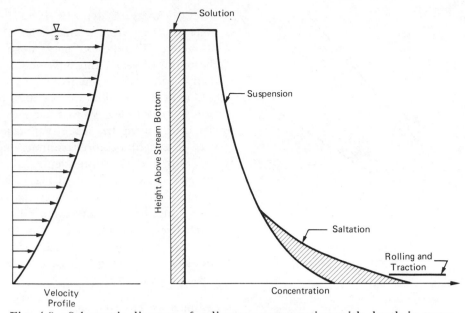

Fig. 4.9   Schematic diagram of sediment concentration with depth in transporting system.

Table 4.4   Comparison of Transporting Agents

Agent	Type of Flow	Approximate Average Velocity	Maximum Size Eroded by Average Velocity	Areas Affected	Max load per m³	Type of Transport	Relative Effect
Streams	turbulent	a few km/hr	Sand	All land	a few tens of kilograms	Bed load, suspended load, solution	1
Waves	turbulent	a few km/hr	Sand	Coastlines	a few tens of kilograms	Same as streams	2
Wind	turbulent	15 km/hr	Sand	Arid, semiarid, beaches, plowed fields	a kilogram	Bed load, suspended load	3
Glaciers	laminar	a few m/yr	Large boulders	High latitudes and altitudes	hundreds of kilograms	Bed load, suspended load, surface load	2
Groundwater	laminar	a few m/yr	Colloids	Soluble material and colloids	a kilogram	Solution	3
Gravity		cm/yr to a few m/sec	Boulders	Steep slopes, sensitive clays, saturated cohesionless soils, unconsolidated rock	2000 kilogram	Bed load	3

Adapted from Garrels (1951).

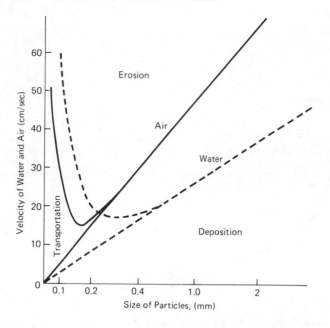

Fig. 4.10 Comparison of erosion and transport curves for air and running water. The air is a slightly more effective erosional agent than streams for very small particles but is ineffective for those larger than sand (Garrels, 1951).

Deposition

The deposition of sediments is controlled by the same laws as their transportation. If the stream velocity and turbulence are reduced below the values needed to hold the particles in suspension or moving with the bed load, then particles will settle. When ice melts, the sediment may be deposited in place or carried away by meltwaters. Materials in solution may precipitate when introduced into conditions of changed temperature or chemical composition. Sediments may be divided into those formed primarily by chemical and biological means and those composed mainly of mineral and rock fragments (detrital deposits). Those whose texture is controlled by the detrital material are termed *clastic*.

Cyclical deposition is the rule rather than the exception and results in a pulsating influx of sediments into any given area. Some causes of cyclic deposition are:

1. Periodic earth movements.
2. Climatic cycles of various lengths, most notably the annual rhythm.
3. Cyclic shifting of tributaries on a delta.
4. Periodic vulcanism.

The thickness of deposits formed during any one cycle may vary from a millimeter or less to hundreds of meters. The period may range from months to thousands of years, and only one type of sediment or many types may be involved.

One of the best known examples of a cyclic deposition is *varved clay*. Varved clays were formed in glacial lakes during the ice-retreat stage. Each layer consists of a lighter-colored summer-deposited silt grading into a dark winter-deposited clay. Spring and summer thaws contributed clay- and silt-laden meltwater to the lake, the coarsest part of which settled out at once to form the summer layer. Because of the much slower settling velocity of the clay, most of it did not settle out until the quiet winter period. Figure 4.11 is a photograph of a typical vertical section through a varved clay.

Sedimentary environments

The environment of deposition is important in determining the properties of sediments. It determines the complex of physical, chemical and biological conditions under which a sediment accumulates and consolidates. A geographic classification and some associated characteristic soil types are as follows:

1. Continental (above tidal reach).
   a. Terrestrial.
      1) Glacial—see Table 4.7.
      2) Aeolian—desert areas. Cobbles and pebbles are found at the base of steep slopes; sand dunes and fines, in basins.

Fig. 4.11 Varved clay from the New Jersey meadowlands (Courtesy of S. Saxena).

3) Alluvial—pluvial (high rain area) and fluvial (river). Lenticular beds are oriented downstream with different size characteristics. Gravel commonly in contact with sand or silt.

b. Paludal (swamp)—silts, muds, clays with high water content, and organic matter.

c. Lacustrine (lake).

1) Fresh water—fine-grained, quiet water deposits except for narrow shore zone of sand.

2) Saline—precipitated salt beds.

2. Mixed continental and marine.

a. Littoral (between tides).

1) Tidal lagoon—fine sand and silt in channels, organic-rich silt and clay in quiet areas. Abundant organic matter and carbonates.

2) Beach—sand.

3) Tidal flats—fine-grained dark muds, with lenses or stringers of sand or gravel, free of intermediate sizes.

b. Delta—coarse and fine material, organic matter and marl. There are alternations between coarse and fine deposits owing to continual shifting of stream. Suspended silt and clay in the main stream is flocculated by salts in sea water to form marine mud on the seaward delta face which is later covered by alluvial, lacustrine and beach deposits as the delta grows.

c. Estuarine—muds with coarser materials in channels.

3. Marine (below tidal limits) (Fig. 4.12).

a. Continental Shelf (neritic)—extends from low tide to a water depth of about 600 ft. The

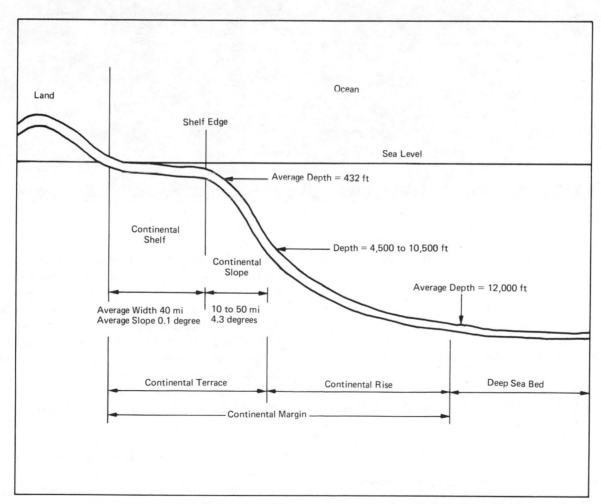

Fig. 4.12 Idealized profile of continental margin (vertical exaggeration) (After Heselton, 1969).

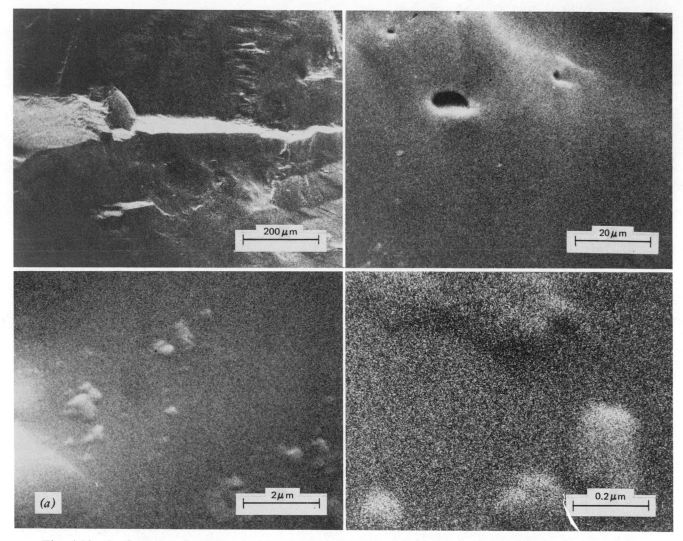

Fig. 4.13   Surface texture of four sands of differing origins (Courtesy of Norris, 1975). (a) River sand.

slope is very gentle, about 0.1 degree. Decrease in particle size and increased influences of chemical and biological factors develop in the seaward direction. Sediment distribution may be irregular. Abundant sandstone, shale and limestone.

b. Continental slope and continental rise (bathyal)—a steeper (4° average) slope leads down to a gentler slope known as the continental rise. Quiet bottom conditions with fine sand, silt, and mud.

c. Deep ocean (abyssal)—water depth averages 12,000 ft. Deposits are mainly oozes and muds.

Effects of transportation on sediment properties

The major effects of transportation processes on the physical properties of sediments are sorting and abra-sion. Sorting may be of two types: a local sorting, which produces layers or lenses with differential grain size distributions, and a longitudinal or progressive sorting, which leads to a progressive variation in particle sizes in the direction of flow. In general, sediments coarsen in the direction of their source, as is strongly evidenced by the rocky surface characteristics of mountainous areas and the fine-grained sediments of old river valleys.

Spherical particles settle from suspension more rapidly than nonspherical particles of the same size; the reverse is true, however, for the traction load. Small grains of high density will settle with large grains of low density. The greater the stream flow, the greater the quantity of sediment that can be carried. The greater the velocity, the larger the particle size that can be moved. On the other hand, recognition of

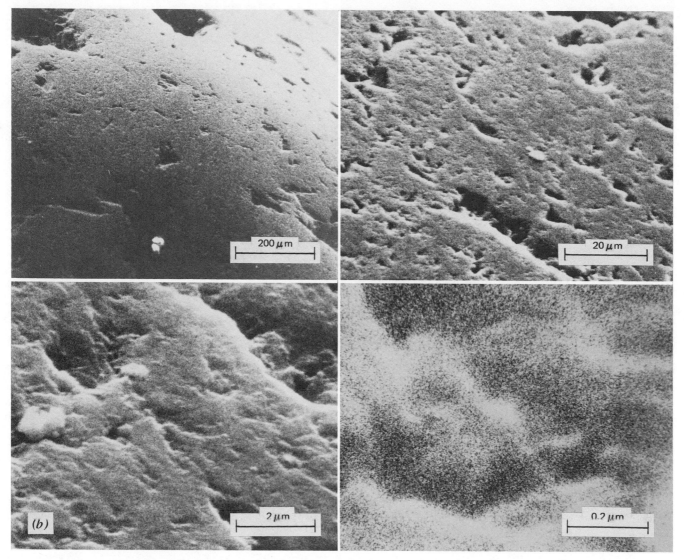

Fig. 4.13 Surface texture of four sands of differing origins (Courtesy of Norris, 1975). (*b*) Beach sand.

these factors does not mean that the sorting in any given area can be reliably predicted. Their interactions coupled with the fact that the intensity of flow of the transporting agent may vary considerably from time to time, for example, the seasonal variation in stream flow, means that sorting at any specific point may be complex.

Particle size and shape may be modified mechanically by grinding, impact, crushing, and blasting. Abrasion affects the shape and size of gravel size particles; whereas, it only serves to modify the shape of sand size and smaller particles. Water working of sands causes rounding and polishing of grains; wind driven impact may cause frosting of grains. All minerals are not equally affected with time. The pyrox-

enes, amphiboles, feldspars, and rock fragments, for example, are rapidly broken down chemically and removed during transport. Chemically stable minerals, such as quartz, are modified primarily by mechanical action but at a much slower rate. Quartz sand grains may survive a number of successive sedimentation cycles with no more than a percent or two loss by weight by chipping during transport by wind, and even less if the grain shape is more rounded. Furthermore, losses during wind transport are hundreds of times greater than during water transport (Kuenen, 1959).

Older studies of natural particle roughness were carried out with the light microscope and were limited by magnification capability. With the advent of

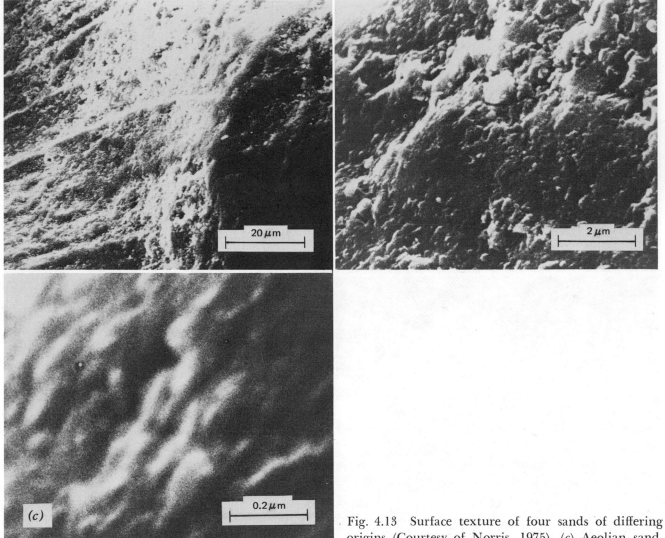

Fig. 4.13   Surface texture of four sands of differing origins (Courtesy of Norris, 1975). (c) Aeolian sand.

electron microscopy, higher magnifications were attained and geologists have been able to correlate patterns of surface roughness with their associated environments. However, such studies were carried out on fixed particle sizes (0.5 to 1 mm) and a set magnification range in order to standardize results for diagnostic use. From an engineering point of view, one is interested in the potential interaction of features at all levels of magnification and the possible variation of patterns of surface roughness characteristic of a given environment with particle size (Fig. 4.13). This study is in progress (Norris, 1975).

Tentative conclusions regarding the association of particle surface roughness and environment can be made. The mechanical and chemical action associated with a beach environment yields a relatively smooth pitted surface texture at all levels of magnification.

Aeolian sands exhibit a rougher surface texture, particularly at the highest level of magnification. River sands are characterized by a very smooth particle surface that reflects the influence of significant chemical action. Sand that has undergone change after deposition and burial is termed diagenetic sand. The surface texture of this type of sand reflects a long and stable period of interaction with the groundwater regime. Under certain chemical conditions very rough surface textures can develop. Ottawa sand, a material widely used for soil mechanics research investigations, is a diagenetic sand. Scanning electron photomicrographs of beach, aeolian, river, and diagenetic sands at several magnifications are compared in Fig. 4.13.

Some of the effects of transportation on sediment properties are listed in Table 4.5. The gradational characteristics of sedimentary materials reflect their

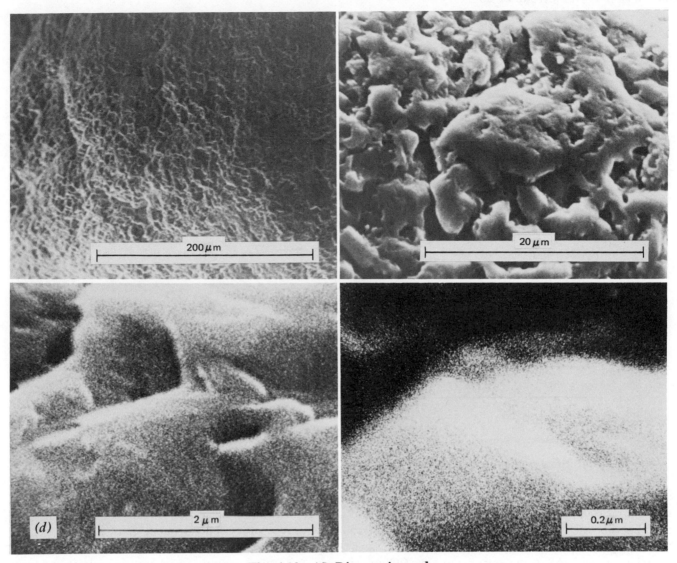

Fig. 4.13   (d) Diagenetic sand.

Table 4.5   Effects of Transportation on Sediments

	Water	Air	Ice	Gravity
Size	Reduction through solution, little abrasion in suspended load, some abrasion and impact in traction load	Considerable reduction	Considerable grinding and impact	Considerable impact
Shape and roundness	Rounding of sand and gravel	High degree of rounding	Angular, soled particles (—)	Angular, non-spherical
Surface texture	Sand: smooth, polished, shiny	Impact produces frosted surfaces	Striated surfaces	Striated surfaces
	Silt: little effect			
Sorting	Considerable sorting	Very considerable sorting (progressive)	Very little sorting	No sorting

Adapted from Lambe and Whitman (1969).

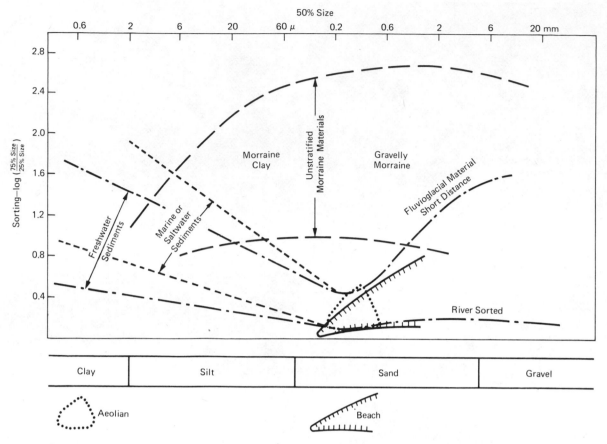

Fig. 4.14   Influence of geologic history on sorting of particle sizes (Adapted from Selmer-Olsen, 1964).

transportational mode as indicated in Fig. 4.14. On this diagram the logarithm of the ratio of 75 percent particle size to 25 percent particle size is shown as a function of median (50 percent) grain size. Sediments of different origin lie within specific zones as indicated on the diagram.

4.11   ALLUVIAL DEPOSITS

Deposition from streams results from a decrease in slope, increased frictional resistance to flow, a decrease in stream discharge, or a discharge into the more quiet water of oceans and lakes. As the slope flattens, the stream loses energy, and all particles above a certain size are dumped in a jumble of large and small particles. The current then slips to one side following the steepest slope. This channel may fill and the stream shifts again. The result is an alluvial fan, a temporary feature, which is a symmetrical pile of material spread out radially from the point of slope change.

In advanced stages of stream development, the stream occupies only a small part of a broad flat valley. As the stream overflows its banks during flood stage, friction against the ground outside the channel decreases the energy of the water, and a layer of material is dropped. Natural levees form in this way. Natural levees consist mainly of sands and gravels.

Gravity settling of sediments discharged into lakes and, more commonly, into oceans may be accelerated by flocculation of clay particles in the changed chemical environment.

Types of alluvial deposits and their complexity are well-illustrated by the sediments of the lower Mississippi River valley. The alluvial valley of the Mississippi River covers some 125,000 km², extending from Cairo, Illinois to the Gulf of Mexico. Virtually all types of deposits from sands to highly plastic clays and organic materials are to be found at some point or other within the valley. The fall of sea level associated with the last stages of glaciation led to scouring of a valley beneath the present flood plain surface.

The rising sea at the end of the glacial period resulted in deposition of sands and gravels in the bottom of the valley followed by finer material above. In the 25,000 yr since the last glaciation, the Mississippi has changed from an overloaded, shallow braided stream to a deep, single-channeled, meandering river.

Table 4.6 lists the different types of deposits to be found presently within the Mississippi River valley and some of their engineering properties. Although the variety of deposits is great and their interrelationships are complex, each can be accounted for in a logical way in terms of the factors governing its deposition. Figure 4.15 illustrates the conditions near Greenville, Mississippi, and Fig. 4.16 shows conditions near New Orleans, Louisiana. The substream materials are the coarser materials laid down initially in the bottom of the valley. Occasional lenses of clay, sandy silt, and silty sand are found in the substratum. The depth to these materials varies from about 3 m in the north to 30 m at the southern part of the river, and the thickness varies from 15 to 125 m in the same direction.

The braided stream deposits are usually remote from present large streams. Most are relatively dense sandy silts and clayey sands. Natural levees rise to 5 m or more above the floodplain and decrease in grain size away from the crest and in a downstream direction.

Point bar deposits composed of silts and silty sands form on the inside of river bends during high water periods. Clayey swales with high organic and water content form between the sand bars and the original river bank. The alternating pervious bars and impervious swales have created underseepage problems in connection with artificial levees. Abandoned sections of the river, left as oxbow lakes, fill with weak and compressible clay and silty clay in thicknesses up to 30 m or more. Abandoned courses many miles long fill with materials similar to those of the oxbow lakes.

Clays of medium to high plasticity, often organic, termed backswamp deposits, form in shallow areas during flood stage. Because of dessication between periods of deposition, the water contents are lower than in abandoned channel deposits.

The complex formations of the Mississippi delta reflect the composite effect of the advancing delta and the encroaching sea. Pleistocene sediments, consisting of high density clays to sands and gravels underlie the delta (Fig. 4.16). The sand and shell beaches, often 5 m high or more, are among the most useful deltaic formations for use as foundations. Some of the most difficult problems are associated with marshes and swamps because of low strength, high compressibility, and marsh gas.

4.12 AEOLIAN DEPOSITS

Wind is most easily able to erode sand. Of the agents of transportation, wind is the only one that can transport material uphill. Wind is not a universal agent of erosion; its effects are restricted to areas of certain climate such as deserts, or to specific places such as beaches and plowed fields. The wind suspension load is carried high above the ground; silt size particles may be transported for great distances. The bed load (moved by saltation and traction) moves slowly and as a unit. Saltation movement is the major wind transport mode.

Deposition from wind occurs with reduction in wind velocity. Chief accumulations are found in the lee of desert areas. Coarser particles (sand) carried by saltation and traction pile in dunes with their long axis parallel to the wind. *Loess* deposits, composed of wind-blown particles of silt size, are of great interest because of their unique properties and are described more fully in Section 4.16.

4.13 GLACIAL DEPOSITS

Ice has the greatest competency for sediment movement of all the transportation agents. There is no limiting size of particles that may be carried. Ice pushes material along in front and erodes the bottom and sides of the valleys through which it flows. In an *active* glacier (Fig. 4.17), there is a continuous erosion and transport of material from the region of ice accumulation to the region of melting. A *dead* glacier is cut off from a feeding ice field.

Deposition from ice results from melting, and a variety of deposits may result, as listed in Table 4.7 and shown in Figs. 4.17 to 4.19. Glacial deposits are dropped directly by the melting ice; the bottom morraine (Fig. 4.18) is an example. Glacio-fluvial deposits are transported from the melting point by flowing meltwater; kames and eskers (Fig. 4.19) are examples. Many lateral morraines and dead ice deposits (Fig. 4.19) are mixed glacial and glacio-fluvial deposits. Glacial lake deposits are quiet water deposits usually composed of fine-grained materials. Varved clays (Fig. 4.11) are an example.

Loess deposits may be found in association with previously glaciated areas. When the ice melts, a

Table 4.6 Typical Properties of Selected Environments of Deposition within the Mississippi Alluvial Valley (Kolb and Shockley (1957)).

Environment	Grain Size and Organic Content[1]	Predominant Soil Texture[2]	Natural Water Content, Per Cent	Liquid Limit	Plasticity Index	Shear Strength[3] Cohesion lb per sq ft	Angle of Internal Friction, Deg
Braided stream[4]		Clay sands (SC) to silty clays (CL)	25–40	30–75	10–55	200–1100	30
		Sands (SP)	—	NP	NP	0	30–40
Natural levee[5]		Clays (CL)	25–35	35–45	15–25	360–1200	0
		Silts (ML)	15–35	NP–35	NP–5	180–700	10–35
Point bar (ridges)[6]		Silts (ML) and silty sands (SM)	25–45	30–55	10–25	0–850	25–35
Abandoned channel[6]		Clays (CL and CH)	30–95	30–100	10–65	300–1200	0
Backswamp[7]		Clays (CH)	25–70	40–115	25–100	400–2500	0
Swamp[8]		Organic clay (OH)	110–265	135–200	100–165	Very low	Very low
Marsh[8]		Peat (PT)	160–465	250–500	150–400	Very low	Very low
Pro-delta[8]		Clay (CL and CH)	20–120	25–95	10–80	175–300	0
Lacustrine[8]		Clay (CH)	45–165	85–115	65–95	75–150	0
Beach[8]		Sand (SP)	Saturated	NP	NP	0	30
Bay-sound[8]		Clay silts and silty clays (CL and CH)	20–70	40–80	25–65	250–700	15–20
Substratum		Sand (SP)	Saturated	NP	NP	0	30–38
Pleistocene[8]		Clays and silty clays (CL and CH)	15–50	25–80	20–75	550–5000	0

(1) Legend:

Legend: Gravel (>2.0 mm), Sand (2.0 to 0.06 mm), Silt (0.002 to 0.06 mm), Clay (<0.002 mm), Organic Material,

Characteristic grain sizes for braided stream, natural levee, point bar, backswamp, and abandoned channel deposits based on Fisk, N.H., *fine-grained alluvial deposits and their effects on Mississippi River activity*, Mississippi River Commission, Vicksburg, Miss., 1947.

(2) Symbols based on unified soil classification system.

(3) Shearing strengths of clays based on unconfined compression tests.

(4) Data other than grain size based on 3 borings at Haynes and Devall Bluff, Ark.

(5) Data other than grain size based on approximately 25 borings in southern portion of valley.

(6) Data other than grain size based on approximately 100 borings in valley south of Memphis, Tenn.

(7) Data other than grain size based on approximately 150 borings south of Natchez, Miss.

(8) Data other than grain size based on approximately 200 borings in New Orleans area.

70

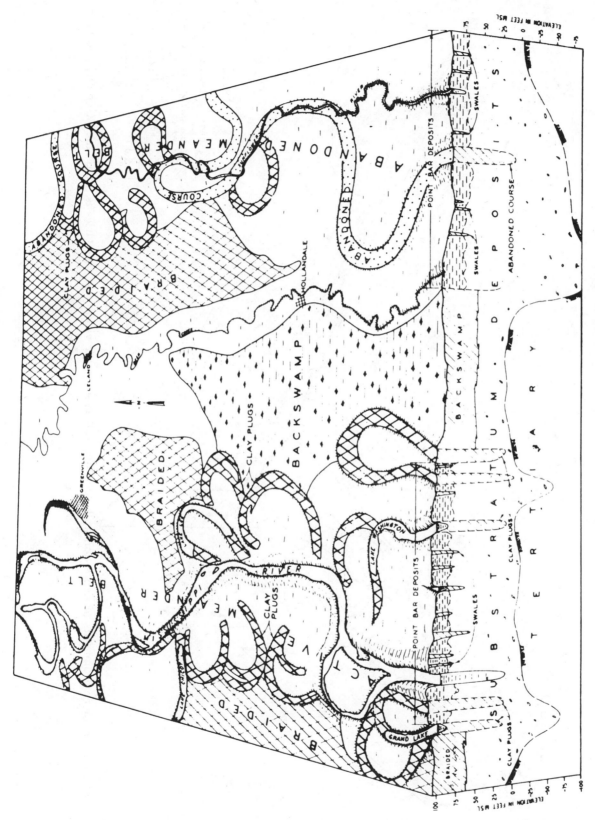

Fig. 4.15 Major environments of deposition and associated soil types in the vicinity of Greenville, Mississippi. On the subsurface profile dots represent sand; slant lines, clay; and dashed lines, silt (Kolb and Shockley, 1957).

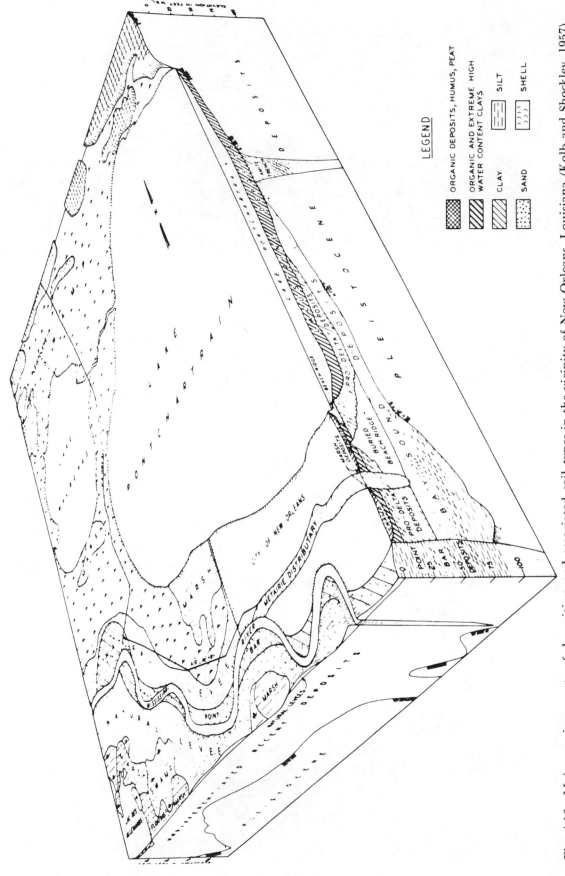

Fig. 4.16 Major environments of deposition and associated soil types in the vicinity of New Orleans, Louisiana (Kolb and Shockley, 1957).

Table 4.7 Deposits of the Glacial Environment

I. Ice Deposited Material
   A. Sediments
      1. Boulder clay or till (drift includes glacial and glacio-fluvial sediments)
      2. Erratics
   B. Structures
      1. Moraines
         a. *Lateral moraine*—ribbon of debris on sides of glacier
         b. *Medial moraine*—merging of inner lateral moraines of two joining glaciers
         c. *Englacial moraine*—material within ice
         d. *Subglacial moraine*—material at sole of glacier
         e. *Ground moraine*—deposited subglacial moraine
         f. *Terminal or end moraine*—ridge of deposits built up at end of glacier
         g. *Recessional moraine*—terminal moraine of receding glacier
      2. Drumlins
         Mounds of boulder clay formed under deep ice

II. Glacio-Fluvial Deposited Material
   A. Sediments
      1. Coarse gravel to clay, progressively sorted dams and deltas
      2. Crudely bedded gravel and sand in kames and eskers
   B. Structures
      1. *Alluvial fans* for glaciers terminating on land
         Outwash plains merged with fans
      2. *Deltas* for glaciers terminating in standing water
      3. *Kettle holes* caused by melting of stranded ice blocks
      4. *Kames*—mounds of crudely bedded sand and gravel caused by stream from melting ice
      5. *Esker*—winding ridge of sand and gravel from meltwater stream in ice tunnel or from receding ice

III. Glacial Lake Deposited Material
   A. Sediments
      1. Sands to clay
      2. Poor sorting and stratification of channel deposits
      3. Excellent stratification of lake floor deposits
   B. Structures
      1. *Overflow channels* where lake water escaped
      2. *Shore line deposits and terraces* from waves and currents
      3. Deltas
      4. *Lake floor sediments* including varved clays

---

vegetation-free area, which is vulnerable to wind erosion, is left behind.

The characteristics of glacial deposits depend on such factors as the erodability of the original material, type, and distance of transportation, gradients, and pressures. As an example, bottom morraines are usually finer-grained and more consolidated than lateral or end morraines. Finely ground (silt and clay size) "rock flour" is produced by the grinding actions of the ice and may be a major constituent of post-glacial marine and lake clays of the type found in Canada and Scandinavia.

4.14 MARINE SEDIMENTS

In shallow water, deposition from waves occurs when the intensity of wave action turbulence diminishes either due to periods of calm water or when sediment is carried seaward by return currents into zones of relatively more quiet water. Characteristic

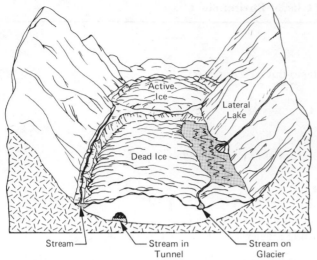

Fig. 4.17 Characteristics of glaciers (Selmer-Olsen, 1964).

shore line deposits are sands on beaches and finer materials in lagoons and on tidal flats.

The area of the continental shelves amounts to about 30 percent of the land area of the earth, and the continental shelves are rich in resources.

Sediments on the continental shelves are generally derived from the continents. Sands, silts, and clays are found, and their distribution is influenced largely by the local geology. The physical characteristics of the continental shelf deposits are essentially the same as those of comparable terrestrial soils (Keller, 1969; Noorany and Gizienski, 1970). Because of the present development of the continental shelf for energy production and storage, terminal facilities, and so on, current study of shelf deposits is intensive.

The deep sea floor is covered mainly by brown clays and calcareous and siliceous oozes, with thicknesses from 300 to 600 m. *Terrigenous* deposits are derived

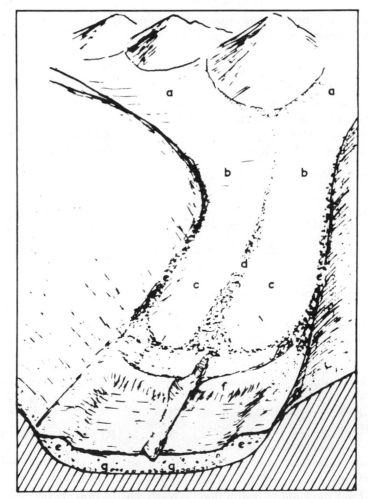

Fig. 4.18   Morraines (Selmer-Olsen, 1964). (*a*) Accumulation. (*b*) Debris on glacier. (*c*) Melting area. (*d*) Middle morraine. (*e*) Lateral morraine. (*f*) End morraine. (*g*) Bottom morraine.

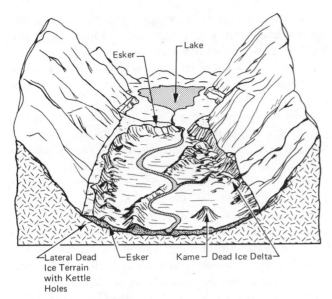

Fig. 4.19 Glacio-fluvial sediments (Selmer-Olsen, 1964).

from land; whereas, *pelagic* sediments settle from the water alone and contain the shells and skeletal remains of tiny marine organisms and plants. Accumulation rates may range from less than a millimeter per 1000 yr in the deep sea (Griffin, Windom, and Goldberg, 1968) to a few tens of centimeters per year in near-shore areas near the mouth of large rivers. *Oozes* contain more than 30 percent biotic material.

Calcareous ooze, composed of empty shells or tests, covers about 35 percent of the sea floor. It is usually nonplastic, cream to white in color, and composed of easily crushed sand-to-silt size particles. Brown clay is found over most of the deeper ocean areas. Its origin is believed to be atmospheric dust and fine material circulated by ocean currents. About 60 percent of the material is finer than 60 μm, and the clay fraction contains various amounts of chlorite, montmorillonite, illite, and kaolinite, with illite often the most abundant. Brown clays are characterized by high-water contents, moderate to high plasticity, and low strength. Siliceous ooze, composed of plant remains, is found mainly in the Antarctic, northeast of Japan, and in some areas of the equatorial Pacific.

Deep sea deposits are normally consolidated and highly compressible. There is an apparent overconsolidation of many of the near-surface materials, which evidently reflects a bonding developed as a result of the extremely slow rate of deposition and physicochemical effects (Noorany and Gizienski, 1970). Most of the published data on the mechanical properties of sea floor soils pertain to the upper 6 m.

4.15 CHEMICAL AND BIOLOGICAL DEPOSITS

Evaporite deposits formed by precipitation from salt lakes and seas as a result of evaporation are sometimes encountered in layers several meters thick. The major chemical constituents of sea water and their relative proportions are listed in Table 4.8. Also listed are some of the important evaporite deposits. In some cases, alternating layers of evaporite and clay or other

Table 4.8 Major Constituents of Sea Water and Evaporite Deposits

Ion	Grams per Liter	Percent by Weight of Total Solids	Important Evaporite Deposits	
Sodium, Na^+	10.56	30.61	Anhydrite	$CaSO_4$
Magnesium, Mg^{2+}	1.27	3.69	Barite	$BaSO_4$
Calcium, Ca^{2+}	0.40	1.16	Celesite	$SrSO_4$
Potassium, K^+	0.38	1.10	Kieserite	$MgSO_4 \cdot H_2O$
Strontium, Sr^{2+}	0.013	0.04	Gypsum	$CaSO_4 \cdot H_2O$
Chloride, Cl^-	18.98	55.04	Polyhalite	$Ca_2K_2Mg(SO_4) \cdot 2H_2O$
Sulfate, SO_4^{2-}	2.65	7.68	Bloedite	$Ma_2Mg(SO_4)_2 \cdot 4H_2O$
Bicarbonate, HCO_3^-	0.14	0.41	Hexahydrite	$MgSO_4 \cdot 6H_2O$
Bromide, Br^-	0.065	0.19	Epsomite	$MgSO_4 \cdot 7H_2O$
Fluoride, F^-	0.001	—	Kainite	$K_4Mg_4(Cl/SO_4) \cdot 11H_2O$
Boric Acid, H_3BO_3	0.026	0.08	Halite	$NaCl$
	34.485	100.00	Sylvite	KCl
			Flourite	CaF_2
			Bischofite	$MgCl_2 \cdot 6H_2O$
			Carnallite	$KMgCl_3 \cdot 6H_2O$

Adapted from data by Degens (1965).

fine-grained clastic sediment are formed, resulting from cyclic wet and dry periods.

Many limestones have been formed by precipitation or from the remains of various organisms. Because of the much greater solubility of limestones than most other rocks, they present special problems caused by the possibilities of solution channels and cavities under foundations. Coral deposits, encountered throughout much of the southwestern Pacific area, are almost entirely composed of calcium carbonate and are formed organically. Carbonate sediments form in areas of high latitude and cold water as well as in warm ocean regions.

More than 12 percent of Canada is covered by a peaty material, composed almost entirely of decaying vegetation, termed *muskeg* (Radforth and MacFarlane, 1957). The engineering properties of this material have been extensively studied in recent years as areas covered by muskeg are developed. Peats and muskegs may have water contents of 1000 percent or more; they are compressible, and they have low strength. Typical engineering property data are given by Adams (1965).

Subsidence of peat bogs and burial under younger sediments may result in the formation of coal. Physical and chemical alterations can cause transition through the coal series from peat through brown coal and bituminous coals to anthracite. In many respects, the behavior of brown coal is similar to that of clays (Trollop, Rosengren, and Brown, 1965).

Chemical sediments and rocks in fresh water lakes, ponds, swamps, and bays are occasionally encountered in civil engineering problems. Marls ranging from relatively pure calcium carbonate to mixtures with mud and organic matter are formed by biochemical processes. Iron oxide is formed in some lakes. Diatomite or diatomaceous earth is essentially pure silica formed from the skeletal remains of small (up to a few tenths of a millimeter) freshwater and saltwater organisms.

4.16 COLLAPSING SOILS

Large areas of the earth's surface, particularly in the midwestern and western United States, parts of of Asia, and southern Africa, are covered by soils that are susceptible to large decreases in bulk volume when they become saturated. Such materials are termed collapsing soils. Collapse may be triggered by water alone or by saturation and loading acting together. Soils with collapsible grain structures may be residual, water deposited, or aeolian (Dudley, 1970). In most cases, the deposits are characterized by loose structures of bulky shaped grains, often in the silt to fine sand size range. In residual soils, collapsible grain structures form as a result of leaching of soluble and colloidal material. Water and wind-deposited collapsing soils are usually found in arid and semi-arid regions.

Debris flows (mudflows and torrential stream deposits) are deposited suddenly and locally, and may form a loose, metastable structure. Torrential stream deposits, in particular, form a loose, poorly graded material. Some amount of clay is present that may serve as a binder for the deposit when dry. Some cementation may also develop, because in the arid climates where such deposits form, water moves upward to evaporate, leaving behind its content of dissolved salts. If subsequently wetted, the loose structure may collapse and cause large settlements.

The San Joaquin Valley–Southern California Aqueduct intersects many debris flow deposits. In order to minimize the potential settlements of the canal and appertenant structures as a result of canal leakage, it was necessary to carry out extensive preponding before constructing the canal. Figure 4.20 is a photograph illustrating the large amount of surface subsidence caused by wetting collapsible soil formations.

Soils susceptible to large collapse can be identified using a density criterion. If the density is sufficiently low to give a void space larger than that needed to hold the liquid limit water content, then collapse problems on saturation are likely (Gibbs and Bara, 1967). If the void space is less than that needed for the liquid limit water content if saturated, then collapse is not likely unless the soil is loaded.

Loess deposits are widespread throughout the midwestern United States and parts of Asia. The material, a windlaid silt, is tan to light-brown in color, crumbly, and essentially devoid of stratification. The particles are predominantly silt-sized and composed of feldspar and quartz. A small amount of clay, usually less than 15 percent, may be present. Montmorillonite is the usual clay mineral. Calcite may be present in amounts up to 30 percent and serve as a weak cement that precipitates along the sides of vertical root holes and at interparticle contacts. Densities of the undisturbed material may be as low as 75 lb/ft^3, and the natural water content of metastable deposits is low, of the order of 10 percent. Most loesses plot near the A line on the plasticity chart.

Because of the vertical root holes formed by gradual burial of grassy plains, the absence of stratification, and light cementation, loess cleaves on vertical planes, and vertical faces cut in loess are quite stable as shown in Fig. 4.21. In fact, if inclined slopes are

Fig. 4.20   Preponding on collapsible soil to reduce future settlements.

cut, they will gradually erode back to a series of steplike vertical faces.

The low density and light cementation of the loess structure makes it susceptible to collapse. When maintained dry, it is reasonably strong and incompressible, as may be seen from Figs. 4.22 and 4.23. The porous structure may persist even under 200 ft of overburden. When saturated, however, loess deposits may lose their stability. Settlement due to saturation alone may be small, but with a surcharge acting, it may be very large as shown in Fig. 4.23. If first satu-

rated and then subjected to shock loading by earthquake or other cause, there may be instantaneous liquefaction. Watering the lawns around houses founded on loess has been known to cause serious settlement. The undisturbed density of loess may be a fair indicator of the potential settlement and loss of strength that may result from saturation.

Detailed studies on the nature and behavior of loess are reported by Krinetzky and Turnbull (1967) and Lutton (1969).

Fig. 4.20 (*continued*).

Fig. 4.21 Loess deposit. Note vertical slopes.

Fig. 4.21 (continued).

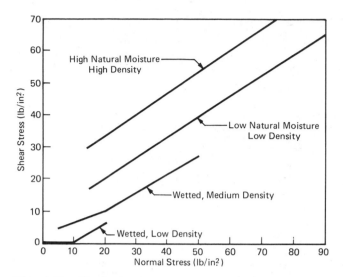

Fig. 4.22 Typical shear envelopes for loess (Clevenger, 1958).

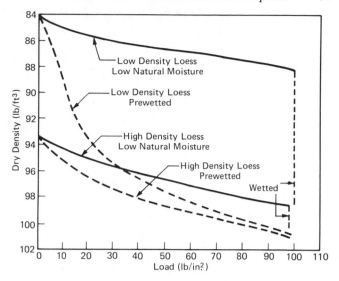

Fig. 4.23 Typical consolidation curves for Missouri River basin loess (Clevenger, 1958).

4.17 POSTDEPOSITIONAL CHANGES IN SEDIMENTS

Between the time a sediment is laid down and the time it is encountered in connection with some activity by man, it may have been altered as a result of the action of any one or more of several postdepositional processes. These processes may be physical, chemical, or biological in character. These changes occur because the sediment is not necessarily stable in its new environment where there are changed conditions of temperature, pressure, and chemistry. A knowledge of the nature of these changes is of great value in understanding properties, interpreting soil profile data, and reconstructing geological history.

Dessication

The exposure of fine-grained sediments to drying is accompanied by shrinkage and possibly cracking. Precompression of the upper portions of clay layers by drying is frequently observed in many of the profiles encountered in practice. Figure 4.24 shows the effects of dessication on the strength and water content variations with depth near the surface in London clay from the Thames estuary. Care must be exercised in interpreting profiles of this type, since drying is only one of several possible causes of apparent overconsolidation. Partial consolidation under increased overburden and the effects of weathering are other important mechanisms.

Exposure to the air and drying may cause oxidation and acceleration of weathering processes. For exam-

ple, a layer of hard yellow clay overlies much of the soft Boston blue clay. The yellow clay is believed to have formed during exposure of the clay during a period of uplift. Many clays and clay shales, particularly those containing the more active clay minerals, may lose much of their stability as a result of drying and slake after continued exposure to air or as a result of rewetting.

Weathering

Gradual compositional changes may result from weathering. Weathering and soil forming processes are active on sedimentary deposits after exposure to the atmosphere just as they are on freshly exposed rock. In some instances, weathering can lead to improvement in properties or protection of underlying material. For example, the weathering of uplifted marine clays may lead to a replacement of sodium by potassium as the dominant exchange cation (Moum and Rosenqvist, 1957). This increases both the undisturbed and remolded strength. Figure 4.25 shows water content and strength data for a Norwegian marine clay profile. It may be seen that the upper 5 m of clay, which have been weathered, have water content and strength variation characteristics similar to those of the Thames estuary clay (Fig. 4.24). That is, the Norwegian clay appears precompressed. It may be noted, however, that the plasticity values of the Norwegian clay have been changed in this zone; whereas, those of the Thames estuary clay have not, thus providing evidence of the changed composition.

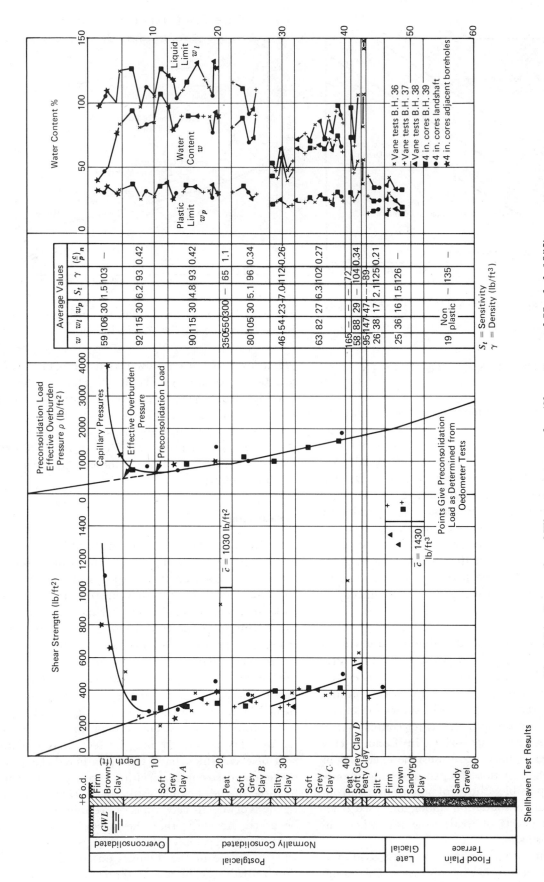

Fig. 4.24 Properties of Thames estuary clay (Skempton and Henkel, 1953).

80

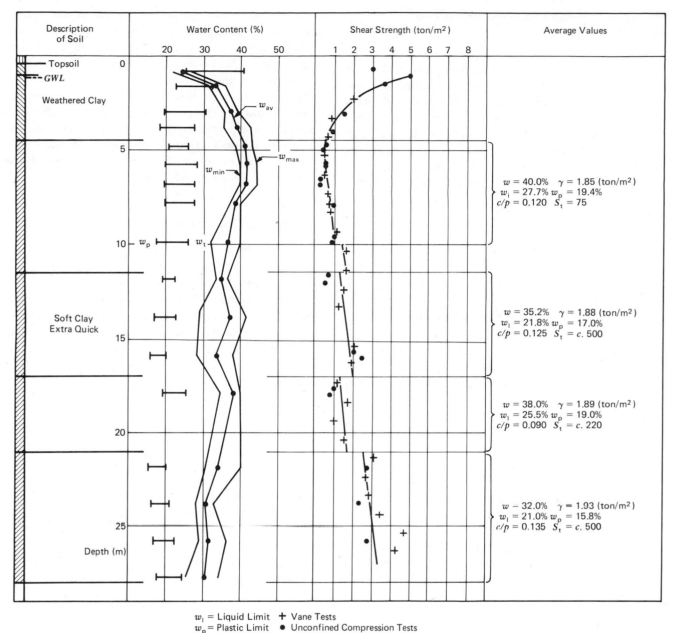

Fig. 4.25 Clay characteristics at Manglerud in Oslo, Norway (Bjerrum, 1954).

The weathering of some loess deposits has resulted in formation of a layer of relatively impervious loam which serves to protect the underlying metastable loess structure from the deleterious effects of water.

Consolidation and densification

Consolidation (termed compaction by geologists) of fine-grained sediments occurs from increased overburden, drying, or changes in ground water table such that the effective stress on the material is increased. Deposits of granular material may be affected to some extent in the same way. More significant densification of cohesionless soil occurs, however, as a result of dynamic loading such as induced by earthquake or the activities of man. The usual effects of consolidation on sediments are to increase strength, decrease compressibility, increase swell potential, and decrease permeability.

Authigenesis, diagenesis, cementation, and recrystallization

Authigenesis is the formation of new minerals in place after deposition. Authigenesis may make the grains more angular, lower the void ratio, and decrease the permeability. Small crystals and rock fragments may grow into aggregates of coarser particles.

Diagenesis refers to such phenomena as changes in particle surface texture, the conversion of minerals from one type to another, and the formation of interparticle bonds as a result of temperature, pressure, and time. With increasing depth of burial in a sedimentary basin, a clayey sediment may undergo substantial transformation. Expansive clay minerals may transfer to a nonexpansive form, (e.g., montmorillonite through mixed layer to illite) as a result of the progressive removal of water layers under pressure (Burst, 1969). Burial depths of 1000 to 5000 m are required, and the transformation process appears thermally activated as a result of the increased temperature at these depths. Chlorite may form in muds and shales during deep burial (Weaver and Pollard, 1973).

Cementation has important effects on the properties and stability of many soil materials. Cementation is not always easily identified, nor are its effects always readily determined quantitatively, although it is known to contribute to clay sensitivity (Mitchell, 1956), and it may result in an apparent preconsolidation pressure. Removal of iron compounds from a very sensitive clay from Labrador, Canada led to a decrease in apparent preconsolidation pressure of 30 ton/m² (Kenney, Moum, and Berre, 1967).

Failure to recognize cementation may result in construction difficulties and disputes. As an example, a soil on a major project was marked simply as glacial till on the contract drawings. It proved to be so hard that it had to be blasted. The contractor and the government became engaged in litigation when the contractor claimed an extra. He contended that the soil was cemented, because during digging failure took place through pebbles as well as the clay matrix. The government contended that this was because the pebbles were weathered.

Leaching, ion exchange, and differential solution

A change in environmental setting, as may take place, for example, by the uplift of a marine clay above sea level followed by the leaching action of percolating fresh groundwater, can lead to both ion exchange and a removal of dissolved salts. Leaching may cause marked changes in the interparticle forces between colloids and has been shown to be a major cause in the formation of quick clays (Lambe, 1953; Bjerrum, 1954; Bjerrum and Rosenqvist, 1956; Mitchell, 1956). These effects are discussed in more detail in Chapters 7 and 11.

Materials may be removed from sediments by differential solution and subsequent leaching. Calcareous sediments are particularly susceptible to solution, resulting in the formation of channels, sink holes, and cavities. The presence of such features under the foundations of structures can be the source of difficulties.

Jointing and fissuring of clay soils

Some normally consolidated clays, almost all flood plain clays, and many preconsolidated clays are weakened by joints. Joints in flood plain clays result from deposition followed by cyclic expansion and contraction from wetting and drying. Joints and fissures in preconsolidated clays may result from unloading or from shrinkage cracking during drying. Closely spaced joints in these types of clays may contribute to slides some years after excavation of cuts. The unloading enables joints to open, water to enter, and the clay to soften.

Fissures have been found in normally consolidated clays at water contents well above the shrinkage limit that cannot have been caused by drying or unloading (Skempton and Northey, 1952). Increased brittleness may develop in clay chunks stored for some time at constant water content (Rosenqvist, 1955). This behavior may have resulted from syneresis—the mutual attraction of clay particles to form closely knit aggregates with fissures between.* It has also been postulated that weathering and the release of potassium may result in fissuring, as may numerous small shear failures during consolidation.

Trees have been observed to induce fissuring and shrinkage of some clays (Barber, 1958). The root systems suck up water causing large capillary pressures in the ground. When rain falls on the crusted deposit of dried up saline lakes, it is absorbed by capillarity; the air inside the clay can become so compressed as to cause tension cracking or actual blowouts in a form similar in appearance to root holes. These sediments may undergo severe cracking apparently as a result of shock. Cracks up to 2 ft wide, of unknown depth and spaced tens of feet apart have caused damage to buildings and highways.

---

* Such behavior is many times observed in gelatine after aging.

4.18 PRACTICAL IMPLICATIONS

A knowledge of the history of a soil deposit aids in anticipating its probable composition, structure, and properties. Effects of processes, such as earth movements, unloading and heating, that occurred many thousands or millions of years ago may remain in the form of joints, fissures, locked-in stress, planes of weakness, clay seams, and other features that may influence design and construction.

An understanding of soil-forming processes and the characteristics of residual soils can be used to anticipate organic and high clay content layers, to infer clay mineral types, to locate borrow material for construction, and to estimate the depth to unaltered parent material. Pedologic data can be used to infer compositions and physical properties.

Transported soils are sorted, abraded, and have particle surface textures that reflect the transporting medium. Conditions of sedimentation and the depositional environment have a large influence on grain size, size distribution, and grain arrangement in any depositional basin. Inferences concerning physical properties can be based on a knowledge of transportational and depositional history. If postdepositional changes are known and understood in any given case, then current behavior and possible future changes in properties can be postulated. Stress history is of particular importance because of the strong dependence of deformation and strength on both past and present state of stress. Changes in the chemical environment may cause changes in interparticle forces in fine-grained soils, thus causing initially stable particle arrangements to become metastable and susceptible to collapse when physically disturbed. The status of homogeneity and anisotropy found in any soil relate to its formational mode and subsequent history.

SUGGESTIONS FOR FURTHER STUDY

Degens, E. T. (1965), *Geochemistry of Sediments*, Prentice-Hall, Englewood Cliffs, N. J.

Dudley, J. H. (1970), "Review of Collapsing Soils," *Journal of the Soil Mechanics and Foundations Division*, A.S.C.E., Vol. 96, No. SM3, pp. 925–947.

Grim, R. E. (1968), *Clay Mineralogy*, 2nd Ed., McGraw-Hill, New York.

Inderbitzen, A. L. (Ed.) (1974), *Deep Sea Sediments,* Plenum, New York.

Jenny, H. (1941), *Factors of Soil Formation*, McGraw-Hill, New York.

Keller, W. D. (1957), *The Principles of Chemical Weathering*, revised ed., Lucas Bros., Columbia, Mo., 111 pp.

Keller, W. D. (1964), "Processes of Origin and Alteration of Clay Minerals," in *Soil Clay Mineralogy* (C. I. Rich and G. W. Kunze, Ed.), University of North Carolina Chapel Hill, pp. 3–76.

Krumbein, W. C., and Sloss, L. L. (1951), *Stratigraphy and Sedimentation*, Freeman, San Francisco.

Millar, C. E., Turk, F. M., and Foth, H. D. (1965), *Fundamentals of Soil Sciences,* 4th ed., Wiley, New York.

Noorany, I., and Gizienski, S. F. (1970), "Engineering Properties of Submarine Soils: State-the-Art Review," *Journal of the Soil Mechanics and Foundations Division*, A.S.C.E., Vol. 96, No. SM5, pp. 1735–1762.

Reiche, P. (1945), "A Survey of Weathering Processes and Products," University of New Mexico, Publication in Geology, No. 1, Albuquerque, N.M.

Weaver, C. E., and Pollard, L. D. (1973), *The Chemistry of Clay Minerals*, Developments in Sedimentology 15, Elsevier, New York, 213 pp.

Woods, K. B., Miles, R. D., and Lovell, C. W., Jr. (1962), "Origin, Formation, and Distribution of Soils in North America," Chapter 1, in *Foundation Engineering* (G. A. Leonards, Ed., McGraw-Hill, New York.

CHAPTER 5

Determination of Soil Composition

5.1 INTRODUCTION

The properties of any soil depend directly on mineralogical and chemical composition, the texture and arrangement of particles, and the effects of environment. However, present knowledge does not permit quantitative expression of properties in terms of these factors, except in the simplest systems.

On the other hand, a knowledge of the mineralogical and noncrystalline chemical composition of a soil can be of great value for understanding engineering properties. The general physical characteristics of the important clay and nonclay minerals are reasonably well known, and the influences of such constituents as organic matter and the noncrystalline oxides of iron, silicon, and aluminum are becoming better understood. Qualitative identification of the minerals, both clay and nonclay, in a soil can ordinarily be made using X-ray diffraction. Simple chemical tests can be used to indicate the presence of organic matter and other constituents. The microscope may be used to identify the constituents of the nonclay fraction. From a knowledge of the grain size distribution and the relative intensities of the different X-ray diffraction peaks a semiquantitative analysis may be made that is adequate for most purposes.

In this chapter, a general approach to determination of soil composition is presented, some of the techniques are described briefly, and criteria for identification of the important soil constituents are given.

5.2 METHODS FOR COMPOSITIONAL ANALYSIS

The methods and techniques that may be employed for determination of soil composition and study of soil grains are as follows:

1. Particle size analysis and separation.
2. Various pretreatments prior to mineralogical analysis.
3. Chemical analyses for determination of free oxides, hydroxides, amorphous constituents, and organic matter.
4. Petrographic microscope study of silt and sand grains.
5. Electron microscope study of the clay phase.
6. X-ray diffraction for identification of crystalline minerals.
7. Thermal analysis for identification of components.
8. Determination of specific surface.
9. Chemical analysis for cation exchange capacity, exchangeable cations, pH, and soluble salts.
10. Staining test for identification of clays.

Procedures for determination of soil composition are described in some detail in the two volume series, *Methods of Soil Analysis,* Monograph Number 9, published by the American Society of Agronomy in 1965. For each method, principles are presented as well as the details of the method. In addition, the interpretation of results is discussed and extensive bibliographies are given.

5.3 ACCURACY OF COMPOSITIONAL ANALYSIS

Although techniques for chemical analysis are generally of a high order of accuracy, this accuracy does not extend to the overall compositional analysis of the soil in terms of components of interest in understanding behavior. The problem arises because knowledge of the chemical composition alone is of limited value. The value of chemical analysis of the solid phase of a soil is limited because it does not indicate the organization of the elements into crystalline and noncrystalline components.

In the case of quantitative mineralogical analysis of the clay fraction, it is generally necessary to assume that the properties of the mineral in the soil are the same as those of a reference mineral. Different samples of any given clay mineral may exhibit significant differences in composition, surface area, particle size and shape, and cation exchange capacity. Thus, selection of "standard" minerals for reference is arbitrary. It is doubtful, therefore, that quantitative clay mineral determinations can be made to an accuracy of more than about plus or minus a few percent without exhaustive chemical and mineralogical tests. Fortunately, for most applications, greater accuracy is not needed.

5.4 GENERAL SCHEME FOR COMPOSITIONAL ANALYSIS

Figure 5.1 provides a general scheme for determination of the components of a soil. Those techniques that are of the most value for qualitative and semiquantitative analysis of the clay minerals are indicated by a double asterisk, and those that may be particularly useful for explaining unusual properties are indicated by a single asterisk. The scheme shown is by no means the only one that could be used; in any given case a feedback approach is desirable, wherein the results of each test are used to plan subsequent tests. Brief discussions of the various techniques listed in Fig. 5.1 are given below. X-ray diffraction analysis is treated in more detail because of its especially great value for the study of soil composition.

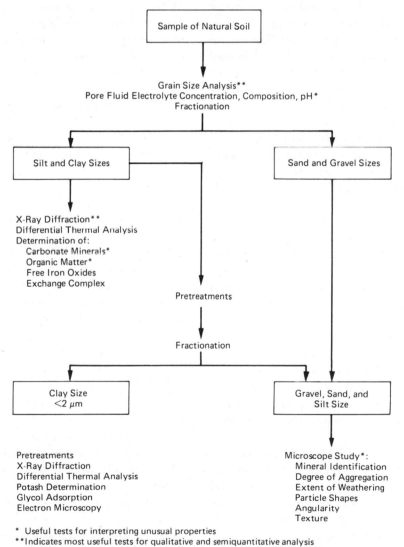

Fig. 5.1 Flow sheet for compositional analysis of soils (Adapted from Lambe and Martin 1954).

Grain size analysis

The determination of particle size and size distribution is most often accomplished using sieve analysis for the coarse fraction (sizes greater than 200 mesh sieve, i.e., 74 μm) and by sedimentation methods for the fine fraction. Details of these methods are presented in standard soil mechanics texts and in the standards of the American Society for Testing and Materials. Determination of sizes by sedimentation is based on the application of Stoke's law for the settling velocity of spherical particles, which states

$$v = \frac{\gamma_s - \gamma_w}{18\mu} D^2 \qquad (5.1)$$

where γ_s = unit weight of particle, γ_w = unit weight of liquid, μ = viscosity of liquid, and D = diameter of sphere. The sizes determined according to Stoke's law are not actual particle diameters but equivalent spherical diameters. Gravity sedimentation is limited to particle sizes in the range of about 0.2 mm to 0.2 μm, the upper bound reflecting the size limit where flow around the particles is no longer laminar, and the lower bound representing a size where Brownian motion keeps particles in suspension indefinitely.

Sedimentation methods call for treatment of a soil–water suspension with a dispersing agent and thorough mixing prior to the start of the test. This leads to the breakdown of aggregates of soil particles, and the degree of breakdown may vary greatly with the method of preparation. For example, the American Society for Testing and Materials standard method of test permits use of either an air dispersion cup or a blender-type mixer. The amount of material less than 2 μm in equivalent spherical diameter may vary by as much as a factor of two by the two techniques. The relationship between the size distribution that results from preparation of the sample to that of the particles and aggregates in the natural soil is unknown. The distribution of the particles and aggregates and the resistance of the aggregates to disruption are important in influencing the mechanical properties.

Occasionally, the optical and electron microscopes are used to study particle sizes and size distributions. Analysis by these methods, while more tedious and only providing data for a small sample, provides information on particle shape, aggregation, angularity, weathering, and surface texture.

Fractionation of samples into different sizes is done by sieve (200 mesh) if separation into coarse (sand and gravel) and fine (silt and clay) fractions is desired. Separation of the clay sizes (finer than 2 μm), or different sizes within the clay fraction, can be done by sedimentation. The centrifuge is useful for accelerating the settlement of small particles and is the most practical means for extracting particles smaller than about a micrometer in size. The times for particles of 2, 5, and 20 μm equivalent spherical diameter to fall through water a distance of 10 cm are about 8 hr, 1.25 hr, and 5 min, respectively, at 20°C. At 30°C the required times are about 6.5 hr, 1 hr, and 4 min.

Pore fluid electrolyte

The total concentration of soluble salts may be determined from the electrical conductivity of extracted pore fluid (American Society for Testing and Materials, 1970). Chemical or photometric techniques may be used to determine the elemental constituents of the extract (Bower and Wilcox, 1965). Removal of excess soluble salts by washing the sample with water or alcohol is desirable before proceeding with subsequent analysis. If they are not removed, the soil may be difficult to disperse, it may be difficult to remove organic matter, reliable cation exchange capacity determinations will be impossible, and mineralogical analyses will be complicated (Kunze, 1965).

pH

The determination of the acidity or alkalinity of a soil in terms of the pH is a relatively simple measurement that may be made using a pH meter or special indicators (Peech, 1965; American Society for Testing and Materials, 1970). As the value obtained decreases as the ratio of soil to water decreases, it is usual to standardize the measurement using a 1 : 1 ratio of soil to water by weight. The pH also decreases with increasing concentration of neutral salts in solution and with increasing amounts of dissolved CO_2.

Carbonates

Carbonates, in the form of calcite ($CaCO_3$), dolomite ($CaMg(CO_3)_2$), marl, and shells are frequently found in soils. Their presence can be readily detected by effervescence when the soil is treated with dilute HCl. Quantitative analysis of well-crystallized carbonate minerals can be made using differential thermal analysis. Poorly crystallized carbonates are determined quantitatively on the basis of the loss in weight resulting from treatment with 2 N acetic acid. Alternative procedures are given by Allison and Moodie (1965).

Organic matter

The presence of organic matter can be readily detected by treatment of the soil with a 15 percent hydrogen peroxide solution. The H_2O_2 and organic matter react to give vigorous effervescence. As organic matter has an aggregating effect, and because its presence may interfere with other mineralogical analyses, it is desirable to remove most of it by digestion with H_2O_2 (Kunze, 1965). Quantitative analysis methods for soil organic matter are described by Broadbent (1965), Stevenson (1965), and the American Society for Testing and Materials (1970).

Free iron oxides

Free iron oxides may occur as discrete particles or as coatings around grains and cementing agents between particles. Appreciable quantities of iron oxides may make dispersion of the soil difficult and interfere with other analysis procedures. Kunze (1965) presents a procedure for iron oxide removal using sodium dithionite-citrate, which is a slight modification of a method presented by Mehra and Jackson (1960). R. V. Olson (1965) presents a method for determination of the quantity of free iron oxide.

Exchange complex

Determination of the cation exchange capacity (expressed in milliequivalents per hundred grams of dry soil) is made after first freeing the soil of excess soluble salts. The adsorbed cations are then replaced by a known cation species, and the amount of the known cation needed to saturate the exchange sites is determined analytically (Chapman, 1965). The composition of the original cation complex can be determined by chemical analysis of the original extract.

Potash

Since the hydrous mica minerals (illites) are the only ones commonly found in the clay size fraction of a soil that contain potassium in their structure, knowledge of the K_2O content is useful for quantitative determination of their abundance. A method for potassium determination is given by Pratt (1965). Well-organized 10 Å illite layers contain 9 to 10 percent K_2O (Weaver and Pollard, 1973). Values less than this indicate the presence of nonillite layers or interlayer cations other than potassium.

Glycol adsorption

Ethylene glycol and glycerol are adsorbed on clay surfaces. As different clay minerals have different values of specific surface, the amount of glycol or glycerol retained under controlled conditions has been used to aid in the quantitative determinations of clay minerals (Martin, 1955; Diamond and Kinter, 1956; Mortland and Kemper, 1965; and American Society for Testing and Materials, 1970). Quantities adsorbed for different clay minerals are indicated in Table 5.1.

Table 5.1 Quantities of Ethylene Glycol and Glycerol Adsorbed by the Different Clay Minerals

Mineral	Adsorbate	Quantity Adsorbed per Gram of Oven-Dry Mineral of Indicated Particle Diameter (g)		
		2 to 1 μm	<2 μm	<0.1 μm
Kaolinite	Glycol	0.011	0.016	0.027
	Glycerol	0.014	0.023	0.033
Halloysite	Glycol	0.0209	0.0342	0.0476
	Glycerol	0.0255	0.0418	0.0580
Illite	Glycol	0.0430	0.0610	0.0790
	Glycerol	0.0500	0.0730	0.0960
Montmoril-lonite	Glycol[a]		0.3117	0.3117
	Glycerol[b]		0.4110	0.4110
Vermiculite	Glycol[c]		0.1951	0.2122
	Glycerol[c]		0.2294	0.2393
Allophane	Glycol	0.1200	0.1300	0.1400
	Glycerol	0.1470	0.1509	0.1700
Quartz	Glycol		0.0040	
	Glycerol		0.0033	

[a] $d_{(001)} = 17.1$ Å.
[b] $d_{(001)} = 18.4$ Å.
[c] $d_{(001)} = 14.3$ Å.
Adapted from Barshad (1965).

Staining tests

Dyes have been used for the rapid identification of clay minerals. The methods are based on clay-organic adsorption reactions. Malachite green and benzidine have been used for identification of montmorillonite. Such tests have not been widely adopted for routine identification of clay minerals, since their reliability has not been established.

5.5 X-RAY DIFFRACTION ANALYSIS

Generation of X-rays

X-ray diffraction is the most widely used method for identification of fine-grained soil minerals and the study of their crystal structure. X-rays are one of the several types of waves in the electromagnetic spectrum and have wave lengths in the range of 0.01 to 100 Å. When high speed electrons impinge on a target material one of two phenomena may occur:

1. The high speed electron may strike and displace an electron from an inner shell of one of the atoms of the target material. An electron from one of the outer shells then falls into the vacancy to lower the energy state of the atom. An X-ray of wave length and intensity characteristic of the target atom and of the particular electronic positions is emitted. Because electronic transfers may take place in several shells and each has a characteristic frequency, the result is a relationship between radiation intensity and wave length such as shown in Fig. 5.2.

2. If the high speed electron does not strike an electron in the target material but slows down in the intense electric fields near atomic nuclei, then the decrease in energy is converted to heat and to X-ray protons. X-rays produced in this way are independent of the nature of the bom-

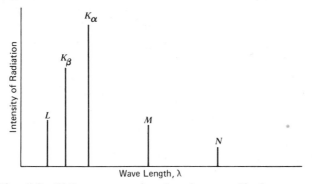

Fig. 5.2   X-Ray generation by electron displacement. Letters designate shells in which electron transfer takes place.

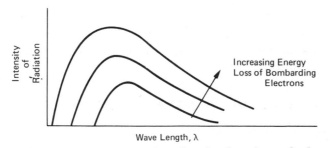

Fig. 5.3   X-Ray generation by deceleration of electrons in an electric field.

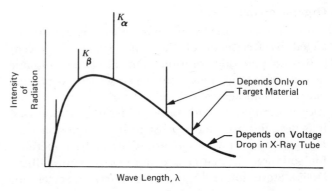

Fig. 5.4   Composite relationship for X-ray intensity as a function of wave length.

barded atoms and appear as a band of continuously varying wave length as shown in Fig. 5.3.

The resultant output of X-rays from these two effects acting together is shown in Fig. 5.4. In practice, X-rays are generated using a tube in which electrons are caused to stream from a filament to a target material across a voltage drop of about 50,000 V. Curved crystal monochrometers can be used to give X-rays of a single wave length. Alternatively certain materials are able to absorb X-rays of different wave lengths so that it is possible to filter the output of an X-ray tube to give rays of only one wave length. The wave lengths of monochromatic radiation (usually K_α, Fig. 5.4) produced from commonly used target materials range from 0.71 Å for molybdenum to 2.29 Å for chromium. Copper radiation, which is the most frequently used type for clay mineral identification, has a wave length of 1.54 Å.

Diffraction of X-rays

It is because wave lengths of about 1 Å are of the same order as the spacings of atomic planes in crystalline materials that makes X-rays useful for analysis of crystal structures. The atomic planes act with respect to X-rays in much the same way as a diffraction grating relative to visible light. When X-rays strike a crystal, they may penetrate to a depth of several million layers before being absorbed. At each atomic plane a minute portion of the beam is absorbed by individual atoms that then oscillate as dipoles and radiate in all directions. Radiation in certain directions will be in phase and can be interpreted in simplistic fashion as a wave resulting from a reflection of the incident beam. In-phase radiations emerge as a coherent beam and can be detected on film or by a radiation counting device. The orientation of parallel atomic

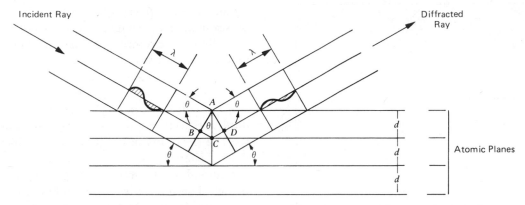

Fig. 5.5 Geometrical conditions for X-ray diffraction according to Bragg's law.

planes, relative to the direction of the incident beam, at which radiations are in phase depends on the wave length of the X-rays and the spacing between atomic planes.

Figure 5.5 shows a parallel beam of X-rays of wave length λ striking a crystal at an angle θ to parallel atomic planes spaced at distance d. If the reflected wave from C is to reinforce the wave reflected from A, then the path differences between the two waves must be an integral number of wave lengths $n\lambda$. From Fig. 5.5, the difference in path lengths is distance $BC + CD$. Thus,

$$BC + CD = n\lambda$$

But from symmetry, $BC = CD$, and by trigonometry, $CD = d \sin \theta$. Thus the necessary condition is given by

$$n\lambda = 2d \sin \theta \qquad (5.2)$$

which is *Bragg's law*. This law forms the basis for identification of crystals using X-ray diffraction. Since no two minerals have the same spacings of interatomic planes in three dimensions, the angles at which diffraction occurs (and the atomic spacings calculated therefrom) are used for identification. X-Ray diffraction is particularly well suited for identification of the clay minerals because the (001) spacing is characteristic for each clay mineral group. The basal planes generally give the most intense reflections of any planes in the crystals because of the close packing of atoms in these planes. The common nonclay minerals occurring in soils are also detectable by X-ray diffraction.

Detection of diffracted X-rays

Because the small size of most soil particles prevents the study of single crystals, use is made of the *powder method* and of oriented aggregates. In the powder method, a small sample giving particles of all possible orientations is placed in a collimated beam of parallel X-rays and diffracted beams of various intensities are either recorded on a film strip placed

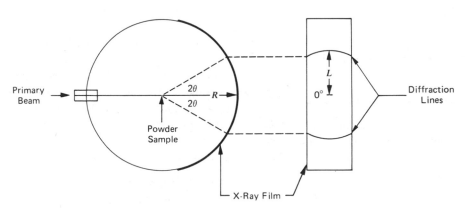

Fig. 5.6 Registration of diffraction lines on film in a cylindrical camera.

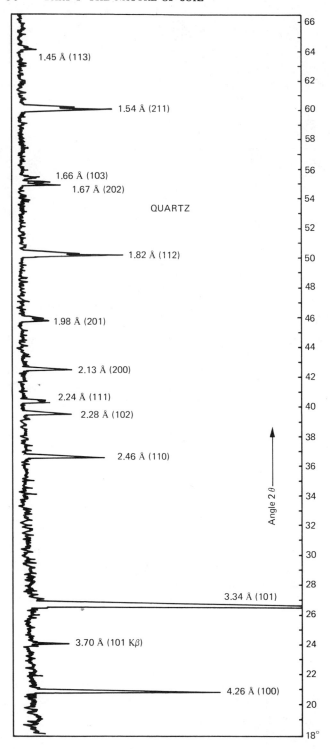

1.45 Å (113)

1.54 Å (211)

1.66 Å (103)
1.67 Å (202)

QUARTZ

1.82 Å (112)

1.98 Å (201)

2.13 Å (200)

2.24 Å (111)
2.28 Å (102)

2.46 Å (110)

Angle 2θ

3.34 Å (101)

3.70 Å (101 Kβ)

4.26 Å (100)

18°

Fig. 5.7 X-Ray diffractometer chart for quartz. Peaks occur at specific 2θ angles, converted to d spacings by Bragg's law. Numbers in parentheses are Miller indices for crystal planes.

around the periphery of a circular camera (Fig. 5.6) or scanned by a Geiger–Muller counting tube and read directly or recorded automatically (Fig. 5.7). The powder method works because the very large number of particles in a powder sample insures that some will be properly oriented to produce a reflection.

All prominent atomic planes in a crystal will produce a reflection when properly positioned with respect to the X-ray beam. Thus, each mineral will produce a characteristic set of reflections at values of θ corresponding to the spacings of the prominent planes. The intensities of the different reflections vary according to the density of atomic packing and other factors. When the oriented aggregate method is used, platy clay particles are precipitated onto a glass slide, usually by drying from a deflocculated suspension. With most particles oriented parallel to the slide, the (001) reflections are intensified; whereas, reflections from (hk0) planes are minimized.

In the Bragg equation, n may assume a value of any whole number. The reflection corresponding to $n = 1$ is termed the first order reflection. If the first order reflection for a numeral gives $d_{(001)} = 10$ Å, then for $n = 2$, there is an apparent reflection at $d = 5$ Å, for $n = 3$, there is an apparent reflection at $d = 3.33$ Å, and so on. It is common to refer to these spacings as due to the (002) plane, the (003) plane, and so on, even though atomic planes do not exist at these spacings. These "spacings" are in reality values of $d/n = \lambda/(2 \sin \theta)$.

Analysis of X-ray patterns

The complete X-ray diffraction pattern, either film or strip chart record (Figs. 5.6 and 5.7), will consist of a series of reflections of different intensities and values of θ. Each reflection must be accounted for in terms of some component of the sample. The first step in the analysis is to determine all values of d/n for the particular type of radiation (which determines λ) using equation (5.2). The test pattern may be compared directly with patterns for known materials. The American Society for Testing and Materials maintains a card file of patterns indexed on the basis of the strongest lines in the pattern. Grim (1968), Brown (1961), and Carroll (1970) present extensive X-ray diffraction data for the clay minerals. The most intense reflections for minerals commonly found in soils are listed in Table 5.2.

Criteria for clays

The different clay minerals are characterized by first order basal reflections at 7, 10, or 14 Å. Positive

Table 5.2 X-Ray Diffraction Data for Clay Minerals and Common Nonclay Minerals

d(Å)	Mineral[b]	d(Å)	Mineral[b]
14	*Mont. (VS) Chl. Verm. (VS)*	2.93–3.00	Felds.
12	*Sepiolite*, heated corrensite	2.89–2.90	Carb.
10	*Illite, Mica(S)*	2.86	Felds.
9.23	Heated Verm.	2.84	Carb. *Chl.*
7	*Kaol.(S). Chl.*	2.84–2.87	*Chl.*
6.90	*Chl.*	2.73	Carb.
6.44	Attapulgite	2.61	Attapulgite
6.39	*Felds.*	2.60	Verm., Sepiol.
4.90–5.00	*10Å (2nd)*	2.56	Illite (VS), Kaol.
4.70–4.79	*Chlor. (S)*	2.53–2.56	Chlor., Felds., Mont.
4.60	Sepiol., Verm. (S)	2.49	Kaol. (VS)
4.45–4.50	Illite (VS)	2.46	Quartz, heated Verm.
4.46	Kaol.	2.43–2.46	Chlorite
4.36	Kaol.	2.39	*Verm.*, Illite
4.26	Quartz (S)	2.38	*Kaol.*
4.18	Kaol.	2.34	Kaol. (VS)
4.02–4.04	Felds.(S)	2.29	Kaol. (VS)
3.85–3.90	Felds.	2.28	Quartz, Sepiol.
3.82	Sepiol.	2.23	Illite, Chl.
3.78	Felds.	2.13	Quartz, Mica
3.67	Felds.	2.05–2.06	Kaol. (WK)
3.58	Carbonate, Chl.	1.99–2.00	*Mica, Illite(S)*, Kaol. Chl.
3.57	*Kaol. (VS)*, Chl.	1.90	Kaol.
3.54–3.56	Verm.	1.83	Carb.
3.50	Felds., Chlor.	1.82	Quartz
3.40	Carb.	1.79	Kaol.
3.34	Quartz (VS)	1.68	Quartz
3.32–3.35	*Illite (VS)*	1.66	Kaolin
3.30	Carb.	1.62	Kaolin
3.23	Attapulgite	1.54B	Verm. (S), Quartz
3.21	Felds.	1.55	Quartz
3.20	Mica	1.58	Chl.
3.19	*Felds. (VS)*	1.53	Verm., Illite (Trioctahed)
3.05	Mont.	1.50	Ill. (S), Kaol.
3.04	Carb. (VS)	1.48–1.50	Kaol. (VS), Mont.
3.02	Felds.	1.45B	Kaol.
3.00	Heated Verm.	1.38	Quartz, Chl.
2.98	Mica (S)	1.31, 1.34, 1.36	Kaol. (B)

[a] *Italics*: *(00l) spacing.*
[b] (B) = broad; (S) = strong; (VS) = very strong; (WK) = weak; Mont. = Montmorillonite; Chl. = Chlorite; Verm. = Vermiculite; Kaol. = Kaolinite; Carb. = Carbonate; Felds. = Feldspar; Sepiol. = Sepiolite.

identification of specific mineral groups ordinarily requires certain pretreatments.

Kaolinite minerals. Kaolinite has a basal spacing of about 7.2 Å, which is insensitive to drying or moderate heating. Kaolinite minerals are destroyed by heating to 500°C. The other clay minerals are not. Hydrated halloysite has a basal spacing of 10 Å, which collapses irreversibly to 7 Å on drying at 110°C. The electron microscope is often needed to distinguish dehydrated halloysite (metahalloysite), with its tubular morphology, from kaolinite.

Hydrous mica (illite) minerals. The illites are characterized by a $d_{(001)}$ of about 10 Å, which remains fixed both in the presence of polar liquids and after drying.

Smectite (montmorillonite) minerals. The expansive character of this group of minerals provides the basis for their positive identification. When air dried, these minerals may have basal spacings of 12 to 15 Å. After treatment with ethylene glycol or glycerol, the smectites expand to a $d_{(001)}$ value of 17 to 18 Å. When oven dried, $d_{(001)}$ drops to about 10 Å as a result of the removal of interlayer water.

Vermiculite. Although an expansive mineral, the greater interlayer ordering in vermiculite results in less variability in basal spacing as a result of heating and drying than occurs in the smectite minerals. When Mg-saturated, the hydration states of vermiculite yield a discrete set of basal spacings, resulting from a changing, but ordered, arrangement of Mg cations and water in the interlayer complex. When fully saturated, the d-spacing is 14.8 Å, which reduces to 11.6 Å when heated at 70°C. All interlayer water can be expelled at 500°C, but rehydration is rapid on cooling. Permanent dehydration and collapse to 9.02 Å can be achieved by heating to 700°C.

Chlorite minerals. The basal spacing of the chlorite minerals is fixed at 14 Å because of the strong ordering of the interlayer complex. Chlorites often have a clear sequence of four or five basal reflections. The third order reflection at 4.7 Å is quite often strong. Iron-rich chlorites have a weak first order reflection but strong second order and, thus, may be confused with kaolinite. The facts that chlorite is destroyed when treated with 1 N HCl at 60°C while kaolinite is unaffected and that kaolinite is destroyed but chlorite may not be affected on heating to 600°C are useful for distinguishing the two clay types.

Table 5.3 is a summarization of basal spacings and effects of glycol treatment and heating which can be used for X-ray identification. Regularly interstratified mixtures of clay minerals are also frequently encountered. Basal spacings useful for identification of some of the common interlayered mixtures are listed in Table 5.4.

Quantitative analysis by X-ray diffraction

Quantitative determinations of the amounts of different minerals in a soil on the basis of simple comparison of diffraction peak heights or areas cannot be made because of differences in mass absorption coefficients of different minerals, in particle orientations, in sample weights, in surface texture of the sample, in crystallinity of the minerals, in hydration, and in other factors. In some cases, estimates based on X-ray data alone are at best semiquantitative; however, in others techniques that account for differences in mass absorption characteristics and utilize comparisons with known mixtures or internal standards may give good results. Soils containing only two or three well-crystallized mineral components are more easily analyzed than those with multimineral compositions and mixed layering.

A theory for quantitative analysis based on linear interaction between mineral pairs is described by Moore (1968). Methods for determination of interaction coefficients are given. Experimental accuracy to within 2 percent appears possible. A method based on the use of at least two synthetic mixtures to bracket the background X-ray intensity and composition of the unknown is described by Buck (1973).

For a more detailed treatment of X-ray diffraction theory, identification criteria, and techniques, particularly as related to the study of clays, the reader is referred to Klug and Alexander (1954), Cullity (1956), Brown (1961), Whittig (1965), Grim (1968), and Carroll (1970).

5.6 DIFFERENTIAL THERMAL ANALYSIS

Principle

Thermal techniques have been used for the study of soils and clay minerals. Differential thermal analysis (DTA) has found the widest application, and consists of simultaneously heating a test sample and a thermally inert substance at constant rate (usually about 10°C/min) to over 1000°C and continuously measuring differences in temperature between the sample and the inert material. Differences in temperature between the sample and the inert substance reflect reactions in the sample brought about by the heating. Thermogravimetric analyses, in which

Table 5.3 X-Ray Identification of the Principal Clay Minerals (<2 μm) In an Oriented Mount of a Separated Clay Fraction from Sedimentary Material

Mineral	Basal d Spacings (001)	Glycolation Effect (1 hr, 60°C)	Heating Effect (1 hr)
Kaolinite	7.15 Å (001); 3.75 Å (002)	No change	Becomes amorphous 550–600°C
Kaolinite, disordered	7.15 Å (001) broad; 3.75 Å broad	No change	Becomes amorphous at lower temperatures than kaolinite
Halloysite, 4H$_2$O (hydrated)	10 Å (001) broad	No change	Dehydrates to 2H$_2$O at 110°C
Halloysite, 2H$_2$O (dehydrated)	7.2 Å (001) broad	No change	Dehydrates at 125–150°C; becomes amorphous 560–590°C
Mica	10 Å (002); 5 Å (004) generally referred to as (001) and (002)	No change	(001) becomes more intense on heating but structure is maintained to 700°C
Illite	10 Å (002), broad, other basal spacings present but small	No change	(001) noticeably more intense on heating as water layers are removed; at higher temperatures like mica
Montmorillonite group	15 Å (001) and integral series of basal spacings	(001) expands to 17 Å with rational sequence of higher orders	At 300°C (001) becomes 9 Å
Vermiculite	14 Å (001) and integral series of basal spacings	No change	Dehydrates in steps
Chlorite, Mg-form	14 Å (001) and integral series of basal spacings	No change	(001) increases in intensity; <800°C shows weight loss but no structural change
Chlorite, Fe-form	14 Å (001) less intense than in Mg-form; integral series of basal spacings	No change	(001) scarcely increases; structure collapses below 800°C
Mixed-layer minerals	*Regular*, one (001) and integral series of basal spacings	No change unless an expandable component is present	Various, see descriptions of individual minerals
	Random, (001) is addition of individual minerals and depends on amount of those present	Expands if montmorillonite is a constituent	Depends on minerals present in interlayered mineral
Attapulgite (palygorskite)	High intensity d reflections at 10.5 Å, 4.5 Å, 3.23 Å, 2.62 Å	No change	Dehydrates stepwise (see description)
Sepiolite	High intensity reflections at 12.6 Å, 4.31 Å, 2.61 Å	No change	Do
Amorphous clay, allophane	No d reflections	No change	Dehydrates and loses weight

Compiled by Carroll (1970).

Table 5.4   X-Ray Diffraction Spacings (Å) Obtained from (001) Planes of Binary, Regularly Alternating, Layer-Silicate Species as Related to Sample Treatment

Interstratified Mixture	Mg-saturated Air-Dried	Mg-saturated Glycerol-Solvated	K-saturated Heated (500°C)
	diffraction spacings (Å)		
Mica–vermiculite	24	24	10
Mica–chlorite	24	24	24
Mica–montmorillonite	24	28	10
Vermiculite–chlorite	28	28	24
Vermiculite–montmorillonite	28	32	10
Montmorillonite–chlorite	28	32	24

Whittig (1965).

changes in weight caused by loss of water or CO_2 or gain in oxygen are determined, are also used to some extent, but will not be discussed herein.

The results of differential thermal analysis are generally presented as a plot of the difference in temperature between sample and inert material (ΔT) versus temperature (T) as indicated in Fig. 5.8. Endothermic reactions are those wherein the sample takes up heat, and in exothermic reactions, heat is liberated. Analysis of test results consists in comparing the sample curve with those for known materials so that each deflection can be accounted for.

Apparatus

Apparatus for DTA consists of a sample holder, usually ceramic, nickel, or platinum; a furnace; a temperature controller to provide a constant rate of heating; thermocouples for measurement of temperature and the difference in temperature between the sample and inert reference material; and a recorder for the thermocouple output. The amount of sample required is of the order of one gram. Figure 5.9 is a schematic diagram of the apparatus. Although the temperatures at which thermal reactions take place are a function only of the sample, the size and shape of the reaction peaks depend also on the thermal characteristics of the apparatus and the heating rate. Thus, there is some difficulty in comparing quantitatively the results from different apparatus.

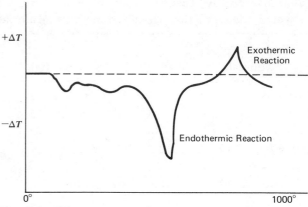

Fig. 5.8   Thermogram of a sandy clay soil.

Reactions giving thermal peaks

The important thermal reactions giving rise to peaks on a thermogram are:

1. *Dehydration.* Water may be present in two forms in addition to free pore water: adsorbed water or water of hydration, which is driven off at 100 to 300°C and crystal lattice water in the form of (OH) ions, the removal of which is termed "dehydroxylation" (Barshad, 1965). Dehydroxylation causes a complete destruction of mineral structures; whereas removal of adsorbed water is a reversible process. The temperature at which

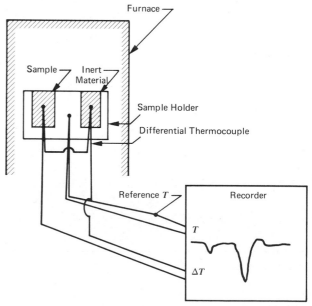

Fig. 5.9   Schematic diagram of apparatus for differential thermal analysis.

the major amount of crystal lattice water is lost is the most indicative property for identification of minerals. Dehydration reactions are endothermic.

2. *Crystallization.* New crystals can form from amorphous materials or from old crystals destroyed at a lower temperature. Crystallization reactions usually are accompanied by an energy loss and thus are exothermic.

3. *Phase changes.* Some crystal structures change from one form to another at a specific temperature and the energy of transformation shows up as a peak on the thermogram. Quartz changes from the α to β form (reversible) at 573°C. The peak for the quartz phase change is sharp (Fig. 5.10), and its amplitude is nearly in direct proportion to the amount of quartz present. The quartz peak is frequently masked within the peak for some other reacting material. The quartz may be readily identified, however, by determining the thermogram during cooling of the sample or by letting it cool first and then rerunning it. The other minerals are destroyed during the initial run while the quartz reaction is reversible.

4. *Oxidation.* Exothermic oxidation reactions may take place such as the combustion of organic matter or the oxidation of Fe^{2+} to Fe^{3+}. Organic matter oxidizes in the range of about 250 to 450°C.

Characteristic thermograms

Thermograms for several clay minerals and quartz are shown in Fig. 5.10. In general, the clay minerals show an endothermic reaction between 100 and 200°C as adsorbed water is lost, dehydroxylation between about 500 to 1000°C, and an exothermic recrystallization reaction between 800 and 1000°C. The thermal behavior of soil clays may differ from that of pure clay deposits (Barshad, 1965); thus, there may be some difficulty in selection of reference standards. In particular, montmorillonite, vermiculite, and illite or other micalike minerals in soil clays may not show any endothermic reaction associated with loss of lattice water.

Beside quartz, the only common nonclay minerals in soils that give reactions with large thermal peaks are carbonates and free oxides such as gibbsite, brucite, and goethite. The carbonates give very large endothermic peaks between about 800 and 1000°C, and the oxides have an endothermic peak between about 250 and 450°C. Thermograms for a large number of clay and nonclay minerals are presented by Lambe (1952).

Quantitative analysis

Theory shows that the area of the reaction peak can be taken as a measure of the amount of mineral present in the sample. For sharp large amplitude peaks, such as the quartz inversion at 573°C and the kaolinite endotherm at 650°C, the amplitude can be used for quantitative analysis. In either case, calibration of the apparatus is necessary, and the overall accuracy is of the order of the determined value plus or minus 5 percent.

Further details of differential thermal analysis theory and techniques are given by, among others, Lambe (1952), Smothers and Chiang (1958), and Barshad (1965).

5.7 OPTICAL MICROSCOPE STUDIES OF SOIL

Both the binocular and petrographic microscopes may be useful for the study of the identity, size, shape, texture, and condition of single grains and aggregates in the silt and sand size range; for study of the fabric, that is, the distribution and interrelationships of the constituents in thin section; and for study of the orientations of groups of clay particles. Because the in-focus depth of field decreases sharply as magnification increases, studies of soil thin sections become impractical at magnifications greater than about a hundred. Thus, individual clay particles cannot be distinguished.

Useful information about the shape, texture, size, and size distribution of silt and sand grains may be obtained directly without formal previous training in petrographic techniques. Some background is needed to identify the various minerals; however, relatively simple diagnostic criteria that can be used for identification of over 80 percent of the coarse grains in most soils have been given by Cady (1965). These criteria are based on such factors as color, refractive index, birefringence, cleavage, and particle morphology. The nature of surface textures, the presence of coatings, layers of decomposition, and so on, are useful both for interpretation of the history of a soil and as a guide to the soundness and durability of the particles.

5.8 ELECTRON MICROSCOPY

The electron microscope is a valuable tool for the study of soils, because with modern electron micro-

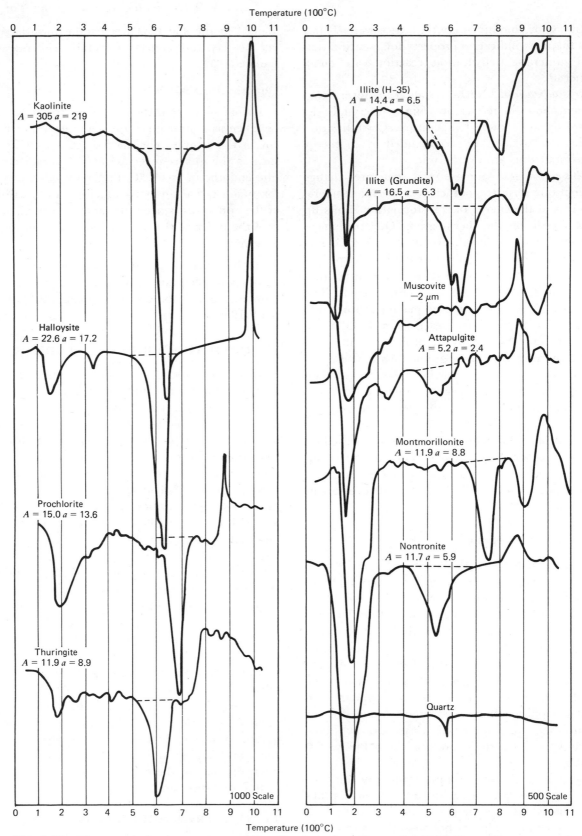

Fig. 5.10 Characteristic thermograms for soil minerals. All samples initially at 50 percent relative humidity, A is the area (cm²), and a is the amplitude (cm).

scopes, it is possible to resolve distances to less than 100 Å, thus making study of small clay particles possible. Electron diffraction study of single particles also is possible. Electron diffraction is similar to X-ray diffraction except an electron beam instead of an X-ray beam is used.

Magnetic lenses, which refract an electron beam, form the basis of the transmission electron microscope optical system. An electron beam is focused on the specimen, which is usually a replica of the surface structure of the material under study. Some of the electrons are scattered from the specimen, and different parts of the specimen appear light or dark in proportion to the amount of scattering. After passing through a series of lenses, the image is displayed on a fluorescent screen for viewing. Probably the most critical aspect of successful transmission electron microscopy is specimen preparation. Some of the available techniques that have been used for study of discrete particles and surface structures of particle assemblages are reviewed by Kittrick (1965).

The scanning electron microscope (SEM) represents a rather recent development. With this instrument, secondary electrons emitted from a sample surface form what appear to be three dimensional images. The SEM has a magnification range of $\times 20$ to $\times 150,000$ and a depth of field some 300 times greater than that of the light microscope. These characteristics, coupled with the fact that clay particles themselves and fracture surfaces through soil masses may be viewed directly, as opposed to the study of replicated surfaces in the transmission electron microscope, have led to almost exclusive use of the SEM for electron microscope study of clays. Examples of electron photomicrographs of clays and soils are given in Chapters 3, 8, and 11. Principles of electron microscopy techniques, and additional examples are presented by McCrone and Delly (1973).

5.9 QUANTITATIVE ESTIMATION OF SOIL COMPONENTS

Qualitative X-ray diffraction and a few simple tests generally suffice to indicate the minerals present in a soil. More data are needed, however, for more precise quantitative estimates. As a rule, the number of different analyses needed is equal to the number of mineral species present (Barshad, 1965). The results of glycol adsorption, cation exchange capacity, X-ray diffraction, differential thermal analysis, and the various chemical tests noted in Section 5.4 all give data that may be used for quantitative estimations. Table 5.5 summarizes some of the pertinent identification criteria and indicates reference values for the clay minerals. Quantitative estimates may be made based on measured quantities and reference values (Lambe and Martin, 1954; Barshad, 1965).

After the quantities of organic matter, carbonates, free oxides, and nonclay minerals have been determined, the percentages of clay minerals are estimated using the appropriate glycol adsorption, cation exchange capacity, K_2O, and DTA data. The nonclays can be identified, and their abundance determined using the microscope, grain size distribution analysis, X-ray diffraction, and DTA. The amount of illite is estimated from the K_2O content since this is the only clay mineral containing potassium. The amount of kaolinite is most reliably determined from the 600°C DTA endotherm amplitude. If the presence of montmorillonite, chlorite, and/or vermiculite has been indicated by X ray, then quantitative estimates are made based on the glycol adsorption and exchange capacity data. The total exchange capacity and glycol retention is ascribed to the clay minerals, and the measured values must be accounted for in terms of proportionate contributions by the different clay minerals present. In some cases these values may be used to provide a check on each other.

As a simple example, assume that quartz, illite, and montmorillonite (smectite) are identified in the -2 μm fraction of a soil. Additional data indicate 4.0 percent K_2O, ethylene glycol retention of 100 mg/g, and a cation exchange capacity of 35 meq/100 g. Then assuming 9 percent as an average value of K_2O for pure illite (Table 5.5), the content of illite is estimated at 4.0/9.0 of 44 percent. Because only the illite and montmorillonite will contribute to the glycol adsorption, the amount of montmorillonite may be estimated

$$0.44 \times 60 + M \times 300 = 110$$

$$\therefore M = \frac{110 - 26.4}{300} = 25\%$$

The remaining 31 percent can be ascribed to quartz and other nonclay components. For this clay mineral composition, the theoretical cation exchange capacity should be, based on the reference values in Table 5.5:

$$0.44 \times 25 + 0.25 \times 85 = 11 + 21 = 33 \text{ meq}/100 \text{ g}$$

Table 5.5 Summary of Clay Mineral Identification Criteria — Reference Data for Clay Mineral Identification (-2μm fraction)

Clay	X-Ray $d(001)$	Glycol mg/g	CEC meq/100 g	K_2O (%)	DTA[b]
Kaolinite	7	16	3	0	End. 500–660°+ Sharp[c] Exo. 900–975° Sharp
Dehydrated halloysite	7	35	12	0	Same as kaolinite but 600 peak slope ratio >2.5
Hydrated halloysite	10	60	12	0	Same as kaolinite but 600° peak slope ratio >2.5
Illite	10	60	25	8–10	End. 500–650° Broad End. 800–900° Broad Exo. 950°
Vermiculite	10–14	200	150	0	
Smectite	10–18	300	85		End. 600–750° End. 900° Exo. 950°
Chlorite	14[a]	30	40	0	End. 610 ± 10° or 720 ± 20°

[a] Heat treatment will accentuate 14 Å line and weaken 7 Å line.

[b] For clays prepared at same relative humidity the size of the 100–300°C endotherm (adsorbed water removal) increases in the order kaolinite–illite–smectite.

[c] For samples started at 50% R.H. the amplitude of 600° peak/amplitude of adsorbed water peak $>>> 1$.

This compares favorably with the measured quantity of 35 meq/100 g. Thus, the composition of the clay size fraction is

Illite	44%
Montmorillonite	25%
Quartz and other nonclays	31%

In most instances, a semiquantitative analysis is sufficient for most applications. This may be done as follows. The silt and sand fraction can be examined by microscope and the approximate proportion of nonclay minerals determined. The amount of clay size material, $-2\,\mu$m, can be determined by grain size distribution analysis. As a first approximation, it may be assumed that the amount of clay mineral equals at least the amount of clay size. This assumption is justified for the following reasons. It is known that nonclay minerals, principally quartz, are found in the clay size fraction. On the other hand, for most soils, the amount of clay mineral exceeds the amount of clay size. This most probably results from cementation of small clay particles into aggregates larger than 2 μm in diameter. The approximate proportions of the different clay minerals in the clay fraction can be estimated from the relative intensities of the X-ray diffraction reflections for each mineral. Greater precision may be obtained by application of one of the techniques noted earlier for quantitative analysis by X-ray diffraction. The presence of organic matter and carbonates can be easily detected using the tests listed in Section 5.4.

5.10 NATURE OF SOIL FINES

In a series of five papers, Lambe and Martin (1953–1957) reported compositional data and their relationships to engineering properties for a large number of soils. These soils were predominantly those from civil engineering projects from all over the world, but the largest group came from the United States. In many instances they were associated with special soil engineering problems. In 137 soils, the percentage of the following minerals was:

Quartz	45%	Feldspars	21%
Illite	66%	Chlorite	21%
Fe_2O_3	58%	Halloysite	10%
Smectite	46%	Gibbsite	10%
Organic matter	45%	Mica	6%
Carbonates	28%	Vermiculite	4%
Kaolinite	23%		

This shows clearly the prevalance of quartz as the dominant nonclay and illite as the dominant clay mineral in soils. More than one of the clay minerals occurred in about 70 percent of the soils studied, and the clay minerals were usually interstratified.

SUGGESTIONS FOR FURTHER STUDY

American Society for Testing and Materials (1970), *Special Procedures for Testing Soil and Rock for Engineering Purposes,* STP 479, 5th Ed., American Society for Testing and Materials.

Black, C. A., Ed. (1965), *Methods of Soil Analysis,* in Two Parts, Agronomy, No. 9, American Society of Agronomy, Madison, Wisc.

Brown, G., Ed. (1961), *The X-Ray Identification and Crystal Structures of Clay Minerals,* The Mineralogical Society of London, London.

Carroll, D. (1970), "Clay Minerals: A Guide to Their X-Ray Identification," Geological Society of America, Special Paper 126.

Grim, R. E. (1968), *Clay Mineralogy,* 2nd ed., McGraw-Hill, New York.

CHAPTER 6

Soil Water

6.1 INTRODUCTION

A saturated soil with a void ratio greater than 1.0 contains a greater volume of water than solids, and void ratios greater than 1.0 are more the rule than the exception in the case of fine-grained soils. Yet, the emphasis in the study of soil composition and properties has been almost entirely on the mineralogy and structure of the solid phase, with very little regard for the properties of the liquid phase. Two reasons might be suggested for this:

1. Classical soil mechanics is founded on the concept of effective stress, which postulates that volume change and strength behavior depend on the stresses carried by the solid grain structure and that the water phase is neutral.
2. The properties of water are known—a clear, colorless, odorless, tasteless liquid that has a density very nearly equal to unity, freezes at 0°C, boils at 100°C, and has quite well-defined viscosity and thermal properties. It is so familiar that it is scarcely given a second thought.

To those who have looked further into the nature of water, however, a third reason might be appropriate:

3. The structure and properties of water in a soil are not known in detail. In fact, there is no rigorous theory available for the structure of pure, liquid water, only hypotheses, some of which explain properties better than others.

Since neither water nor soil surfaces are inert chemically, water and soil particles interact with each other. These interactions can be expected to influence the physical and physico-chemical behavior of the material. Although details of these interactions and their consequences cannot be stated with certainty, some things are known, and these considerations form the subject of this chapter.

6.2 THE NATURE OF ICE AND WATER

The water molecule (H_2O) is composed of a V-shaped arrangement of atomic nuclei, with an average H–O–H angle of slightly less than 105°. An equilibrium state, point charge model for the molecule as proposed by Pople (1951) is shown in Fig. 6.1. The outer shell electronic charges, six from the oxygen and one from each hydrogen, are distributed in the form of electron pairs, as shown. The resulting configuration is that of a tetrahedron with two positive corners that are the sites of the hydrogen protons and two negative corners that are located above and below the plane of the atomic nuclei. Bond energy considerations (Pauling, 1960) indicate that the H–O bond is about 40 percent ionic and 60 percent covalent, thereby accounting for the directionality of the bond and producing a permanent dipole.

In water and ice, the positive corner of one molecule attracts the negative corner of another. The proton is shared by the two oxygens, which results in hydrogen bonding and a tendency for each molecule to bond to four neighboring molecules, which surround it tetrahedrally. It has been established (Frank, 1958) that the lone pair electrons are distorted in the electrical field, and thus the intermolecular bond is partly covalent. In ice I, the stable crystalline state of water at temperatures less than 0°C and atmospheric pressure, a hexagonal network structure is formed, as shown schematically in Fig. 6.2. Three molecules of the hexagon are in one plane, and three are in another. The oxygen-to-oxygen distance between hydrogen bonded molecules is 2.76 Å, with the hydro-

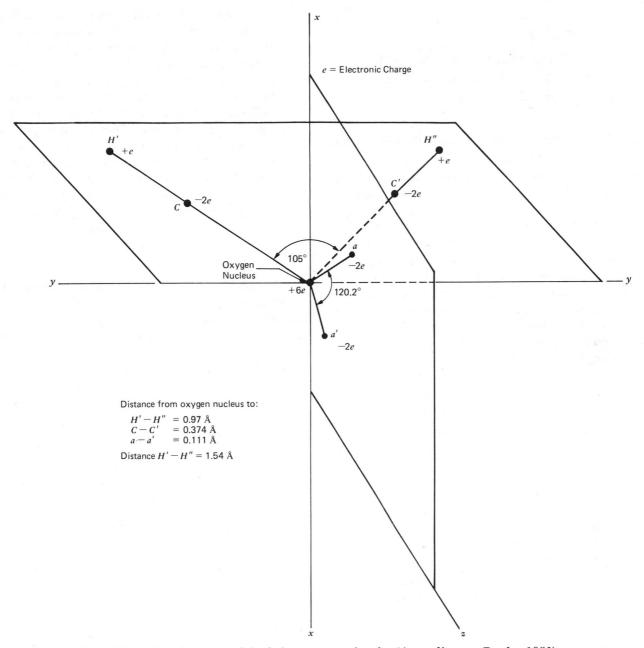

e = Electronic Charge

H' $+e$

H'' $+e$

C $-2e$

C' $-2e$

a $-2e$

105°

Oxygen Nucleus $+6e$

120.2°

a' $-2e$

Distance from oxygen nucleus to:

$H' - H''$ = 0.97 Å
$C - C'$ = 0.374 Å
$a - a'$ = 0.111 Å

Distance $H' - H''$ = 1.54 Å

Fig. 6.1 Point charge model of the water molecule (According to Pople, 1951).

gen located (statistically) 1.00 Å from one oxygen and 1.76 Å from the other. The distance between molecules in a given plane is 4.5 Å. The strength of the bonds in ice is about 4.5 kcal/mole. As the melting point is approached, the number of broken or distorted hydrogen bonds increases. This accounts for the lower strength and higher creep rates in ice and frozen soil as the temperature increases.

At temperature above 0°C (under atmospheric pressure conditions), enough hydrogen bonds are rup-

tured or bent so that ice loses its rigidity, and water exists in the liquid state. Rupture of 16 percent of the possible hydrogen bonds would account for the heat of fusion of ice.

The structure of liquid water is not definitely known, except that it is clear that some hydrogen bonding and structure remain. If they did not, then each water molecule would have 12 nearest neighbors, and the density would be 1.84. The high melting point, boiling point, heat of fusion, heat of

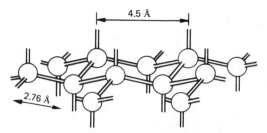

Fig. 6.2 The arrangement of water molecules in ice I.

vaporization, specific heat, dielectric constant, and viscosity of water relative to similar compounds are ascribed to the energy needed to break hydrogen bonds. The 9 percent volume decrease on melting results from the fact that each molecule has somewhat more than four nearest neighbors, even though the average intermolecular distance increases to about 2.9 Å as a result of the increased thermal agitation. Evidently, bond rupture and distortion predominate at temperatures less than 4°C; whereas, increased intermolecular distances prevail above, thus giving maximum density at 4°C.

Various models have been proposed for the structure of water in the light of available physical and chemical evidence (Eisenberg and Kauzman, 1969).

1. *Mixture models* assume a mixture (at any instant) of a small number of distinguishable species of water molecules. One proposal is that there are clusters of hydrogen bonded molecules and non-H-bonded molecules. Increased temperature leads to an increase in the concentration of non-H-bonded molecules.

2. *Interstitial models* are a special class of mixture models, wherein one species of molecule forms an H-bonded framework containing cavities in which non-H-bonded molecules reside.

3. In the *distorted hydrogen bond model* (Pople, 1951), the majority of hydrogen bonds are assumed to be distorted rather than broken. A bond is said to be distorted when either the lone pair electrons or the O—H—O bond direction depart from the line between two oxygen centers. Bending of hydrogen bonds permits some of the second and third neighbors to penetrate into regions near the central molecule, which results in an apparent number of nearest neighbors greater than four.

4. The *random network model* was proposed by Bernal (1960, 1964) and is an extension of the distorted H-bond model. Each water molecule is H-bonded to others, although the bonds may

be distorted. The result is an irregular network of rings rather than an ordered lattice as in ice. It is postulated that many rings contain five molecules, because the H–O–H angle is close to the 108 degree angle of a five-membered ring, but others may contain four, six, seven, or more molecules. According to Bernal, in a noncrystalline, irregular structure, fivefold arrangements are, for geometrical reasons, likely to be the rule. Five-coordinated structures cannot exist in crystalline solids because regular, repeating patterns cannot be formed. The liquid state could be explained precisely because of this, and solids and liquids can be distinguished as coherent materials with and without regular structure, respectively.

Although there is not yet a rigorous theory for liquid water, Eisenberg and Kauzman (1969) conclude that mixture models are not supported by the data, but that the distorted H-bond models (including random networks) seem to accord with most of what is known about water from experiment. No fundamental distinction is seen between broken and highly distorted hydrogen bonds in liquid water.

6.3 THE INFLUENCE OF DISSOLVED IONS

As a result of the uneven charge distribution and dipolar character of water molecules, they are attracted to ions in solution, leading to ion hydration. Positive ions attract the negative corners of water molecules, and vice versa. Water molecules will move

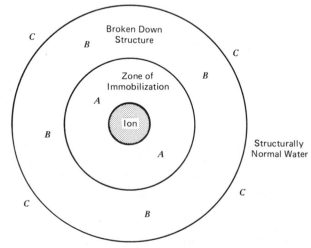

Fig. 6.3 Ion–water interaction as postulated by Frank and Wen (1957).

from their normal structure into positions in the hydration shell of an ion provided the energy is less as water of hydration than as normal water. Not all ions hydrate, although cations common in soils do.

Ions disrupt normal water structure, whether they hydrate or not. Those dissolved ions that do not hydrate still occupy space; those that do hydrate attract only those corners of water molecules of oppo-

site electrical charge; whereas, in normal water there is an alternating character to the directions of positive and negative corners.

A model for ion–water interaction is shown schematically in Fig. 6.3 (Frank and Wen, 1957). Region A is a zone of immobilization. Water molecules are strongly oriented in the field of the ion and have little kinetic energy. In region B the water structure is

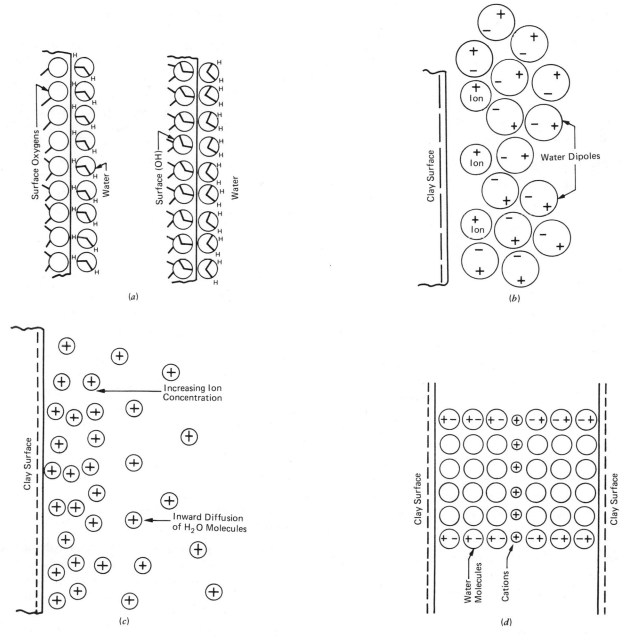

Fig. 6.4 Possible mechanisms of water adsorption by clay surfaces. (*a*) Hydrogen bonding. (*b*) Ion hydration. (*c*) Attraction by osmosis. (*d*) Dipole attraction.

broken down and more random than in normal water. Region C contains water with normal structure, but which is polarized by the ionic field.

6.4   POSSIBLE MECHANISMS OF SOIL–WATER INTERACTION

There is ample evidence to show that water is attracted to soil minerals, particularly to clays. Dried clays will adsorb water from the atmosphere at low relative humidities, many soils swell when given access to water, and temperatures above 100°C are needed to remove all the water from a soil. In fact it is not always evident just what is meant by a dry soil (Lambe, 1949). The nature of soil–water interaction and the interpretations of its consequences, however, are not clear.

Possible mechanisms for clay–water interaction include the following (Low, 1961) and are shown schematically in Fig. 6.4.

Hydrogen bonding

Because the surfaces of the soil minerals are usually composed of a layer of either oxygens or hydroxyls, hydrogen bonding could easily develop, with oxygens attracting the positive corners and hydroxyls the negative corners of water molecules (Fig. 6.4a). Early concepts of the structure of adsorbed water suggested an icelike character because of the similarity between the hexagonal symmetry of the oxygens and hydroxyls in clay surfaces and the structure of ice. Subsequent work has shown, however, that the structure cannot be that of ice.

The formation of H bonds with particle surfaces would alter the electron distribution from that in normal water (Low, 1961), thus making it easier for bonded molecules to form additional bonds with molecules in the same and next layer. The directional properties of the bonds would induce a tetrahedral arrangement of water molecules which would become less rigid with distance from the surface due to a decrease in the surface force fields and an increase in the force fields of normal water structure.

Hydration of exchangeable cations

Since cations are attracted to negatively charged clay surfaces, their water of hydration will also be (Fig. 6.4b). This mechanism would be most important at low water contents.

Attraction by osmosis

The concentration of cations increases as negatively charged clay surfaces are approached, Fig. 6.4c, as discussed in Chapter 7. Because of this increased concentration and the restriction on diffusion of ions from the vicinity of the surface, as a result of electrostatic attraction, water molecules tend to diffuse towards the surface in an attempt to equalize concentrations.

Charged surface–dipole attraction

Clay particles can be viewed as negative condenser plates, with an electrical field strength that decreases with distance from the surface because of the presence of positive charges. Water dipoles would then orient with their positive poles directed toward the negative surfaces and with the degree of orientation decreasing with distance from the surface. At the midplane between parallel plates there would be structural disorder, however, because like poles would be adjacent to each other. Ingles (1968) suggests that because of the high hydration number and energy of aluminum in the clay structure, water is strongly attracted to the surfaces and interposes itself between the surfaces and the counterions, with the counterions removed as far as possible from the surface, that is, to the midplane between opposing parallel sheets. With this model the structure shown in Fig. 6.4d can be conceived. The same type of arrangement could result simply from ion hydration. In a dry clay, adsorbed cations occupy positions in holes on the clay surfaces. On hydration they surround themselves with water and move to the central region between clay layers.

Attraction by London Dispersion Forces

van der Waals attractive forces could bond water molecules to clay surfaces. In-phase fluctuations of electron clouds form temporary dipoles and induce displacements in neighboring molecules so that dipole–dipole attraction occurs. Because such bonds would be nondirectional, the water structure would be close packed and more fluid than the H-bonded structure.

6.5   EVIDENCE ON THE STRUCTURE AND PROPERTIES OF ADSORBED WATER

Different arrangements of water molecules are associated with each of the postulated mechanisms for soil–water interaction. It would be anticipated that different water structures would give different properties. Evidence concerning the validity of the different models has come from a variety of sources, including X-ray and electron diffraction data, density measurements, dielectric and fluid flow measurements, and

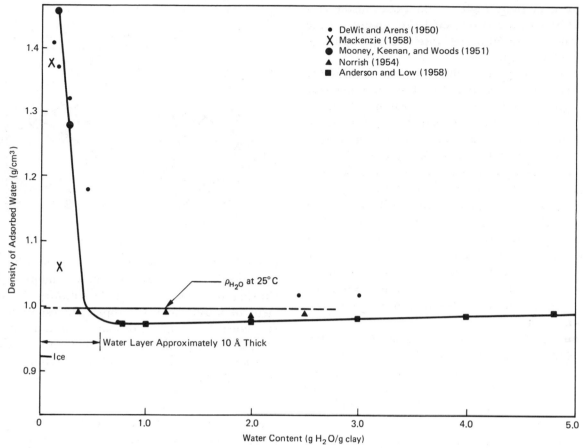

Fig. 6.5 Density of adsorbed water (From Martin, 1960).

behavior on freezing. Unfortunately, almost all of the data or their interpretations may be questioned for various reasons, and unequivocal support for any one model is not yet available. Nonetheless, a review of some of the findings does serve to provide further insight into the most probable characteristics of water in soils.

Density of adsorbed water

Data are summarized in Fig. 6.5 (Martin, 1960) for the density of adsorbed water on sodium montmorillonite as a function of water content. The determinations were made using pycnometer and X-ray diffraction techniques. The data points indicate either the average density of the water for the water content shown or, in the case of Anderson and Low's data, possibly the density at the midplane between particles, because an incremental measurement technique was used. The curve cannot be interpreted to represent the variation in density as a function of distance from the clay surface; however, there is nothing in the data

to preclude variable density with distance from the surface. For water contents less than needed to give about three molecular layers on the clay surfaces (a 10 Å thick layer), the density is greater than that of normal water; whereas, at greater water contents, it is less. At no water content is the density the same as that of ice. The data in Fig. 6.5, or portions of the data covering specific ranges of water content, can be used in support of close-packed liquid models and open H-bond models, and no clear preference can be made for either.

X-Ray evidence

X-Ray analyses of frozen clay pastes of montmorillonite, halloysite, and kaolinite at low water contents were made by Anderson and Hoekstra (1965). Water films were 5 to 10 Å in thickness. In samples containing some unfrozen water, the ice was separated from the clay surfaces by liquid water, in which the adsorbed cations were also located. Normal ice structure was observed; however, the c-axes of the ice crystals tended to be at right angles to the *c*-axes of the clay plates. This is rather strong evidence against a rigid icelike structure for adsorbed water with an epitaxial fit to the particle surfaces.

In a recent study of swelling of montmorillonites, Ravina and Low (1972) found that as the water content of a given montmorillonite increased, so also did the *b* dimension, as may be seen in Fig. 6.6. Differences in the *b* dimensions of different montmorillonites in the dry state are a consequence of differences in isomorphous substitution, which cause small rotations of the silica tetrahedra in alternating clockwise and counterclockwise directions. When each clay was completely swollen, the *b* dimension was 9.0 Å. Whatever the water structure at the end of swelling, it is in energy equilibrium with normal water. At smaller values of the *b* dimension, the water structure must have been so as to give an energy level lower than that of free water, or swelling would not have been spontaneous.

The finding that a change in *b* dimension accompanied a change in water content at first seems surprising. It would perhaps be more surprising if such an interaction did not occur, since both the clay and the water have preferred internal structures, and each

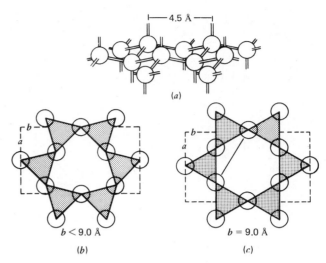

Fig. 6.7 Arrangement of water molecules in ice, dry montmorillonite, and idealized montmorillonite (Ravina and Low, 1972). (*a*) Diagram of the arrangement of water molecules (open circles) in ice I. (*b*) Ditrigonal symmetry of the surface of a dry montmorillonite (O—oxygen atoms). (*c*) Regular hexagonal symmetry of the surface of idealized montmorillonite (O—oxygen atoms).

should exert an influence on the other as long as there is attraction between the two phases. The situation is analogous to many types of soil-structure interaction that have become so important in soil engineering and design.

The misfit between the oxygen configuration on dry montmorillonite surfaces and water structure may be sufficiently small that the hydrogen bonding between them would be strong enough, in relation to their elastic moduli, to result in pseudomorphism (Ravina and Low, 1972). Pseudomorphism is the straining of an overgrowth structure to match a substrate structure. The similarity between the configurations of water molecules in ice I and the surface oxygents is shown by Fig. 6.7. The bases of the silica tetrahedra are not arranged in a perfectly hexagonal array in dry montmorillonite because of strains induced by interconnection with the octahedral sheet of the mineral structure.

At very low water contents, the water conforms more to the montmorillonite structure; whereas, at high water contents, the surface structure conforms more to that of the water. The water molecules at the higher water contents are assumed by Ravina and Low to have a configuration similar, but not identical, to that in ice I.

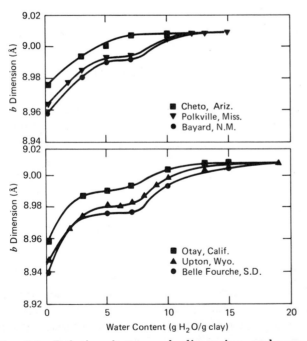

Fig. 6.6 Relation between *b* dimension and water content for six Na-saturated dioctahedral montmorillonites. (From Ravina and Low, 1972) (Reproducibility of *b* dimension = ± 0.001 Å at low water content, ± 0.003 Å at high water content.)

Theoretical analyses (Lahay and Bresler, 1973) suggest also that change in the *b* dimension with change in the water content may be caused by variation in the penetration of cations into the hexagonal holes on the silica surfaces of the montmorillonite structure.

Diffusion, viscosity, and fluid flow

If the bonding and structure in adsorbed water are different than in normal water, then there should also be differences in viscosity and in ion mobility, which would be reflected by measurements of diffusion and flow properties. Thus, the results of measurements of hydraulic flow rates, diffusion coefficients, and activation energy for flow and diffusion have been used to infer details of water structure in clays.

Low (1961) arrived at the conclusion that available data supported a quasicrystalline structure for adsorbed water that should exhibit non-Newtonian flow, higher than normal viscosity, and a threshold gradient below which flow would not develop. Martin (1960) concluded, however, that fluid flow and diffusion experiments provided no clear evidence about the nature of adsorbed water, because difficulties in both experimentation and analysis of data make the interpretation of results uncertain. The interpretation of most measurements requires assumptions concerning tortuosity of flow path. Changes in fabric and/or chemical environment during measurements, bacterial growth, and electro-kinetic and chemical coupling effects (see Chapter 15) all may affect the results.

A supposed failure of Darcy's law was taken as evidence of abnormal water properties in soils. Because Darcy's law is the commonly used relationship between flow rate *q* and hydraulic gradient *i* in soil mechanics analyses of seepage and consolidation, that is,

$$q = kia \qquad (6.1)$$

where *k* is the hydraulic conductivity (permeability) and *a* is the cross-sectional area, possible deviations are of particular concern to geotechnical engineers. Current evidence (Olsen, 1965, 1969; Gray and Mitchell, 1967; Miller, Overman, and Peverly, 1969) suggests, however, that Darcy's law is obeyed exactly in saturated clays, but that apparent deviations may arise as a result of particle migrations, electro-kinetic effects and chemical concentration gradients.

It would appear, therefore, that fluid flow data indicate that adsorbed water, whatever its structure, behaves as a Newtonian liquid.

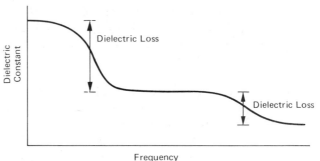

Fig. 6.8 Dependence of dielectric constant on frequency.

Dielectric and magnetic properties

The dielectric properties of a material depend on the ease with which its constituent molecules can be polarized. There are two contributions to the polarization that develop in an electrical field: distortion polarization, which is instantaneous, and orientation polarization, which is time-dependent. In a low frequency AC field, dipoles can rotate with the changes in current direction. As the frequency increases, however, less time is available for dipole rotation. Above some frequency the molecules will no longer be able to follow the field, and the dielectric constant will drop. The relationship between dielectric constant and frequency for a material might appear as in Fig. 6.8. The dielectric losses and the frequencies at which they occur are dependent on the intermolecular bond types and strengths. Hence, the results of dielectric constant measurements have been used to infer details of water structure.

Measured values of dielectric constant for clay–water systems are less than for normal water, but evidently they are consistent with either an ordered water structure or a "two-dimensional liquid" model (Martin, 1960). A two-dimensional fluid would consist of water molecules that are bonded strongly to the surfaces of clay particles, but that could slide along the surface with relative ease. The result would be that the two-dimensional fluid would resist significant normal stress, but it would shear easily along planes parallel to the surface.

Nuclear magnetic resonance (NMR) measurements have been used to study adsorbed water (Pickett and Lemcoe, 1959; Wu, 1964; Graham, Walker, and West, 1964). The results of these measurements provide information on the time for a molecule to move through a distance comparable with its dimensions (about 3×10^{-12} sec for normal water at 20°C), magnetic field inhomogeneity over the sample, proton mobility, and the number of resonating nuclei. Interpretation of NMR data must be done with care, because anisotropic susceptibility of clay crystals and the effects of particle size influence the results (Graham, Walker, and West, 1964).

Nuclear magnetic resonance data confirm that water is tightly held by clay, the structure is not that of ice, and the mechanism of attraction is not simple dipole adsorption. Some interpretations have led to the conclusion that adsorbed water viscosity is higher than that of normal water. The data can be explained in terms of either an ordered, three-dimensional water structure or a two-dimensional liquid (Martin, 1960).

Some NMR data (Graham et al., 1964) suggest two regions of swelling in layer silicate minerals. In region I a limited number of water layers exist between the silicate layers, resulting in distinct phases or crystalline hydrates. Region I is a comparatively ordered phase, with proton mobility decreasing with decreasing water content. In Region II, at higher water contents, the interlayer water quantity is continuously variable and the interlayer cations form a diffuse double layer (Chapter 7). Both kinds of structure probably coexist between Regions I and II.

Molecular orientation results from general attraction between protons and the negative silicate layers. Extended covalent or hydrogen bonding does not seem to develop, and the postulated structure can be described in terms of a two-dimensional liquid. The evidence suggests very little, if any, special water structure in the osmotic swelling region, and the molecular motion may not be much different than in free water.

This general picture is in accord with the density data in Fig. 6.5—namely, at water contents less than that needed for surface and cation hydration, amounting to only a few molecular layers, the organization of water molecules is quite unlike that in normal water, being dominated by the silicate surfaces and cations. At higher water contents, the influence of the water dominates. The structure would not be identical to that of free, pure water because of the presence of cations in solution and the clay particles.

Supercooling and freezing of adsorbed water

The water in clay–water systems will both supercool* and exhibit freezing point depression.† Thus, the

* *Supercooling* is the reduction of temperature below the normal freezing point without the initiation of freezing.
† *Freezing point depression* is a reduction in the usual freezing temperature of a liquid, such as that caused by dissolving salt in water.

adsorbed water cannot have the structure of ice, because if it did, it would act as a crystallization nucleus and there could be neither effect. Both supercooling and freezing point depression could result from more or less order in the water relative to normal water. If there were more order, then it would be more difficult to rearrange the water molecule into the ice structure. With a less ordered water structure, more energy would need to be removed to initiate freezing.

Thermodynamic considerations

Evaluations of the thermodynamic properties of soil water have been made by several investigators, and many of the results are discussed by Martin (1960) and Low (1961). Moisture adsorption from the vapor phase has been the most widely used technique, although calorimetric measurements of heat of wetting, and measurements of free energy changes by different methods have also been used. Thermodynamic measurements provide information on changes in properties from one state to another; they do not provide specific information on the mechanisms responsible for the changes.

The heat of adsorption can be determined using the Clausius–Clapeyron equation,

$$\ln\left(\frac{p_2}{p_1}\right) = \frac{\Delta \bar{H}}{R}\frac{T_2 - T_1}{T_1 T_2} \qquad (6.2)$$

where p_1 and p_2 are equilibrium vapor pressures above the clay at temperature T_1 and T_2 and constant soil water content, R is the gas content, and $\Delta \bar{H}$ is the change in partial molar heat content of the water on adsorption from the vapor state.

At equilibrium, the partial molar free energy of the water in the clay is the same as that in the vapor. The change in partial molar free energy on adsorption, $\Delta \bar{F}$, is defined by

$$\Delta \bar{F} = \Delta \bar{H} - T\Delta \bar{S} \qquad (6.3)$$

where $\Delta \bar{S}$ is the change in partial molar entropy. As $\Delta \bar{F} = 0$,

$$\Delta \bar{S} = \frac{\Delta \bar{H}}{T} \qquad (6.4)$$

The entropy is a measure of molecular randomness, so $\Delta \bar{S}$ is presumed to reflect the change in disorder of water molecules in going from the vapor to the adsorbed state. Measurements show that $\Delta \bar{H}$ is greater for water vapor adsorption on clays than for the condensation of pure water. Thus, by equation (6.4) the

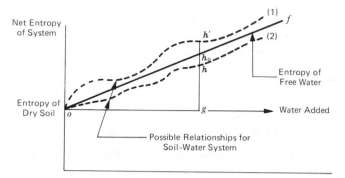

Fig. 6.9 Differential and integral entropy. Differential entropy is the slope at any point (if less than slope *of*, then conclude there is more order in adsorbed water). Integral entropy of soil–water system = gh/og or gh'/og. Integral entropy of system (1) greater than for water; therefore, it has less order in adsorbed water.

decrease in partial molar entropy is greater on adsorption to the clay than on condensation, and it is concluded that adsorbed water is more ordered than free water. The nature of the order cannot be specified from the measurement or the theory.

This interpretation has been challenged (Martin, 1960; Cary, Kohl, and Taylor, 1964; and others) on the basis that integral thermodynamic functions should be used for analysis of the systems studied (Hill, 1950) rather than differential thermodynamic functions. The difference between differential and integral entropy is illustrated by Fig. 6.9. From this figure it may be seen that the integral entropy of adsorbed water at a given water content could be greater than that of free water even if the differential entropy is less. This is in fact the case for adsorption on kaolinite and there are discrepancies in the data on montmorillonite (Martin, 1960).

Thus we are left with a rather confused picture of adsorbed water based on the entropy data. Less order than normal water would certainly not be inconsistent with the density data, Fig. 6.5, the conclusions of Graham, Walker, and West (1964) based on NMR for Region I swelling, and the unfrozen zones observed by Anderson and Hoekstra (1965). It is important to keep in mind, however, that all these observations pertain to the first few molecular layers adsorbed onto a dry clay.

Measurements of the heat of wetting (heat liberated as a result of wetting a clay with liquid water) have shown that the heat of wetting is almost zero after formation of the first few water layers, a large

amount of energy is released on adsorption of the first one or two molecular layers, and exchangeable cations influence the amount of heat released.

Values of specific heat as great as 1.1 cal/g have been measured (Oster and Low, 1964) at very low water contents, decreasing to values close to unity at higher water contents. There is evidence (referred to by Oster and Low, 1964) that the heat capacity of the water in a clay increases with time of rest after remolding.

Moisture tension increases (pore pressure decreases) with time after disturbance of a thixotropic clay–water mixture (Kolaian and Low, 1960; Mitchell, 1960; and Ripple and Day, 1966). This effect could be a reflection of a time-dependent change in the free energy and structure of the water as a new, at-rest equilibrium condition is developed. Because the moisture tension depends on the free energy of the whole clay–water system, however, and time-dependent changes in fabric have been observed to accompany thixotropic hardening, the data cannot be accepted as proof of a change in water structure. Nonetheless, it does not seem unreasonable to expect that mechanical disturbance can change the structural state of water in a clay–water system.

Infrared data

Recent measurements by infrared spectroscopy (Low and White, 1970) show that hydrogen bonds in clay adsorbed water are weaker than those in bulk water. Weak hydrogen bonds do not necessarily imply disorder in the water, nor do the data mean that there cannot be as many or more H bonds developed. Surface constraint and steric considerations can be used as a basis for an argument that there should be fewer, however.

6.6 SUMMARY AND CONCLUSIONS

What can be concluded from this rather cursory review of the nature of water in soils, and what are the implications relative to the understanding of soil behavior in geotechnical engineering?

Some things appear well established about water and adsorbed water in soils. These include:

1. The volume of water in a soil may be equal to or greater than the volume of solid material.
2. The electron distribution on the oxygen in a water molecule gives a tetrahedral geometry, with the hydrogen protons at two of the corners, that in turn makes possible a variety of

geometrical configurations held together by hydrogen bonds.

3. The structure of ice is known, but that of liquid water is not. A distorted hydrogen bond model with random networks and an appreciable number of five-coordinated rings of molecules appears most compatible with available data.

4. Dissolved ions disrupt water structure.

5. Adsorbed water structure differs from that of ice and normal water.

6. Adsorbed water exhibits both supercooling and freezing point depression.

7. Energy is released by the adsorption of water by clays.

8. Hydrogen bonds in adsorbed water are weaker than in normal water, and they may be fewer in number in the vicinity of the clay surface.

9. Time-dependent increases in the moisture tension of water are evident after mechanical disturbance of the at-rest structure of a clay–water system.

10. The interaction of water with clay surfaces causes a change in the b-dimension of the clay lattice. A mutual adjustment of both the clay and water structures develops during adsorption.

11. Most data on adsorbed water are interpretable in more than one way.

These considerations have led to a number of hypotheses about the nature and behavior of adsorbed water in soils.

1. Possible mechanisms for clay–water interaction include:
   a. Hydrogen bonding.
   b. Cation hydration—important at low water content.
   c. Attraction by osmosis—important for water contents above that needed for the first few molecular layers.
   d. Dipole orientation in an electric field—NMR data argue against this.
   e. Attraction by London dispersion forces—some contribution from this effect is probable.

2. A water structure similar to that of ice I may exist in fully swollen clays (Ravina and Low, 1972). The properties are not those of ice, however.

3. There is no strong evidence for abnormal water viscosity or failure of Darcy's law in clays.

4. The structure and properties of water in clays at low water contents may differ from those at high water contents.

5. Adsorbed water structure may vary with distance from the clay surface in a given layer. The fact that ionic distribution varies with distance from the surface (Chapter 7) is by itself an argument to support this conclusion.

Many of the apparently contradictory hypotheses concerning adsorbed water structure could perhaps be reconciled by recognizing that at low water contents, characteristic of those developed in adsorption studies, the water structure could be quite different than that in wet or saturated clays. The density data in Fig. 6.5 and the NMR results of Graham, Walker, and West (1964) argue in favor of this, as do some of the thermodynamic properties, that show sharp gradients at low water content, but only small changes at high water content. At very low water contents, competition for water molecules between the adsorbed cations and the surfaces is keen. A strong attraction to the surface with disorder in the arrangement of water molecules and high lateral mobility are reasonable consequences.

At higher water contents, cations are diffused from the surface and normal water exerts a strong influence on the adsorbed water. If the random network model for water structure (Bernal, 1960, 1964) is correct, then it seems reasonable that the development of networks that could both adapt to the clay surfaces and incorporate adsorbed cations should be possible. Because both the surfaces and cations would influence the nature of the networks developed, it would be expected that the bond strengths and thermodynamic properties would differ from those in normal water. A gradual transition from one structure at the surface to another in the bulk water seems reasonable.

6.7   PRACTICAL IMPLICATIONS

The fact that water is strongly adsorbed by the clay minerals has important consequences as regards the swelling and compression behavior of soils. The structure of adsorbed water and its thermodynamic properties play a role in heat and water flow through soils. Because the energy status of adsorbed water is different than that of normal water, there may be consequences in terms of measured pore pressures, and time-dependent changes in water structure may influence physical properties. These effects are considered further in subsequent chapters.

SUGGESTIONS FOR FURTHER STUDY

Bernal, J. D. (1960), "The Structure of Liquids," *Scientific American*, Vol. 203, pp. 124–134.

Bernal, J. D. (1964), "The Structure of Liquids," The Bakerian Lecture, 1962, *Proceedings of the Royal Society London*, A230, pp. 299–322.

Low, P. F. (1961), "Physical Chemistry of Clay-Water Interaction," *Advances in Agronomy*, Vol. 13, pp. 269–327, Academic, New York.

Martin, R. T. (1960), "Adsorbed Water on Clay: A Review," *Clays and Clay Minerals*, Vol. 9, pp. 28–70, Pergamon, New York.

CHAPTER 7

Clay–Water–Electrolyte System

7.1 INTRODUCTION

In Chapter 3 the composition, structure, and some of the characteristics of soil minerals were described. In Chapter 6 consideration was given to water and its interaction with soil particles. Interactions between soil particles, adsorbed cations, and water arise because there are unbalanced force fields at the interfaces between the constituents. When two particles are brought into close proximity, their respective force fields begin to overlap and may influence the behavior of the system if the magnitudes of these forces are large relative to the weights of the particles themselves. Clay particles, because of their small size and large surface area, are well known to be susceptible to such effects.

The effects of surface force interactions and small particle size are manifested by a variety of interparticle attractive and repulsive forces, which, in turn, influence or control the flocculation–deflocculation behavior of clays in suspension and the volume change and strength properties of clays at void ratios common to clay deposits encountered in engineering practice. Since the fabric acquired by a clay deposit at the time of formation may have a profound influence on its subsequent engineering properties, an understanding of factors influencing flocculation–deflocculation behavior is of considerable usefulness. Furthermore, postdepositional or postconstruction changes in engineering properties may be brought about by changes in physico-chemical forces of interaction.

Colloid chemistry provides a means for description of interactions in the clay–water–electrolyte system. In this chapter, the distributions of cations and anions adjacent to clay surfaces are considered, and the consequences in terms of particle interactions in suspensions are examined. No attempt is made to give a rigorous treatment of colloid chemistry, emphasis is on the development of an understanding and appreciation for interactions in the systems of interest to geotechnical engineers.

One point that is sometimes confusing is that clays are usually treated as lyophobic (liquid-hating) or hydrophobic (water-hating) colloids rather than as lyophilic or hydrophilic colloids, even though water wets clays and is adsorbed on particle surfaces. This has resulted (van Olphen, 1963) from the need, historically, to distinguish colloids, such as clay, from colloids already termed hydrophilic, such as gums, which exhibit such an affinity for water that they spontaneously form a colloidal solution. Hydrophobic colloids are now considered to be those which are liquid dispersions of small solid particles; are two-phase systems with a large interfacial area; have a behavior dominated by surface forces; and can flocculate in the presence of small amounts of salt. Clay–water–electrolyte systems satisfy all these criteria.

7.2 ION DISTRIBUTIONS IN CLAY–WATER SYSTEMS

In a dry clay, adsorbed cations are tightly held by the negatively charged clay surfaces. Cations in excess of those needed to neutralize the electro-negativity of the clay particles and their associated anions are present as salt precipitates. When the clay is placed in water the precipitated salts go into solution. Because the adsorbed cations are responsible for a much higher concentration near the surfaces of particles, there is a tendency for them to diffuse away in order to equalize concentrations throughout. Their freedom to do so, however, is restricted by the negative electric field originating in the particle surfaces. The escaping

112

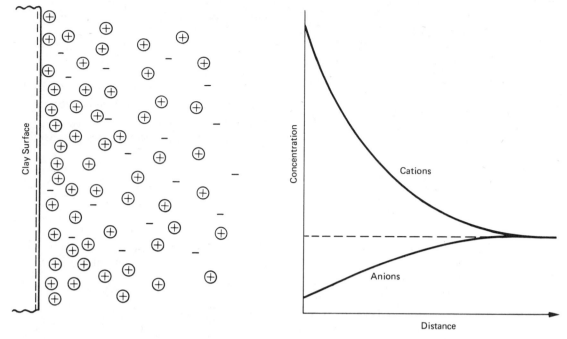

Fig. 7.1 Distribution of ions adjacent to a clay surface according to the concept of the diffuse double layer.

tendency due to diffusion and the opposing electrostatic attraction lead to a cation distribution adjacent to a clay particle in suspension as shown in Fig. 7.1. The distribution of cations adjacent to a clay is analogous to that of the air in the atmosphere, where the escaping tendency of the gas is countered by the gravitational attraction of the earth.

Anions, however, are excluded from the negative force fields of the particles, giving the distribution shown in Fig. 7.1. This phenomenom is sometimes termed negative adsorption. The negative surface and the distributed charge in the adjacent phase are together termed the *diffuse double layer*. Several theories have been proposed for the description of ion distributions adjacent to charged surfaces in colloids. The Gouy–Chapman theory of the diffuse double layer (Gouy, 1910; Chapman, 1913) has received the greatest attention, and it has been applied to the behavior of clays with varying degrees of success.

7.3 DOUBLE LAYER EQUATIONS

The mathematical description of the diffuse double layer has been developed for the cases of both planar and spherical surfaces. Only the planar (one-dimen-

sional) case is treated here. The following idealizing assumptions are made:

1. Ions in the double layer can be considered point charges, and there are no interactions between them.
2. Charge on the particle surface is uniformly distributed.
3. The particle surface is a plate that is large relative to the thickness of the double layer (one-dimensional condition).
4. The static dielectric constant* of the medium is independent of position.

* As noted in Chapter 6, the dielectric constant is a measure of the ease with which molecules can be polarized and oriented in an electric field. Quantitatively, the static dielectric constant is defined by D in Coulomb's equation for the force of electrostatic attraction F, between two charges, Q and Q^1, separated by a distance, d; that is,

$$F = \frac{Q\,Q^1}{Dd^2}$$

The dielectric constant is also given by the ratio of the electrostatic capacity of condenser plates separated by the given material to that of the same condenser with vacuum between the plates. The dielectric constant of water at 20°C is about 80. Because the presence of dissolved ions and the structure imparted to the adsorbed water are likely to influence the arrangement and mobility of water molecules, it does not follow that the dielectric constant of the water in clay is also about 80 or that it is independent of position. There are some data, summarized by Low (1961), to indicate it may be much less than for pure water.

Fig. 7.2 Variation of electrical potential with distance from a charged surface. Except in very unusual cases ψ in soils is negative.

The concentration of ions of type i, n_i, in a force field at equilibrium is given by the Boltzmann equation

$$n_i = n_{io} \exp\left(\frac{E_{io} - E_i}{kT}\right) \qquad (7.1)$$

where the subscript o represents the reference state, conveniently taken to be the conditions at a large distance from the surface, E is the potential energy, T is temperature (°K), and k is the Boltzmann constant (1.38×10^{-16} erg/°K). Concentrations n_i and n_{io} are expressed as ions/cm³.

For ions in an electric field, the potential energy is given by

$$E_i = v_i \epsilon \psi \qquad (7.2)$$

where v_i is the ionic valence, ϵ is the unit electronic charge (16.0×10^{-20} coulomb or 4.8×10^{-10} esu), and ψ is the electrical potential at the point.* Potential varies with distance from a charged surface in the manner shown by Fig. 7.2. In clays, ψ is negative because of the negative surface charge. The potential at the surface is designated as ψ_o. As $E_{io} = 0$, because $\psi = 0$ at a large distance from the surface,

$$E_{io} - E_i = -v_i \epsilon \psi$$

* The electrical potential ψ is defined as the work to bring a positive unit charge from the reference state to the specified point in the electric field.

and the Boltzmann equation becomes

$$n_i = n_{io} \exp\left(\frac{-v_i \epsilon \psi}{kT}\right) \qquad (7.3)$$

Equation (7.3) relates concentration to potential, as illustrated by Fig. 7.3. For negatively charged clay particles $n_i^+ > n_{io}$ and $n_i^- < n_{io}$.

The Poisson equation relates potential, charge, and distance

$$\frac{d^2\psi}{dx^2} = -\frac{4\pi\rho}{D} \qquad (7.4)$$

where x is distance from the surface, ρ is charge density, and D is dielectric constant. The charge density in the double layer is contributed by the ions so that

$$\rho = \epsilon \Sigma v_i n_i \qquad (7.5)$$

Substitution for n_i from equation (7.3) gives

$$\rho = \epsilon \Sigma v_i n_{io}\, e^{(-v_i \epsilon \psi / kT)} \qquad (7.6)$$

which when substituted into equation (7.4) yields

$$\frac{d^2\psi}{dx^2} = -\frac{4\pi\epsilon}{D} \Sigma\, v_i n_{io}\, e^{(-v_i \epsilon \psi / kT)} \qquad (7.7)$$

Equation (7.7) is the general differential equation for the electric double layer adjacent to a planar surface. Its solution provides a basis for computation of electrical potential and ion concentrations as a function of distance from the surface.

For the case of a single cation and anion species of equal valence, that is, $i = 2$,

$$|v_+| = |v| \equiv v, \qquad n_o^+ = n_o^- \equiv n_o$$

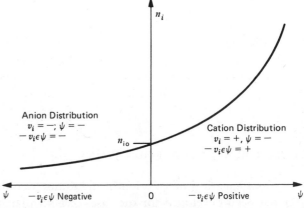

Fig. 7.3 Ionic concentrations in a potential field according to the Boltzmann equation.

equation (7.7) simplifies to the Poisson–Boltzmann equation

$$\frac{d^2\psi}{dx^2} = \frac{8\pi n_0 v\epsilon}{D} \sinh \frac{v\epsilon\psi}{kT} \qquad (7.8)$$

for which solutions are available (Verwey and Overbeek, 1948; Bolt and Peech, 1953; Bolt, 1955a, 1956; Babcock, 1963; van Olphen, 1963).

As the influence of ions of the same sign as the surface charge (termed *co-ions*) is not very important (van Olphen, 1963, p. 252), the solutions can be applied to a good approximation when $i = 2$, but $|v_+| \neq |v_-|$. In the case where there is more than one *counterion* species (ions of the opposite sign as the surface charge, cations in the case of clays) the analysis is more complex, but some solutions are available. A few of the solutions are given here (Bolt, 1955a, 1956; van Olphen, 1963, and Collis-George and Bozeman, 1970).

There is a difference between the boundary conditions assumed for many classical colloid chemical treatments of the double layer and those appropriate for clays. In many systems, behavior is dictated in terms of a constant surface potential, which is controlled by the concentration of "potential-determining ions" in solution. A similar condition may hold for double layers that form on the edges of clay particles, as the solution characteristics control the association and dissociation of alumina in the edges of the octahedral sheets. However, the clay surfaces, which provide by far the major part of the surface area, exhibit a constant surface charge, determined by the isomorphous substitution in the clay structure. Solutions appropriate to the case of constant surface charge are given here; van Olphen (1963) compares the two solutions for some simple cases.

Single diffuse double layer

Solutions are commonly given in terms of the dimensionless quantities

$$y = \frac{v\epsilon\psi}{kT}$$

$$z = \frac{v\epsilon\psi_0}{kT} \qquad (7.9)$$

and

$$\xi = Kx$$

where

$$K^2 = \frac{8\pi n_0 \epsilon^2 v^2}{DkT} \quad (\text{cm}^{-2}) \qquad (7.10)$$

In terms of these variables equation (7.8) becomes

$$\frac{d^2 y}{d\xi^2} = \sinh y \qquad (7.11)$$

Boundary conditions for the first integration are that for $\xi = \infty$, $y = 0$, and $dy/d\xi = 0$. Thus,

$$\frac{dy}{d\xi} = -2 \sinh \left(\frac{y}{2}\right) \qquad (7.12)$$

The boundary condition for the second integration is that for $\xi = 0$, $y = z$ ($\psi = \psi_0$), which leads to

$$e^{y/2} = \frac{e^{z/2} + 1 + (e^{z/2} - 1) e^{-\xi}}{e^{z/2} + 1 - (e^{z/2} - 1) e^{-\xi}} \qquad (7.13)$$

Equation (7.13) describes a roughly exponential decay of potential with distance from a surface at given potential ($y_0 \sim \psi_0$).

In the case where the surface potential is small (less than about 25 mV), $v\epsilon\psi/kT \ll 1$, and equation (7.8) may be approximated by

$$\frac{d^2\psi}{dx^2} = K^2\psi \qquad (7.14)$$

where

$$\psi = \psi_0 e^{-Kx} \qquad (7.15)$$

and the potential decreases purely exponentially with distance. In this case, the center of gravity of the diffuse charge is located at a distance $x = 1/K$ from the surface. Consequently, the quantity $(1/K)$ is often referred to as the "thickness" of the double layer.

The double layer charge is given by

$$\sigma = -\int_0^\infty \rho \, dx \qquad (7.16)$$

and, using the Poisson equation, we obtain

$$\sigma = \frac{D}{4\pi}\int_0^\infty \frac{d^2\psi}{dx^2} \, dx = -\frac{D}{4\pi}\left(\frac{d\psi}{dx}\right)_{x=0} \qquad (7.17)$$

The slope of the potential function at the surface $(d\psi/dx)_{x=0}$ can be found from equation (7.12), leading to

$$\sigma = \left(\frac{2n_0 DkT}{\pi}\right)^{1/2} \sinh \frac{z}{2} \qquad (7.18)$$

for the general case, and to

$$\sigma = \left(\frac{DK}{4\pi}\right)\psi_0 \qquad (7.19)$$

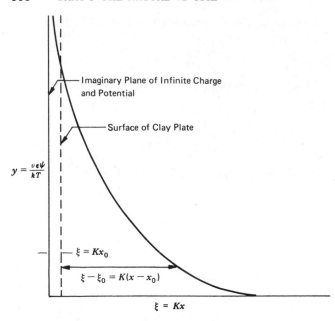

Fig. 7.4 Potential vs. distance relationship used to develop equation (7.20).

for the case of small ψ_0. With the relationship between concentration and potential defined by equation (7.3), that between potential and distance by equation (7.13) or (7.15), and charge and surface potential related by equation (7.18) or (7.19), the double layer is completely described, provided the solution con-

centration n_0, cation valence v, and either the surface charge σ or surface potential ψ_0 are known.

A somewhat simpler relationship between potential and distance than equation (7.13) can be obtained (Bolt, 1955a) by choosing the zero point for the ξ axis at an imaginary plane, a distance x_0 behind the true surface, where the potential reaches infinity (Fig. 7.4). This gives

$$\exp(-y) = \frac{\cosh \xi + 1}{\cosh \xi - 1} = \coth^2\left(\frac{\xi}{2}\right) \quad (7.20)$$

$$\xi_0 = Kx_0 = \frac{4vc_0}{K\Gamma} \quad (7.21)$$

where c_0 is the concentration at a large distance from the surface in mmole/cm³ ($= 1000n_0/N$ where N is Avogadro's number) and Γ is the surface charge density in meq/cm² (cation exchange capacity per specific surface). The relationship between the dimensionless potential function y and the dimensionless distance function ξ, as given by equation (7.20), is plotted in Fig. 7.5. Equation (7.21) is an approximation that is valid except for very high c_0 or very low Γ. Another form of equation (7.21) is

$$x_0 = \frac{4}{v\beta\Gamma} \quad (7.22)$$

where $\beta = 8\pi F^2/1000DRT \approx 1 \times 10^{15}$ cm/mmole at 20°C, F is the Faraday constant, and R is the gas con-

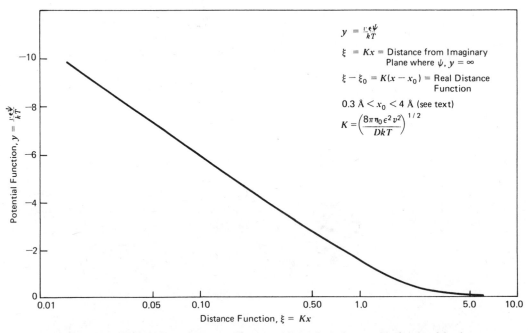

Fig. 7.5 Potential function vs. distance function for a single double layer.

stant-kN. According to Bolt (1956), x_0 can be approximated by $1/v$ Å for illite, $2/v$ Å for kaolinite, and $4/v$ Å for montmorillonite. It is evident from these values that the correction required to the distance functions to account for the distance to the imaginary plane where $\psi = \infty$ is small.

Although the case of a single diffuse double layer is not representative of the physical conditions in most actual clay systems because double layers of adjacent particles will usually overlap, the solutions are useful in practice to provide estimates of thickness of diffuse layer, surface potentials, and effects of changes in solution composition.

Interacting double layers

The potential and charge distributions for the case of interacting double layers from parallel flat plates, separated at distance $2d$ are shown in Fig. 7.6. The potential function at the midplane, $y = v\epsilon\psi_d/kT$, is denoted by u, and the integration boundary condi-

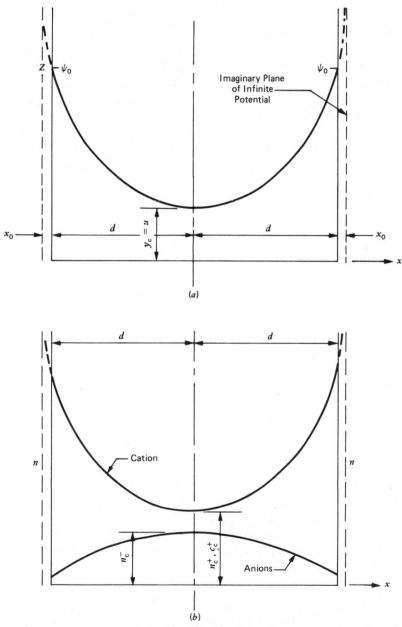

Fig. 7.6 Potential and charge distributions for interacting double layers from parallel flat plates. (*a*) Potential. (*b*) Charge.

tions for equation (7.11) are that for $\xi = Kd$, $y = u$, and $dy/d\xi = 0$. The second integration using the boundary condition imposed on equation (7.12) leads to

$$\int_z^u (2 \cosh y - 2 \cosh u)^{-1/2} dy = -\int_0^d d\xi = -Kd$$

(7.23)

The integral in equation (7.23) can be evaluated using tables to give the midplane potential (from u) for given values of surface potential $2d$ and K. Corresponding values of Kd, z, and u are given by Verwey and Overbeek (1948), which are useful for computation of midplane potentials as a function of variations in plate distance and electrolyte concentration for the case of constant surface potential. Convenient tabulations for conditions of constant surface charge are given by van Olphen (1963).

For small interactions, that is, for large values of Kd, as would be the case for large plate separation, high n, high v, or small ψ_0, the midplane potential can be taken as the sum of the double layer potentials at distance d based on the solutions for a single plate, that is,

$$u = 2y_d$$

(7.24)

Concentrations at the midplane can be obtained from Boltzmann's equation:

$$n_- = n_0 e^u \quad \text{and} \quad n_+ = n_0 e^{-u}$$

(7.25)

Alternatively, midplane concentrations c_c may be computed using the following equation (Bolt, 1956):

$$v(\beta c_0)^{1/2}(x_0 + d) = 2\left(\frac{c_0}{c_c}\right)^{1/2} \int_{\phi=0}^{\pi/2} \frac{d\phi}{(1 - (c_0/c_c)^2 \sin^2 \phi)^{1/2}}$$

(7.26)

where all terms are previously defined. Values for the complete elliptic integral are given in standard tables.

The preceeding solutions all have been developed for systems containing a single cation species; however, as in the case of the single double layer, they are applicable to systems where $v_+ \neq v_-$, because the co-ions have little influence on the results. For systems containing mixed valent cations (e.g., Na+ and Ca2+), the equations are not applicable.

Collis-George and Bozeman (1970) have developed a double layer equation for heterovalent systems applicable to the case of constant surface charge. It shows that in a clay containing both monovalent and divalent cations there is a much greater concentration of divalent cations than monovalent cations near the particle surface even if the concentration of monovalent ions is much greater in the bulk solution.

Since the overlap of double layers of the same sign is potentially the source of an interparticle repulsion, it is reasonable to inquire if double layers in typical soils are sufficiently thick that interactions between those of adjacent particles can occur. If the quantity $1/K$ is taken as the "thickness" of the double layer, then values of 10 Å in a 0.1 M solution of monovalent cation, increasing to 100 Å in a 0.001 M solution are obtained. For water distributed uniformly on surfaces of clay particles, the water layer thickness is equal to half the particle spacing or d in Fig. 7.6. This thickness is given by the water content (cm³/g) divided by the surface area (cm²/g). For a water content of 50 percent and a specific surface in the range of 50 to 300 m²/g, values of d from 17 to 100 Å are obtained, indicating that spacings are likely to be well within the range where interactions are important. In fact, in many cases, spacings may be small due to nonparallelism of particles, and the effects of other types of interparticle forces, including physical interactions between particles, may override the influences of double layer interactions.

Nonetheless, relationships of the type developed in this section are useful for understanding long range forces that control flocculation–deflocculation behavior, for study of swelling properties, and for analysis of some aspects of ion exchange.

7.4 INFLUENCES OF SYSTEM VARIABLES ON THE DOUBLE LAYER ACCORDING TO THE GOUY THEORY

Computations of ionic and potential distributions adjacent to charged surfaces show that the thickness of the double layer is sensitive to variations in surface charge density σ or surface potential ψ_0, electrolyte concentration n_0, cation valence v, dielectric constant of the medium D, and temperature T. An approximate quantitative indication of the influences of these factors can be seen in terms of the "thickness" of the double layer as given by

$$\frac{1}{K} = \left(\frac{DkT}{8\pi n_0 \epsilon^2 v^2}\right)^{1/2}$$

(7.27)

From this relationship it may be noted that the thickness decreases inversely as the valence and the square root of the concentration and that it increases with the square root of dielectric constant and temperature, other factors remaining constant.

Since the long range interparticle repulsive force depends on the amount of overlap or interaction between adjacent double layers, one can estimate the probable influences on behavior that result from changes in the system variables. In general, the thicker the double layer the less the tendency for particles in suspension to flocculate and the higher the swelling pressure in cohesive soils.

A numerical example is helpful for the quantitative illustration of the influence of system variables. Concentrations and potential distributions in single double layers will be developed using the relationships given in the previous section, and midplane concentrations and potentials are computed using Table III-2 of van Olphen (1963). The following conditions are assumed for the example: a montmorillonite clay with a specific surface of 800 m²/g and a cation exchange capacity of 83 meq/100 g, constant surface charge, and solutions of 0.83×10^{-4} M NaCl, 0.83×10^{-2} M NaCl, and 0.83×10^{-4} CaCl$_2$ in water and 0.83×10^{-4} M NaCl in ethyl alcohol.

The following constants are needed: N = Avogadro's number = 6.02×10^{23} ions/mole; $k = 1.38 \times 10^{-16}$ ergs/°K; $\epsilon = 4.80 \times 10^{-10}$ esu; $D_{\text{water}} = 80$; and $D_{\text{ethyl alcohol}} = 24.3$.

The surface charge density is

$$\Gamma = \frac{0.83 \text{ meq/g}}{800 \text{ m}^2/\text{g}}$$

$$= \frac{0.83}{800} \ (\text{meq/m}^2) \times (\text{m}^2/10^4 \text{ cm}^2)$$

$$= 0.1035 \times 10^{-6} \ (\text{meq/cm}^2)$$

$$\sigma = \Gamma \times \frac{F}{1000} = 0.1035 \times 10^{-6} \ (\text{meq/cm}^2)$$

$$\times \ (96.5 \text{ coulomb/meq})$$

$$= 10.0 \times 10^{-6} \text{ coulomb/cm}^2$$

$$\sigma = 10.0 \times 10^{-6} \text{ coulomb/cm}^2 \div 3.33$$

$$\times 10^{-10} \text{ coulomb/esu}$$

$$= 3.0 \times 10^4 \text{ esu/cm}^2$$

Effect of electrolyte concentration

A concentration change from 0.83×10^{-4} M NaCl to 0.83×10^{-2} M NaCl in water solution is considered. n_0 is the molarity $\times$ N $\times$ 10^{-3} ions/cm³ and is equal to 6.02×10^{20} M ions/cm³; kT is equal to

0.4×10^{-13} ergs at 290°K (17°C). For 0.83×10^{-4} M NaCl,

$$n_0 = 5.0 \times 10^{16} \text{ ions/cm}^3$$

$$K^2 = \frac{8\pi n_0 \epsilon^2 v^2}{DkT} = \frac{8\pi \times 5.0 \times 10^{16} \times (4.8 \times 10^{-10})^2 \times 1^2}{80 \times 0.4 \times 10^{-13}}$$

$$= 9.0 \times 10^{10}$$

$$K = 3.0 \times 10^5 \text{ cm}^{-1}$$

$$\frac{1}{K} = \text{``thickness'' of double layer} = 3.33 \times 10^{-6} \text{ cm}$$

$$= 333 \text{ Å}$$

and for 0.83×10^{-2} M NaCl,

$$n_0 = 5.0 \times 10^{18} \text{ ions/cm}^3$$

$$K = 3.0 \times 10^6 \text{ cm}^{-1}$$

$$\frac{1}{K} = \text{``thickness''} = 3.33 \times 10^{-7} \text{ cm} = 33.3 \text{ Å}$$

The surface potential is given by equation (7.18).

$$\sigma = \left(\frac{2n_0 DkT}{\pi}\right)^{1/2} \sinh\left(\frac{z}{2}\right)$$

$$\sinh\left(\frac{z}{2}\right) = 3.0 \times 10^4 \left(\frac{\pi}{2n_0 \times 80 \times 0.4 \times 10^{-13}}\right)^{1/2}$$

$$= \frac{3.0 \times 10^4}{(n_0)^{1/2}} (49.1 \times 10^{10})^{1/2}$$

$$\sinh\left(\frac{z}{2}\right) = \frac{2.10 \times 10^{10}}{(n_0)^{1/2}}$$

For 0.83×10^{-4} M NaCl

$$\sinh\left(\frac{z}{2}\right) = \frac{2.10 \times 10^{10}}{(5.0 \times 10^{16})^{1/2}} = \frac{2.10 \times 10^{10}}{2.23 \times 10^8}$$

$$= 0.94 \times 10^2 = 94$$

$$\frac{z}{2} = \frac{v\epsilon\psi_0}{2kT} = 5.235$$

$$z = 10.47$$

$$\psi_0 = \frac{kTz}{v\epsilon} = \frac{0.4 \times 10^{-13}}{1 \times 4.80 \times 10^{-10}}$$

$$\times 300^* \times 10^3 \times 10.47 = 262 \text{ mV}$$

The corresponding values for 0.83×10^{-2} M NaCl are

$$\sinh\left(\frac{z}{2}\right) = 9.4$$

---

* 1.0 esu of electrical potential is 299.79 V.

$$\frac{z}{2} = 2.937$$

$$z = 5.874$$

$$\psi_0 = 147 \text{ m}v$$

The decay of potential with distance from the surface may be determined using equations (7.20) and (7.21)

$$\exp(-y) = \frac{\cosh \xi + 1}{\cosh \xi - 1} = \coth^2\left(\frac{\xi}{2}\right) \quad (7.20)$$

$$\xi_0 = Kx_0 = \frac{4vc_0}{K\Gamma} \quad (7.21)$$

or with the aid of Fig. 7.5. For 0.83×10^{-4} M NaCl $= 0.83 \times 10^{-4}$ mmole/cm³,

$$K = 3.0 \times 10^5 \text{ cm}^{-1}$$

$$x_0 = \frac{4 \times 1 \times 0.83 \times 10^{-4}}{(3.0 \times 10^5)^2 \times 0.1035 \times 10^{-6}}$$

$$= 3.5 \times 10^{-8} \text{ cm} \approx 4 \text{ Å}$$

The same value of x_0 is obtained for 0.83×10^{-2} M NaCl.

The potential and concentrations n^+ and n^- may now be determined using Fig. 7.5 and the relationships $n^+ = n_0 \exp(-y)$ and $n^- = n_0 \exp(+y)$. The results are shown in Fig. 7.7 for potential as a function of distance from the surface. The concentration variations with distance are shown in Fig. 7.8. Figure 7.7 indicates that not only does the increase in electrolyte concentration reduce the surface potential for the condition of constant surface charge, but also the decay of potential with distance is much more rapid.

Concentrations in Fig. 7.8 are shown to a logarithmic scale to enable comparison between the two systems. Again, the strong suppression of the double layer as a result of the increase in electrolyte concentration is evident. For each case the number of excess cations contained in region ABD (or $A'B'D'$) plus the number of excluded anions in region BCD (or $B'C'D'$) together must give a charge equal to the surface

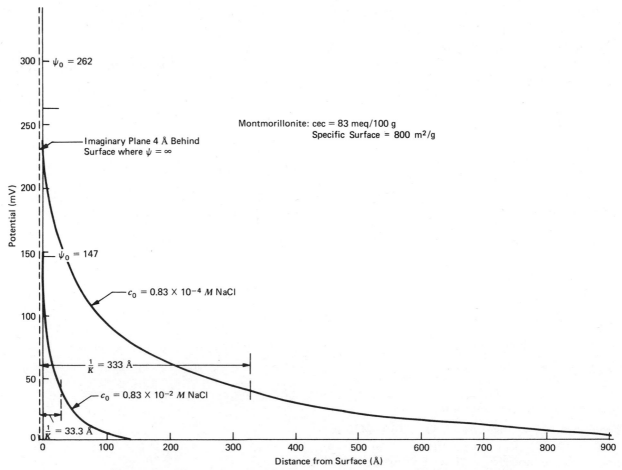

Fig. 7.7   Effect of electrolyte concentration on double layer potential.

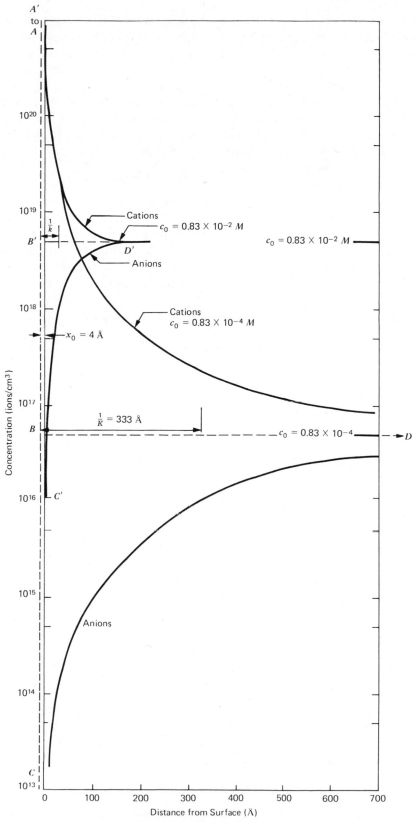

Fig. 7.8 Effect of electrolyte concentration on ion distributions in the double layer.

charge density. Furthermore, because we are dealing with a constant surface charge, the number of excess cations plus deficient anions in region ACD must equal that in region $A'C'D'$.

When the same systems are analyzed in terms of interacting parallel plates using van Olphen's (1963) tabulations, the results for surface and midplane potentials and concentrations as a function of plate spacing ($2d$) are as shown in Figs. 7.9 and 7.10. Values of water content are shown corresponding to half the plate spacings. The average value of half the plate spacing is given by the thickness of the water layer on each surface, or the water content divided by the specific surface, as noted earlier.

Figure 7.9 shows that interparticle spacing has essentially no influence on the surface potential except for half distances less than about 20 Å. The mid-

plane potential, however, is very sensitive to both spacing and concentration. Figure 7.10 indicates that interactions extend to much greater particle spacings for the low electrolyte concentration system. One mechanism of clay swelling is related to double layer interactions (Chapter 13), and higher swell pressures are associated with greater interactions. Thus, swelling behavior depends in part on the electrolyte concentration.

Effect of cation valence

For solutions of the same molarity and a constant surface charge, a change in cation valence affects both the surface potential and the thickness of the double layer. The surface potential function z computed from equation (7.18) is independent of valence. Thus, for the example considered (0.83×10^{-4} M NaCl and

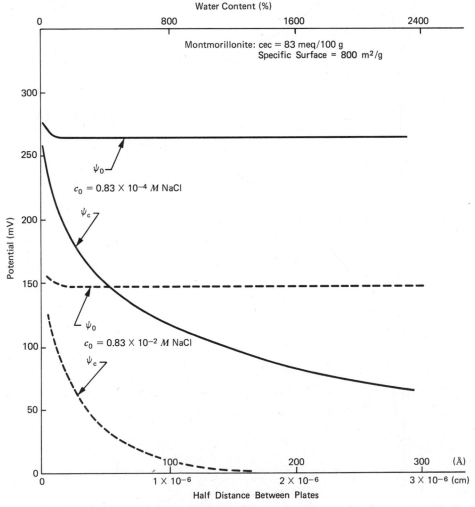

Fig. 7.9 Effect of electrolyte concentration on surface and midplane potentials for interacting parallel plates.

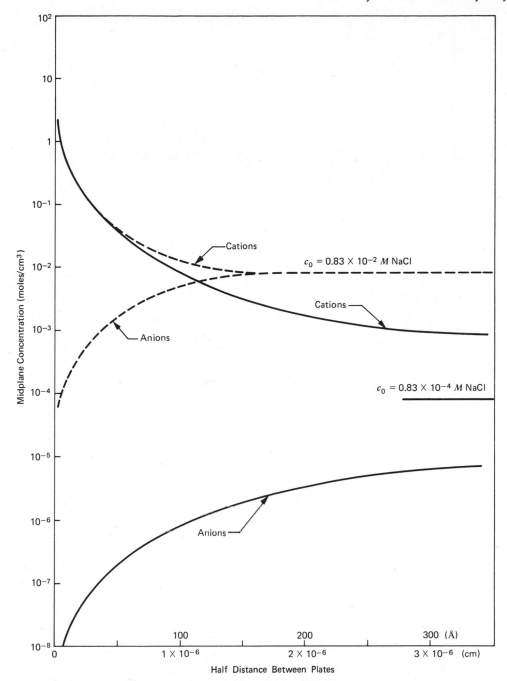

Fig. 7.10 Effect of electrolyte concentration on midplane concentration—parallel plates.

$CaCl_2$, $n = 5 \times 10^{16}$ ions/cm³), z is equal to 10.47, as computed previously. As ψ_0 varies inversely as v, the surface potential in the $CaCl_2$ system will be half that of the NaCl system, that is, $(\psi_0)_{Na} = 262$ mV, $(\psi_0)_{Ca} = 131$ mV. The effect of valence on double layer thickness can be seen through its effect on $1/K$ from equation (7.10)

$$\frac{1}{K} \propto \frac{1}{v}$$

The distributions of Na$^+$ and Ca^{2+} are shown in Fig. 7.11. It also follows that an increase in valence will suppress the midplane concentrations and potential between interacting plates, thus leading to a decrease in interplate repulsion.

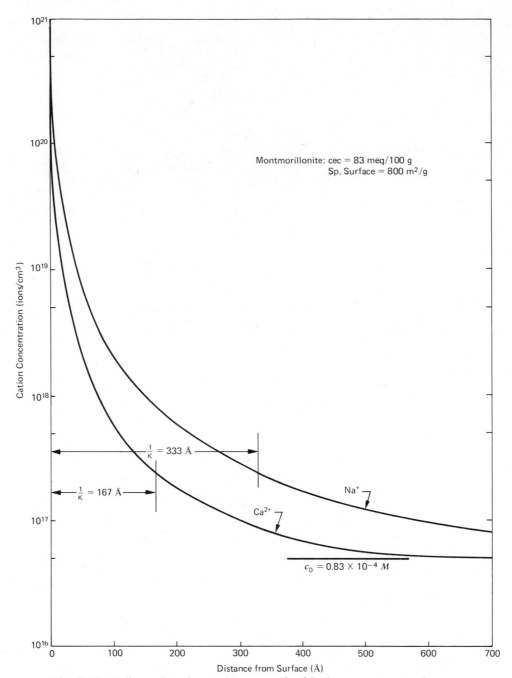

Fig. 7.11　Effect of cation valence on double layer concentrations.

The preferential adsorption of multivalent ions, which is a well-established experimental fact, also predicted by the theory of Collis–George and Bozeman cited previously, means that even relatively small amounts of di- or trivalent cations added to clay–water–monovalent electrolyte systems can have a significant influence on physical properties.

Effect of dielectric constant

The dielectric constant of the electrolyte–pore fluid system should influence both the surface potential and the double layer thickness. In dealing with a constant surface charge, the surface potential function will increase as D decreases, according to

$$\sinh\left(\frac{z}{2}\right) = \left(\frac{\pi}{2n_0 D k T}\right)^{1/2} \sigma \qquad (7.28)$$

For a $0.83 \times 10^{-4}\ M$ solution of NaCl in ethyl alcohol ($D = 24.3$) as opposed to water ($D = 80$),

$$\sinh\left(\frac{z}{2}\right) = 94 \times \left(\frac{80}{24.3}\right)^{1/2} = 170$$

$$\frac{z}{2} = \frac{v\epsilon\psi_0}{kT} = 5.83$$

$$z = 11.66$$

and

$$\psi_0 = 292\ \text{mV}$$

are obtained as compared with 262mV with water.

The effect of dielectric constant on thickness of the double layer is given by

$$\frac{1}{K} \propto D^{1/2}$$

so that with alcohol as the dielectric the layer will be reduced in thickness by a factor of $(24.3/80)^{1/2}$ or 0.55.

Detailed consideration of the influences of dielectric constant may seem academic because the pore fluid in soils usually is water, but there may be special instances where liquids such as oil and waste chemicals are the pore fluid. The dielectric constant of these materials is different from that of normal water.

Effect of temperature

According to equations (7.18) and (7.27), an increase in temperature causes an increase in double layer thickness and a decrease in surface potential for a constant surface charge all other factors constant.

However, an increase in temperature results also in a decrease in the dielectric constant. The following tabulation shows the nature of the variation for water.

$T(°C)$	$T(°K)$	Dielectric Constant (D)	DT
0	273	88	2.40×10^4
20	293	80	2.34×10^4
25	298	78.5	2.34×10^4
60	333	66	2.20×10^4

The small variation of the product DT with change in temperature means, theoretically, that the double layer should not be influenced greatly. This assumes, of course, that the values of the dielectric constant are unaffected by particle surface forces and ionic concentration. It also accounts, in part, for apparent contradictory findings reported in the literature on the effects of temperature change on such soil properties as strength, compressibility, and swelling.

7.5 SHORTCOMINGS OF DOUBLE LAYER THEORY AND ADDITIONAL FACTORS INFLUENCING BEHAVIOR

The validity of the Gouy–Chapman theory may be examined in at least two ways: in terms of its completeness in taking into account factors likely to influence behavior, and in terms of its ability to describe observed behavior. The first of these is considered here, and the second is examined in several places in later chapters.

Effects of secondary energy terms

Several energy terms should be important in real systems (Bolt, 1955a). These are the effect of the electrical field strength on the dielectric constant, coulombic interaction between the ions, and short range repulsion between the surface and ions. The results from a corrected theory are almost the same as those from the simple theory for double layer interaction, although a corrected theory is more suitable for description of ion exchange.

Adsorbed water effects

The effects of the electric field and water structure on the water pressure in the double layer may be important (Low, 1961; Ravina and Zaslavsky, 1972), and this may influence behavior. Some revision of the theory has been proposed to take these effects into account.

The effect of ion size—The Stern layer

The application of the relationships presented in the last section to some systems may lead to impossibly high ion concentrations next to the surface. This is because the theory deals with point charges; whereas, ions are of finite size. Thus, the actual concentration adjacent to the surface will be less than predicted. The hydrated radii of some cations in soils are:

Ion	Hydrated Radius (Å)
Li^+	7.3–10.0
Na^+	5.6– 7.9
K^+	3.8– 5.3
NH^+	5.4
Rb^+	3.6– 5.1
Ca^+	3.6– 5.0
Mg^{2+}	10.8
Ca^{2+}	9.6
Sr^{2+}	9.6
Ba^{2+}	8.8

The Gouy–Chapman theory has been corrected for this effect (Stern, 1924), and the theory is summarized by van Olphen (1963) for both single and interacting flat double layers. The so-called "Stern layer" is assumed to consist of counterions in a closely packed layer close to the surface with an adjacent diffuse layer extending into the solution. The equations enable computation of the charge in each layer and the potential at their interface.

The physical consequences of the development of a Stern layer in terms of potential and cation distributions are shown in Fig. 7.12. From a particle interaction standpoint, the larger the ion size the thicker the layer required to accommodate the necessary number of cations, and hence the greater the repulsion.

The effect of pH

Clay particles may have hydroxyl (OH) exposed on their surfaces and edges. The tendency for the hydroxyl to dissociate,

$$SiOH \xrightarrow{H_2O} SiO^- + H^+$$

is strongly influenced by the pH[*]; the higher the pH, the greater the tendency for the H^+ to go into solu-

[*] $pH = \dfrac{1}{\log_{10} H^+ \text{ concentration}}$; pH < 7—acid—high H^+ concentration, and pH > 7—basic—low H^+ concentration.

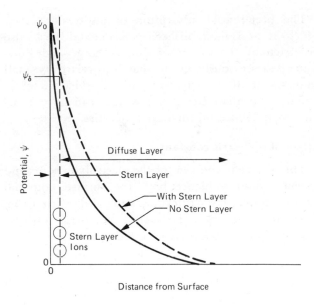

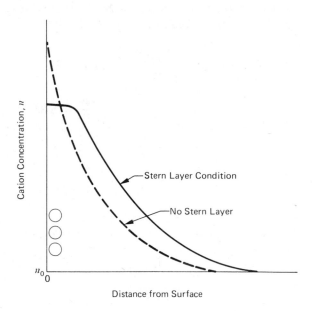

Fig. 7.12 Effect of Stern layer on cation concentration and double layer potential.

tion, and the greater the effective negative charge of the particle.

In addition to this, alumina, which is exposed at the edges of clay particles, is amphoteric and ionizes positively at low pH and negatively at high pH. As a consequence, positive double layers may develop at the edges of some clay particles in an acid environment. Such double layers are of the constant surface potential type, as opposed to constant surface charge, with H^+ serving as the potential determining ion.

Because of both of these considerations, pH plays a very important role in the behavior of clay suspensions. A low pH promotes a positive edge to negative surface interaction, often leading to flocculation from suspension. Stable suspensions or dispersions of clay particles often require high pH conditions. While a proper understanding of the behavior of many clay–water systems cannot be obtained without consideration of the influences of pH, quantitative relationships for doing so do not appear to be available at present.

Effects of anion adsorption

The theory as developed assumes attraction of cations to the negatively charged clay surfaces and repulsion of all anions, leading to concentration distributions of the type shown in Fig. 7.8. There may be situations, however, where specific anions are attracted to and essentially become part of particle surfaces or edges, thereby increasing the electronegativity.

Attraction of anions, particularly those such as phosphate, arsenate, and borate, which have about the same size and geometry as the silica tetrahedron, appears well established. Phosphates, in particular, seem strongly attracted, as certain of the phosphate compounds are among the most effective deflocculating agents for soil suspensions. Little is known about whether or not clays can have anion exchange spots on basal surfaces, although replacement of $(OH)^-$ is a plausible mechanism.

7.6 ENERGY AND FORCE OF REPULSION

Knowledge of the potential and charge distributions between interacting double layers can be used to compute the energy and force of electrostatic repulsion per unit area of plate. The repulsive energy V_R is given by

$$V_R = 2(F_d - F_\infty) \qquad (7.29)$$

where F_d is the free energy of the double layer per unit area at a plate spacing of *2d*, and $F\infty$ is the free energy of a single noninteracting double layer. Tabulations of F_d and $F\infty$ for the constant surface potential case are available (Verwey and Overbeek, 1948).

The repulsive force per unit area (repulsive pressure) can be computed as the difference in osmotic pressure (see Chapter 13) midway between plates relative to that in the equilibrium solution. The osmotic pressure difference depends directly on the difference in numbers of ions in the two regions, that is,

$$p \propto n_c^+ + n_c^- - (n_0^+ + n_0^-)$$

$$\therefore p \propto n_0 e^u - n_0 + n_0 e^{-u} - n_0 = 2n_0 (\cosh u - 1)$$

The resulting equation is (van Olphen, 1963)

$$p = 2n_0 kT (\cosh u - 1) \qquad (7.30)$$

which is valid for both constant charge and constant potential surfaces, with u computed for the appropriate condition. A corresponding equation may be written using Bolt's formulation or equation (7.26). These relationships have been used, with varying degrees of success, as a basis for the physico-chemical description of clay swelling, and are considered further in Chapter 13.

7.7 LONG RANGE ATTRACTION

Fluctuating dipole bonds (van der Waals forces) act between all units of matter and are the source of attraction between colloidal particles (Section 2.4). The attractive energy between pairs of molecules V_A (London, 1937) was extended by Casimir and Polder (1948) to obtain the attraction between parallel plates by assuming interaction to be additive. The following equation was obtained:

$$V_A = -\left[\frac{A}{48\pi} \frac{1}{d^2} + \frac{1}{(d+\delta)^2} - \frac{2}{(d+\delta/2)^2} \right] \qquad (7.30)$$

where d is the half distance between plates measured from the plane of surface layer atoms, δ is the thickness of the plate measured between the same planes, and A is the van der Waals constant, which is in the range 10^{-11} to 10^{-14} ergs. van Olphen (1963) suggests A to be of the order of 10^{-12}; whereas, Derjaguin (1960) considers 5×10^{-14} a more reasonable value. Equation (7.30) has been widely used in conjunction with relationships for repulsion as a function of distance for the development of net curves of interaction between parallel plates (Verwey and Overbeek, 1948).

Later work, e.g., Lifshitz (1955), Derjaguin (1960), has shown that the van der Waals forces are electromagnetic, and that the instantaneous electric moment is frequency dependent. As a result the Casimir-Polder theory is invalid but is a good approximation for particle separations less than about 1000 Å. The Lifshitz theory (Lifshitz, 1955; Dyzaloshinskii et al., 1961) applies for all particle separations. The general results of these theories (Ingles, 1962) are that the attractive forces are dependent on distance according to

Casimir-Polder Theory $F_1 \propto \dfrac{Ak}{d^3}$

Lifshitz Theory $F_2 \propto \dfrac{Bk'}{d^4}$

where A, B, k and k' are constants. A is of the order of 10^{-13} erg and B, an analytically deducible constant, is of the order of 10^{-19} erg-cm. Black et al (1960) determined a force per cm² between parallel quartz plates given by

$$F_2 = \frac{C}{d_\mu^{\;4}} \text{ dyne/cm}^2 \qquad (7.31)$$

where d_μ is separation distance in μm and

$C = 1.0 \times 10^{-3}$ to 2.0×10^{-3} (experimental)

$C = 0.6 \times 10^{-3}$ to 1.6×10^{-3} (theoretical—
Lifshitz theory)

The close agreement between theory and experiment is encouraging, as past results were often in conflict.

For the general case of two bodies separated by a medium, the complex dielectric constant must be known for the full range of frequencies. Although the resulting equations are complicated, some simplification is possible in certain cases (Dzyaloshinskii et al, 1961). For two clay particles having static dielectric constants D_{10}, separated at distance d by a pore fluid having a static dielectric constant D_{30}, the following relationship is obtained

$$F_2 = \frac{\pi}{480} \frac{hc}{d^4} \frac{1}{\sqrt{D_{30}}} \frac{D_{10} - D_{30}}{D_{10} + D_{30}} \; \Phi\left(\frac{D_{10}}{D_{30}}\right) \quad (7.32)$$

where h is Planck's constant and c is the velocity of light. The function $\Phi\,(D_{10}/D_{30})$ is shown in Fig. 7.13. The static dielectric constant of clay particles D_{10}, is approximately 4.0.

Evaluation of the coefficient of $1/d^4$ in equation (7.32) (which is the same as C in equation (7.31)), as a function of the dielectric constant of the medium separating particles D_{30}, for d in μm, gives the result shown in Fig. 7.14. Thus, according to the Lifshitz theory the dielectric constant of the pore fluid in a clay should influence the attractive forces in the manner shown.

As interparticle repulsions due to double layer interactions increase monotonically ($1/K \propto D^{1/2}$) with increasing dielectric constant, the net force of interaction should vary in the manner shown in Fig. 7.15 (Moore and Mitchell, 1974). Figure 7.16 shows the

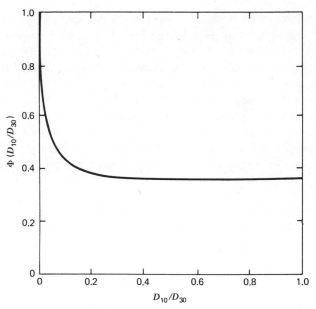

Fig. 7.13 Dielectric constant function for attractive force calculation by Lifshitz theory (After Dzyaloshinskii, Lifshitz, and Pitaevskii, 1961).

variation of undrained strength (normalized to that with water as the pore fluid) with dielectric constant of the pore fluid for three series of tests on kaolinite. In each case, samples were consolidated initially in water, followed by leaching with water-miscible pore fluids. Since it seems reasonable that strength should depend on the net attractive force between particles, and the variation in strength with D_{30} is similar to

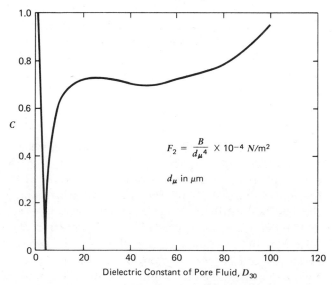

Fig. 7.14 Effect of dielectric constant of the pore fluid on van der Waals attraction.

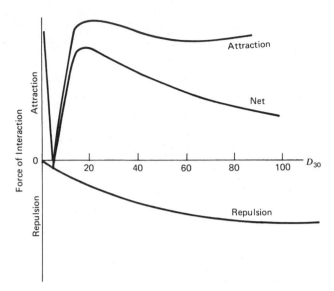

Fig. 7.15 Combined forces of interaction.

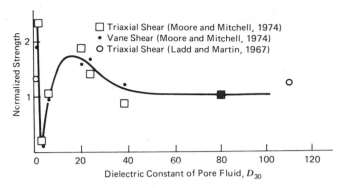

Fig. 7.16 Effect of the dielectric constant of the pore fluid on the strength of Peerless #2 kaolinite.

that of the van der Waals attraction (Figs. 7.14 to 7.16), there appears some support for the use of the Lifshitz theory for van der Waals forces in clay systems.

Although a condition of perfectly parallel plates is not a probable particle arrangement in a clay soil, and particle corners, edges, and asperities are likely to be in contact with adjacent particles, the average particle surface to particle surface distance should be great enough (a few tenths of a micrometer or more) to justify application of the Lifshitz theory. It should be noted also that for the case of interaction of a flat plate and one or two spheres the attractive force decays as $1/d^3$ instead of $1/d^4$ (Ingles, 1962).

7.8 NET ENERGY AND FORCE OF INTERACTION

Double layer repulsions and van der Waals electromagnetic attractions combine in the manner shown schematically in Fig. 7.17. The energy of repulsion is sensitive to changes in electrolyte concentration, cation valence, dielectric constant, and pH; whereas, the attractive energy is sensitive only to changes in the dielectric constant and temperature. In cases where the net curve of interaction exhibits a high repulsive energy barrier, particles in suspension are prevented from close approach, and the suspension will be stable. In cases where the repulsive energy barrier does not exist, particles are drawn into close proximity and flocculation results, represented by

the minima in the energy curves. In this case, flocs of several particles settle together from suspension.

The character of the net curve of interaction may have a major influence on the particle arrangements in sedimented deposits of clay. Changes in system chemistry, which in turn can cause changes in the net curve of interaction, may have important consequences in terms of the behavior of the soil when disturbed or subjected to the action of flowing water.

7.9 CATION EXCHANGE—GENERAL CONSIDERATIONS

From the considerations in Section 7.5, it is apparent that the type and amount of different cations in a clay–water–electrolyte system have a major influence on double layer interactions. Changes in these interactions may lead to changes in physical and physicochemical properties. Under a given set of environmental conditions (temperature, pressure, pH, total electrolyte concentration), a clay adsorbs cations with a fixed total charge. Exchange reactions involve replacement of these ions with a group of different ions having the same total charge. The exchange of ions of one type by ions of another type does not affect the structure of clay particles themselves.

Common ions in soils

The most commonly found cations in soils are calcium (Ca^{2+}), magnesium (Mg^{2+}), sodium (Na^+), and potassium (K^+), usually in that decreasing order of abundance for residual and nonmarine sedimentary soils. The commonest anions in soils are sulfate (SO_4^{2-}), chloride (Cl^-), phosphate (PO_4^{3-}) and nitrate (NO_3^-).

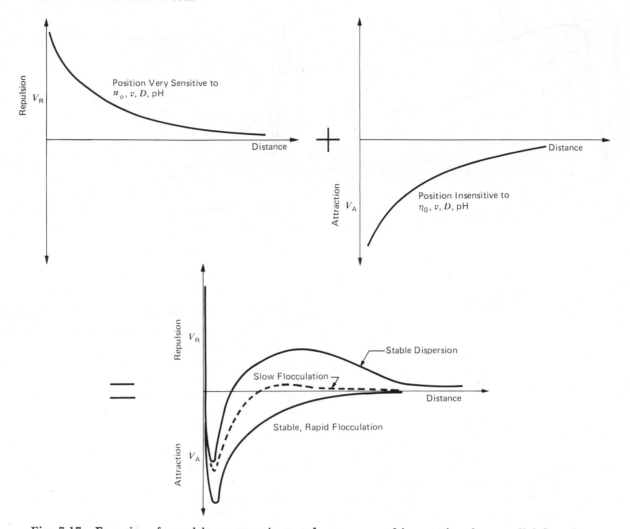

Fig. 7.17   Energies of repulsion, attraction, and net curves of interaction for parallel flat plates.

Sources of exchange capacity

Three sources of exchange capacity in clays have been identified (Grim, 1968):

1. **Isomorphous substitution.*** Al^{3+} for Si^{4+} in the silica sheet and Mg^{2+} for Al^{3+} in the tetrahedral sheet are most common forms. Balancing cations are attracted to cleavage surfaces. This is the major source of exchange capacity, except possibly for the kaolin minerals.
2. **Broken bonds.**   Particle edges and noncleavage surfaces. The importance of this contribution increases with decreasing particle size. It may be the major cause in kaolinite, and contribute 20 percent to the total in smectite.

---

* See Sections 2.8 and 3.4 for discussion of the meaning of the term isomorphous substitution.

3. **Replacement** of the hydrogen of an exposed hydroxyl.

Exchange capacities of the clay minerals

Because the amount of exchange capacity contributed by each of these effects depends on various compositional and environmental factors, a given clay mineral does not have a fixed, single value of exchange capacity. Typical ranges of exchange capacity are summarized in Table 3.3.

Cation replaceability

Ions of one type can be replaced by ions of another type; for example, Ca^{2+} for Na^+, Na^+ for Ca^{2+}, Fe^{3+} for Mg^{2+}, and so on. The ease with which an ion of one type can replace an ion of another type depends mainly on the valence, relative abundance of the different ion types, and ion size. Other things being equal, trivalent cations are held more tightly than divalent, and divalent more tightly than monovalent. Ordinarily small cations tend to displace large cations. A typical replaceability series is

$$Na^+ < Li^+ < K^+ < Rb^+ < Cs^+ < Mg^{2+} < Ca^{2+}$$
$$< Ba^{2+} < Cu^{2+} < Al^{3+} < Fe^{3+} < Th^{4+}$$

It is possible, however, to displace a cation of high replacing power, such as Al^{3+}, by one of low replacing power, such as Na^+, by mass action, wherein the concentration of sodium in solution is made very high relative to that of aluminum.

Rate of exchange

The rate of exchange varies with clay type, solution concentrations, temperature, and so on. In general, however, exchange reactions in the kaolin minerals are almost instantaneous. In illites, a few hours may be needed for completion, because a small part of the exchange capacity may be between unit layers. A longer time is required in the smectites, because the major part of the exchange capacity is located in the interlayer region.

Organic ions

Organic anions and cations may be adsorbed by clays as well as inorganic ions, and both exchange adsorption and adsorption onto particle surfaces in place of previously adsorbed water molecules may occur (van Olphen, 1963). Although clay–organic reactions have not been studied extensively by engineers, they are of importance in the technology of drilling fluids, in stability control of clay suspensions, in the manufacture of lubricants as flocculation aids, in soil stabilization, and in soil conditioners for agricultural purposes.

7.10 STABILITY OF ADSORBED ION COMPLEXES ON CLAYS

Ion exchange reactions usually occur in aqueous environment; however, clays can also take ions from trace concentrations in solution of rather insoluble substances by exchange and adsorption reactions, even when little water is present. The deterioration of clay samples after prolonged storage in Shelby tubes is a good example of this. The process, sometimes termed galvanic action, involves adsorption by the clay minerals of iron ions as soon as they pass into solution from the tube. Because the clay takes up the iron ions quickly there remains a tendency for further solution

of the metal tube, so the process continues. The result is that after some weeks or months the clay in contact with the sample tube is altered, and no longer representative of the *in situ* material. Hence, some organizations now use stainless steel, brass, or tube liners to minimize this corrosive process.

Other examples may be cited of how a clay of one ionic form may alter to another over relatively short time periods, depending on the disequilibrium with the environment. The results of a number of studies have been published in which a clay was reported as a "hydrogen clay," in that special treatments had been used in an attempt to make the clay homoionic in hydrogen. The behavior of such clays was generally observed to be characteristic not of a clay containing monovalent adsorbed cations, but more nearly like that of a material with di- or trivalent cations. Subsequent work has shown that after preparation of "hydrogen clays," aluminum ions move from octahedral lattice positions and displace the hydrogen ions from the exchange sites, producing an aluminum clay.

Findings of this type underscore difficulties in attempting to study homoionic clay systems. They also raise questions relative to proper interpretation of clay behavior under changed environmental conditions. The selectivity of clay surfaces for different ions in mixed ion systems is temperature-dependent. Bischoff, Greer, and Luistro (1970) found that the composition of the interstitial waters of a clayey marine sediment was altered as a result of a change in temperature from 5°C at the ocean bottom to 22.5°C in the laboratory, as shown in Fig. 7.18. As the free pore water concentrations of potassium and chlorine were increased by 13 and 1.4 percent, respectively, whereas magnesium and calcium were depleted by 2.5 and 4.9 percent, it can be inferred that the proportions of these constituents in the exchange complex were changed in the opposite direction.

Fanning and Pilson (1971) found that the interstitial silica concentration in water squeezed from a marine sediment was 51 percent higher after warming to a temperature 20°C greater than the *in situ* value, and the pH was also slightly higher. It would appear, therefore, that until the effects of temperature change on the chemistry of fine-grained soils are better understood, testing at *in situ* temperature would be desirable.

7.11 THEORIES OF ION EXCHANGE

A number of theories have been proposed for the quantitative description of the equilibrium concen-

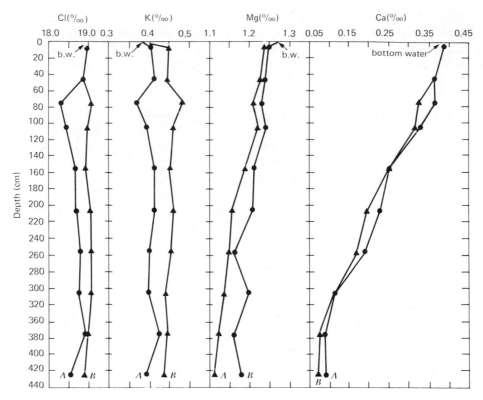

Fig. 7.18 Effect of temperature change on composition of interstitial water of a marine sediment (Bischoff, Greer, and Luistro, 1970). Relationship of depth to concentrations (in parts per thousand) of chloride, potassium, magnesium, and calcium in interstitial water from the San Pedro basin; *A* refers to samples squeezed at *in situ* temperature 5.0°C), and *B* refers to those equilibrated to laboratory temperature (22.5°C) for 1 hr prior to squeezing. Core location, 33°28′12″N, 118°20′09″W; water depth, 950 m.

trations of different cations in the adsorbed layer in terms of concentrations in the bulk solution. Approaches to the problem have included application of the mass law, kinetic theory, Donnan theory, and double layer theory. None has been completely satisfactory because of the great complexity of the system and the large number of variables involved.

Approaches based on double layer theory (Bolt, 1955*b*; van Olphen, 1963; Bolt and Page, 1965) and on the mass law appear to have given the best results. In the case of mixed cations of the same valence, double layer theory predicts the ratio of counterion concentrations in the double layer to be the same as in the equilibrium solution. Because of differences in ion size and specific interaction energies, however, substantial deviations may exist.

From double layer theory, it may be derived (Collis-George and Bozeman, 1970) that in monovalent–divalent systems the ratio of divalent to monovalent

cations is much higher in the double layer than in the equilibrium solution. This accords well with experimental observations. For practical purposes the *Gapon equation* is useful for assessing the proportions of monovalent and divalent ions, except in highly acid soils.

If the subscript *s* refers to the exchange complex of the soil, the subscript *e* refers to the equilibrium solution, *M* and *N* are monovalent ion concentrations, and *P* refers to the concentration of divalent ions, then

$$\left(\frac{M^+}{N^+}\right)_s = k_1 \left(\frac{M^+}{N^+}\right)_e \qquad (7.33)$$

$$\left(\frac{M^+}{P^{2+}}\right)_s = k_2 \left[\frac{M^+}{(P^{2+})^{1/2}}\right]_e \qquad (7.34)$$

where k_1 and k_2 are selectivity constants. The relationships given by equations (7.33) and (7.34) are incorporated in the statement of *Schofield's ratio law,* which is: A change in bulk solution concentration will not change the adsorbed ion equilibrium if the concentration of monovalent ions is changed in one ratio, the concentration of divalent ions in the square of that ratio, and the concentration of trivalent ions in the cube of that ratio.

As an example, if the proportions of Na^+ and Ca^{2+} in the adsorbed complex are equal, and the concentration of Na^+ in the free solution is doubled, then the concentration of Ca^{2+} in the free solution must be quadrupled if the proportions of adsorbed ions are to remain the same.

A useful practical form of the Gapon equation is

$$\left(\frac{Na^+}{Ca^{2+} + Mg^{2+}}\right)_s = k\left[\frac{Na^+}{[(Ca^{2+} + Mg^{2+})/2]^{1/2}}\right]_e \quad (7.35)$$

where concentrations are in milliequivalents per liter. The quantity

$$\left[\frac{Na^+}{[(Ca^{2+} + Mg^{2+})/2]^{1/2}}\right]_e = SAR \ (meq/liter)^{1/2} \quad (7.36)$$

is termed the *sodium adsorption ratio* (SAR). It can be determined by chemical analysis of the pore water (saturation extract). The selectivity constant k has been found to have a value of 0.017 for a wide range of soils. Thus, for these soils, if the composition of the pore fluid is known, the relative amounts of monovalent and divalent ions in the adsorbed cation complex can be estimated.

The proportion of sodium in the adsorbed layer has an important bearing on the structural status of a soil, and is often described in terms of the *exchangeable sodium percentage* (ESP), defined by

$$ESP = \frac{(Na^+)_s}{\text{total exchange capacity}} \times 100 \quad (7.37)$$

The results of a number of studies (e.g., Richards, 1954; Aitchison and Wood, 1965; Ingles, 1968; Ingles and Aitchison, 1969; Sherard, Decker, and Ryker, 1972a; and Arulanandan et al., 1973) have indicated that the ESP and the SAR provide a good indication of the stability of clay soil structure to breakdown and particle dispersion, at least in the case of non-marine clays. Soils with ESP greater than about 2 percent are susceptible to spontaneous dispersion in water. Both acid and alkaline soils may be dispersive, as well as many soils with very high porewater salt contents, up to and exceeding 100 meq./liter.

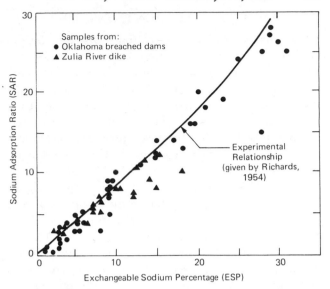

Fig. 7.19 Correlation between exchangeable sodium percentage and sodium adsorption ratio (Sherard, Decker, and Ryker, 1972).

For most soils the SAR and ESP are uniquely related, as may be seen by manipulation of equations (7.36) and (7.37), and from the test data (Sherard, Decker, and Ryker, 1972a, and Richards, 1954) shown in Fig. 7.19. Since the SAR is more easily determined than is the ESP, it is more widely used in practice.

7.12 ANION EXCHANGE

Clays may contain anion exchange sites, although little is known about them. Certainly, in low pH systems where particle edges are charged positively, anion adsorption would be expected. Tannates are widely used to improve the stability of drilling fluids (van Olphen, 1963). The tannate anions are adsorbed at particle edges by complexing with the exposed octahedral aluminum ions. This leads to a reversal in edge charge from positive to negative and prevents edge-to-face flocculation.

Sodium polyphosphates are effective soil dispersants, which have had application in soil engineering (Lambe, 1954b) for the construction of high density, low permeability clay blankets for use as storage reservoir liners. A major factor controlling the effectiveness of these materials is the adsorption of the negative phosphate radical on particle edges.

Anions, particularly bicarbonate, appear to have played a role in erosion (by dispersion) in Australia (Ingles, 1972) in soils of low to moderate sodium content.

7.13 PRACTICAL IMPLICATIONS

The analysis of clays as colloids provides insight into long-range repulsive and attractive forces and the influences of system variables, such as pore solution chemistry, type of adsorbed cations, and type of clay on them. The theory accounts for flocculation and deflocculation phenomena, both of which play an important part in the development of soil fabric and structure, which in turn exert a major influence on mechanical properties (Chapters 11 and 12). Long range repulsive forces caused by overlapping double layers have been used to describe the compression and swelling behavior of some clays (Chapter 13), and dispersion phenomena due to changes in the relative proportions of monovalent and divalent cations in the double layer can be used to account for erosion of clays and tunneling failures in dams (Chapter 11).

Some methods of soil stabilization depend, at least in part, on changes in double layer interactions and ion exchange to alter interparticle forces and particle arrangements in order to achieve a soil structure with properties suited for a particular application. One example is the use of dispersing chemicals to produce a high density and deflocculated particle arrangements to insure a very low permeability. Another is the use of a flocculating chemical to give a higher strength or more open particle arrangement to facilitate drainage.

Ion exchange reactions are important in the properties and stability control of drilling muds and slurries. Groundwater quality, underground waste disposal, and pollutant transport through soils are all influenced by the ion exchange properties of the soils. As the composition of the exchange complex relative to that of the pore water is influenced by both temperature and composition of the free water and as mechanical properties are influenced by the adsorbed ion types, suitable controls on laboratory testing conditions are needed if properties are to be correctly measured.

SUGGESTIONS FOR FURTHER STUDY

Ingles, O. G. (1968), "Soil Chemistry Relevant to Engineering Behavior of Soils," Chapter 1 in *Soil Mechanics —Selected Topics* (I. K. Lee, Ed.,) American Elsevier, New York.

van Olphen, H. (1963), *An Introduction to Clay Colloid Chemistry, Wiley Interscience,* New York.

Verwey, E. J. W., and Overbeek, J. Th. G. (1948), *Theory of the Stability of Lyphobic Colloids,* Elsevier, Amsterdam, New York, London.

CHAPTER 8

Soil Fabric and Its Measurement

8.1 INTRODUCTION

Although soils are composed of discrete soil particles and particle groups, a soil mass is almost always treated as a continuum for purposes of analysis and design. This is not surprising, because adequate theories of particulate mechanics are not available for direct characterization of properties such as strength, permeability, compressibility, and deformation modulus. Nonetheless, the values of properties that are chosen for use in continuum theories of soil mechanics are controlled directly by the particle characteristics, their arrangements, and the forces between them. Thus, to *understand* a property requires a consideration of these factors.

The particle arrangements in soils remained largely unknown until development of suitable optical, X-ray diffraction, and electron microscope techniques made direct observations possible starting in about the mid 1950's. Then interest centered mainly on clay particle arrangements and their relationships to mechanical properties. In the late 1960's, research accelerated significantly, sparked by improved techniques of sample preparation and study, with a major contribution arising from the development of the scanning electron microscope.

Starting in about 1970, attention was also directed at particle arrangements in cohesionless soils, which had not been much studied by engineers previously. From this work has come a realization that characterization of the properties of sands and gravels cannot be done in terms of density or relative density alone, as had once been thought. Particle arrangements and stress history must be considered as well.

The term "fabric" refers to the arrangement of particles, particle groups, and pore spaces in a soil. The term "structure" is used by some interchangeably with the term "fabric." Herein, however, the term "structure" is taken to have the broader meaning of the combined effects of fabric, composition, and interparticle forces. Emphasis is on the *microfabric,* which is the level of fabric requiring at least an optical microscope for study.

Macrofabric, that is, those features that can be seen with the unaided eye or a hand lens, is also of great importance. Some macrofabric features, such as stratification, fissuring, voids, and large scale inhomogeneities, may be determinative in an analysis of stability, settlement, or seepage (Rowe, 1972). The search for such features is of major concern in the soil investigation for any project.

8.2 DEFINITIONS OF FABRICS AND FABRIC ELEMENTS

As a result of both the large number of fabric studies in recent years and the variety and complexity of fabrics that have been observed, there has been a proliferation of terms for the description of fabrics and fabric features. It is convenient to describe fabric features in terms of the form and function of the units.

Particle associations in clay suspensions

Particle associations in clay suspensions can be described as follows (van Olphen, 1963):

1. *Dispersed.* No face-to-face association of clay particles.
2. *Aggregated.* Face-to-face association of several clay particles.
3. *Flocculated.* Edge-to-edge or edge-to-face association of aggregates.
4. *Deflocculated.* No association between aggregates.

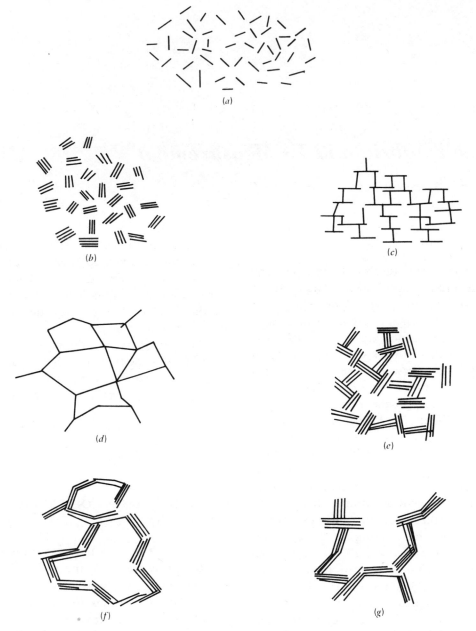

Fig. 8.1 Modes of particle association in clay suspensions (After van Olphen, 1963). (*a*) Dispersed and deflocculated. (*b*) Aggregated but deflocculated. (*c*) Edge-to-face flocculated but dispersed. (*d*) Edge-to-edge flocculated but dispersed. (*e*) Edge-to-face flocculated and aggregated. (*f*) Edge-to-edge flocculated and aggregated. (*g*) Edge-to-face and edge-to-edge flocculated and aggregated.

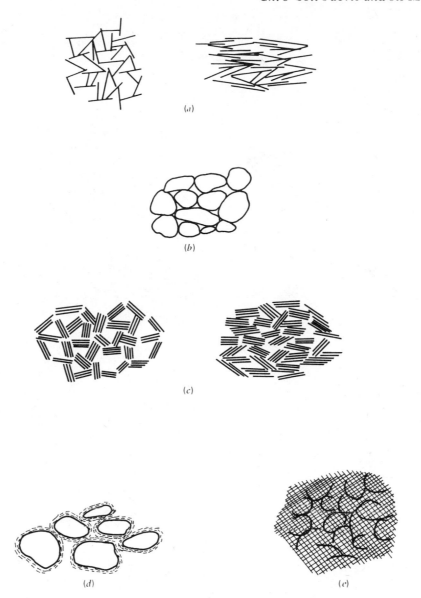

Fig. 8.2 Schematic representations of elementary particle arrangements (Collins and McGown, 1974). (*a*) Individual clay platelet interaction. (*b*) Individual silt or sand particle interaction. (*c*) Clay platelet group interaction. (*d*) Clothed silt or sand particle interaction. (*e*) Partly discernable particle interaction.

Various modes of association and the appropriate terminology are shown in Fig. 8.1. Evidence for the edge-to-face flocculated but dispersed fabric from the results of light scattering measurements on dilute clay suspensions is available (Schweitzer and Jennings, 1971). A similar fabric has been observed in a Norwegian quick clay (Rosenqvist, 1959). In most sediments and other soils of engineering interest, however, individual particle associations are quite rare, and aggregates of several clay plates are the more usual fabric forming units.

Particle associations in soils

Particle associations in sediments, residual soils, and compacted clays assume a variety of forms, but most of them are related to the configurations shown in Fig. 8.1 and reflect the difference in water content between a suspension and a denser soil mass. Three

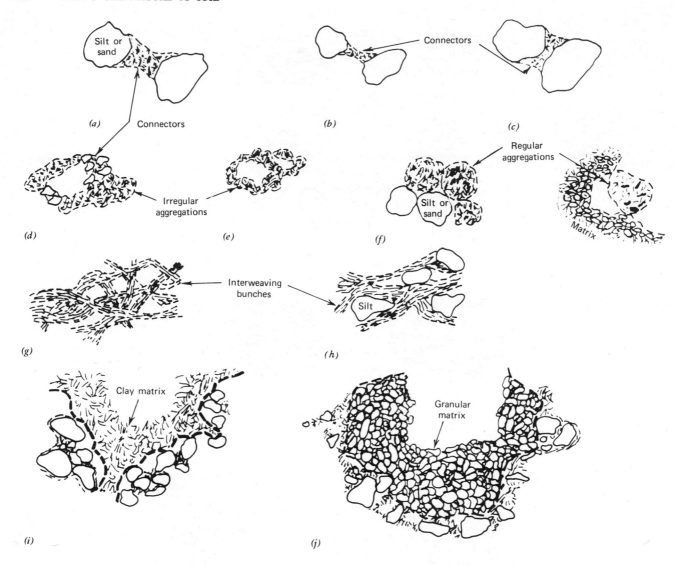

Fig. 8.3   Schematic representations of particle assemblages (Collins and McGown, 1974). (*a*) Connectors. (*b*) Connectors. (*c*) Connectors. (*d*) Irregular aggregations by connector assemblages. (*e*) Irregular aggregations in a honeycomb. (*f*) Regular aggregation interacting with particle matrix. (*g*) Interweaving bunches of clay. (*h*) Interweaving bunches of clay with silt inclusions. (*i*) Clay particle matrix. (*j*) Granular particle matrix.

main groupings may be identified (Collins and McGown, 1974):

1. **Elementary particle arrangements.** Single forms of particle interaction at the level of individual clay, silt, or sand particles.
2. **Particle assemblages.** Units of particle organization having definable physical boundaries and a specific mechanical function, and which consist of one or more forms of Elementary Particle Arrangements or smaller particle assemblages.

3. **Pore spaces.**

Schematic representations of the fabric features in each of these three classes are shown in Figs. 8.2 through 8.4. Electron photomicrographs illustrating some of the features are shown in Fig. 8.5. Figure 8.6 shows the overall microfabric of undisturbed Tucson silty clay, a fresh water alluvial deposit.

The features shown in Figs. 8.1 through 8.6 should be sufficient to describe most fabrics. A number of additional terms have been suggested. Those listed

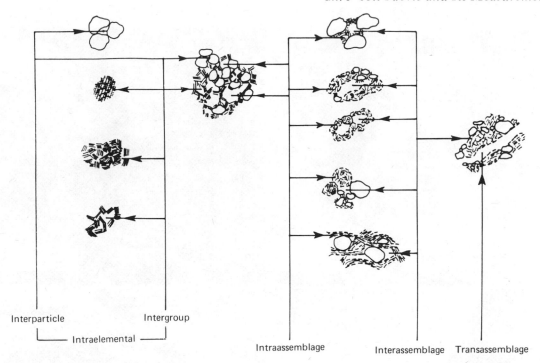

Interparticle   Intergroup

Intraelemental

Intraassemblage   Interassemblage   Transassemblage

Fig. 8.4   Schematic representation of pore space types (Collins and McGown, 1974).

below pertain mainly to monomineralic clay fabrics. *Cardhouse* (Goldschmidt, 1926) is an edge-to-face arrangement forming an open fabric similar to the edge-to-face flocculated but dispersed arrangement of Fig. 8.1c. *Domain* (Aylmore and Quirk, 1960, 1962) is an aggregate of parallel clay plates. An array of such aggregates is termed a *turbostratic fabric* and is similar to the interweaving bunches of Fig. 8.3h. *Packets or books* (Sloane and Kell, 1966) are fabric units consisting of parallel clay plates that is, a domain. An edge-to-face association of packets or books is termed a *bookhouse*. This arrangement is similar to the edge-to-face flocculated and aggregated arrangement of Fig. 8.1e. *Stair-stepped cardhouse* (O'Brien, 1971) is a three-dimensional network of twisted chains of clay platelets having a stepped face-to-face interaction. *Cluster* (Olsen, 1962; Yong and Sheeran, 1972) is a grouping of particles or aggregates into larger fabric units. *Ultramicroblocks* (Bochko, 1973) are face-to-face associations of clay plates. *Microblocks* are assemblages of oriented ultramicroblocks. *Micro-aggregates* are accumulations of ultramicroblocks and small microblocks without definite orientation.

Terms that have been used mainly for the description of soils containing more than one mineral and a range of particle sizes include the following: *Aggre-*

gation (Pusch, 1973a) is a fabric unit within which individual clay particles of various sizes interact. Aggregations correspond to the clay platelet group interactions of Fig. 8.2. *Ped* (Brewer, 1964) is "an individual soil aggregate consisting of a cluster of primary particles, and separated from adjoining peds by surfaces of weaknesses." *Intrapedal pores* are pores within peds and *interpedal pores* occur between peds. *Transpedal pores* refer to pores that transverse the soil beyond the limits of a single ped, but do not have specific relationships to peds. *Aggregated clay grains* (Dudley, 1970) are sand grains with surface coatings of clay particles in closely packed parallel arrangement. *Ring buttresses* are formed by sand grains buttressed by clay particles in the interparticle contact zone. *Particle intergrowth in domains* (Smart, 1971) are particle domains wherein the particles are intergrowing and in intimate contact.

8.3   SINGLE GRAIN FABRICS

Although in clays the smallest fabric element is the clay plate group or aggregate, the particles in sand are sufficiently large and bulky that they generally behave as independent units. Attempts have been made, for example, Deresiewicz (1958), Rowe (1962, 1973), Horne (1965), to describe the stress-deformation

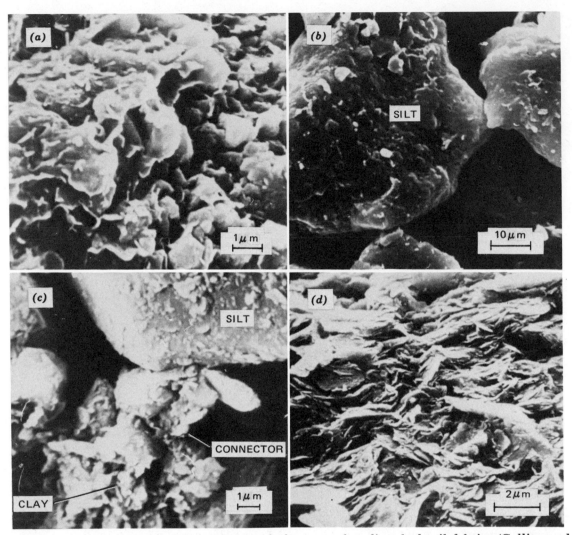

Fig. 8.5 Scanning electron photomicrograph features of undisturbed soil fabrics (Collins and McGown, 1974). (*a*) Partly discernable particle systems in Lydda silty clay, Israel (freshwater alluvial deposit). (*b*) Grain–grain contacts in Ford silty loess, England (aeolian deposit). (*c*) Connector assemblages in Breidmerkur silty till, Iceland (glacial ablation deposit). (*d*) Particle matrix assemblage in Immingham silty clay, England (estuarine deposit). (*e*) Regular aggregation assemblage in Holon silty clay, Israel (consisting of elementary particle arrangements interacting with each other and silt) (freshwater alluvial deposit). (*f*) Interweaving bunch assemblage in Hurlford organic silty clay Scotland (freshwater lacustrine deposit). (*g*) Irregular aggregation assemblage in Sundland silty clay, Norway (marine deposit).

behavior of granular soils using particulate mechanics theories. Some of these theories are based on elastic distortion of particles and the sliding and rolling of particles in regular packings of equal-sized spheres. In real granular soils the irregular particle shapes and distribution of sizes invalidate the assumption of uniform spheres, and packings are usually far from regular. Nonetheless, the theories do provide some

insight into behavior, and a knowledge of the characteristics of ideal systems can be useful for interpreting data on real soils.

Packings of equal-sized spheres

Analysis of the possible regular packings of spheres of the same size provides insight into the maximum and minimum possible densities, porosities, and void

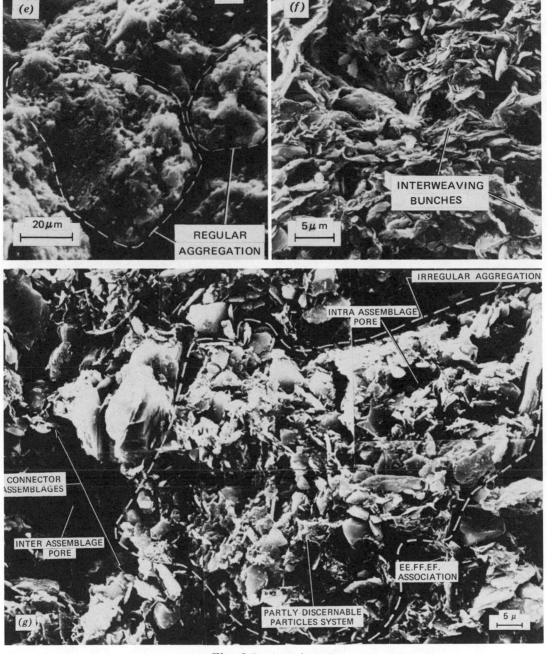

Fig. 8.5 *(continued)*

ratios that are likely in single grain fabrics. Figure 8.7 shows five different possible packings, and properties of these arrangements are listed in Table 8.1. The range of possible porosities is from 25.95 to 47.64 percent, and the range of void ratios is from 0.34 to 0.91.

Random packings of equal spheres can be con-

sidered to be composed of clusters of simple packings, each present in appropriate proportion to give the observed porosity. The relationship between coordination number N and porosity n in such systems is

$$N = 26.486 - \frac{10.726}{n} \qquad (8.1)$$

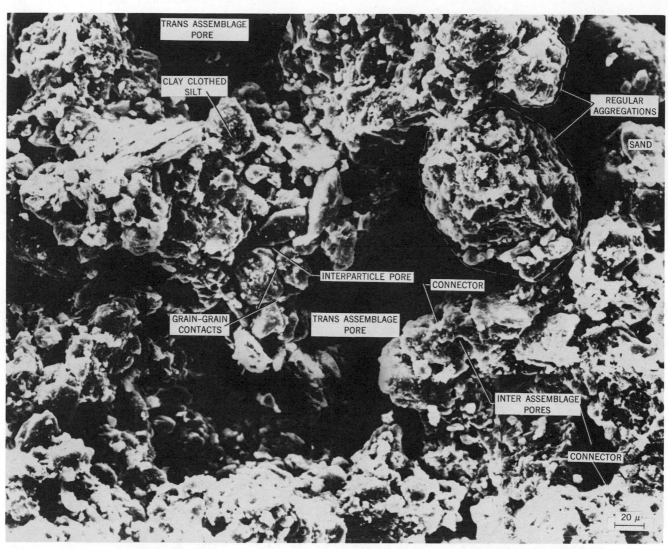

Fig. 8.6 Overall microfabric in Tucson silty clay, U.S.A.—(freshwater alluvial deposit) (Collins and McGown, 1974).

Table 8.1 Properties of Ideal Packings of Uniformly Sized Spheres

Type of Packing	Coordination Number	Layer Spacing (R = radius)	Volume of Unit	Porosity (%)	Void Ratio
Simple Cubic	6	$2R$	$8R^3$	47.64	0.91
Cubical-tetrahedral	8	$2R$	$4\sqrt{3}\,R^3$	39.54	0.65
Tetragonal-sphenoidal	10	$R\sqrt{3}$	$6R^3$	30.19	0.43
Pyramidal	12	$R\sqrt{2}$	$4\sqrt{2}\,R^3$	25.95	0.34
Tetrahedral	12	$2R\sqrt{2/3}$	$4\sqrt{2}\,R^3$	25.95	0.34

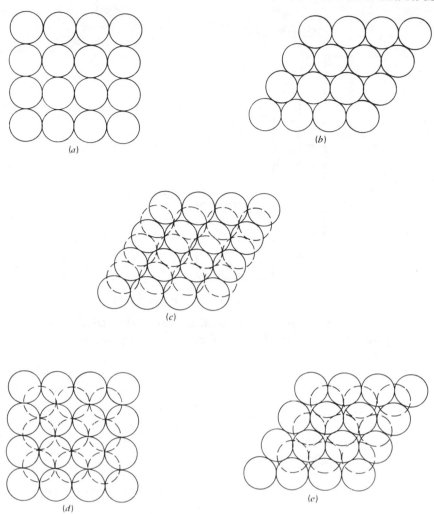

Fig. 8.7 Ideal packings of uniform spheres. (*a*) Simple cubic. (*b*) Cubical tetrahedral. (*c*) Tetragonal sphenoidal. (*d*) Pyramidal. (*e*) Tetrahedral.

Glass balls allowed to fall freely form an anisotropic assembly, with balls tending to arrange in chains (Kallstenius and Bergau, 1961). The number of balls per unit area in contact with a vertical plane can be different from the number in contact with a horizontal area.

Packings in granular soils

In soils all particles are not the same size, and as a result, smaller particles can occupy pore spaces between larger particles. This leads to a tendency towards higher densities and lower porosities and void ratios than for the case of uniform spheres. On the other hand, irregular particle shapes result in a tendency towards lower densities and higher porosities and void ratios. The net result is that the range of porosities and void ratios encountered in real soils with single grain fabrics is not too different than that for uniform spheres. The difference between the maximum and minimum values for a given soil generally is considerably less than the full range shown in Table 8.1, however.

Many recent studies have shown that a given cohesionless soil may have different fabrics at the same void ratio or relative density. Characterization of the fabric of a cohesionless soil can be done in terms of

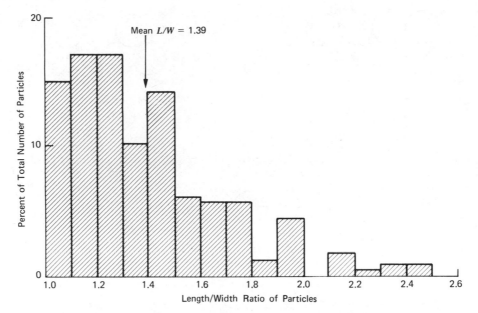

Fig. 8.8   Grain shape distribution of Monterey no. 0 sand. Results based on study of 277 particles $d_{50} = 0.43$ mm, $C_\mu = 1.4$ (Mahmood, 1973).

grain shape factors, grain orientations, and interparticle contact orientations (Lafeber, 1966; Oda, 1972a; Mahmood, 1973).

Most of the nonclay minerals in soils, with the exception of mica, occur as bulky particles when present in silt size or larger.* Most of these particles are not equidimensional, however, and are at least slightly elongate or tabular. Figure 8.8 shows a frequency histogram of the particle length to width ratio (L/W) for Monterey No. 0 sand. This sand, which is composed mainly of quartz with some feldspar, is a well-sorted beach sand with rounded particles. It may be seen that fewer than 50 percent of the particles have L/W ratios less than 1.39. This distribution is typical of that determined for several different sands and silty sands.

The orientation of grains in a sand can be described in terms of the inclination of the particle axes to a set of reference axes. For example the orientation of the particle shown in Fig. 8.9 can be expressed in terms of the angles α and β. In most studies, however, a thin section is studied to give the orientations of apparent long axes.† In this case the long axis of par-

ticles can be referred to a single reference axis by angle θ as shown in Fig. 8.10. The spacial orientation of the thin section itself with respect to the sample and field deposit is also an essential part of the fabric description.

The results of measurement of the orientations of long axes for a large number of grains can be ex-

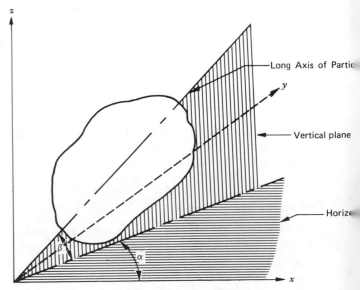

Fig. 8.9   Three-dimensional orientation of a sand particle.

---

* Krinsley and Smalley (1973) present evidence that quartz particles become flatter with decreasing size, and may develop a platy morphology when subdivided to a fineness approaching clay size.

† The method underestimates the value of L/W for elongate particles whose long axis is out of the plane of the thin section.

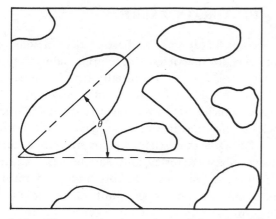

Fig. 8.10 Characterization of particle orientation in thin section.

For these histograms the orientation of each grain was assigned to one of the 15-degree intervals between 0 and 180 degrees. Evaluation of orientation is facilitated by use of a photographic enlargement of the thin section. The V-section refers to a thin section from a vertical plane (oriented parallel to the cylinder axis). The H-section refers to orientations in the horizontal plane.

The orientations of long axes in the vertical plane of two samples of a well-graded crushed basalt (mean $L/W = 1.64$) are shown by the rose diagrams in Figs. 8.12 and 8.13 (Mahmood, 1973). In this study, the orientations of at least 400 grains were measured for each sample, and the orientation of each was assigned to one of the eighteen 10-degree intervals between 0 and 180 degrees. A completely random particle distribution would yield the dashed circles shown in the figures. Figure 8.12 shows strong preferred orientation in the horizontal direction for the sample prepared by pouring. Dynamic compaction, however, resulted in a more nearly random fabric (Fig. 8.13).

pressed by histogram or rose diagram. Figure 8.11 shows a frequency histogram for a sand (with mean axial ratio equal to 1.65) that was placed by tapping the side of a vertical, cylindrical mold (Oda, 1972a).

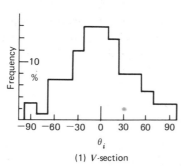
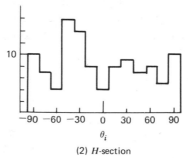

(1) *V*-section   (2) *H*-section

Fig. 8.11   Frequency histogram of long particle axis orientations in two planes for a uniform fine sand (Oda, 1972a).

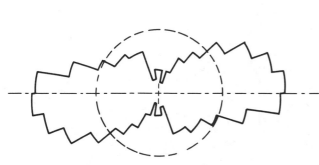

Fig. 8.12 Particle orientation diagram for crushed basalt. Vertical section in sample prepared by pouring. Density was 1.60 gm/cm³ and the relative density was 62 percent.

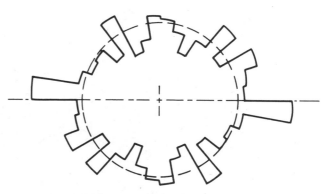

Fig. 8.13 Particle orientation diagram for crushed basalt. Vertical section in a dynamically compacted specimen. Density was 1.84 gm/cm³ and the relative density was 90 percent.

The distribution of interparticle contact orientations is not accounted for by grain orientation alone. These orientations can be described in terms of a perpendicular N_i to the tangent plane at the point of contact. The orientation of N_i is defined by angles α' and β' as shown in Fig. 8.14. A procedure for determination of the angular distributions of normals $E(\alpha', \beta')$ as a function of α' and β' is given by Oda (1972a). For a fabric with axial symmetry around the vertical axis Z, the function $E(\alpha', \beta')$ becomes independent of α, and so the distribution of $E(\beta')$ as a function of β' can be used to characterize the distribution of interparticle contact normals. Such distributions for four sands placed into a cylindrical mold by tapping are shown in Fig. 8.15. The horizontal dashed line represents the distribution for an isotropic fabric. In each case, there is a greater proportion of contact plane normals in the near-vertical direction; that is, there is a preferred orientation of contact planes towards the horizontal.

Other methods for characterizing particle packings in granular soils are possible. For example, Marsal (1973) has analyzed the mechanical properties of rockfill using considerations of grain geometry, grain concentrations, number of contacts per particle, and contact force distributions.

8.4 MULTIGRAIN FABRICS

In Section 8.2, it was emphasized that in soils containing clay size particles, single grain fabrics are exceedingly rare. This is also usually true in the case of silts (particle sizes in the range of 2 to 74 μm). For example, experiments have shown that silt-size quartz particles sedimented in water can have a void ratio as high as 2.2. Quartz particles in this size range can be somewhat platy (Krinsley and Smalley, 1973), which would account for a part of this high void ratio in comparison with an upper limit of about 1.0 for single-grain assemblages of bulky particles. It is also probable that multigrain arrangements form in such materials during slow sedimentation, because particles in this size range are sufficiently small that behavior can be to some extent influenced by surface forces. A "honeycomb" type of arrangement as shown schematically in Fig. 8.16 is thought to exist in some silts (Terzaghi, 1925a). Loose fabrics such as this are usually metastable and may be subject to sudden collapse or liquefaction under the action of rapidly applied stresses.

Multigrain fabrics involving clays and clay–nonclay mixtures form because of the importance of particle surface forces in relation to particle weight; the affi-

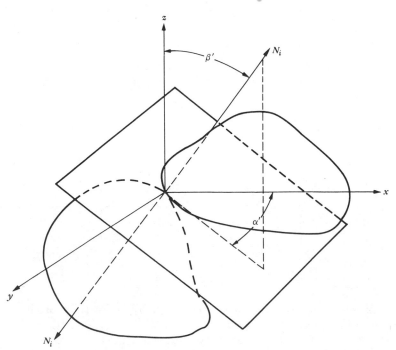

Fig. 8.14   Characterization of contact orientation.

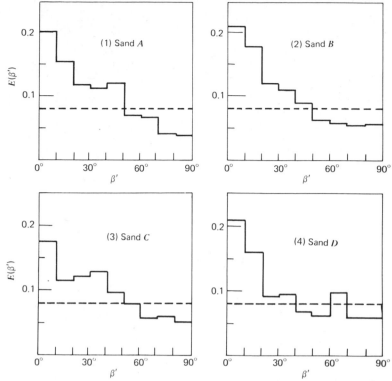

Fig. 8.15 Distribution of interparticle contact normals for four sands (Oda, 1972a). The dotted lines represent theoretical lines for an isotropic assembly. The solid line that shows the distribution of N_i of sand A deviates from the dotted line as follows: $0 \leq \beta < 50$, or $130 < \beta \leq 180 \ldots E(\beta) \geq \frac{1}{4}\pi$ and $50 \leq \beta \leq 90$, or $90 \leq \beta \leq 130 \ldots E(\beta) < \frac{1}{4}\pi$.

nity of nonclay particle surfaces for adsorption of clays; and because of the chemical reactivity of clay surfaces. In addition, clay plate groups in many soils may represent remnants of a preexisting sedimentary rock from which the soil was derived.

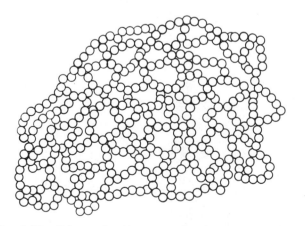

Fig. 8.16 Schematic diagram of a honeycomb fabric in a silt.

8.5 FABRIC DETERMINATION IN FINE-GRAINED SOILS

A variety of methods, both direct and indirect, has been used to study the fabric and fabric features in soils, as listed in Table 8.2. Figure 8.17 is an illustrative schematic diagram prepared by Professor R. N. Yong that summarizes methods for analysis of soil composition and fabric that utilize various parts of the electromagnetic spectrum.

Of the methods listed in Table 8.2, optical and electron microscopy, X-ray diffraction, and pore size distribution offer the advantages of providing direct, (usually) unambiguous data on specific fabric features, provided the samples studied are representative and the sample preparation method has not destroyed the original fabric. On the other hand, these techniques are limited to the study of small samples, and they are destructive of the samples examined. While the other techniques are nondestructive and can, at least in principle, be used for the study of soil fabric *in situ*

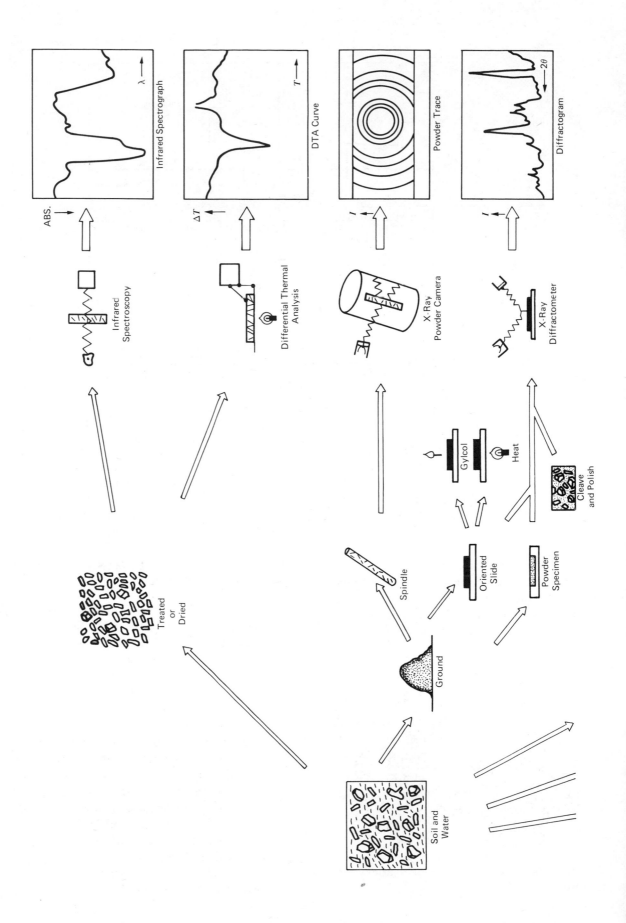

Infrared Spectrograph

$\lambda \longrightarrow$

ABS. $\longrightarrow$

Infrared Spectroscopy

DTA Curve

$T \longrightarrow$

$\Delta T \longleftarrow$

Differential Thermal Analysis

Powder Trace

X-Ray Powder Camera

Diffractogram

2θ

X-Ray Diffractometer

Gylcol

Heat

Cleave and Polish

Treated or Dried

Spindle

Oriented Slide

Powder Specimen

Ground

Soil and Water

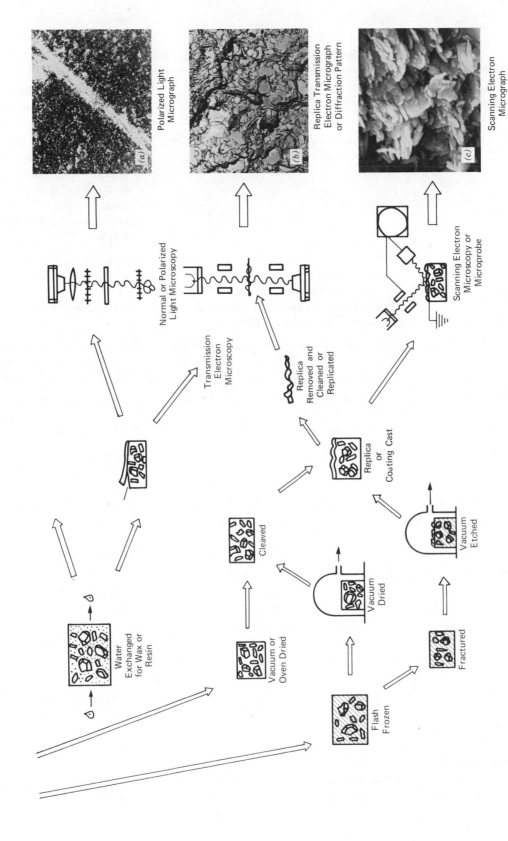

Fig. 8.17 Methods of examining mineralogy, fabric, and structure of soils using parts of the electromagnetic spectrum. (Prepared by R. N. Yong: McGill University Soil Mechanics Laboratory.)

Table 8.2 Techniques for Study of Soil Fabric

Method	Basis	Scale of Observations and Features Discernable
Optical Microscope (Polarizing)	Direct observation of fracture surfaces or thin sections	Individual particles of silt size and larger, clay particle groups, preferred orientation of clay, homogeneity on a millimeter scale or larger, large pores, shear zones Useful upper limit of magnification about 300 ×
Electron Microscope	Direct observation of particles or fracture surfaces through soil sample (scanning electron microscope—SEM) observation of surface replicas (transmission electron microscope—TEM)	Resolution to about 100 Å; Large depth of field with SEM; direct observation of particles; particle groups and pore space; details of microfabric
X-Ray Diffraction	Groups of parallel clay plates produce stronger diffraction than randomly oriented plates	Orientation in zones several square millimeters in area and several micrometers thick; best in single mineral clays
Pore Size Distribution	(1) Forced intrusion of nonwetting fluid (usually mercury) (2) Capillary condensation	(1) Pores in range from $\sim$0.01 to $>$10 μm (2) 0.1 μm maximum
Acoustical Velocity	Particle alignment influences velocity	Anisotropy; measures microfabric averaged over a volume equal to sample size[a]
Dielectric Dispersion and Electrical Conductivity	Variation of dielectric constant and conductivity with frequency	Assessment of anisotropy; flocculation and deflocculation; measure microfabric averaged over a volume equal to sample size[a]
Thermal Conductivity	Particle orientations influence thermal conductivity	Anisotropy; measures microfabric averaged over a volume equal to sample size[a]
Magnetic Susceptibility	Variation in magnetic susceptibility with change of sample orientation relative to magnetic field	Anisotropy; measures microfabric averaged over a volume equal to sample size[a]
Mechanical Properties Strength Modulus Permeability Compressibility Shrinkage and swell	Properties reflect influences of fabric See Chapter 12	Microfabric averaged over a volume equal to sample size[a]; anisotropy; macrofabric features in some cases

[a] For a homogeneous sample. Discontinuities, stratification, and so on, on a macroscale can override effects of microfabric.

or for study of changes in fabric in samples subjected to shear, consolidation, and so on, interpretation of the data is seldom straightforward or conclusive. In most studies, only a single method for fabric determination has been used. It may be useful to use several methods together in some studies, because not only will it be possible to obtain information on different levels of detail (Tchalenko, 1968a; Gillott, 1970; McKyes and Yong, 1971) but also comparison of results may enable assessment of the reliability of data from each.

8.6 SAMPLE PREPARATION FOR FABRIC ANALYSIS

Acoustical, dielectric, thermal, and magnetic measurements can be made directly on samples in their undisturbed, wet state. Study using optical and electron microscopes, X-ray diffraction, or porosimetry requires that the pore fluid be removed or replaced. To do this without disturbance of the original fabric is difficult.

Pore fluid removal

Air drying may be suitable for stiff soils, partly saturated soils, and other soils that do not undergo significant shrinkage. For soft samples at high water content, oven drying may cause less change in fabric than air drying, evidently because the relatively long (24 hr) time needed for air drying allows for greater particle rearrangement (Tovey and Wong, 1973). On the other hand, the stresses induced during oven drying may result in some particle breakage.

Freeze drying has been used successfully. Small samples, of the order of 10 mm in diameter and 5 mm thick, are quickly frozen using liquid nitrogen or propane cooled by liquid nitrogen. It is desirable that freezing be done at temperatures less than $-130°C$ to avoid formation of crystalline ice. Sublimation of the frozen water is then carried out at temperatures between -50 and $-100°C$. Times ranging upward from 24 hr are required, with the time increasing with decreasing temperature. For temperatures less than $-100°C$ the vapor pressure of the ice ($\sim 10^{-5}$ Torr) may be less than the capability of the vacuum system. Considerations in the preparation of samples using freeze drying are discussed by Tovey and Wong (1973) and Tovey, Frydman, and Wong (1973).

Critical point drying involves raising the temperature and pressure of the sample above the critical values, which for water are $374°C$ and 217.7 atm, respectively. In the critical range, the liquid and vapor phases are indistinguishable, and the pore water may be distilled away without the presence of air–water interfaces that lead to shrinkage. The high critical temperature and pressure may change the clay particles, however. To avoid this, replacement by carbon dioxide (critical $T = 31.1°C$, critical $p = 71$ atm) has been used. This requires prior impregnation of the sample with acetone (Tovey and Wong, 1973), which may cause swelling in partly saturated and expansive soils.

Both critical point and freeze drying cause less disturbance to the sample than do air or oven drying. On the other hand, both are more difficult and time consuming.

Pore fluid replacement

In cases where thin sections are required, as for optical microscopy or where drying shrinkage must be minimized, but the presence of a material in pore spaces is not objectionable, replacement of the pore water may be required. Various resins and plastics have been used for this purpose. Carbowax 6000 is one of the most widely used materials. Carbowax is a high molecular weight ethylene glycol that is miscible with water in all proportions. It melts at 55°C but is solid at lower temperature.

Carbowax impregnated samples are prepared by immersing an undisturbed cube sample, 1 to 2 cm on a side, of the soil in melted carbowax at 60 to 65°C. The top surface of the specimen should be left exposed to vapor for the first day of immersion to allow escape of trapped gases and to prevent specimen rupture. The Carbowax should be changed after 2 or 3 days, to insure water-free wax in the pores of the sample. Replacement of the water by the Carbowax is usually complete in a few days. At the end of the replacement period, the impregnated sample is removed from the melted wax and allowed to cool.

Thin sections are prepared by grinding using emery cloth or abrasive powders and standard thin section techniques. Care must be taken that neither heat, water, nor other water-soluble liquids are used at any stage of the grinding or mounting process. Measurements using X-ray diffraction (Martin, 1966) have shown that Carbowax replacement of water has essentially no effect on the fabric of wet kaolinite.

Gelatins or water soluble resins may be used in lieu of Carbowax (Tovey 1973a), or the sample may be impregnated with methanol or acetone before replacement with resins or plastics such as Araldite AY 18, Durcupan, or Vestopal W. These latter materials appear best suited for preparation of ultrathin sections using a microtome; whereas, the Carbowax method is more used for preparation of normal thin sections for optical microscopy and for X-ray diffraction studies. A method for stabilization of the fabric for preparation of thin sections from moist organic soils is described by Langston and Lee (1965).

Preparation of surfaces for study

Care must be taken to insure that the surfaces chosen for study reflect the original fabric of the soil and not the preparation method. Grinding or cutting air-dried and Carbowax treated samples may result in substantial particle rearrangement at the surface, thus making them of little value for study using the electron microscope. Successive peels using adhesive tape from the surface of dried specimens will insure exposure of the original fabric (Barden and Sides, 1971). Alternatively, the surface may be coated with a resin solution that partly penetrates the sample. After hardening, the resin is peeled away revealing an undisturbed fabric (Tovey and Wong, 1973). A comparison between surfaces before and after treatment by this method is shown in Fig. 8.18.

The disturbed zone at the surface of Carbowax treated samples extends to a maximum depth of 1 μm in the case of kaolinite (Barden and Sides, 1971). As thin sections used for polarizing microscope study are of the order of 30 μm thick, this disturbed zone is of little consequence. This disturbance also is insignificant in the case of X-ray diffraction studies.

Fracture surfaces in dried specimens are sometimes studied as representative of the undisturbed fabric. Probably additional preparation, such as gentle blowing of the surface or peeling, is needed following fracture, because (1) there may be loose particles on the surface and (2) a fracture surface may be more representative of a plane of weakness than the sample as a whole.

Sections as thin as 300 to 500 Å for study in the transmission electron microscope can be obtained using a microtome (Pusch, 1966a, 1966b, 1970, 1973a). The presence of coarse particles may be the source of some disturbance in this method.

The method of sample preparation should be selected after consideration of scale of fabric features of interest, method of observation to be used, and the soil type and state as regards water content, strength, disturbance, and so on. With these factors in mind, the probable effects of the preparation methods on the fabric can be assessed.

8.7 FABRIC STUDY USING THE POLARIZING MICROSCOPE

In the case of thin sections and ground surfaces of soils composed of silt and sand-size particles, individual soil grains can be seen, and the sizes, orientations, and distributions of particles and pore spaces can be described systematically as discussed in Section 8.2. For a two-dimensional analysis of pore spaces, a

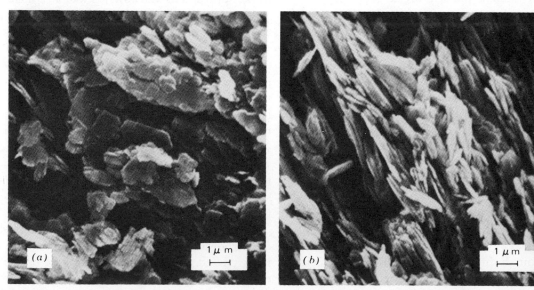

Fig. 8.18   Effect of surface preparation on fabric seen by the scanning electronic microscope (From Tovey and Wong, 1973). × 5000. (a) Before peeling. (b) After one peeling.

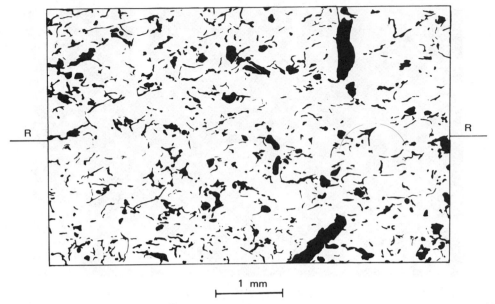

Fig. 8.19   Pore pattern (pores in black) of a section from a stony tableland soil from Woomera, Australia (From Lafeber, 1965).

thin section or polished surface can be used. Three-dimensional analysis requires a series of parallel cross sections (Lafeber, 1965). A photographic technique may aid in the identification of features of interest (Lafeber and Kurbanovic, 1965). Approximately 5 percent by weight of anthracene (a flourochrome) is added to Carbowax before replacement of the pore water in the soil. The pores then appear as bright bluish-white when viewed under bright ultraviolet light. Photographic prints are obtained in which the pores appear white, the clay matrix in black, and large grains, except for those cut perpendicular or nearly perpendicular to their optical axes also in black, but surrounded by a fine white line marking the particle boundaries.

Two-dimensional planar pore patterns can be represented by rose diagrams. Figure 8.19 shows the pore pattern in a section of a stony desert tableland soil from near Woomera, Australia, which suggests some degree of preferred orientation. Figure 8.20 shows the same cross section as Fig. 8.19, but here the pores are shown in white, the clay in gray, and the silt and sand grains in black. Figure 8.21 shows rose diagrams of pore orientation (white figure) and silt and sand grain orientation (black figure). There is considerable preferred orientation of both pores and particles.

Three-dimensional pore patterns are somewhat more difficult to characterize. A procedure for equal area stereo net projection that can be used for study and comparison of large amounts of data is given

by Lafeber (1965). Two- and three-dimensional representations of pore patterns are useful in that they reveal orientations and distributions not readily visible in the fabric cross section.

A technique was also developed by Lafeber (1967) for determination of the three-dimensional orientations of platy clay minerals and mineral aggregates in thin sections using a polarizing microscope, universal stage, and Berek compensator. This method is based on the geometrical relationships between the basal cleavage that is, cleavage along (001) planes, and the maximum or minimum apparent birefringence* in an arbitrary cross section of a platy mineral. By this method, it is unnecessary to know or to determine the thickness of thin section, refractive indices, or actual birefringence of the mineral.

It is not usually possible to see individual clay particles with the polarizing microscope, because of optical limitations in resolving power and depth of field. Practical resolution is to a few micrometers using magnifications up to about ×300. If, however, a group of clay plates are aligned parallel to each other in a face-to-face arrangement, then they behave optically as one large particle with definite optical properties.

The optical axes and the crystallographic axes of the clay minerals are almost coincident. For plate-shaped particles the refractive indices in the a and b

---

* The birefringence of a crystal is an optical property determined by the difference in refractive indices along different optical axes.

1 mm

Fig. 8.20   The same section as in Fig. 8.19, but showing pores in white, clay matrix in gray, and silt and sand grains in black (From Lafeber, 1965).

directions are approximately equal, but different from that in the c direction. For example, in kaolinite the refractive indices are 1.559 to 1.569, 1.560 to 1.570, and 1.553 to 1.563 in the a, b, and c directions, respectively. If a group of parallel particles is viewed in plane polarized light looking down the c-axis, a uniform field is observed as the group is rotated around the c-axis. If the same particle group is viewed with the c-axis normal to the direction of light, no light is transmitted when the basal planes are parallel to the direction of polarization, and a maximum is transmitted when they are at 45 degrees to it. Thus, there are four positions of extinction and illumination when the sample is viewed using light passed through crossed nicols and the microscope stage is rotated

through 360 degrees. For rod-shaped particles in parallel orientation, a uniform field is observed looking down the long axis; whereas, illumination and extinction are seen when looking normal to this axis. Use of a tint plate in the microscope is often helpful, because the resulting retardation of light waves results in distinct different colors for extinction and illumination.

If particle orientation is less than perfect, or if the c-axis direction of a group of parallel plates is other than normal to the direction of light, then the minimum intensity is finite and the maximum intensity is less than for perfect orientation. The ratio of minimum intensity I_{min} to maximum intensity I_{max} is called the birefringence ratio β.

Fig. 8.21   Cumulative lengths, in percent of total length, of elongated pores (white figure) and of elongated skeleton grains (black figure) vs. orientation direction in polar coordinates (5 degree intervals) for the pattern in Fig. 8.20. Note that for this figure the broken circle represents an even distribution of lengths over all directions (2.77 percent). s_1 and s_2 are the major maxima of the elongated pores, L_1 is the major maxima of elongated grains, and R is the reference direction. (From Lafeber, 1965).

Photometric measurements of the birefringence ratio can be used (Wu, 1960, and Morgenstern and Tchalenko, 1967a) to quantify clay orientation. Although there may be difficulties in photometric methods when dealing with other than monomineralic materials with singular orientations of particles (Lafeber, 1968), the semiquantitative scale proposed by Morgenstern and Tchalenko (1967c) given in Table 8.3 may be useful.

Table 8.3 Orientation Scale for Clay Aggregates Viewed in Plane Polarized Light

Birefringence Ratio	Particle Parallelism
1.0	Random
1.0–0.9	Slight
0.9–0.5	Medium
0.5–0.1	Strong
0	Perfect

Morgenstern and Tchalenko (1967c).

Figure 8.22 shows a thin section taken on a vertical plane through a varved clay. The upper half shows the winter-deposited clay varve, and the lower half, the summer-deposited silt varve. Strong preferred orientation of the clay is evident by comparison of illumination on the left and extinction on the right. Were the clay plates free of orientation, the thin section would have had the same appearance at both orientations. The upper portion of the silt varve is also seen to contain some zones of well-oriented clay. A series of planar pores are also visible, which probably were developed during impregnation of the sample or preparation of the thin section.

Optical microscope study of fabric provides a view of some features that are too small to be seen by eye, too large to be appreciated using electron microscopy, but important as regards soil behavior. Some features that may be of interest include distributions of silt and sand grains, silt and sand particle coatings, homogeneity of fabric and texture, discontinuities of various types, and shear planes (Mitchell, 1956; Morgen-

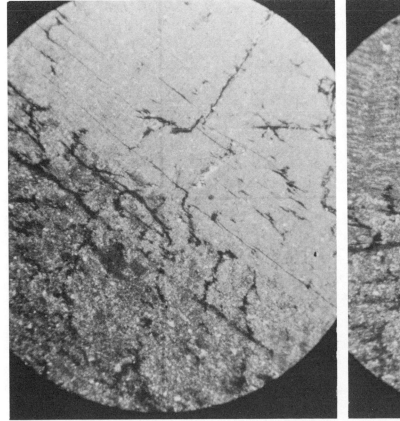

Fig. 8.22 A thin section of varved clay under polarized light (Courtesy of Division of Building Research, National Research Council, Canada).

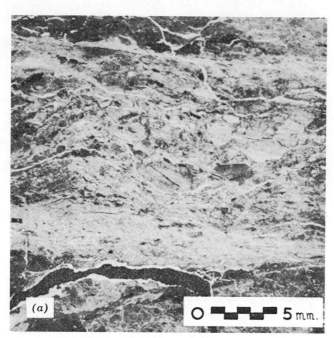

Fig. 8.23a Photograph of Fiddler's ferry shear zone (Morgenstern and Tchalenko, 1967).

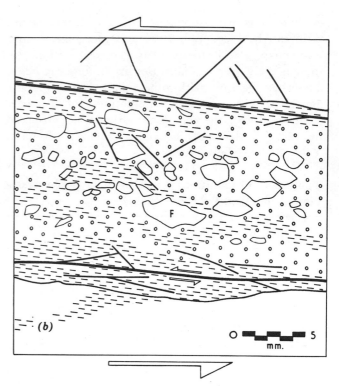

Fig. 8.23b Details of Fiddler's ferry shear zone (Morgenstern and Tchalenko, 1967). F is the fragment of ambient material; the hatched areas indicates the shear matrix where $\beta = 0.45$; and the direction of hatching is the average particle orientation over the stippled areas where $\beta = 1.00$.

stern and Tchalenko, 1967b, 1967c; McKyes and Yong, 1971). A thin section from a shear zone through a soft silty clay at the site of a foundation failure under an embankment on the flood plain of the Mersey River (England) is shown in Fig. 8.23. Details of the shear zone deduced from the photomicrograph (Morgenstern and Tchalenko, 1967c) are shown in Fig. 8.23b.

8.8 FABRIC STUDY USING THE ELECTRON MICROSCOPE

Of the techniques available, the electron microscope is the only one that can reveal particles and particle arrangements directly. The operation of the transmission electron microscope (TEM) and the scanning electron microscope (SEM) are described briefly in Section 5.8. More detailed treatments are given by Kimoto and Russ (1969), Gillot (1969), Tovey (1971), and McCrone and Delly (1973). The practical limit of resolution using the TEM is about 20 Å and using the SEM is about 100 Å; thus, detail at the individual clay particle level is easily seen. Major advantages of the SEM over the TEM are the very great depth of field (some 300 times greater than with the light microscope), the wide range of magnifications available (about ×20 to ×20,000, continuous), and sur-

faces may be studied directly; whereas, either surface replicas or ultrathin sections are needed for the TEM. The main advantage of the TEM is a higher limit of resolution.

Both types of electron microscopy require an evacuated sample chamber (1×10^{-5} Torr), so wet soils cannot be studied directly. It is usually necessary to coat SEM sample surfaces with a conducting film to prevent surface charging and consequent loss of resolution. Gold placed in a very thin layer (200 to 300 Å) in a vacuum evaporator is often used.

The main difficulty in the use of the electron microscope for study of fabric is the preparation of sample surfaces, surface replicas, or ultrathin sections that retain the undisturbed fabric of the original soil. In general, the higher the water content and void ratio of the original sample, the greater the likelihood of disturbance. Soils containing expansive clay minerals may also pose problems because removal of interlayer water may cause undetected changes in microfabric, or there may be excessive shrinkage. As noted pre-

Fig. 8.24 Spestone kaolin remolded at 160 percent water content and freeze dried. Note the open structure, the importance of the larger particles, and small particles clinging to the surfaces of larger particles. Picture width is 16 μm (Tovey, 1971).

Fig. 8.25 Undisturbed Drammen clay from Konnerudgt, Norway. Air dried. Picture width 20 μm (Tovey, 1971).

viously, ground surfaces are not suitable. The dry–fracture–peel technique appears the best of the available alternatives.

That careful techniques are successful in preserving delicate fabrics is evidenced by Fig. 8.24 that shows the fabric of a kaolinite remolded at 160 percent water content, a value at least twice the liquid limit. The fabric was preserved by freeze-drying. A flocculated–aggregated or bookhouse type of particle arrangement is clearly visible, with the larger clay plates serving as the structural framework. The small particles are shown clinging in face-to-face arrangement to the surface of the larger particles. The disposition of these particles in the original saturated sample possibly could have been different.

Undisturbed Drammen clay (Norway) is shown in Fig. 8.25. This material consists of silt-sized particles of a variety of shapes with clay particles adhering to their surfaces. The usefulness of the SEM is particularly apparent when dealing with soils containing a range of particle sizes. The large particles would interfere with the preparation of ultrathin sections or surface replicas for the TEM, and the clay particles are too small to be seen using the optical microscope.

Although the field of view at high magnifications is limited, mosaics of photomicrographs may be prepared to show larger fabric features. Such a composite is shown in Fig. 8.6. Another example is given in Fig. 8.26, which shows the failure zone in a triaxial compression test specimen of kaolinite.

Recent developments in electron microscopy include *cathodoluminescent* observations using the SEM to provide internal crystallographic information on sand grains (Tovey, 1973a). This information may be related to the stress history of the material. The SEM can be used in a *scanning transmission* mode. This appears to have advantages over the conventional TEM in that thicker sections can be studied, and contrast can be improved. Operation of the SEM in the *X-ray* mode may have application for detection of mineralogic details.

High voltage transmission electron microscopy (HVEM) makes use of thicker sections possible as well as the direct study of wet samples in a special cell. It is also possible to obtain information on internal structure using HVEM. For example, Smalley, Cabrera, and Hammond (1973) studied dislocation networks in nonclay particles from a quick clay. These patterns reflect stress history and method of particle production.

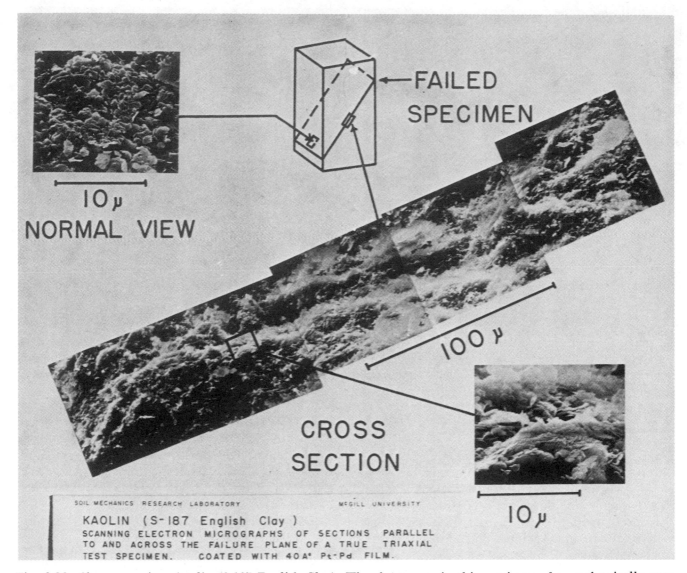

Fig. 8.26 Shear zone in a kaolin (S-187 English Clay). The shear zone in this specimen of a mechanically compacted true triaxial test specimen consists of slip planes of well-oriented particles within a shear zone of width 40 to 80 μm. Some local conjugate planes can be seen (Scanning electron micrographs of sections coated with 40 Å Pt-Pd film). (R. L. Sloane and S. E. McKyes.)

8.9 FABRIC STUDY BY X-RAY DIFFRACTION

For a given mineral the intensity of X-ray diffraction from a particular crystallographic plane depends on (1) the mineral concentrations in the volume irradiated, and (2) the orientation of the mineral particles. In the case of the clay minerals, parallel orientation of clay plates enhances basal reflections, but decreases the intensity of reflection from lattice planes oriented in other directions. Thus, the intensity of the (001) reflections in a given soil provides a measure of clay particle orientation.

The relative heights of basal peaks for different samples of the same material can be used as a measure of particle orientation differences (Meade, 1961). A Fabric Index (FI) based on areas of diffraction peaks can be defined (Gillott, 1970):

$$FI = \frac{V}{(P + V)} \tag{8.2}$$

where V is the area of the basal peak in a section cut perpendicular to the orientation plane and P is the area of the same peak from a section cut parallel to the plane of parallel orientation of particles. The value of Fabric Index ranges from zero for perfect preferred orientation to 0.5 for perfectly random orientation.

A similar procedure, which, while retaining the concept of peak area, does not quite require direct measurement of peak areas, is (Yoshinaka and Kazama, 1973)

$$FI = \frac{H_v W_v}{H_h W_h + H_v W_v} \qquad (8.3)$$

where H_h and H_v are basal reflection intensities in horizontal and vertical sections, respectively, and W_h and W_v are the width of diffraction peaks at half intensity of horizontal and vertical sections, respectively, as illustrated by Fig. 8.27. The direction of preferred orientation may not always be parallel to either the horizontal or vertical. A more general form of equation (8.3) would be based on sections cut parallel and perpendicular to the orientation plane, as is the case for equation (8.2), provided the direction of orientation is known.

The peak ratio (PR), which is the amplitude of the (002) reflection divided by the amplitude of the (020) reflection can be used as a measure of orientation (Martin, 1966). The *PR* has the advantage of being independent of particle concentration and minimizing the effects of mechanical and instrumentation variables. For kaolinite the PR of a sample with completely random particle orientations is about 2.0. Maximum obtainable orientation is characterized by

a PR of 200. The reasons for choosing the (020) and (002) reflections, at least for kaolinite are:

1. They are strong reflections.
2. They are distinct from other reflections.
3. They are close enough together ($2\theta = 19.9$ and 24.9 degrees for (020) and (002) in kaolinite, respectively, using copper radiation) that approximately the same volume will be irradiated for determination of both peaks.

A pole figure device can be used to determine particle orientation at any angle to the particle surface. In this way particle orientations and their variations about the axes in the three orthogonal directions can be described quantitatively. The pole figure device offers a practical method by which parallel particle assemblages of a volume of the order of that sampled by the X-ray beam, but oriented in a direction other than parallel to the sample surface, can be detected. An apparatus and technique for automatic measurement and plotting of orientation data using a pole figure device is described by Borodkina and Osipov (1973).

X-Ray diffraction methods for the study of soil fabric offer the advantage of quantification of the data in a way that is not possible with optical and electron microscope techniques. On the other hand, X-ray methods suffer some disadvantages which include:

1. The interpretation of results may be difficult in materials containing more than one mineral.
2. The volume sampled by the X-ray beam, which depends on sample characteristics, X-ray tube voltage and current, X-ray tube slit openings, and diffraction angle, may contain discontinuities and smaller fabric elements that will not be detected. It is conceivable that the same X-ray peak heights could be produced by different microfabrics. For example, on a millimeter scale, all the arrangements shown in Fig. 8.1 are random; whereas, on a micrometer scale they are quite different.
3. The results are weighted in favor of the fabric nearest the sample surface.

Thus, X-ray diffraction would appear best suited for fabric analysis of monomineralic clays in which particle orientations over regions the size of the X-ray beam (a few millimeters) are of interest, or in conjunction with other methods that can provide detail on the character of the microfabric. A volume of sample of the order of a few cubic millimeters is irra-

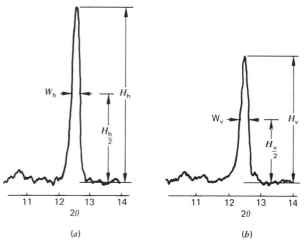

Fig. 8.27 Quantities needed for evaluation of fabric index according to Yoshinaka and Kazama (1973). (*a*) Horizontal section. (*b*) Vertical section.

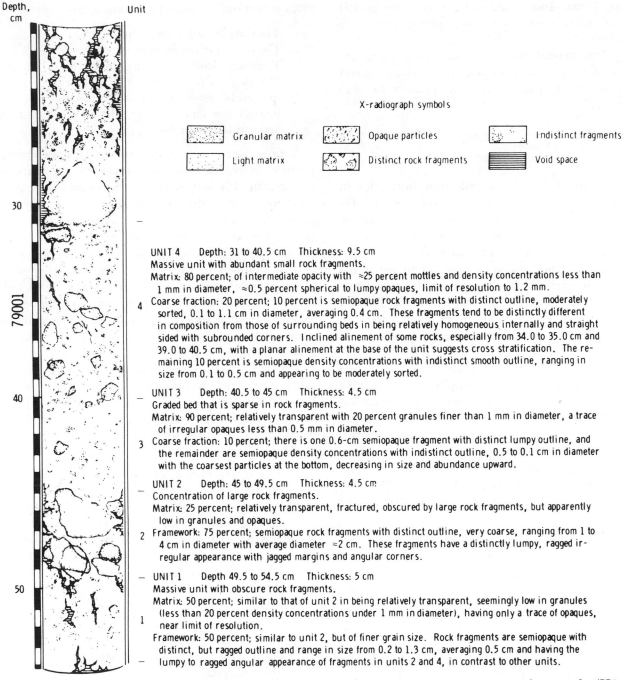

X-radiograph symbols

Granular matrix	Opaque particles	Indistinct fragments
Light matrix	Distinct rock fragments	Void space

UNIT 4   Depth: 31 to 40.5 cm   Thickness: 9.5 cm
Massive unit with abundant small rock fragments.
Matrix: 80 percent; of intermediate opacity with ≈25 percent mottles and density concentrations less than 1 mm in diameter, ≈0.5 percent spherical to lumpy opaques, limit of resolution to 1.2 mm.
Coarse fraction: 20 percent; 10 percent is semiopaque rock fragments with distinct outline, moderately sorted, 0.1 to 1.1 cm in diameter, averaging 0.4 cm. These fragments tend to be distinctly different in composition from those of surrounding beds in being relatively homogeneous internally and straight sided with subrounded corners. Inclined alinement of some rocks, especially from 34.0 to 35.0 cm and 39.0 to 40.5 cm, with a planar alinement at the base of the unit suggests cross stratification. The remaining 10 percent is semiopaque density concentrations with indistinct smooth outline, ranging in size from 0.1 to 0.5 cm and appearing to be moderately sorted.

UNIT 3   Depth: 40.5 to 45 cm   Thickness: 4.5 cm
Graded bed that is sparse in rock fragments.
Matrix: 90 percent; relatively transparent with 20 percent granules finer than 1 mm in diameter, a trace of irregular opaques less than 0.5 mm in diameter.
Coarse fraction: 10 percent; there is one 0.6-cm semiopaque fragment with distinct lumpy outline, and the remainder are semiopaque density concentrations with indistinct outline, 0.5 to 0.1 cm in diameter with the coarsest particles at the bottom, decreasing in size and abundance upward.

UNIT 2   Depth: 45 to 49.5 cm   Thickness: 4.5 cm
Concentration of large rock fragments.
Matrix: 25 percent; relatively transparent, fractured, obscured by large rock fragments, but apparently low in granules and opaques.
Framework: 75 percent; semiopaque rock fragments with distinct outline, very coarse, ranging from 1 to 4 cm in diameter with average diameter ≈2 cm. These fragments have a distinctly lumpy, ragged irregular appearance with jagged margins and angular corners.

UNIT 1   Depth 49.5 to 54.5 cm   Thickness: 5 cm
Massive unit with obscure rock fragments.
Matrix: 50 percent; similar to that of unit 2 in being relatively transparent, seemingly low in granules (less than 20 percent density concentrations under 1 mm in diameter), having only a trace of opaques, near limit of resolution.
Framework: 50 percent; similar to unit 2, but of finer grain size. Rock fragments are semiopaque with distinct, but ragged outline and range in size from 0.2 to 1.3 cm, averaging 0.5 cm and having the lumpy to ragged angular appearance of fragments in units 2 and 4, in contrast to other units.

Fig. 8.28   Stratigraphic sketch prepared from X-radiographs of an Apollo 16 core tube sample (Hörz et al., 1973).

diated when conventional diffraction apparatus is used. Sampled volumes as small as a cubic micrometer are possible using special high intensity tubes.

8.10 APPLICATIONS OF TRANSMISSION X-RAY (RADIOGRAPHY)

Transmission X-ray is a useful nondestructive method for study of both soil macrofabric and deformation phenomena. X-Radiographs of samples while still in sample tubes can provide useful information on stratigraphy, texture, and disturbance (Kenney and Chan, 1972). Many of the core tube samples obtained from the surface of the moon during the Apollo missions have not yet (1975) been opened; however, X-radiographs have been useful to deduce fabric and particle size details (Hörz et al., 1973; Mahmood, 1973). Figure 8.28 is an example. In addition to details of macrofabric X-radiographs may be reveal density variations within samples of an apparently homogeneous material.

In one technique for study of deformation patterns, lead shot is placed in regular patterns in blocks of soil used for model tests. Determination of the position of the shot by means of X-radiographs before and after the test allows for computation of displacements and strains. The technique offers an easy means for locating shear planes in soil samples (Sopp, 1964).

8.11 PORE SIZE DISTRIBUTION ANALYSIS

Most emphasis in soil fabric studies has been on the orientation and arrangement of particles and particle groups. The nature of the pore spaces is also an important element of fabric. The permeability of a soil depends directly on pore sizes and size distributions; deformation behavior depends in part upon the pore spaces available to accommodate displaced particles; and planar pores may be responsible for anisotropic properties. Procedures for characterizing orientations of pores visible in thin sections are referred to in Section 8.7. Similar methods can be applied to thin sections seen in the electron microscope (Pusch, 1966*b*).

Determinations of pore size distribution are possible in two other ways: (1) by forced intrusion of a nonwetting fluid and (2) a capillary condensation method based on the interpretations of adsorption and desorption isotherms. The maximum pore size that can be measured using the capillary condensation method is 1000 Å (0.1 μm). As many of the pores in soils are larger, this method is of limited usefulness.

The intrusion method, however, is useful for measurement of pore sizes ranging from 0.01 μm to several tens of micrometers. A mercury intrusion method has been applied to clays (Diamond, 1970; Sridharan, Altschaeffl, and Diamond, 1971; Mahmood, 1973; and Ahmed, Lovell, and Diamond, 1974). The basis of the method is that a nonwetting fluid (fluid to solid contact angle > 90°) will not enter pores without application of pressure. For pores of cylindrical shape

$$d = - \frac{4 \tau \cos \theta}{p} \qquad (8.4)$$

where d is the diameter of pore intruded, τ is the surface tension of the intruding liquid, θ is the contact angle, and p is the applied pressure. The volume of mercury intruded into an evacuated dry sample of the order of 1 g in size using successively higher pressures is measured. The total volume of mercury intruded at each pressure gives the total volume of pores with an equivalent diameter larger than that corresponding to that pressure. The surface tension of mercury is 4.84×10^{-4} N/mm at 25°C. The contact angle θ has been taken as 130 or 140 degrees in some studies; measurements by Diamond (1970) gave 139 degrees for montmorillonite and 147 degrees for other clay mineral types.

Some limitations of the method are:

1. Pores must be dry initially.
2. Isolated pores are not measured.
3. Pores accessible only through smaller pores will not be measured until the smaller pore is penetrated.
4. The apparatus may not have the pressure capacity to intrude the smallest pores in a sample.

Apparatus for porosimetry measurements is available commercially with different pressure capacities.

Some examples of pore size distributions are shown in Figs. 8.29 and 8.30. The heavy lines on the ordinate axis in these plots represent the total pore volume in the sample. That pores may cover a range of sizes and that changes in density and sample preparation technique may lead to a change in pore size distributions is evident.

8.12 INDIRECT METHODS FOR FABRIC CHARACTERIZATION

As virtually all physical properties of a soil depend in part on fabric, the measurement of almost any

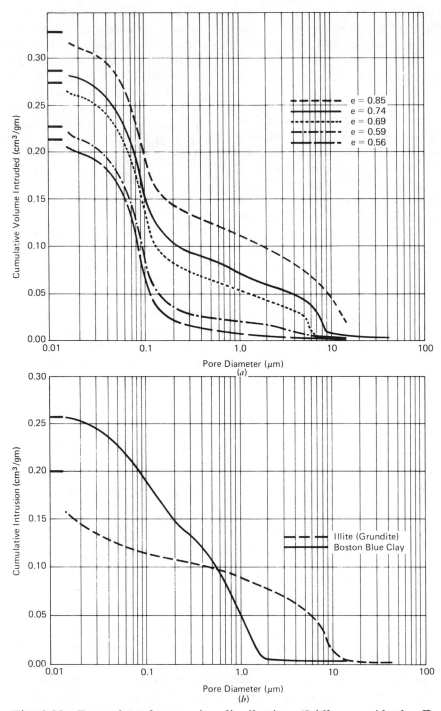

Fig. 8.29 Examples of pore size distribution (Sridharan, Altschaeffl, and Diamond, 1971). (a) Effect of degree of compaction—statically compacted kaolinite with 21.0 percent water content. (b) Statically compacted illite (w = 14 percent, e = 0.57) and Boston blue clay (w = 15 percent, e = 0.70).

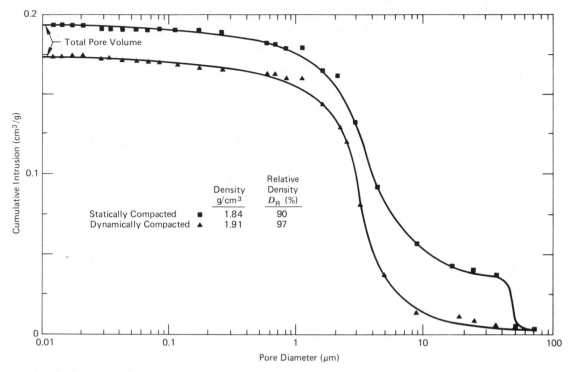

Fig. 8.30   Pore size distributions in **crushed** basalt as affected by compaction method.

physical property serves as an indirect measure of fabric. Some of the measurements that are particularly useful or promising are listed in Table 8.2 and are discussed further in this section.

Acoustical velocity

Particle alignment influences the velocity of sound waves through a soil. For example, it has been found that the ratio of velocity in a horizontal direction to that in a vertical direction increases with consolidation pressure for an ocean bed sediment (Tovey, 1973a). Presumably this reflects an increase in preferred orientation under the higher pressures. In the interpretation of wave propagation data, the effects of pressure must be considered, because velocity depends on the square root of deformation modulus (shear, bulk, or Young's), and modulus depends on confining pressure.

Dielectric dispersion and electrical conductivity

A measurement of the direct current resistance or conductivity of a soil yields values dependent in part on the sample fabric. Such measurements are complicated, however, by the fact that prolonged application of a dc field causes electrokinetic coupling phenomena, such as electro-osmosis and electro-chemical effects (Chapter 15) that may cause irreversible

changes in the system. Problems of this type can be minimized by use of alternating current. In this case, however, it is found that the measured responses depend on frequency.

Measured values of capacitance C and resistance R can be converted into a dielectric constant ϵ', which is defined as C/C_0 where C_0 is the capacitance when vacuum occupies the space between electrodes and conductivity σ using

$$\epsilon' = \frac{CL}{A\epsilon_r} \qquad (8.5)$$

$$\sigma = \frac{L}{RA} \qquad (8.6)$$

where L is the sample length, A is the cross-sectional area, and ϵ_r is the dielectric constant of vacuum (8.85×10^{-14} farads/cm). In fine grained materials there is a concentration of electrical charges adjacent to particle surfaces. When an AC field is applied, the charges move back and forth with an amplitude that depends on such factors as type of charge, association of charge with surfaces, particle arrangement, and strength and frequency of field. Those charges, usually in the form of ions that are free to drift through the material and discharge at electrodes, produce a

DC conductivity. Others can oscillate in an AC field but are not free to break away completely from their associated particles. These oscillating charges contribute to a polarization current that can be measured. The number of charges per unit volume times the average displacement is the polarizability. The magnitude of the polarizability is determined by the composition and structure of the material and is reflected by the dielectric constant.

Phenomena contributing to polarization include dipole rotation, accumulation of charges at interfaces between particles and their suspending medium, ion atmosphere distortion, coupling of flows, and distortion of a molecular system. The extent to which any polarization can develop depends on ease of charge movement and time available for displacement. With increase in frequency, the dielectric constant may fall, and the conductivity may rise. These changes are termed *anomalous dispersion*. For a given material, several regions of anomalous dispersion may be observed over the frequency range from zero to microwave (Fig. 8.31). Different polarization mechanisms cease to be effective at different frequency ranges, thus accounting for the several ranges of anomalous dispersion in Fig. 8.31. Electrolyte solutions alone do not exhibit conductivity or dielectric dispersion effects at frequencies up to 10^8 Hz. Figure 8.32 shows the conductivity and dielectric dispersion behavior of saturated illite in the radiofrequency range.

Electrical response characteristics in the low frequency range depend on particle size and size distribution, water content, direction of current flow relative to direction of preferred particle orientation, type and concentration of electrolyte in the pore fluid, particle surface characteristics, and sample disturbance. The dielectric and conductivity dispersions in the radio frequency range can be described by a three-element model (Arulanandan and Smith, 1973) based on a division of electrical current into flow paths: through pore fluid alone, through solid alone; and through solid and pore fluid. This model can be used to obtain quantitative values for compositional and geometrical properties of a soil.

Techniques for measurement of quantities needed for obtaining conductivity and dielectric dispersion curves and interpretations in terms of factors relating to soil fabric and structure are given by Arulanandan and Mitchell (1968); Arulanandan, Smith, and Spiegler (1973); and Arulanandan and Smith (1973).

Thermal conductivity

Preferred orientation of platy particles causes anisotropic thermal conductivity. Ratios of thermal conductivity in a horizontal direction k_h to that in a vertical direction k_v for three clays with preferred orientations in the horizontal direction were in the range of 1.05 to 1.70, depending on the clay type, consolidation pressure, and disturbance (Penner, 1963a).

A simple transient heat flow method, shown schematically in Fig. 8.33, can be used for determination of k_h and k_v. A line heat source in the form of a needle probe is inserted in a cube sample (about 10 cm on a side) as shown. The probe contains both a heater coil and a temperature thermocouple. When heat is introduced through the probe, the temperatures T_2 and T_1 at times t_2 and t_1 are related to the thermal conductivity k, according to

$$k = \frac{4}{Q} \pi \frac{\ln t_2 - \ln t_1}{T_2 - T_1} \qquad (8.7)$$

where Q is the heat input between t_1 and t_2.

For the probe in the vertical position, assuming a cross anistropic fabric, the value of k determined from equation (8.7) is k_h. For the probe in the horizontal position, a value of k_1 is measured which is related to k_v and k_h according to (Carlslaw and Jaeger, 1957)

$$k_v = \frac{k_i^2}{k_h} \qquad (8.8)$$

Magnetic susceptibility

The magnetic susceptibility of an anisotropic soil varies with direction in a magnetic field (Osipov and

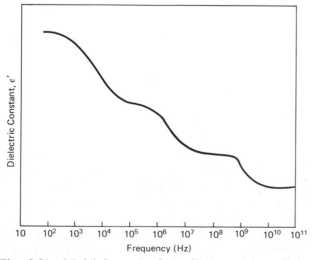

Fig. 8.31 Multiple anomolous dispersion in a dielectric material.

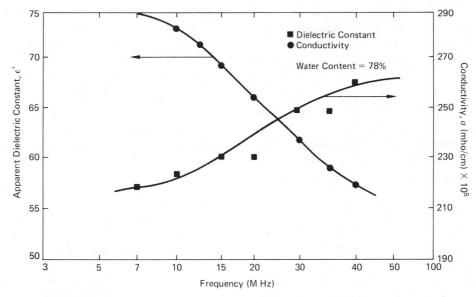

Fig. 8.32 Dielectric and conductivity dispersion characteristics of saturated illite (grundite) (Arulanandan, Smith, and Spiegler, 1973).

Sokolov, 1972). It is greatest in a direction parallel to bedding and least in directions normal to bedding. Measurements in various directions are used to obtain the magnitudes and directions of the principle axes of anisotropy.

Mechanical properties

The mechanical properties of a soil, including strength, compressibility, stress–strain characteristics, and permeability, depend strongly on fabric in ways that are reasonably well understood (Chapter 12). It follows, therefore, that measurements of mechanical properties can be used to deduce information about fabric.

8.13 QUANTIFICATION OF FABRIC

There is a natural desire on the part of most geotechnical engineers to quantify studies whenever possible. Fabric measurements are no exception. A number of quantitative techniques have been discussed in this chapter, including particle and contact orientation distributions in single grain fabrics, photometric measurements of clay orientation using polarizing microscope, peak heights and ratios using X-ray diffraction, and pore size distributions. In addition, quantification of fabric data from electron microscopy is treated by Foster (1973), Tovey (1973b), Smart (1973), and Matsuo and Kamon (1973).

Unquestionably, quantification of fabric has value as a research tool and for comparisons by statistical analysis. Use of a specific quantification procedure

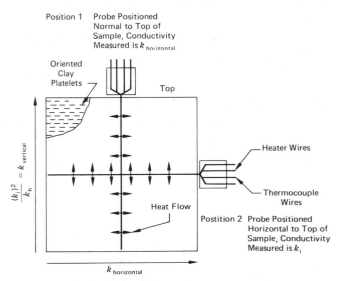

Fig. 8.33 Method for determination of thermal conductivity of cube samples (Penner, 1963a). Schematic drawing of 4 in. cube sample containing clay platelets with the long axis horizontally oriented to ground surface. The position of the probe and the direction in which conductivity is measured are also known. Measurements are taken on the center line of the cube.

makes the interpretation of fabric data less arbitrary and less subject to interpretation by preconception. Furthermore, development of quantitative interrelationships between fabric and properties is a useful research goal. From the standpoint of routine engineering application, however, quantitative values of fabric are of limited use at the present time, and direct measurement of properties is still required. Nonetheless, a knowledge of what the fabric is and how it relates to engineering properties can be valuable in understanding how the soil may behave.

8.14   PRACTICAL APPLICATIONS

Fabric analyses are useful in research to reveal how the mechanical properties relate to particle associations and arrangements. Fabric information can be used to deduce details of the depositional and post-depositional history of a deposit. The effects of different sampling methods can be assessed through study of fabric changes. Insights into the mechanics of strength mobilization, the nature of peak and residual strengths, and the stress–strain behavior of soils can be obtained through fabric analyses.

A problem of great practical importance at the present time when dealing with granular soils is the proper simulation of field conditions using laboratory samples. Fabric analyses offer one technique for determination of appropriate procedures for sample preparation. Some of the indirect methods described in Section 8.12 may be suitable for use in the field for determination of properties, homogeneity, and anisotropy *in situ*.

SUGGESTIONS FOR FURTHER STUDY

Brewer, R. (1964), *Fabric and Mineral Analysis of Soils*, Wiley, New York.

Collins, K., and McGown, A. (1974), "The Form and Function of Microfabric Features in a Variety of Natural Soils," *Geotechnique*, Vol. 24, No. 2.

Kenney, T. C., and Chan, H. T. (1972), "Use of Radiographs in a Geological and Geotechnical Investigation of Varved Soil," *Canadian Geotechnical Journal*, Vol. 9, pp. 195–205.

Lafeber, E. (1966), "Soil Structural Concepts," *Engineering Geology*, Vol. 1, No. 4, pp. 261–290.

Martin, R. T. (1966), "Quantitative Fabric of Wet Kaolinite," *Fourteenth National Clay Conference*, pp. 271–287.

Mitchell, J. K. (1956), "The Fabric of Natural Clays and its Relation to Engineering Properties," *Proceedings of the Highway Research Board*, Vol. 35, pp. 693–713.

Morgenstern, N. R., and Tchalenko, J. S. (1967a), "The Optical Determination of Preferred Orientation in Clays and its Application to the Study of Microstructure in Consolidated Kaolin," *Proceedings of the Royal Society, London*, A 300, I, pp. 218–234 and II, pp. 235–250.

Oda, M. (1972a), "Initial Fabrics and Their Relations to Mechanical Properties of Granular Material," *Soils and Foundations*, Vol. 12, No. 1, pp. 17–36.

Rowe, P. W. (1972), "The Relevance of Soil Fabric to Site Investigation Practice," *Geotechnique*, Vol. 22, No. 2, pp. 195–300.

Sridharan, A., Altschaeffl, A. G., and Diamond, S. (1971), "Pore Size Distribution Studies," *Journal of the Soil Mechanics and Foundations Division*, A.S.C.E., Vol. 97, No. SM 5, pp. 771–787.

Tovey, N. K. (1971), "A Selection of Scanning Electron Micrographs of Clays," CUED/TR - SOILS/TR 5a, Cambridge Univ., Department of Engineering.

Tovey, N. K., and Wong, K. Y. (1973), "The Preparation of Soils and other Geological Materials for the S.E.M.," *Proceedings of the International Symposium of Soil Structure, Gothenburg, Sweden*, pp. 59–67.

PART II Soil Behavior

CHAPTER 9

Soil Composition and Engineering Properties

9.1 INTRODUCTION

The engineering properties of a soil depend on the composite effects of several interacting and often interrelated factors. These factors may be divided into two groups; compositional and environmental.

Compositional factors determine the potential range of values for any property. They can be studied using disturbed samples. Included in this group are:

1. Type of minerals.
2. Amount of each mineral.
3. Type of adsorbed cations.
4. Shape and size distribution of particles.
5. Pore water composition.

Environmental factors determine the actual value for any property. Undisturbed samples or *in situ* measurements are required for this study. They include:

1. Water content.
2. Density.
3. Confining pressure.
4. Temperature.
5. Fabric.
6. Availability of water.

Quantitative prediction of soil behavior completely in terms of compositional and environmental factors is an unrealistic objective for several reasons:

1. Most natural oil compositions are complex.
2. Determination of soil composition is difficult.
3. Physical and chemical interactions may occur between different phases and constituents.
4. The determination and quantitative expression of soil fabric is difficult.
5. Past geologic history and the present *in situ* environment are difficult to simulate in the laboratory.
6. Physico-chemical theories for relating quantitatively composition and environment to properties are inadequate.

In spite of these problems, compositional data are useful for development of a better understanding of properties and establishment of qualitative guidelines for how real soils behave. In this chapter, some relationships between compositional factors and engineering properties are summarized. Major emphasis is given to the clay minerals, because the clay phase dominates the behavior of most fine-grained soils. The purpose is to help provide a "feel" for the values of important properties. Detailed study of the three most important property classes for most engineering problems—volume change, strength and deformation, and conductivity characteristics—are presented in Chapters 13, 14, and 15, respectively.

9.2 APPROACHES TO THE STUDY OF COMPOSITION AND PROPERTY INTERRELATIONSHIPS

The general problem of the study of composition and its relationship to soil properties may be approached in two ways. In the first, natural soils are used, the composition and engineering properties are determined, and correlations are made. This method has the advantage that measured properties are those of the soil in nature. Disadvantages, however, are that compositional analyses are difficult and time consuming, and that in soils containing several minerals or other constituents, such as organic matter, silica, alumina, and iron oxide, the influence of any one constituent may be difficult to isolate.

In the second approach, the engineering properties of synthetic soils are determined. Soils of known composition can be prepared by blending different commercially available clay minerals of relatively high purity with each other and with silts and sands. Although this approach is much easier, it has the disadvantages that the properties of the pure minerals may not necessarily be the same as those of the minerals in the natural soil. Whether or not the influences of such things as organic matter, oxides, and cementation and other chemical effects can be successfully studied using this approach has not yet been thoroughly investigated.

Regardless of the approach used, at least two difficulties arise. One is that in many instances variability of both composition and properties in any one deposit may be great. Thus, care must be taken in the selection of representative samples for study. It has yet to be established that variations in mineralogy do not occur in sediments within distances as small as a few centimeters (Weaver and Pollard, 1973).

The other difficulty is that the different constituents of a soil may not influence properties in direct or even (in many cases) predictable proportion to the quantity present. Both physical and physico-chemical interactions are responsible for this. As an example of physical interactions, it would be found that blending equal amounts of a uniform sand and a clay, each having a compacted density of 18 kN/m³ would not yield a mixture also with a density of 18 kN/m³ after compaction. The resulting density could be as much as 20 kN/m³ or more because the clay could fill void spaces between sand particles.

Figure 9.1 illustrates physico-chemical interaction between clay minerals. Mixtures of bentonite (sodium montmorillonite) and kaolinite, and bentonite and a commercial illite (containing about 40 percent illite clay mineral with the rest silt sizes) were prepared and the liquid limits determined. The dashed line shows the liquid limit values to be expected if each mineral contributed in proportion to the amount present. The solid lines show the actual values observed. Although the bentonite–kaolinite mixture gave values close to theoretical, the liquid limit values for the bentonite–illite mixtures were much less than predicted. It has been established that this resulted from excess salt in the illite that, when mixed with the bentonite, prevented full expansion of the montmorillonite in the presence of water.

9.3 THE DOMINATING INFLUENCE OF THE CLAY PHASE

In general, the greater the quantity of clay mineral in a soil, the higher the plasticity, the greater the potential shrinkage and swell, the lower the permeability, the higher the compressibility, the higher the true cohesion, and the lower the true angle of internal friction. Surface forces and their range of influence are small relative to the weight and size of silt and sand particles; whereas, the behavior of small and flaky clay mineral particles is strongly influenced by surface forces.

The tremendous range in particle sizes that may be found in a soil is illustrated by Fig. 9.2, where different sizes are shown to the same scale. The largest size shown represents fine sand. It may be recalled that particles finer than about 0.06 mm cannot be seen by the naked eye. The orders of magnitude difference in particle sizes found in any one soil is perhaps better appreciated from a representation such as Fig. 9.2 than by the usual size distribution curve where particle diameters are shown to a logarithmic scale.

Whether or not the clay or granular phase will tend to dominate the behavior of a soil can be anticipated to some extent in terms of the amount of clay in the soil. Water is strongly attracted to clay mineral sur-

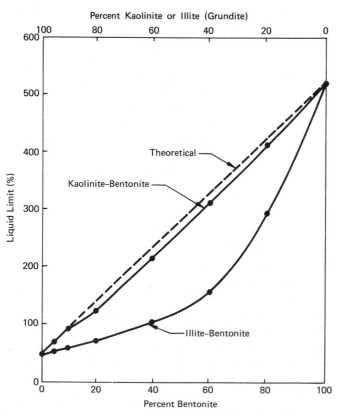

Fig. 9.1 Interactions between clay minerals as indicated by liquid limit (Data from Seed, Woodward, and Lundgren, 1964a).

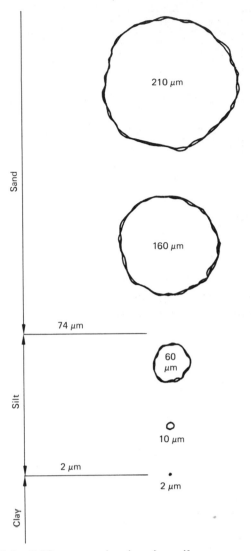

Fig. 9.2 Different grain sizes in soils.

faces (Chapter 6) and results in plasticity; whereas, the nonclay minerals have little affinity for water and do not develop significant plasticity, even when in finely ground form. It can be assumed, therefore, as a first approximation, that all of the water in a soil is associated with the clay phase (Seed, Woodward, and Lundgren, 1964*b*). On this basis, it is possible to estimate the amount of clay required to fill the voids of the granular phase and prevent direct contact between granular particles for any water content. Figure 9.3 shows schematically the weight and volume relationships for the different phases of a saturated soil, from which

$$\frac{w}{100} + \frac{C}{100\,G_{sC}} = \left(1 - \frac{C}{100}\right)\frac{e_G}{G_{sG}} \qquad (9.1)$$

where w is the percent water content; C is the percent dry clay which, when mixed with the water, will fill the voids in the granular phase; e_G is the void ratio of the granular phase; and G_{sC} and G_{sG} are the specific gravities of the clay and granular particles, respectively.

The void ratio of a granular material composed of bulky particles is of the order of 0.9 in the loosest possible state. The specific gravity of the nonclay fraction of most soils is about 2.67 and that of the clay fraction is about 2.75. Inserting these values in equation (9.1) gives

$$C = 48.4 - 1.42w \qquad (9.2)$$

This relationship is shown in Fig. 9.4. It may be seen that, at water contents usually encountered in practice, only a maximum of about one-third of the soil solids need be clay in order to have a condition where the clay is likely to dominate the behavior by preventing direct interparticle contact of the granular particles. Since there is a tendency for clay particles to coat the granular particles in many soils, the clay may exert a significant influence on properties even when present in quantities less than shown by Fig. 9.4.

9.4 ENGINEERING PROPERTIES OF THE CLAY MINERALS*

The different groups of clay minerals exhibit a wide range of engineering properties. Within any one group, the range may also be great and is a function of such factors as particle size, degree of crystallinity, type of adsorbed cations, and type and amount of free electrolyte in the pore water. In general, the importance of these factors increases in the order kaolin < hydrous mica (illite) < smectite. The chlorites exhibit characteristics in the kaolin–hydrous mica range. Vermiculites and the relatively uncommon attapulgite type of mineral behave in the hydrous mica–smectite range.

Because of the influence of the above compositional factors, as well as the effects of environmental influences, only ranges of values for the various engineering properties of the different minerals can be given.

Atterberg limits

The plasticity characteristics of different clay minerals are listed in Table 9.1 in terms of ranges in the

---

* Typical values of properties are presented in this section. Factors that determine these values are analyzed more completely in Chapters 12 through 15.

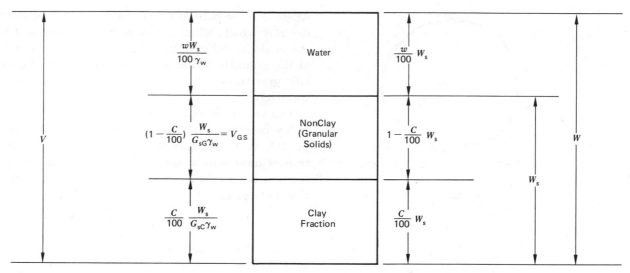

Fig. 9.3 Weight–volume relationships for saturated clay–granular soil mixture. C, percent clay by weight; G_{sC}, specific gravity of clay particles; w, water content in percent; γ_w, unit weight of water; and G_{sG}, specific gravity of granular particles. Volume voids in granular phase, $V_{GS}e_G$ is

$$e_G V_{GS} = \left(1 - \frac{C}{100}\right)\frac{W_s}{G_{sG}\gamma_w}\, e_G$$

where e_G is the void ratio of granular phase.

$$\text{Vol. water} + \text{vol. clay} = \frac{w}{100}\frac{W_s}{\gamma_w} + \frac{C}{100}\frac{W_s}{G_{sC}\gamma_w}$$

If clay and water fill voids in the granular phase, then

$$\frac{w}{100}\frac{W_s}{\gamma_w} + \frac{C}{100}\frac{W_s}{G_{sC}\gamma_w} = \left(1 - \frac{C}{100}\right)\frac{W_s}{G_{sG}\gamma_w}\, e_G$$

$$\frac{w}{100} + \frac{C}{100G_{sC}} = \left(1 - \frac{C}{100}\right)\frac{e_G}{G_{sG}}$$

Atterberg limit values. Most of the values were determined using samples composed of particles finer than 2 μm. Several general conclusions can be made concerning the Atterberg limits of the clay minerals.

1. The liquid and plastic limit values for any one clay mineral species may vary over a wide range, even for a given adsorbed cation type.
2. For any given clay mineral, the range in liquid limit values is greater than the range in plastic limit values.
3. The variation in liquid limit among different clay mineral groups is much greater than the variation in plastic limits.
4. The type of adsorbed cation has a much greater influence on the high plasticity minerals, for example, montmorillonite, than on the low plasticity minerals, for example, kaolinite.

5. Increasing cation valence decreases the liquid limit values of the expansive clays, but tends to increase the liquid limit of the nonexpansive minerals.
6. Hydrated halloysite has an unusually low plasticity index.
7. The greater the plasticity the greater the shrinkage on drying (lower the shrinkage limit).

Particle size and shape

The different clay minerals tend to occur in somewhat different size ranges (Table 3.3). Most reported size analyses indicate the amount finer than 2 μm but do not consider the distribution of sizes within this fraction. Table 9.2 shows, however, that there is some concentration of different clay minerals within different size ranges.

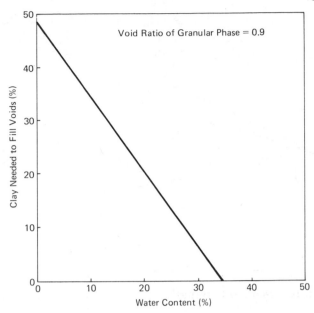

Fig. 9.4 Relationship between water content and amount of clay required to fill voids in a granular soil.

Table 9.2 Mineral Composition of Different Particle Size Ranges in Soils

Particle Size (μm)	Predominating Constituents	Common Constituents	Rare Constituents
0.1	Montmorillonite Beidellite	Mica intermediates	Illite (traces)
0.1– 0.2	Mica intermediates	Kaolinite Montmorillonite	Illite Quartz (traces)
0.2– 2.0	Kaolinite	Illite Mica intermediates Micas Halloysite	Quartz Montmorillonite Feldspar
2.0–11.0	Micas Illites Feldspars	Quartz Kaolinite	Halloysite (traces) Montmorillonite (traces)

Soveri (1950).

Table 9.1 Atterberg Limit Values for the Clay Minerals

Mineral[a]	Liquid Limit (%)	Plastic Limit (%)	Shrinkage Limit (%)
Montmorillonite (1)	100–900	50–100	8.5–15
Nontronite (1)(2)	37–72	19–27	
Illite (3)	60–120	35–60	15–17
Kaolinite (3)	30–110	25–40	25–29
Hydrated Halloysite (1)	50–70	47–60	
Dehydrated Halloysite (3)	35–55	30–45	
Attapulgite (4)	160–230	100–120	
Chlorite (5)	44–47	36–40	
Allophane (undried)	200–250	130–140	

[a] (1) Various ionic forms. Highest values are for monovalent; lowest are for di- and trivalent. (2) All samples 10% clay, 90% sand and silt. (3) Various ionic forms. Highest values are for di- and trivalent; lowest are for monovalent. (4) Various ionic forms. (5) Some chlorites are nonplastic.
Data Sources: Cornell University (1951), Samuels (1950), Lambe and Martin (1955), Birrell (1952), Gradwell and Birrell (1954), White (1955), Warkentin (1961), Grim (1962).

The shape of all the common clay minerals is platy, except for halloysite which occurs as tubes (Fig. 3.12). Particles of kaolinite are relatively large, thick, and stiff (Fig. 3.11). The smectites are composed of small, very thin, and filmy particles (Fig. 3.16). Illites are intermediate between kaolinite and smectite (Fig. 3.20) and often are terraced and thin at the edges.

Volume change characteristics

In general, the swelling and shrinking properties of the clay minerals follow the same pattern as their plasticity properties; the more plastic the mineral, the more the potential swell and shrinkage. The actual amount of swell or shrinkage observed as a result of wetting or drying depends on factors in addition to mineralogy, such as particle arrangement, initial water content, and confining pressure. The shrinkage limit values in Table 9.1 provide a measure of the potential shrinkage of the different minerals as they are dried to low water contents. The potential expansion of the different clay minerals follows closely their respective plasticity indexes.

Hydraulic conductivity (permeability)

In addition to mineralogical composition, particle sizes and size distribution, void ratio, fabric, and pore

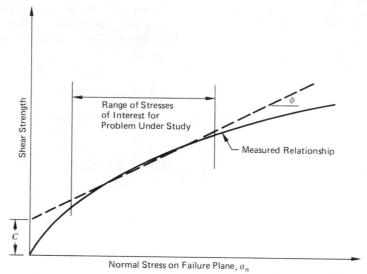

Fig. 9.5 Failure envelopes to represent soil strength.

fluid characteristics all influence the hydraulic conductivity (permeability). Over the normal range of water contents (plastic limit to liquid limit), the hydraulic conductivity of all the clay minerals is less than about 1×10^{-5} cm/sec and may range to values less than 1×10^{-10} cm/sec for some of the monovalent ionic forms of the smectite minerals. The usual range observed for natural clay soils is 1×10^{-6} to 1×10^{-8} cm/sec. For clay minerals compared at the same water content, the hydraulic conductivities are in the order smectite (montmorillonite) < attapulgite < illite < kaolinite.

Shear strength

There are many different ways in which the shear strength of a soil may be measured and expressed. In most cases, a Mohr envelope, where shear strength (usually peak or residual) is plotted as a function of direct stress on the failure plane, or a modified Mohr diagram, where shear strength is plotted versus the average of the major and minor principal stresses at failure, is used. Total and/or effective values of normal stress can be used. A straight line is then fit to the resulting curve over the normal stress range of interest (Fig. 9.5) so that the shear strength τ is given by an equation of the form

$$\tau = c + \sigma_n \tan \phi \qquad (9.3)$$

where σ_n is the normal stress on the shear plane, c is the intercept for σ_n equals zero, often called cohesion, and ϕ is the slope, often called the friction angle.

From the standpoint of relating strength to composition, effective stress failure envelopes are useful. Zones within which effective stress failure envelopes (based on peak strength for pure clay minerals and quartz) fall are shown in Fig. 9.6. The curves show that the increase in shear strength with increase in effective stress is greatest for the nonclay mineral, quartz, followed, in descending order, by kaolinite, illite, and montmorillonite. Ranges in the position of the failure envelope for a given mineral type can be ascribed to differences in such factors as fabric, adsorbed cation type, pH, and overconsolidation ratio.

From many studies (Hvorslev, 1937, 1960; Gibson, 1953; Trollope, 1960; Schmertmann and Osterberg,

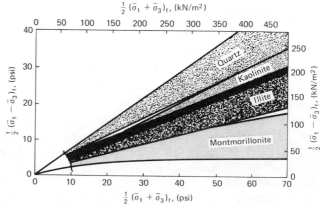

Fig. 9.6 Ranges in effective stress failure envelopes for pure clay minerals and quartz (Olson, 1974).

1960; and others), it has been believed that the total strength of a clay was actually composed of two distinct parts: a cohesion that depends only on void ratio (water content) and a fractional contribution dependent only on normal effective stress. Evaluation of these two parts was done by measurement of the strength of two samples both at the same void ratio but at different effective stress.* The strength parameters determined in this way, sometimes termed the Hvorslev parameters or "true cohesion" and "true friction," showed increasing cohesion and decreasing friction with increasing plasticity and activity (Section 9.7) of the clay.

However, two samples at the same void ratio but different effective stress are known to have different structure (Chapter 12). Furthermore, tests over large ranges of effective stress show that the actual effective stress failure envelope is curved in the manner of Fig. 9.5 and that cohesion is either zero or very small, even for heavily overconsolidated clays. Thus, a significant true cohesion if defined as strength present at zero effective stress, does not exist in the absence of chemical bonding (cementation). These considerations are treated in more detail in Chapter 14.

Even the largest of the friction angle values for clay minerals is significantly less than peak value for cohesionless soils, where values of ϕ_d are generally in the range of 30 to 50 degrees. The residual strengths of some quartz–clay mixtures are shown in Fig. 9.7. If each mineral were equally important, then the curve for a given mixture type should be symmetrical about the 50 percent point, as appears to be the case for kaolinite and hydrous mica with no salt in the pore water. In the other mixtures, however, the clay phase begins to dominate at clay percentages less than 50. This results from the fact that in the case of expansive clay minerals (montmorillonite) or flocculated fabrics (30 g salt/liter) the ratio of volume of wet clay to volume of quartz is greater than the ratio of dry volumes. It is further illustration of the dominating influence of the clay phase.

Compressibility

As might be anticipated, the compressibility of saturated specimens of the clay minerals increases in the order kaolinite < illite < montmorillonite. Values of the compression index C_c have been reported (Cornell University, 1951) in the ranges 0.19 to 0.28

* This condition is obtained by using one normally consolidated and one overconsolidated sample.

for kaolinite, 0.50 to 1.10 for illite, and 1.0 to 2.6 for montmorillonite for different ionic forms. The more compressible the material is, the more pronounced the effect of adsorbed cation type and electrolyte concentration is.

Since both compressibility and hydraulic conductivity are strong functions of soil composition, the coefficient of consolidation c_v should also be related to composition, because c_v is directly proportional to hydraulic conductivity and inversely proportional to the coefficient of compressibility. Values of c_v determined in one study (Cornell, 1951) were in the range 0.06×10^{-4} to 0.3×10^{-4} cm²/sec for montmorillonite, 0.3×10^{-4} to 2.4×10^{-4} cm²/sec for illite, and 12×10^{-4} to 90×10^{-4} cm²/sec for kaolinite. Coefficients of consolidation for kaolinite, illite, montmorillonite, halloysite, and two-mineral mixtures of these materials range in another study from 1×10^{-4} cm²/sec for pure montmorillonite to 378×10^{-4} cm²/sec for pure halloysite (Kondner and Vendrell, 1964). Individual minerals did not influence the coefficient of consolidation of mixtures in direct proportion to the amount present.

9.5   THE EFFECTS OF ORGANIC MATTER

The presence of large amounts of organic matter in soils is usually undesirable from an engineering standpoint. It is known, for example, that organic matter may cause high plasticity, high shrinkage, high compressibility, low permeability, and low strength. Unfortunately few detailed quantitative analyses of the influences of organic matter on engineering properties have been made and present knowledge permits only a few observations.

Organic matter in soils is complex both chemically and physically and varies with age and origin (Mielenz and King, 1955; Kononova, 1961; Schmidt, 1965). It may occur in any of five groups: carbohydrates; proteins; fats, resins and waxes; hydrocarbons; and carbon. Cellulose ($C_6H_{10}O_5$) is the main organic constitutent of soil. In residual soils organic matter is most abundant in the surface horizons. Organic particles may range down to 0.1 μm in size. The specific properties of the colloidal particles vary greatly depending upon parent material, climate, and stage of decomposition.

The humic fraction is gel-like in properties and negatively charged (Marshall, 1964). Organic particles may be strongly adsorbed on mineral surfaces, and

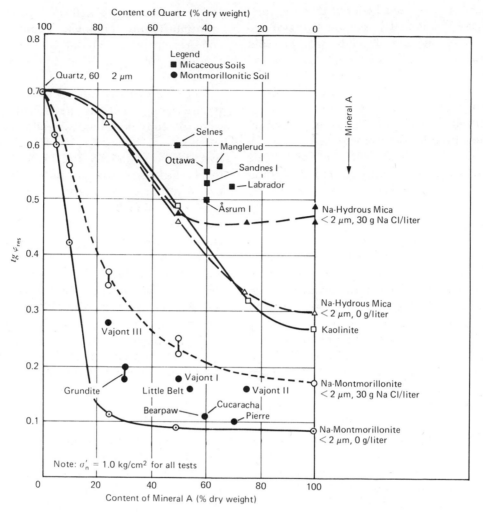

Fig. 9.7   Residual friction angles for clay–quartz mixtures and natural soils (Kenney, 1967).

this adsorption modifies both the properties of the minerals and the organic material itself. The adsorption of organic matter on mineral surfaces and edges may lead to interparticle bonding or the coating of large particles.

At high moisture contents, decomposed organic matter may behave as a reversible swelling system. At some critical stage during drying this reversibility ceases. This often is manifested by a large decrease in Atterberg limits as a result of drying.

A quantitative indication of the effects of organic carbon on plasticity characteristics, in comparison with the influences of percentage of clay and the percentage montmorillonite present in a soil, is given in Fig. 9.8 for more than 70 soil samples from Illinois

(Odell, Thornburn, and McKenzie, 1960). The data show that increasing the organic carbon content by only 1 or 2 percent may increase the limits by as much as an increase of 10 to 20 percent in the amount of material finer than 2 μm or in the amount of montmorillonite.

The effect of organic matter on compaction and strength was studied by Franklin, Orozco, and Semrau (1973). The compaction and strength characteristics on mechanical mixtures of inorganic soils and peat is compared with that of natural soil samples of the same organic content in Fig. 9.9. The effect of organic content on maximum dry density decreases with increased organic content. Increased organic content also causes an increase in optimum water content, and

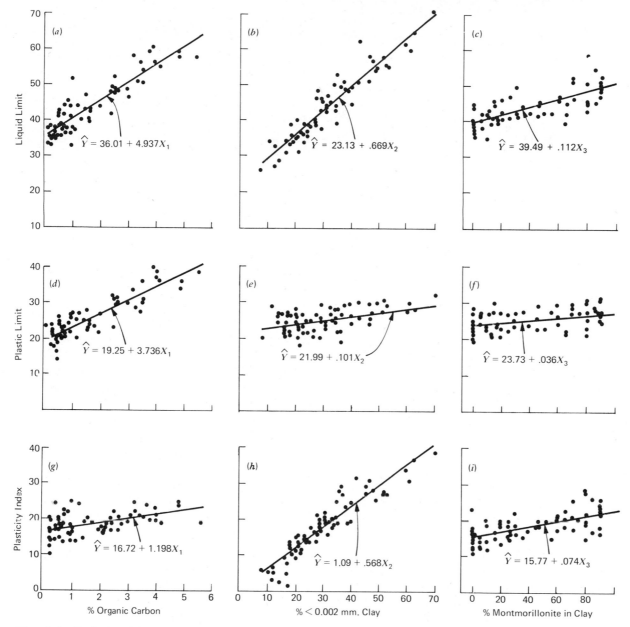

Fig. 9.8   Relationships between Atterberg limits and content of organic carbon, less than 0.002-mm clay, and montmorillonite in clay (Odell, Thornburn, and McKenzie, 1960).

leads to a decrease in maximum unconfined compressive strength, although the relationship is not well defined, as may be seen in Fig. 9.10.

9.6   ATTERBERG LIMITS

The potential value of the Atterberg limits for use in soil mechanics was first indicated by Terzaghi (1926) when he noted, "The results of the simplified soil tests (Atterberg limits) depend precisely on the same physical factors which determine the resistance and the permeability of soils (shape of particles, effective size, uniformity) only in a far more complex manner."

Casagrande (1932) developed a standard device for determination of the liquid limit and noted that the

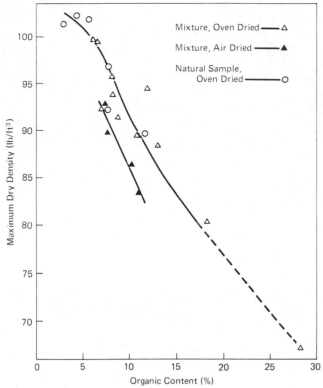

Fig. 9.9 Maximum dry density vs. organic content (From Franklin, Orozco, and Semrau, 1973).

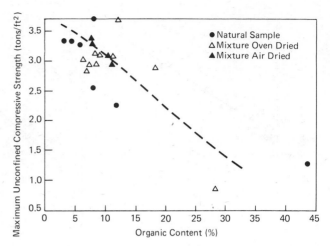

Fig. 9.10 Maximum unconfined compressive strength vs. organic content (From Franklin, Orozco, and Semrau, 1973).

nonclay minerals quartz and feldspar did not develop plastic mixtures with water, even when ground to sizes less than 2 μm. Further studies led to the formulation of a classification system based on the Atterberg limits for identification of cohesive soils (Casagrande, 1948). This classification system was adopted with minor modifications as a part of the Unified Soil Classification System. The plasticity chart (plasticity index vs. liquid limit) shown in Fig. 9.11 forms an essential part of the classification system for fine-grained soils. The characteristic positions of the organic and inorganic silts and clays with reference to the A-line on this chart have been well established.

Although both the liquid and plastic limits are easily determined quantities, and their qualitative correlations with soil composition and physical properties have been quite well established, fundamental physical interpretations of the limits and quantitative relationships between their values and compositional factors are more complex.

Of the two limits, the liquid limit is the more easily interpreted. The liquid limit test is analogous to a shear test. Casagrande (1932) deduced that the liquid limit corresponded approximately to a water content at which a soil has a shear strength of about 2.5 kN/m²; whereas, Norman (1958) reported a strength of the order of 2.0 kN/m² at the liquid limit. The pore water tension at the liquid limit is about 0.4 kN/m² (Croney and Coleman, 1954). For the effective stress corresponding to this negative pore pressure and an angle of friction of 30 degrees, the shear strength due to internal friction would be only 0.25 kN/m². Thus, net interparticle attractive forces must account for the greatest proportion of strength at the liquid limit (Seed, Woodward, and Lundgren, 1964b). The net attractive forces between clay particles are in turn related to the surface activity of the components. This concept can account for the liquid limit of the nonexpansive clay minerals. In the case of the expansive minerals, an appreciable portion of the water content at the liquid limit may be interlayer water which is effectively immobilized. The amount of interlayer water in these minerals is greatly dependent on the type of interlayer cation, thus accounting in part for the great dependence of the liquid limit of expansive soils on cation type.

The plastic limit is less easily interpreted. Terzaghi (1926) suggested that it corresponds to the water content below which the physical properties of the water no longer correspond to those of free water. Yong and Warkentin (1966) interpret the plastic limit as the lowest water content at which the cohesion between particles, or groups of particles is sufficiently low to allow movement, but sufficiently high to allow particles to maintain the new molded positions. Whatever the structural status of the water and the nature

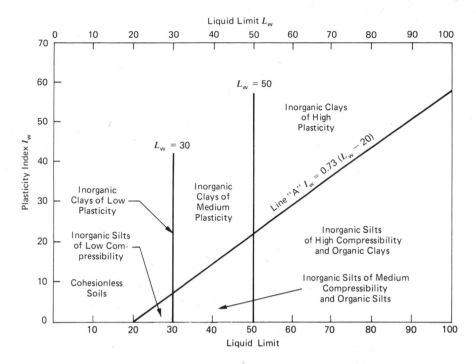

Fig. 9.11 Plasticity chart.

of the interparticle forces, the plastic limit is the lower boundary of the range of water contents within which the soil exhibits plastic behavior; that is, above the plastic limit, the soil can be deformed without volume change or cracking and will retain its deformed shape.

Theoretical relationships between the liquid and plastic limits and the clay content can be derived (Seed, Woodward, and Lundgren, 1964*b*) on the basis that if the volume of the clay-water fraction of a soil is greater than that needed to fill the voids in the nonclay fraction, then the limit of the mixture is determined by the water content of the clay alone. The basic assumption is that the water in a soil is associated entirely with the clay phase. At clay contents less than those needed to fill the voids with the clay at a water content equal to its limit, the limit is essentially independent of the clay content. The relationships that have been verified experimentally are shown in Fig. 9.12. In this figure, *C* is the clay content in percent, and w_{CLL} is the water content of the clay at the liquid limit. With these relationships, it may be shown (Seed, Woodward, and Lundgren 1964*b*) that inorganic clays must lie within a small zone of the plasticity chart, and organic clays must lie in a larger but lower zone of the chart, as is known to be the case (Fig. 9.11).

9.7 ACTIVITY

Both the type and amount of clay in a soil influence the properties, and the Atterberg limits reflect both of these factors. To separate them the ratio of the plasticity index to the clay fraction (percentage by weight of particles finer than 2 μm), termed the *activity*, can be used (Skempton, 1953). For many clays, a plot of plasticity index versus clay content yields a straight line passing through the origin, as shown in Fig. 9.13. The slope of the line for each clay gives the activity. Approximate values for the activities of different clay minerals are listed in Table 9.4.

Table 9.4 Activities of Various Minerals

Mineral	Activity[a]
Smectites	1–7
Illite	0.5–1
Kaolinite	0.5
Halloysite ($2H_2O$)	0.5
Halloysite ($4H_2O$)	0.1
Attapulgite	0.5–1.2
Allophane	0.5–1.2

[a] Activity, $A = \dfrac{\text{Plasticity Index}}{\% < 2\ \mu\text{m}}$.

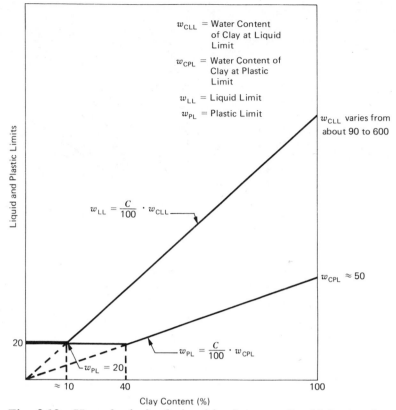

Fig. 9.12 Hypothetical relationships between liquid limit, plastic limit and clay content for inorganic clays (Seed, Woodward, Lundgren, 1964a).

The higher the activity of a soil, the more important the influence of the clay fraction on properties and the more susceptible their values to change due to such factors as variation in type of exchangeable cation and pore fluid composition. For example, the activity of Belle Fourche montmorillonite varies from 1.24 with magnesium as the exchangeable cation to 7.09 for sodium saturation. On the other hand, the activity of Anna kaolinite only varies from 0.30 to 0.41 for six different cation forms (White, 1955).

Studies of the plasticity characteristics of a large number of artificially prepared sand–clay mineral mixtures (Seed, Woodward, and Lundgren, 1962, 1964a, 1964b) show that while the relationship between plasticity index and percentage clay sizes is linear it does not always pass through the origin, but may intersect the percentage clay size axis at values as high as 10 percent, Fig. 9.14. Thus, activity can be redefined as

$$A = \frac{\text{PI}}{C - n} \qquad (9.4)$$

where PI is the plasticity index, C is the percentage of material less than 2 μm, and n is equal to 5 for natural soils or 10 for artificial mixtures. Careful study of factors controlling the Atterberg limits (Seed, Woodward, and Lundgren, 1964b) indicates that the true relationship between plasticity index and clay content is not linear, but it can be approximated by two straight line segments; one applicable for clay contents over about 40 percent, which projects back through the origin; the other applicable for clay contents between about 10 and 40 percent, which is somewhat steeper and intersects the clay content axis at about 10 percent.

Interactions between clay components of a soil are reflected by activity. Figure 9.15 shows the variation in activity of kaolinite–bentonite and illite–bentonite mixtures with the amount of bentonite. Even though illite is considerably more active than kaolinite, when mixed with bentonite it reduced the activity of the mixture considerably below that of corresponding kaolinite–bentonite mixtures, probably because of the presence of excess salt in the illite used. Similar inter-

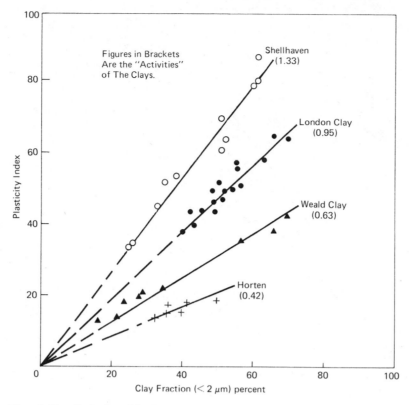

Fig. 9.13   Relationship between plasticity index and clay fraction
(Skempton, 1953).

actions for natural illite–montmorillonite mixtures, as shown in Fig. 9.16, have been observed (Lambe and Martin, 1957). Cementation of clay size particles and interstratification of the clay minerals are probable explanations for this behavior.

9.8 INFLUENCES OF EXCHANGEABLE CATIONS AND pH

In general, the influence of cations on properties increases with increasing activity of the clay. The most important characteristics of the cations are their valence and size. These factors exhibit a major influence on the interparticle physico-chemical forces of interaction discussed in Chapters 7 and 10.

In soils containing expansive clay minerals, the type of exchangeable cation exerts a controlling influence over the amount of expansion that takes place in the presence of water. For example, sodium and lithium montmorillonite can exhibit almost unrestricted interlayer swelling provided water is available, the confining pressure is small, and the excess electrolyte

concentration is low. Di- and trivalent forms of montmorillonite do not expand beyond a basal spacing of about 18 Å regardless of the environment.

In soils composed mainly of nonexpansive clay minerals, the type of adsorbed cation is of greatest importance in influencing the behavior of the material in suspension and the nature of the fabric in sediments formed therefrom. Monovalent cations, particularly sodium and lithium, promote deflocculation; whereas, clay suspensions ordinarily flocculate in the presence of di- and trivalent cations. The role of pH in influencing interparticle repulsions and the existence of positive edge charges in .low pH environments are considered in Section 7.5. These effects are of greatest importance in kaolinite, lesser importance in illite, and relatively unimportant in montmorillonite. In kaolinite the pH may be the single most important factor controlling the fabric of sediments formed from suspension.

The consequences of these cation and pH effects in terms of volume change and strength behavior are examined further in Chapters 13 and 14, respectively.

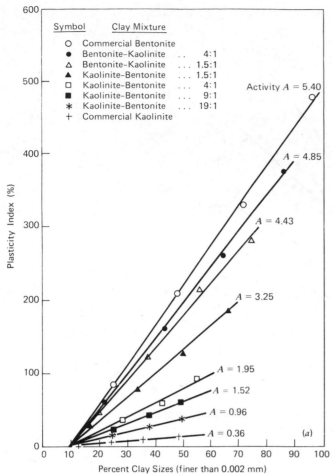

Fig. 9.14a Relationship between plasticity index and percent clay sizes for kaolinite–bentonite clay mixtures (Seed, Woodward, and Lundgren, 1964a).

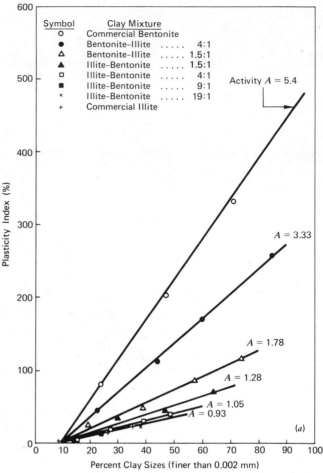

Fig. 914b Relationship between plasticity index and percent clay sizes for illite–bentonite mixtures (Seed, Woodward, and Lundgren, 1964a).

9.9 SHRINKAGE AND SWELLING

Although the actual amount of volume change in response to a change in stress depends on the environmental factors listed in Section 9.1, as well as cation type and electrolyte type and amount, the potential swell or shrinkage is controlled by the type and amount of clay. From a consideration of the clay mineral structures and interlayer bonding (Chapter 3), it would be expected that montmorillonites and vermiculites undergo greater volume changes on wetting and drying than do kaolinites and hydrous micas. Experience clearly indicated this to be the case.

Because of the problems encountered in the performance of structures founded on high volume change soils, several attempts have been made to develop reliable methods for their identification. The most successful of these are based on the determination of some factor that is directly related to clay

mineral composition such as shrinkage limit, plasticity index, and percentage less than 1 μm and activity (Holtz and Gibbs, 1956; Seed, Woodward, and Lundgren, 1962).

Swell tests on artificial sand–clay mineral mixtures were used as the basis for development of one method (Seed, Woodward, and Lundgren, 1962). The standard for comparison was the expansion of laterally confined specimens, prepared at optimum water content using standard AASHO compactive effort, and allowed to swell under a surcharge of 1 psi. An excellent correlation was found between the expansion and compositional factors that expressed both the type and amount of clay, according to

$$S = 3.6 \times 10^{-5} A^{2.44} C^{3.44} \tag{9.5}$$

where S is the percent swell of sample prepared and tested under the specified conditions, A is the percent

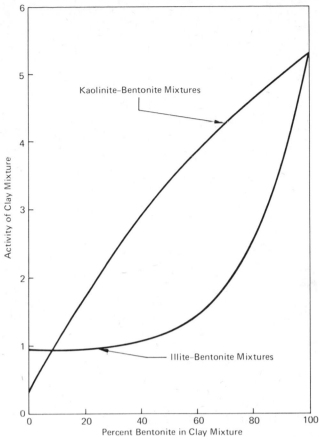

Fig. 9.15 Activities of clay mixtures (Seed, Woodward, and Lundgren, 1964a).

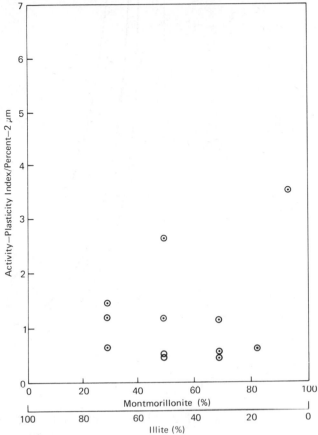

Fig. 9.16 Activity vs. clay composition (Lambe and Martin, 1957).

activity* and C is the percent clay size ($<2\mu$m).

A proposed chart for classification of swelling potential defined in this way is shown in Fig. 9.17.

Further analysis of the relationships indicated that for natural soils the swelling potential could be related to the plasticity index with an accuracy of ± 35 percent, (Seed, Woodward, and Lundgren, 1962) according to

$$S = 2.16 \times 10^{-3}(\text{PI})^{2.44} \qquad (9.6)$$

The approximate relationships between total expansion, swelling potential, and plasticity index are given in Table 9.5.

A slightly different relationship (Ranganathan and Satyanarayana, 1965) has been found to better classify the swell potential of some soils, especially the black cotton soils of India. The swell activity is used to express compositional factors relating to swelling and is defined by

$$SA = \frac{\text{SI}}{C} \qquad (9.7)$$

* With activity defined either by equation (9.4) or by $\Delta\text{PI}/\Delta C$.

Table 9.5 Approximate Relationship Between Total Expansion, Swelling Potential, and Plasticity Index

Plasticity Index (%)	Swelling Potential[a] (%)	Total Expansion[b] (%)
10	0.4– 1.5	4.5–10.0
20	2.2– 3.8	13.5–18.7
30	5.7–12.2	21.4–18.0
40	11.8–25.0	28.0–35.0
50	20.1–42.6	33.0–40.0

[a] For samples compacted at optimum water content using the standard AASHO procedure and allowed to swell under a surcharge of 1 psi.
[b] Air dry to saturation under 1 psi surcharge.
Seed, Woodward, and Lundgren (1962).

where SI is the shrinkage index, given by the liquid limit minus the shrinkage limit. The resulting relationship is

$$S = 41.13 \times 10^{-5}\,\text{SI}^{2.67} \qquad (9.8)$$

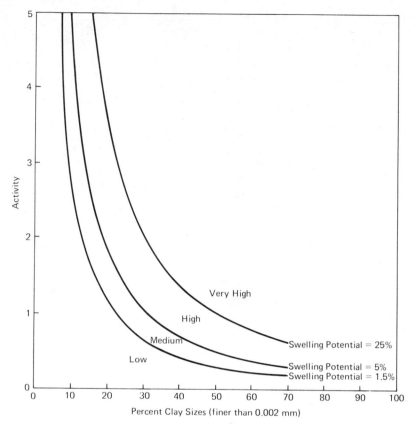

Fig. 9.17 Classification chart for swelling potential (Seed, Woodward, and Lundgren, 1962).

Other relationships of the same general type have also been proposed by Nayak and Christensen (1971).

9.10 VISCOELASTIC BEHAVIOR

Different soil types exhibit varying amounts of time-dependent deformations and stress variations as exhibited by secondary compression, creep, and stress relaxation. The potential for these phenomena depends on compositional factors; whereas, the amount exhibited in any case depends on environmental factors. For example, it is known that retaining walls with wet clay backfills must be designed for at rest earth pressures because of the tendency for stresses to relax along a potential failure plane and thereby increase pressure on the wall. On the other hand, if dry clay is used, and if it is maintained dry, then designs based on active pressures are possible, because time-dependent changes will be negligible.

In general, the greater the organic content and the more plastic the clay, the more pronounced viscoelastic behavior is. Both the type and amount of clay are important as indicated, for example, by the variation of creep rate with clay content for three different clay mineral–sand mixtures shown in Fig. 9.18. In these tests, environmental factors were held constant by preparing all specimens to the same initial conditions (isotropic consolidation of saturated samples to 20 kN/m² and application of a creep stress equal to 90 percent of the strength as determined by a normal compressive strength test. Figure 9.19 shows the variation in creep rate for these specimens as a function of plasticity index. The correlation is reasonably unique, because plasticity index reflects both the type and amount of clay.

9.11 PRACTICAL IMPLICATIONS

A knowledge of composition is useful to indicate probable ranges of geotechnical properties and their variability and sensitivity to changes in environmental conditions. Although quantitative values of properties for analysis and design cannot be derived from compositional data alone, information on composition can be helpful for explaining unusual behavior, identification of expansive soils, the design of testing programs, selection of sampling and sample handling procedures to aid in choice of soil stabiliza-

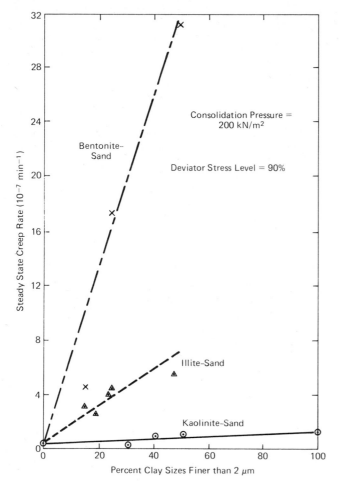

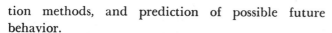

Fig. 9.18 Effect of amount and type of clay on steady-state creep rate.

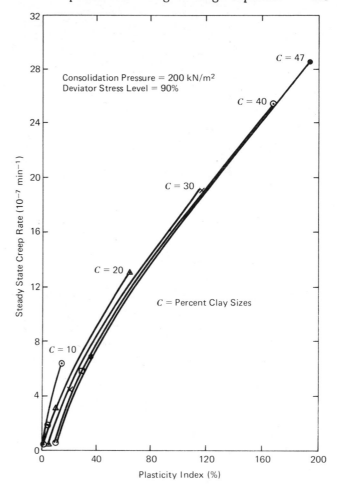

Fig. 9.19 Relationship between clay content, plasticity index, and creep rate.

tion methods, and prediction of possible future behavior.

For example, if it is known that a soil to be used in earthwork construction contains either hydrated halloysite, organic matter, or expansive minerals, then air drying laboratory samples prior to testing could yield totally erroneous data on mechanical properties. If a soil contains a large amount of active clay mineral, then it can be anticipated that properties will be sensitive to changes in chemical environment. Compositional data on the pore water can be used to estimate the dispersion and erosion potential of a soil (Chapter 12).

In many cases, the effects of composition are reflected adequately for practical purposes through information on the particle size, shape, and size distribution of the coarse fraction, and the Atterberg limits of the fine fraction. On large projects and whenever unusual behavior is encountered, however, com-

positional data are useful aids for interpretation of observations.

SUGGESTIONS FOR FURTHER STUDY

Casagrande, A. (1948), "Classification and Identification of Soils," *Transactions*, A.S.C.E., Vol. 113, pp. 901–991.

Grim, R. E. (1962), *Applied Clay Mineralogy*, McGraw-Hill, New York.

Lambe, T. W., and Martin, R. T. (1953–1957), "Composition and Engineering Properties of Soil," *Highway Research Board Proceedings:* I, 1953; II, 1954; III, 1955; IV, 1956; V, 1957.

Low, P. F. (1958), "Mineralogical Data Requirements in Soil Physical Investigations," *Mineralogy in Soil Science and Engineering, Soil Science Society of America*, Special Publication No. 3, pp. 1–34.

Seed, H. B., Woodward, R. J., and Lundgren, R. (1964b), "Fundamental Aspects of the Atterberg Limits," *Journal Soil Mechanics and Foundations Division*, A.S.C.E., Vol. 90, No. SM 6, pp. 75–105.

CHAPTER 10

Effective, Intergranular and Total Stress

10.1 INTRODUCTION

The compressibility and strength of a soil mass depend on the effort required to distort or displace particles or groups of particles relative to each other. In most engineering materials, the resistance to deformation is provided by chemical and physico-chemical forces of interaction. Such forces play a role in the behavior of soils as well. In addition, the compression and strength properties of soils depend on the effects of gravity through self weight and on the stresses applied to the soil mass. The state of a given soil mass, as indicated, for example, by its water content, density, or void ratio, reflects the influences of stresses applied in the past, and this too distinguishes soils from most other engineering materials, which for practical purposes do not change density when loaded or unloaded.

Because of the stress dependencies of the state and the properties, a given soil at different densities could be treated as an infinite number of different materials, which would complicate greatly the treatment of engineering problems. Fortunately, however, the state and properties are not independent, and the relationships between normal stress and volume change and between normal stress and strength can be expressed in terms of definable soil parameters such as compressibility and friction angle. In soils whose properties are influenced significantly by chemical and physicochemical forces of interaction, other parameters such as cohesion may be needed.

Successful treatment of most problems involving volume change and strength requires consideration of that portion of the stress applied to a soil that is carried through the grain assemblage and that portion carried by the fluid phases. This distinction is essential because the grain assemblage can resist both normal and shear stress but the fluid phases (usually water and air) can carry normal stress but not shear stress. Furthermore, whenever the total pressure in the fluid phases within the soil mass differs from that outside the soil mass there will be fluid flow into or out of the soil mass until equality is reached.

In this chapter, the relationships between stresses in a soil mass are examined with particular reference to stress carried by the assemblage of soil particles and stress carried by the pore fluid. Interparticle forces of various types are examined, the nature of effective stress is considered and physico-chemical effects on pore pressure are explored.

10.2 THE PRINCIPLE OF EFFECTIVE STRESS

The principle of effective stress is recognized as a keystone of modern soil mechanics. Development of this principle was begun by Terzaghi about 1920 and extended for several years (Skempton, 1960a). A lucid statement of the principle was given by Terzaghi (1936) on the occasion of the First International Conference on Soil Mechanics and Foundation Engineering. He wrote:

> The stresses in any point of a section through a mass of soil can be computed from the *total principal* stresses, σ_1, σ_2, σ_3, which act in this point. If the voids of the soil are filled with water under a stress u, the total principal stresses consist of two parts. One part, u, acts in the water *and* in the solid in every direction with equal intensity. It is called the *neutral stress* (or the pore water pressure). The balance $\sigma_1' = \sigma_1 - u, \sigma_2' = \sigma_2 - u$ and $\sigma_3' = \sigma_3 - u$ represents an excess

over the neutral stress u, and it has its seat exclusively in the solid phase of the soil.

This fraction of the total principal stresses will be called the *effective principal stresses*. . . . A change in the neutral stress u produces practically no volume change and has practically no influence on the stress conditions for failure. . . . Porous materials (such as sand, clay, and concrete) react to a change of u as if they were incompressible and as if their internal friction were equal to zero. All the measurable effects of a change of stress, such as compression, distortion and a change of shearing resistance are *exclusively* due to changes in the effective stresses σ_1', σ_2' and σ_3'. Hence every investigation of the stability of a saturated body of soil requires the knowledge of both the total and the neutral stresses.

In simplest terms the principle asserts that the effective stress controls volume change and strength and the effective stress is given by $\sigma' = \sigma - u$ for a saturated soil. There is ample experimental evidence to show that these statements are essentially correct for soils; however, this simple definition for σ' is not generally valid for saturated rocks (Skempton, 1960b).

Of the terms in the effective stress equation, only the total stress σ can be directly measured. A pore water pressure, denoted herein by u_0, can be measured at a point remote from the interparticle zone. We know that at equilibrium the total potential of the water at the two points must be equal, as discussed in Section 10.5, but this does not insure that $u = u_0$. The effective stress is a deduced quantity, which in practice is taken as $\sigma' = \sigma - u_0$.

The term "intergranular pressure" has become synonomous with effective stress. Whether or not the intergranular pressure σ_1' is indeed equal to $(\sigma - u)$ cannot be ascertained without a more detailed examination of all the interparticle forces in a soil mass.

10.3 INTERPARTICLE FORCES

Attention in Chapter 7 focused on long range particle interactions associated with electrical double layers and van der Waals forces. These interactions bear on the flocculation–deflocculation behavior of clay particles in suspension. In denser soil masses, other forces of interaction become important as well. Not only do these forces influence the intergranular stresses, but they also control the strength at interparticle contacts, which in turn controls resistance to compression and strength. In a soil mass at equilibrium, there must be a balance among all interparticle forces, the pressure in the water and the applied boundary stresses.

Interparticle repulsive forces

Electrostatic forces. At very small spacings between particles in face-to-face association, or between parts of particles in edge-to-face and edge-to-edge association, a very high repulsion, the *Born repulsion,* develops. This repulsion results from the overlap between electron clouds, and it is sufficiently great to prevent the interpenetration of matter.

At separation distances beyond the region of direct physical interference between adsorbed ions and between hydration water molecules, double layer interactions provide the major source of interparticle repulsion. The theory for quantitative evaluation of these forces is given in Chapter 7. As noted there, this repulsion is very sensitive to cation valence and electrolyte concentration.

Surface and ion hydration. The hydration energy of particle surfaces and interlayer cations causes large repulsive forces at small distances between unit layers (clear distance between surfaces up to about 20 Å). The net energy required to remove the last few layers of water when clay plates are pressed together may amount to 0.05 to 0.1 joule/m². The corresponding pressure required to remove one monomolecular layer of water may be as much as 4×10^5 kN/m² (4000 atm) (van Olphen, 1963).

Thus, pressure alone is not likely to be sufficient to squeeze out all the water between parallel surfaces in naturally occurring clays. Experience shows that heat and/or high vacuum are needed to remove all the water from a fine-grained soil. This does not mean, however, that all the water may not be squeezed from between interparticle contacts. In the case of interacting particle edges and faces of interacting asperities, the contact stress may be several thousand atmospheres (c.f., Lambe and Whitman, 1969, p. 243). This is because the interparticle contact area is only a very small proportion ($<< 1\%$) of the total soil cross-sectional area in most cases. The exact nature of an interparticle contact remains largely a matter for speculation; however, there is evidence (Chapter 14) that it is effectively solid to solid.

Hydration repulsions decay rapidly with separation distance, varying inversely as the square of the distance or less.

Interparticle attractive forces

Electrostatic attractions.   When particle edges and surfaces are oppositely charged, there can arise an attraction due to interactions between double layers of opposite sign. Presumably, the magnitude of this attraction could be calculated using the double layer equations; however, such a calculation has evidently not yet been carried out.

It has been postulated (McEwan, 1953) that when particle separations exceed 10 Å the adsorbed cations divide into groups to form a network linking particles by electrostatic attraction, perhaps somewhat in the manner of Fig. 6.4d. Attractions of this type might be important up to particle separations of 30 Å.

Fine soil particles are often observed to adhere when dry. Electrostatic attraction between surfaces at different potentials has been suggested as a cause. When the gap between parallel particle surfaces at potentials V_1 and V_2, separated by distance d is conductive, there is an attractive force per unit area given by (Ingles, 1962)

$$F = \frac{4.4 \times 10^{-6}(V_1 - V_2)^2}{d^2} \text{ N/m}^2 \quad (10.1)$$

where F is the tensile strength σ_T, and d is measured in micrometers and V_1 and V_2 are measured in millivolts. This force is independent of particle size and becomes significant (greater than 7 kN/m² or 1 psi) for separation distances less than 25 Å.

Electromagnetic attractions

Electromagnetic attractions caused by frequency dependent dipole interactions (van der Waals forces) are described in Section 7.7. Evidence indicates that they are significant long-range attractive forces in clays. Direct computation of the magnitude of these forces is not presently possible; however, they vary inversely as the fourth power of the separation distance between particles.

Primary valence bonding and cementation.   Chemical interactions between particles and between the particles and adjacent liquid phase can only develop at short range. Covalent and ionic bonds occur at spacings less than 3 Å. Cementation involves chemical bonding and can be treated as a short-range attraction.

Whether primary valence bonds, or possibly hydrogen bonds, could develop at interparticle contacts without the presence of cementing agents is largely a matter of speculation. The very high contact stresses between particles could squeeze out adsorbed water and cations and cause mineral surfaces to come close together, perhaps providing opportunity for cold welding. The activation energy for soil deformation is high, in the range characteristic for rupture of chemical bonds, and strength behavior appears in reasonable conformity with the adhesion theory of friction (Chapter 14). Thus, there is reason to suspect interatomic bonding between particles. On the other hand, the absence of cohesion in overconsolidated silts and sands argues against such pressure-induced bonding, because in these soils the individual interparticle contact forces should be orders of magnitude greater than in clays. This is because the larger particle sizes mean many times fewer contacts per unit volume in the coarser soils, with a correspondingly higher normal force per contact for a given effective stress.

Cementation may occur naturally from precipitation of calcite, silica, alumina, or iron oxides or possibly other inorganic or organic compounds. The addition of stabilizers such as cement and lime to a soil may also lead to interparticle cementation.

An analysis of the strength of cemented bonds should consider two cases: failure in the cement or in the particle and failure at the cement-particle interface. The following equation can be derived (Ingles, 1962) for the tensile strength σ_T per unit area of soil cross section:

$$\sigma_T = Pk\left(\frac{1}{1+e}\right)\frac{n}{\sum_1^n A_i} \quad (10.2)$$

where P is the bond strength per contact zone, k is the mean coordination number of a grain, e is the void ratio, n is the number of grains in an ideal breakage plane at right angles to the direction of σ_T, and A_i is the total surface area of the ith grain. It has been shown experimentally that $ke/(1+e) \approx 3.1$.

For a random and isotropic assembly of spheres of diameter d, equation (10.2) becomes

$$\sigma_T = \frac{Pk}{3.1\ d^2(1+e)} \quad (10.3)$$

For a random and isotropic assembly of rods of length l and diameter d

$$\sigma_T = \frac{Pk}{3.1\ d(l + d/2)(1+e)} \quad (10.4)$$

P is evaluated in the following way (Fig. 10.1) for two cemented spheres of radius R. It may be shown that

$$\alpha \cosh \frac{(R - \cos \psi)}{\alpha} = R \sin \psi \qquad (10.5)$$

so for known ψ, α can be computed. Then, for cement failure,

$$P = \sigma_c \times \pi \alpha^2 \qquad (10.6)$$

where σ_c is the tensile strength of the cement; for sphere failure,

$$P' = \sigma_s \times \pi (\alpha')^2 \qquad (10.7)$$

where $\alpha' = R \sin \psi$, and σ_s is the tensile strength of the sphere, and for failure at the interface

$$P'' = \sigma_i \times \frac{\sin \psi}{\psi} \times 2\pi R^2 (1 - \cos \psi) \qquad (10.8)$$

where σ_i is the tensile strength of the interface bond. In principle, equation (10.6), (10.7), or (10.8) can be used to obtain a value for P in equation (10.2) enabling computation of the tensile strength σ_T of a cemented soil.

Capillary stresses. Because water is attracted to soil particles and because water can develop surface tension, capillary menisci form between particles in a partially saturated soil mass. The curved air–water interface causes a pore water tension which, in turn, generates an effective compressive stress between particles. The magnitudes of the tensile strengths developed for certain conditions are considered in some detail by Ingles (1962). The results of his analysis lead to

$$\sigma_T = \frac{a}{d} \frac{kN}{m^2} \qquad (10.9)$$

where d is the particle diameter measured in millimeters and a is in the range of 250 to 650.

Capillary stresses are usually considered responsible for an apparent and temporary cohesion in soils; whereas, the other attractive forces considered produce a true cohesion.

10.4 INTERGRANULAR PRESSURE

A large number of possible intergranular forces has been described in the previous section. Quantitative expression of the interactions of these forces in a soil is beyond the present state of knowledge. Nonetheless, their existence bears directly on the magnitude of intergranular pressure and the relationship between intergranular pressure and effective stress as defined by $\sigma' = \sigma - u$.

A simplified development of the intergranular stress in a soil may be made in the following way. Figure 10.2 shows a horizontal surface through a soil at some depth. Since the stress conditions at contact points, rather than within particles, are of primary concern, a wavy surface that passes through contact points (Fig. 10.2c) is of interest. The proportion of the total wavy surface area that is comprised of intergrain contact area is very small.

The two particles in Fig. 10.2 that contact at point A are shown in Fig. 10.3, along with the forces that act in a vertical direction. Complete saturation is assumed. Vertical equilibrium across wavy surface X-X is considered. The effective area of interparticle contact is a_c; its average value along the wavy surface equals the total mineral contact area along the surface divided by the number of interparticle contacts. Define area a as the average total cross sectional area served by the contact along a horizontal plane. It equals the total area divided by the number of interparticle contacts along the wavy surface. The forces acting on area a in Fig. 10.3 are:

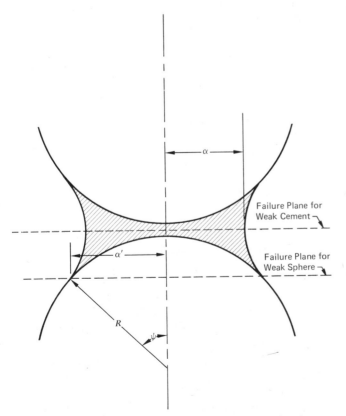

Fig. 10.1   Contact zone failures for cemented spheres.

1. σa, the force transmitted by the applied stress σ.

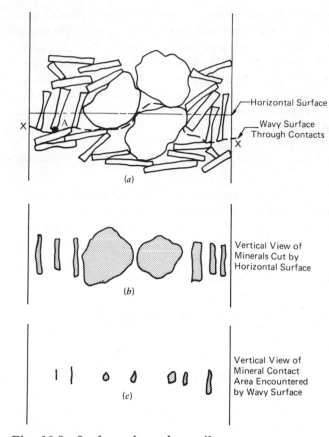

Fig. 10.2   Surfaces through a soil mass.

2. $u(a-a_c)$, the force carried by the hydrostatic pressure u. Because $a >> a_c$ and a_c is very small, the force may be taken as ua. Long range, double-layer repulsions are included in ua.

3. $A(a-a_c) \approx Aa$, the force caused by the long range attractive stress A, that is, van der Waals and electrostatic attractions.

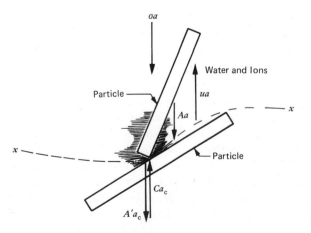

Fig. 10.3   Forces acting on interparticle contact A.

4. $A'a_c$, the force developed by the short range attractive stress A', resulting from primary valence (chemical) bonding and cementation.

5. Ca_c, the force developed by the short range repulsive stress C, resulting from hydration and Born repulsion.

Vertical equilibrium of forces requires that

$$\sigma a + Aa + A'a_c = ua + Ca_c \qquad (10.10)$$

Division of all terms by a converts the forces to stresses per unit area of cross section,

$$\sigma = (C - A')\frac{a_c}{a} + u - A \qquad (10.11)$$

The term $(C - A')\, a_c/a$ represents the net force across the contact divided by the total cross sectional area (soil plus water) that is served by the contact. In other words, it is the intergrain force divided by the gross area, or the *intergranular pressure* in common soil mechanics usage. Designation of this term by σ_i' gives

$$\sigma_i' = \sigma + A - u \qquad (10.12)$$

It should be noted that equations analogous to equations (10.10), (10.11), and (10.12) can be developed for the case of a partly saturated soil. To do so requires consideration of the pressures in the water u_w and in the air u_a, and the proportions of area a contributed by water a_w and by air a_a with the condition that

$$a_w + a_a = a;\ \text{i.e.,}\ a_c \to 0$$

The resulting equation is

$$\sigma_i' = \sigma + A - u_a - \frac{a_w}{a}(u_w - u_a) \qquad (10.13)$$

In the absence of significant long range attractions, this equation is similar in form to that proposed by Bishop in 1955 (Bishop, 1960) for partially saturated soils,

$$\sigma' = \sigma - u_a + \chi(u_a - u_w) \qquad (10.14)$$

where $\chi = a_w/a$. Although it is clear that for a dry soil $\chi = 0$ and for a saturated soil $\chi = 1$, the usefulness of equation (10.14) has been somewhat limited in practice because of uncertainties about χ for intermediate values of saturation.

Limiting the discussion to saturated soils, two questions arise:

1. How does the intergranular pressure σ_i' relate to the effective stress as implied by most analyses; that is, $\sigma' = \sigma - u$?

2. How does the intergranular pressure σ_i' relate to the measured quantity, $\sigma_m' = \sigma - u_o$, that is taken as the effective stress, recalling (Section 10.2) that pore pressure can only be measured at points outside the true interparticle zone?

Answers to these questions require a more detailed consideration of the meaning of fluid pressures in soils.

10.5 WATER PRESSURES AND POTENTIALS

Pressures in the pore fluid of a soil can be expressed in several ways, and the total pressure may involve several contributions. In hydraulic engineering, problems are often analyzed using Bernoulli's equation for the total heads and head losses associated with flow between two points; that is,

$$Z_1 + \frac{p_1}{\gamma_w} + \frac{v_1^2}{2g} = Z_2 + \frac{p_2}{\gamma_w} + \frac{v_2^2}{2g} + \Delta h_{1-2} \quad (10.15)$$

where Z_1 and Z_2 are the elevations of points 1 and 2, p_1 and p_2 are the hydrostatic pressures at points 1 and 2, v_1 and v_2 are the flow velocities at points 1 and 2, γ_w is the unit weight of water, g is the acceleration due to gravity, and Δh_{1-2} is the loss in head between points 1 and 2. The total head H (dimension L) is

$$H = Z + \frac{p}{\gamma_w} + \frac{v^2}{2g} \quad (10.16)$$

Flow results only from differences in total head; conversely, if the total heads at two points are the same, there can be no flow, even if $Z_1 \neq Z_2$ and $p_1 \neq p_2$. If there is no flow, there is no head loss and $\Delta h_{1-2} = 0$.

The flow velocity through soils is low, and as a result, $v^2/2g \rightarrow 0$ and may be neglected. The relationship

$$Z_1 + \frac{p_1}{\gamma_w} = Z_2 + \frac{p_2}{\gamma_w} + \Delta h_{1-2} \quad (10.17)$$

is the basis for evaluation of pore pressures and uplift pressures from flow net solutions to seepage problems.

Although the absence of velocity terms is a factor that seems to simplify the analysis of flows and pressures in soils, there are other considerations that tend to complicate the problem. These include:

1. The use of several terms to describe the status of water in soils (e.g., potential, head, pressure).
2. The possible existence of tensions in the pore water.
3. Compositional differences in the water from point-to-point, and adsorptive force fields from particle surfaces, both of which may influence the energy status of the water (Bolt and Miller, 1958; Low, 1961; Mitchell, 1962).

Some formalism in definition and terminology is necessary to avoid confusion. The status of water in a soil can be expressed in terms of the free energy relative to free, pure water (Aitchison, Russam, and Richards, 1965). The free energy can be (and is) expressed in different ways including

1. *Potential* (Dimensions—ML^2T^{-2}: Joules/kg).
2. *Head* (Dimensions—L: m, cm, ft).
3. *Pressure* (Dimensions—$ML^{-1}T^{-2}$: kN/m², dyne/cm/², tons/m², atm, bar, psi, psf).

In cases where the free energy is less than that of pure water under atmospheric pressure, the terms *suction* and *negative pore water pressure* are often used. The pF is sometimes used as a measure of the suction.

$$pF = \log_{10}\left(\frac{\text{suction in cm water}}{1 \text{ cm}}\right) \quad (10.18)$$

The *total potential (head, pressure)* of soil water is the potential (head, pressure) in pure water that will cause the same free energy at the same temperature as in the soil water. The sign convention is:

+   soil tends to expel water
−   soils tends to absorb water

An alternative definition of total potential is the work done per unit quantity to transport reversibly and isothermally an infinitesimal amount of pure water from a pool at a specified elevation at atmospheric pressure to the pont in soil water under consideration.

The selection of the components of the total potential ψ (head H, pressure p) is somewhat arbitrary (Bolt and Miller, 1958); however, the following appear to have gained acceptance for geotechnical work (Aitchison, Russam, and Richards, 1965):

1. *Gravitational potential* ψ_g (head Z, pressure p_Z) corresponds to elevation head in normal hydraulics usage.
2. *Matrix or capillary potential* ψ_m (head h_m, pressure p) is the work per unit quantity of water

to transport reversibly and isothermally an infinitesimal quantity of water to the soil from a pool containing a solution identical in composition to the soil water at the same elevation and external gas pressure as that of the point under consideration in the soil.

This component corresponds to the pressure head in normal hydraulic usage. It results from that part of the boundary stresses that is transmitted to the water phase from pressures generated by capillary menisci and from water adsorption forces exerted by particle surfaces. A piezometer measures the matrix potential if it contains fluid of the same composition as the soil water.

3. *Osmotic (or solute) potential* ψ_s (head h_s, pressure p_s) is the work per unit quantity of water to transport reversibly and isothermally an infinitesimal quantity of water from a pool of pure water at a specified elevation at atmospheric pressure to a pool containing a solution identical in composition to the soil water but in all other respects identical to the reference pool.

This component is, in effect, the osmotic pressure of the soil water, and it depends on the composition and the ability of clay particles to restrain the movement of adsorbed cations. The osmotic potential is negative; that is, water tends to flow in the direction of increasing concentration.

The total potential, head, and pressure then become

$$\Psi = \psi_g + \psi_m + \psi_s \tag{10.19}$$

$$H = Z + h_m + h_s \tag{10.20}$$

$$P = p_z + p + p_s \tag{10.21}$$

At equilibrium and no flow there can be no variation in $\Psi, H,$ or P within the soil.

10.6 WATER PRESSURE EQUILIBRIUM IN SOIL

Consider a saturated soil mass as shown in Fig. 10.4. Conditions at several points will be analyzed in terms of heads for simplicity, although potential or pressure could also be used with the same result. At point 0, a point inside a piezometer introduced to measure pore pressure, $Z = 0$, $h_m = h_{m_0}$, and $h_{s_0} = 0$ if pure water is used in the piezometer. Thus,

$$H_0 = 0 + h_{m_0} + 0 = h_{m_0}$$

It follows that

$$P_0 = h_{m_0}\gamma_w = u_0 \tag{10.22}$$

the measured pore pressure.

Point 1 is at the same elevation as point 0, except inside the soil mass and midway between two clay particles. At this point, $Z_1 = 0$ but $h_s \neq 0$ because the electrolyte concentration is not zero. Thus,

$$H_1 = 0 + h_{m_1} + h_{s_1}$$

If no water is flowing, $H_1 = H_0$, and

$$h_{m_1} + h_{s_1} = h_{m_0}.$$

Also because $p_1 = p_0 = u_0$

$$u_0 = h_{m_1}\gamma_w + h_{s_1}\gamma_w \tag{10.23}$$

At point 2, which is a point between the same two clay particles as point 1 but closer to a particle surface, there will be a different ion concentration than at 1. Thus at equilibrium, and assuming $Z_2 \approx 0$

$$h_{m_2} + h_{s_2} = h_{m_1} + h_{s_1} = h_{m_0} = \frac{u_0}{\gamma_w}$$

A similar analysis could be applied to any point in the system. If point 3 were midway between two clay particles spaced the same distance apart as the particles on either side of point 1, then $h_{s_3} = h_{s_1}$, but $Z_3 \neq 0$. Thus,

$$\frac{u_0}{\gamma_w} = Z_3 + h_{m_3} + h_{s_3} = Z_3 + h_{m_3} + h_{s_1} \tag{10.24}$$

A partially saturated system can also be analyzed, but the influences of curved air–water interfaces must be taken into account in the development of the h_m terms (Bolt and Miller, 1958).

The conclusions that result from the above analysis of component potentials are:

1. As the osmotic and gravitational components vary from point to point in a soil at equilibrium, the matrix or capillary component must also vary to maintain equal total potential. The concept that hydrostatic pressure must vary with elevation to maintain equilibrium is intuitive; however, the idea that this pressure must vary also in response to compositional differences is less easy to visualize. Nonetheless, this underlies the whole concept of water flow by chemical

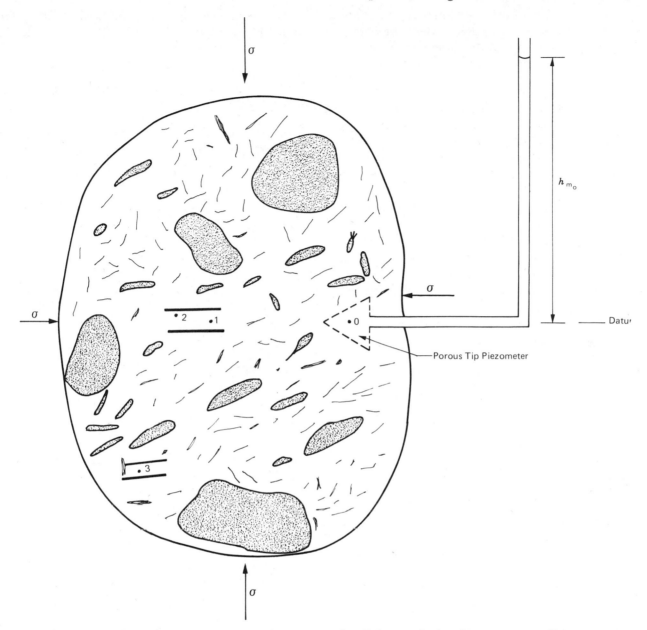

Fig. 10.4 Schematic representation of a saturated soil for analysis of pressure conditions.

osmosis (discussed further in Chapters 13 and 15).

2. The total potential, head, and pressure are measurable and separation into components is possible experimentally, although it is difficult.

3. A pore pressure measurement using a piezometer containing pure water will give a pressure $u_0 = \gamma_w h$, where h is the pressure head at the piezometer. When referred back to points between soil particles, u_0 is seen to include contributions from osmotic pressures as well as matrix pres-

sures. Since osmotic pressures are the cause of long-range repulsions due to double layer interactions, measured pore water pressures may include contributions from long-range interparticle repulsive forces.

10.7 MEASUREMENT OF PORE PRESSURES IN SOILS

A number of techniques for the measurement of pore water pressures have been developed. Some are best suited for laboratory use; whereas, others are intended for use in the field. Some yield the pore pressure or suction by direct measurement, while others require deduction of the value using thermodynamic relationships.

1. Piezometers of various types.   Water in the piezometer communicates with the soil through a porous stone or filter. Pressures are determined from the water level in a standpipe, by a manometer, by a pressure gauge, or by an electronic pressure transducer. A piezometer used to measure pressures less than atmospheric is usually termed a tensiometer.

2. Gypsum block.   The electrical resistance across a specially prepared gypsum block is measured. The water held by the gypsum block determines the resistance, and the moisture tension in the surrounding soil determines the amount of moisture in the gypsum block. This technique is used for measurement of pore pressures less than atmospheric.

3. Pressure–membrane devices.   An exposed soil sample is placed on a membrane in a sealed chamber. Air pressure in the chamber is used to push water from the pores of the soil through the membrane. The relationship between water content and pressure is used to establish the relationship between soil suction and water content.

4. Consolidation tests.   The consolidation pressure on a sample at equilibrium is the soil water suction. If the consolidation pressure were instantaneously removed, then a negative water pressure of the same magnitude would be needed to prevent water movement.

5. Vapor pressure methods.   The relationship between relative humidity and water content is used to establish the relationship between pF and water content.

6. Osmotic pressure methods.   Soil samples are equilibrated with solutions of known osmotic pressure to give a relationship between water content and water suction.

Piezometer methods are used when positive pore pressures are to be measured, as is usually the case in dams, slopes, and foundations on soft clays. The other methods are suitable for measurement of negative pore pressures or suction. Pore pressures are often negative in expansive and partly saturated soils. More detailed descriptions of these and other methods are

given by Croney, Coleman, and Bridge (1952), Aitchison, Russam, and Richards (1965), and Richards (1967, 1969, 1974).

10.8 EFFECTIVE AND INTERGRANULAR PRESSURE

In Section 10.4, it was shown that the intergranular pressure is given by

$$\sigma_i' = \sigma + A - u \qquad (10.12)$$

where u is the hydrostatic pressure between particles (or $h_m\gamma_w$ in the terminology of Section 10.6). Generalized forms of equation (10.24) are

$$u_0 = Z\gamma_w + h_m\gamma_w + h_s\gamma_w \qquad (10.25)$$

and

$$u = h_m\gamma_w = u_0 - Z\gamma_w - h_s\gamma_w \qquad (10.26)$$

Thus, equation (10.12) becomes, for the case of no elevation difference between a piezometer and the point in question,

$$\sigma_i' = \sigma + A - u_0 + h_s\gamma_w. \qquad (10.27)$$

Because the quantity $h_s\gamma_w$ is an osmotic pressure and the salt concentration between particles will invariably be greater than at points away from the soil (such as in a piezometer), $h_s\gamma_w$ will be negative. This pressure reflects double-layer repulsions. It has been termed R in some previous studies (Lambe, 1960; Mitchell, 1962; Chattopadhyay, 1972). If $h_s\gamma_w$ in equation (10.27) is replaced by the absolute value of R, we obtain

$$\sigma_i' = \sigma + A - u_0 - R \qquad (10.28)$$

From equation (10.12), it was seen that the intergranular pressure was dependent on long range interparticle attractions as well as on the applied stress and the pore water pressure between particles. Equation (10.28) indicates that if intergranular pressure is to be expressed in terms of a measured pore pressure u_0, then the long-range repulsion must also be taken into account. The actual hydrostatic pressure between particles, $u = u_0 + R$ includes the effects of long-range repulsions as required by the condition of constant total potential for equilibrium.

In the general case, therefore, the true intergranular pressure, $\sigma_i' = \sigma + A - u_0 - R$, and the conventionally defined effective stress, $\sigma = \sigma - u_0$, differ by the net interparticle stress due to the physico-chemical environment,

$$\sigma_i' - \sigma' = A - R. \tag{10.29}$$

When A and R are both small, as would be true in granular soils, silts, and clays of low plasticity, or in cases where $A \approx -R$, the intergranular and effective stress would be approximately equal. Only in cases where either A or R is large, or both are large but of significantly different magnitude, would the intergranular and effective stress be significantly different. Such a condition appears not to be common, although it might be of importance in a well dispersed sodium montmorillonite, where compression behavior can be accounted for reasonably well in terms of double layer repulsions (Chapter 13).*

10.9   ASSESSMENT OF TERZAGHI'S EQUATION

None of the preceding establishes that the simple Terzaghi equation is indeed the "effective stress" that governs consolidation and strength behavior of soils. Its usefulness has been established from the experience of many years of successful application in practice. Skempton (1960b) has shown, however, that the Terzaghi equation does not give the true effective stress, but gives an excellent approximation for the case of saturated soils.

Long range forces of interaction A and R were not considered by Skempton in his detailed analysis of effective stress. Three relationships for effective stress in saturated soils were proposed and examined using available data to see which, if any, were valid. The relationships proposed were:

1. $$\sigma' = \sigma - (1 - a_c) u \tag{10.30}$$

   This is the true intergranular pressure for the case when $A - R = 0$.

2. The solid phase is treated as a real solid that has a compressibility C_s and a shear strength given by

   $$\tau_i = k + \sigma \tan \psi \tag{10.31}$$

   where ψ is an intrinsic friction angle and k is a true cohesion. The following relationships were derived: For shear strength,

* A detailed analysis of effective stress in clays is presented by Chattopadhyay (1972) which leads to similar conclusions, including equation (10.28). Chattopadhyay termed σ_i' the "true effective stress" and showed that it governed the volume change behavior of Na-montmorillonite.

$$\sigma' = \sigma - \left(1 - \frac{a_c \tan \psi}{\tan \phi'}\right) u \tag{10.32}$$

where ϕ' is the effective stress angle of shearing resistance. For volume change,

$$\sigma' = \sigma - \left(1 - \frac{C_s}{C}\right) u \tag{10.33}$$

where C is the soil compressibility.

3. The solid phase is a perfect solid so that ψ and $C_s = 0$. This gives

$$\sigma' = \sigma - u \tag{10.34}$$

To test the three theories, available data were studied to see which related to the volume change of a system acted upon by both a total stress and a pore water pressure according to

$$\frac{\Delta V}{V} = -C\Delta\sigma' \tag{10.35}$$

and which satisfied the Coulomb equation for drained shear strength τ_d

$$\tau_d = c' + \sigma' \tan \phi' \tag{10.36}$$

when both a total stress and a pore pressure are acting. It may be noted that this approach assumes that the Coulomb strength equation is valid *a priori*.

The results of Skempton's analysis showed that equation (10.30) was not a valid representation of effective stress. Equations (10.32) and (10.33) give the correct results for soils, concrete, and rocks. Equation (10.34) accounts well for the behavior of soils, but not for concrete and rock. The reason for this latter observation is that in soils C_s/C and $a_c \tan \psi/\tan \phi'$ approach zero, and, thus, equations (10.32) and (10.33) reduce to equation (10.34). In rocks and concrete, however C_s/C and $a_c \tan \psi/\tan \phi$ are too great to be neglected. Tan $\psi/\tan \phi$, may range from 0.1 to 0.3, a_c clearly is not negligible, and C_s/C may range from 0.1 to 0.5 (Table 10.1).

Terzaghi's effective stress equation, while not rigorously correct, serves as an excellent approximation in almost all cases for saturated soils.

10.10   PRACTICAL APPLICATIONS

The concepts in this chapter provide insight into the meanings of intergranular, effective, and pore water pressure and factors controlling their values. Because soils behave as particulate materials and not

Table 10.1 Compressibility Values for Soil, Rock, and Concrete

Material	Compressibility[a] per $kN/m^2 \times 10^{-6}$		
	C	C_s	C_s/C
Quartzitic Sandstone	0.059	0.027	0.46
Quincy Granite (30 m deep)	0.076	0.019	0.25
Vermont marble	0.18	0.014	0.08
Concrete (approx.)	0.20	0.025	0.12
Dense sand	18	0.028	0.0015
Loose sand	92	0.028	0.0003
London clay (over cons.)	75	0.020	0.00025
Gosport clay (normally cons.)	600	0.020	0.00003

[a] Compressibilities at $p = 98$ kN/m^2 (1 kgf/cm^2); Water $C_w = 0.49 \times 10^{-6}$ per kN/m^2 (48 $\times$ 10^{-6} per kgf/cm^2).
After Skempton (1960).

as continua, knowledge of these stresses and of the factors influencing them are a necessary prerequisite to the study of compressibility, deformation, and strength.

An understanding of the components of pore water pressure is important to the proper measurement of pore pressure and interpretation of the results.

SUGGESTIONS FOR FURTHER STUDY

Aitchison, G. D., Russam, D., and Richards, B. G. (1965), "Engineering Concepts of Moisture Equilibrium and Moisture Changes in Soils," pp. 7–21. *Moisture Equilibrium and Moisture Changes in Soils Beneath Covered Areas,* Butterworths, Sydney, Australia.

Ingles, O. G. (1962), "Bonding Forces in Soils, Part 3: A Theory of Tensile Strength for Stabilized and Naturally Coherent Soils," *Proceedings of the First Conference of the Australian Road Research Board,* Vol. I, pp. 1025–1947.

Skempton, A. W. (1960a), "Significance of Terzaghi's Concept of Effective Stress," in *From Theory to Practice in Soil Mechanics,* pp. 42–53, Wiley, New York.

Skempton, A. W. (1960b), "Effective Stress in Soils, Concrete, and Rocks," *Proceedings of the Conference on Pore Pressure and Suction in Soils,* Butterworths, London, pp. 4–16.

CHAPTER 11

Soil Structure and Its Stability

11.1 INTRODUCTION

Long before confirmation of specific fabric features was possible, a number of hypotheses were advanced concerning soil fabrics, their formation, and their stability. They were developed in an attempt to account for such phenomena as sensitivity of clays to remolding, differences in properties of soils deposited in different environments, creep and secondary compression, pore pressure phenomena, anisotropy, thixotropic hardening, and the mobilization of friction and cohesion.

Two soils can have the same fabrics but still exhibit different properties if the forces between particles and particle groups are not the same in each material. This would be a case of identical fabrics but different fabric stabilities. The fabric stability of fine-grained soils is sensitive to changes in chemical environment. The stability of coarser materials is influenced by the presence or absence of cementing materials at interparticle contacts. To take account of both fabric and its stability, the term *structure* is used. Structure is determined by particle arrangements (fabric) and interparticle forces.

Both early theories and present understanding of structure are presented. The consequences of structural instability are illustrated through a study of sensitivity and by analysis of spontaneous dispersion of fine-grained soils leading to erosion and to tunneling failures in earth dams.

11.2 STRUCTURE DEVELOPMENT

Early concepts

Early concepts of fabric and structure concentrated on sedimentary clay deposits and were largely speculative, because techniques for direct observation of clay

particles had not yet been developed. There was particular concern for development of a suitable explanation of the loss of strength suffered by most natural clays when remolded at constant water content. This *sensitivity** of the undisturbed structure can be great enough to result in a strength loss due to remolding as great as that illustrated by Fig. 11.1.

Terzaghi (1925) theorized that adsorbed water layers had a high viscosity near particle surfaces and was, thus, responsible for a strong adhesion between mineral grains at the points of contact between particles. Disturbance of the clay causes particle contacts to rupture, more water to fill in around the old contact points, and the strength to drop. Different adsorbed ions were also recognized as possibly responsible for differences in strength and sensitivity (Terzaghi, 1941). The concept that particles in a sensitive clay are arranged in a "cardhouse" that collapses on remolding was advanced by Goldschmidt (1926).

A load-carrying skeleton consisting of highly compressed "bond clay" trapped between silt and fine sand particles, as shown schematically in Fig. 11.2, was hypothesized as responsible for marine clay sensitivity by Casagrande (1932). Such a fabric is assumed to form as a consequence of simultaneous deposition of flocculated clay particles and silt and sand grains in the salt water environment. The clay deposited in the interstices between the elements of the skeleton, termed "matrix clay," is only partly consolidated and remains at high water content. Remolding mixes the matrix and bond clays, thus destroying the primary load carrying structure and causing a reduction in strength. There are some similarities between the grain arrangements in Fig. 11.2 and those shown in Figs. 8.5 and 8.6.

---

* Sensitivity is defined as the ratio of undisturbed to remolded strength.

197

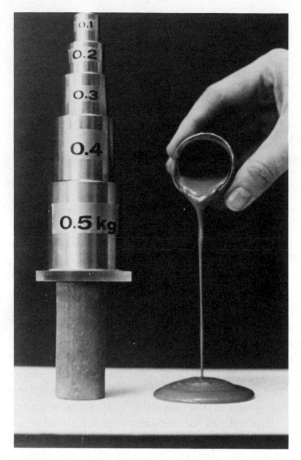

Fig. 11.1 Strength loss of a clay that is extremely sensitive to remolding. A clay that becomes fluid on remolding is termed a "quick" clay (Photograph courtesy of Haley and Aldrich, Inc.).

Winterkorn and Tschebotarioff (1947) suggested that sensitivity resulted from a cementation similar to that in loess and sandstone. This cementation was believed to derive from slow recrystallization or formation of cementing materials from inorganic substances of low solubility. The role of thixotropic hardening in the development of sensitivity was investigated by Skempton and Northey (1952). Both particle arrangements and interparticle forces as influenced by chemical conditions associated with deposition and in the pore water are recognized as important (Lambe, 1953, 1958; Rosenqvist, 1955).

General considerations in structure development

The structure of a soil is composed of a fabric and interparticle force system that reflect, to some degree, all facets of the soil's composition and history. Structure-determining factors and processes are sum-marized in Fig. 11.3. Although it was thought at one time that distinct fabrics, for example, flocculated or dispersed, were associated uniquely with specific sedimentary environments, recent studies show that this is not necessarily the case. The structures of clays formed in salt, brackish, or freshwater environments differ, however, because of differences in electrochemical environment, rate of accumulation, turbulence during sedimentation, postdepositional changes, and other factors.

Initial conditions dominate the structure of young deposits at high porosity; whereas, older soils at lower porosity reflect the postdepositional changes more. In the case of clays, a major consideration appears to be the degree to which a pressure resistant open metastable fabric can develop, and both clay type and chemical environment are important in this regard. For example, the structure of Champlain (Leda) clay, which contains illite as the dominant clay mineral, is relatively little influenced by preconsolidation pressures of up to 500 kN/m² (Gillott, 1970). On the other hand, a one-dimensional consolidation pressure of 10 kN/m² is sufficient to develop intense preferred orientation in a flocculated sediment of kaolin (Morgenstern and Tchalenko, 1967a). Chemical cementation is important in the Champlain clay but is absent in the kaolin.

Single grain fabrics are quite uncommon in soils containing clay, and there may be no relationship between the size and degree of openness of aggregates and the depositional environment. Figures 8.5d and 8.6 are examples of this; the Sundland silty clay (Fig. 8.5d) is a marine deposit and the Tucson silty clay (Fig. 8.6) is a freshwater alluvial deposit, yet their fabrics have many similarities. Complex structural units of micrometer to millimeter size or greater consisting of skeleton grains, clay aggregates and pores are characteristic of most fabrics. Some features may be unique to certain types of soils, however, and these are considered in the remainder of this section.

Residual soils

Relatively few studies have been made of the fabrics of residual soils in relation to geotechnical problems. In the case of the weathering of crystalline rocks in place, the texture of the resulting soil may be quite similar to that of the parent rock. Clay particles may form coatings over silt and sand grains as a result of repeated wetting and drying. Various soil forming processes can lead to open, porous fabrics in some zones and dense, low-porosity fabrics in others.

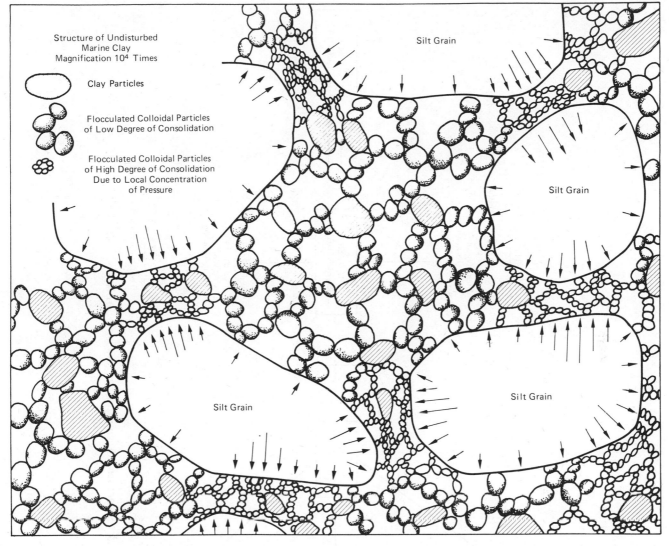

Fig. 11.2 Structure of sensitive clay suggested by Casagrande (1932).

In tropical and subtropical soils, the intense weathering and leaching, coupled with an abundance of aluminum and iron oxides, results in fabrics and textures ranging from open granular to dense and clayey. Concretions and nodules are common in some of these materials. A red kaolinitic clay from Kenya (Barden, 1973) is composed of "crumbs" made up of "subcrumbs" that can in turn break up into "sub-subcrumbs" that contain a random arrangement of individual particles. Pores are in two classes: irregular pores of about 1 μm and very small pores of about 50 Å. Iron oxide particles in such soils tend to aggregate with each other (Smart, 1973).

The fabric of a residual soil may be anisotropic and lead to anisotropic mechanical properties. A horizontally directed transportation component such as creep movement can influence the fabric so that the fabric and properties are not the same in all vertical planes, as is commonly assumed (Lafeber and Willoughby, 1971). In a residual silty clay (70 percent clay size) the only vertical plane of fabric symmetry was in the direction of the dip of the local terrain slope.

Alluvial soils

Alluvial soils can be deposited in marine, brackish water, or freshwater basins. Single particle behavior is rare, and flocculated fabrics of particle groups can form in water over the full range of salinities. Edge-to-face flocculated and aggregated arrangements (similar to Fig. 8.1e) are more common in marine clays,

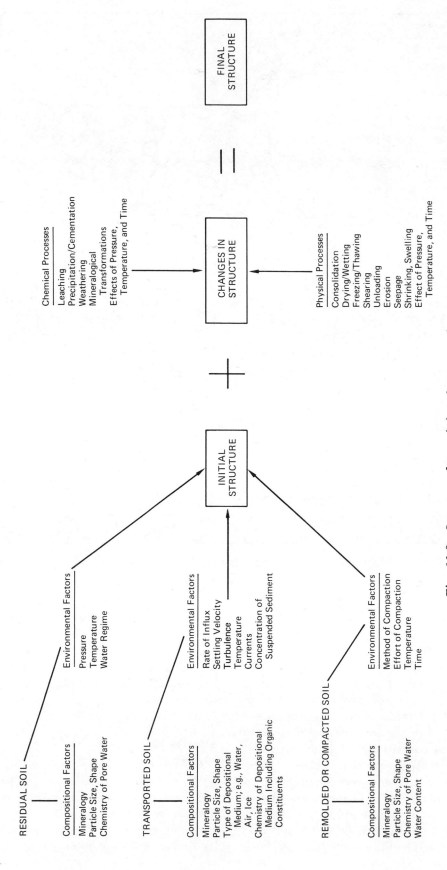

Fig. 11.3 Structure-determining factors and processes.

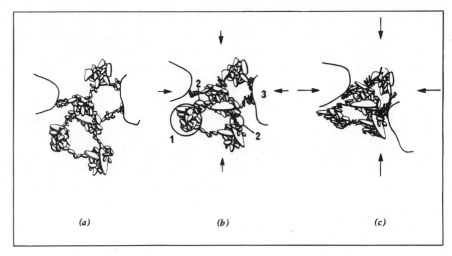

Fig. 11.4   Fabric formation in an aggregated clay (Pusch, 1973).

with dispersed groups and turbostratic groups (similar to interweaving bunches, Fig. 8.3*h*) found mainly in brackish water clays (Collins and McGown, 1974).

Silt and sand grains are reasonably evenly distributed, except in varved or stratified clays, and the larger grains are not usually in contact with each other. Open initial fabrics are characteristic of water-laid sediments, with the degree of openness dependent on clay mineralogy, particle size, and the water chemistry, including both the total salt content and the monovalent/divalent cation ratio. As the intensity of flocculation may be less in brackish and freshwater deposits, subsequent consolidation may lead to development of greater preferred orientation of platy particles and groups than in salt water clays. The rate of sediment accumulation may be an important factor in the development of fabric in different aqueous environments, although specific evidence of this is not available. The very slow rate of accumulation of marine deposits may allow for development of more stability in open fabrics than is the case where deposits thicken rapidly, thereby causing more rapid increase in vertical stress.

The effects of organic matter as well as salinity in the depositional environment may be important. One model (Pusch, 1973*a*, 1973*b*) that seems reasonably well-supported by electron photomicrographs, holds that aggregates of closely packed particles are connected by bridges or links of particles (irregular aggregations linked by connector assemblages according to the terminology of Fig. 8.3) as shown in Fig. 11.4*a*. For clays deposited in freshwater, the aggregates are small, relatively porous and separated by small voids.

In marine clays, the aggregates are large, dense, and separated by large voids (Fig. 11.5). The aggregates are much stronger than the connectors. When a shear stress or consolidation pressure greater than the pre-consolidation stress is applied, the connector links break down and tend toward parallel particle group-

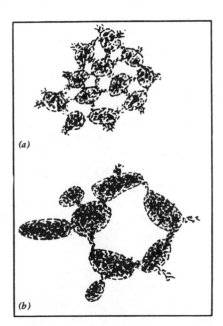

Fig. 11.5 Particle arrangements in fresh and salt water clays (Pusch, 1973). Schematic clay particle arrangement. (*a*) Clay deposited in fresh water, having small, relatively porous aggregates and small voids. (*b*) Marine clay with large, dense aggregates separated by large voids.

ings as aggregates move into more stable positions (Fig. 11.4*b* and 11.4 *c*).

In illitic clays, the aggregates may be composed of particle arrangements ranging from random to booklike. Booklike aggregates are most common in kaolinites, and in smectites the concentration and type of adsorbed cation may control the fabric. A Na-montmorillonite tends to separate into unit layers, and an interwoven network of generally parallel particles may form. Evidence for Ca-montmorillonite is for more irregular arrangements of aggregates containing several unit layers. Some heavily consolidated montmorillonites exhibit surprisingly little preferred orientation.

In soft marine and brackish water illitic clays, there is little or no preferred particle orientation, except within aggregates; whereas, in soft freshwater clays, particles larger than 0.5 μm may align with their long axes normal to the direction of the consolidation pressure. In clay sediments derived from pre-existing shales, the aggregates may themselves be small rock fragments within which the clay plates are intensely oriented.

There is evidence (Krinsley and Smalley, 1973; Smalley, Cabrera and Hammond, 1973) that the open packing of many sensitive postglacial clays may be due in part to the presence of very small quartz particles of platy morphology. Below a critical size of about 100 μm a cleavage mechanism may exist so that platy particles of quartz, and possibly other nonclay minerals, can be formed as a result of glacial grinding.

Organic matter in the form of microscopic animal and plant fragments, micro-organisms, and organic compounds may have a profound effect on the structure and geotechnical properties of postglacial clays (Söderblom, 1966; Pusch, 1973*a*, 1973*b*). The number of bacteria in the oceans at the present time is from 1×10^3 to $3 \times 10^5/cm^3$ at depths of 10 to 50 m below the surface (Reinheimer, 1971). It is probable that micro-organisms were prevalent in the ocean at the time the glacial clays were formed as well. As organic material and clay surfaces interacted, organic matter was attached to the sedimentary aggregates. In the new environment most of the organisms died or became dormant because of the absence of nutrients, subsequently contributing humic acids, fulvic acids, and humus. Either aggregating or dispersing tendencies can result, depending on the environment. Electron micrographs of ultrathin sections show organic matter both as fluffy bodies and as distinct objects associated with aggregates (Fig. 11.6).

Fabric anisotropy is likely as a result of one-dimensional compression after deposition, and this will ordinarily result in some anisotropy of mechanical properties. It was noted earlier that in the case of a residual soil a down-hill transportation component of fabric existed and that a horizontal plane of isotropy was absent. Similar observations have been made for water-laid materials. An example for a beach sand is shown in Fig. 11.7. Here it may be seen that for more or less platy grains, one of the two long axes is parallel to the coastline and the other dips landward at an angle of approximately 10 degrees.

Aeolian soils

Wind-deposited soils, such as loess, are characterized by particles in the silt and fine sand size ranges, although small amounts of clay are sometimes present. These deposits, which are usually partly saturated in nature, are often subject to collapse on saturation. Clean silt or sand grain-to-grain contacts may be common, although clay coatings on the coarser particles are also found. Both granular particle matrices (Fig. 8.31) and partly discernible particle systems (Fig. 8.2*e*) are common. The overall microfabric can be described as "bulky granular."

Directional, preferred orientation in Vicksburg (Mississippi) loess and pre-Vicksburg loess has been observed (Matalucci, Shelton, and Abdel-Hady, 1969). It was found that long axes of grains concentrated in an azimuth direction of 285 to 289 degrees, with an inclination of 3 to 8 degrees. A prevailing wind direction of 290 degrees at the time of deposition was deduced from the thinning pattern of the loess in the area. Such a condition would be expected to result in three-dimensional anisotropy as opposed to simple cross anisotropy.

Glacial deposits

The wide range of particle sizes within and among glacial soils, as well as the widely varying rates of deposition from meltwaters, are responsible for a range of possible fabric types, including clean grain-to-grain contacts, connector assemblages, and regular aggregations (Collins and McGown, 1974). The possible presence of small platy quartz particles in soils derived by glacial grinding was noted earlier.

Many silty and sandy ablation tills have a multimodal grain size distribution, with coarser particles distributed through a fine particle matrix (McGown, 1973). The fabric of the matrix is variable both within and between the tills. Many of the fabric forms are open and similar to those observed (Barden, McGown,

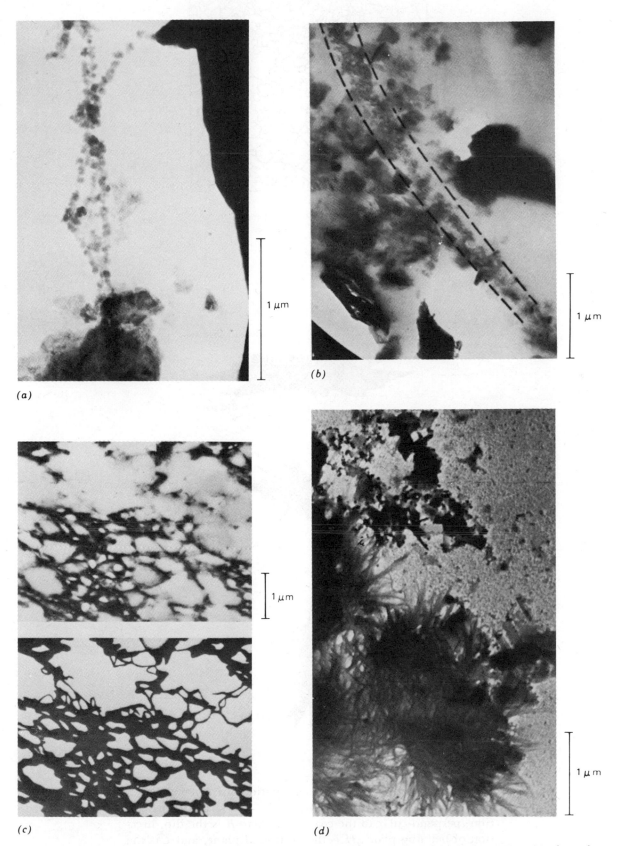

Fig. 11.6 Various types of organic substance in soft glacial and postglacial clays. For (c) the schematic interpretation is shown below the micrograph (Pusch, 1973).

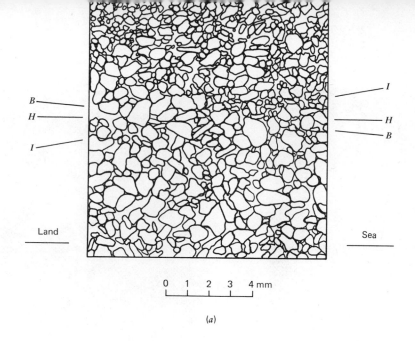

Land

Sea

0 1 2 3 4 mm

(a)

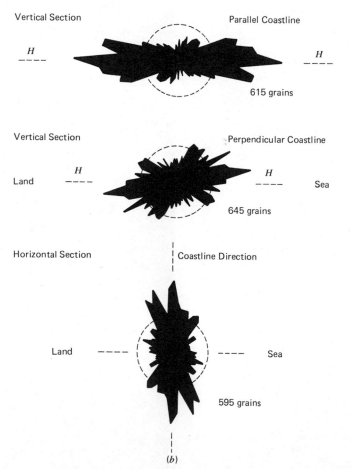

Vertical Section   Parallel Coastline

H   H

615 grains

Vertical Section   Perpendicular Coastline

Land   H   H   Sea

645 grains

Horizontal Section   Coastline Direction

Land   Sea

595 grains

(b)

Fig. 11.7 Fabric and particle orientations in Portsea Beach sand (Lafeber and Willoughby, 1971). (*a*) Vertical cross-section (perpendicular to the coastline) where *B* is the dip direction of bedding plane, *H* is the horizontal plane, and *I* is the dip direction of imbrication plane. (*b*) Distribution of long axis orientations.

Fig. 11.8   Photomicrographs of some glacial tills (McGown, 1973). (*a*) General arrangement of Antarctic Till. (*b*) Open structure of clay sizes in Antarctic Till. (*c*) Area rich in clay sizes in Norwegian Till. (*d*) Unstable clay-silt buttresses in Icelandic Till.

and Collins, 1973) in collapsing soils. Some examples of till microfabrics are shown in Fig. 11.8.

Boulder clays differ from soft, sedimentary clays in that they contain a much wider range of grain sizes, with sizes extending into the gravel to boulder range, and they are much denser. Many boulder clays have

been subjected to high vertical and tangential stresses as a result of readvancing ice sheets. Poor sorting and the presence of a large number of different mineral types is characteristic of these materials.

Well-developed domains of clay are common, although many small local regions of soft clay may be

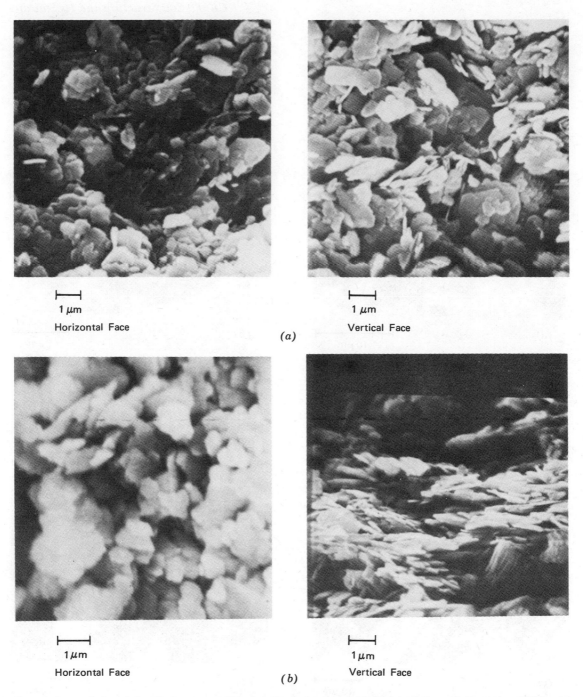

1 μm

Horizontal Face

1 μm

Vertical Face

(a)

1 μm

Horizontal Face

1 μm

Vertical Face

(b)

Fig. 11.9 Fabric of kaolinite consolidated from a slurry (Kirkpatrick and Rennie, 1973). (*a*) Isotropically consolidated sample when the consolidation stress is 420 kN/m² (*b*) K_0 consolidated sample when the major consolidation stress is 420 kN/m².

found (Pusch, 1973*c*). These soft zones are attributed to the arching action of coarse grains, which serve to bridge over the clay in small pores. The high past stresses on some boulder clays have developed features of macrofabric such as shear zones and planes that may influence behavior:

Remolded and compacted soil fabrics

The fabric immediately after remolding or compaction depends on a number of factors, including strength of pre-existing fabric units, compaction method, and compaction or remolding effort. The general effects of disturbance and remolding at constant water content are to break down flocculated aggregations, destroy shear planes, eliminate large pores, and produce a more homogeneous arrangement (on a macroscale) or more dispersed fabric units. Whether or not there will be an overall preferred orientation depends on the methods used. When well-defined shear planes are developed, there tends to be an alignment of platy particles or particle groups along the shear plane.

Under anisotropic consolidation conditions, plates tend to align with their long axes in the plane acted on by the major principal stress. An isotropic (hydrostatic) consolidation stress produces an isotropic fabric, provided the fabric was isotropic at the start of consolidation. These effects are illustrated by Fig. 11.9 for kaolinite consolidated from a slurry at an initial water content of 1.5 times the liquid limit. Working of material such as shown in Fig. 11.9 by kneading, rolling, or extrusion will result in still different fabrics (Weymouth and Williamson, 1953).

Compaction of soils may be done by several different methods, including impact, kneading, vibratory, and static. The method used or the initial state of the soil can have a profound effect on fabrics in both sands (Mahmood, 1973; Oda, 1972a) and clays (Yoshinaka and Kazama, 1973). In clays the water content has an important influence, as it controls the ease with which particles and particle groups can be rearranged under the compaction effort used.

A major consideration in the formation of fabric in a compacted fine-grained soil is whether or not large shear strains are induced by the compaction rammer. If the hammer (impact compaction), tamper (kneading compaction), or piston (static compaction) does not penetrate the soil, as is usual for compaction dry of optimum water content, then there may be a general alignment of particles or particle groups in horizontal planes. For wet of optimum water content, if the compaction effort is high enough, the compaction rammer penetrates the soil surface as a result of a bearing capacity failure under the rammer face. This leads to alignment of particles along the failure surfaces. A series of such zones is developed as a result of successive rammer blows, and a folded or convoluted fabric may result, as shown, for example, by Fig. 11.10.

Effects of postformational changes

There are a large number of different factors that may result in changes in structure subsequent to formation of the initial structure, as listed in Fig. 11.3.

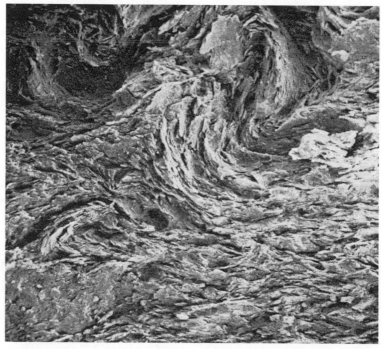

Fig. 11.10 Microfabric of Takahata kaolin compacted wet of optimum using impact compaction (Yoshinaka and Kazama, 1973). × 1000.

Time. Diffusional processes and chemical reactions are time dependent. Thus, following deposition, remolding, or compaction, the fabric and interparticle forces can change simply as a result of time-dependent adjustments including pore pressure redistributions in the new environment. Thixotropic hardening (Mitchell, 1960) can be accounted for in this way.

Seepage and leaching. The flow of water through a deposit can do at least four things:

1. Move particles; see Hansbo (1973).
2. Cause compression due to seepage forces.
3. Remove chemicals, colloids, and micro-organisms.
4. Introduce chemicals, colloids, and micro-organisms.

The change in fabric is usually small except in collapsing soils, but the change in interparticle forces can be great.

Precipitation/cementation. Precipitation of materials onto particle surfaces and at interparticle contacts may develop a cementation. A fabric of partly discernible particles and particle groups may develop.

Weathering. In the zone of weathering, some materials are broken down and others are formed. Changes in pore water chemistry may result that influence the interparticle forces and flocculation–deflocculation tendencies. Weathering can cause significant disruption of the original soil fabric (Chandler, 1972).

Cyclical wetting and drying and freezing and thawing disrupt weak particle assemblages and intergroup associations. Wetting generally means weakening and may lead to collapse of some structures, particularly those with open fabrics where particles are only weakly bonded, such as in loess. Shrinkage associated with drying may cause collapse of open particle arrangements and the development of domain type aggregates in some soils and the formation of tension cracks in others. Drying may concentrate clay around sand and silt particles and between their contact points. Ice-lense formation in frost–susceptible soils can be responsible for the opening of cracks and fissures, followed by later collapse on thawing.

Pressure and consolidation. Consolidation accompanying pressure increases usually strengthens the structure through development of a more stable fabric and stronger interparticle contacts. Platy particles commonly align with long axes normal to the major principal stress direction.

Temperature. Transformations of structure associated with leaching, precipitation/cementation,

weathering, pressure increase, and so on develop more rapidly at high temperature than at low temperature.

Shearing. Shearing can result in a complete collapse of some structures; whereas, in others, such as heavily overconsolidated clay, it may cause a significant change in structure only in the immediate vicinity (within a few millimeters) of the shear plane.

Unloading. Stress relief as a result of unloading can be responsible for elastic rebound of particles and particle groups and the onset of swelling.

11.3 SENSITIVITY

The early concepts of soil fabric and structure, referred to in Section 11.2, were developed at least in part to explain the loss of undrained strength that results when an undisturbed clay is remolded. Although virtually all normally consolidated clays and lightly overconsolidated clays exhibit some amount of sensitivity, with few exceptions the most sensitive large deposits of "quick" clay (a clay that turns to a viscous fluid on remolding, Fig. 11.1) are found in previously glaciated areas of North America and Scandinavia.

The ratio of peak undisturbed strength to remolded strength, as determined by the unconfined compression test was used initially as a quantitative measure of sensitivity (Terzaghi, 1944). The remolded strength of some clays is so low, however, that unconfined compression test specimens cannot be formed, so the vane shear test is often used, both in the field and in the laboratory, as is also the Swedish fall-cone test (Swedish State Railways, 1922; Karlsson, 1961).

Several classifications of clays with respect to sensitivity have been proposed; one of them is given in Table 11.1. It is important to note that clays become quick not because the undisturbed strength becomes

Table 11.1   Classification of Sensitivity Values

	S_t
Insensitive	~1.0
Slightly sensitive clays	1–2
Medium sensitive clays	2–4
Very sensitive clays	4–8
Slightly quick clays	8–16
Medium quick clays	16–32
Very quick clays	32–64
Extra quick clays	>64

From Rosenqvist (1953)

very high, but because the remolded strength becomes very low.

Composition of sensitive clays

Clays of greatly differing composition may exhibit very high sensitivities. Quick clays may not differ from clays of low sensitivity with respect to mineralogical composition, grain size distribution, or fabric. However, the facts that there are small grain sizes and platy particles are important.

Most quick clays are postglacial deposits, with the mineralogy of the clay fraction dominated by illite and chlorite and the nonclay fraction by quartz and feldspar. The activity of quick clays is usually less than 0.5. Pore fluid composition is of paramount importance, particularly differences that develop between the time of deposition and the present. The type and amount of electrolyte, organic compounds, and small quantities of surface active agents may be controlling factors. Changes in one or more of these components are invariably associated with the development of quick clays.

Fabric of sensitive clays

With the possible exception of soils with strongly cemented particles, the undisturbed fabric of sensitive clays is composed of flocculated assemblages of particles or aggregates (Rosenqvist, 1959; Pusch, 1966b, 1970, 1973a; Tovey, 1971; Collins and McGown, 1974). Electron photomicrographs show bookhouse and stepped-flocculated particle arrangements in medium sensitive to quick clays. Dispersed arrangements are sometimes seen in medium sensitive soils. Very sensitive to quick clays contain irregular aggregations linked by connector assemblages. Connector assemblages evidently are not common in soils of low to medium sensitivity. Interassemblage pores are larger in the clays of high sensitivity than in the more insensitive clays. The contribution of fabric to high sensitivity is through the open networks of irregular aggregation assemblages, linked by unstable connector assemblages (Collins and McGown, 1974). The fabric of undisturbed sensitive Drammen clay from Konnerudgt, Norway is shown in Fig. 8.25. Undisturbed Leda clay from Quebec is shown in Fig. 11.11. A very wide range of particle sizes may be seen.

The microfabric of quick clay and that of adjacent zones of much less sensitive clay may be the same; thus, while an open flocculated fabric is necessary, it is not a sufficient condition for quick clay development. Some preferred orientation may develop in

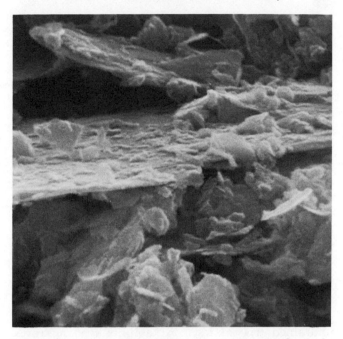

Fig. 11.11 Photomicrograph of undisturbed Leda clay. Air dried. Picture width is 8 μm (Tovey, 1971).

quick clays as a consequence of delayed compression (Kazi and Moum, 1973). Delayed compression (Bjerrum, 1967) can be accelerated as a result of leaching of salts during the formation of quick clay (Torrance, 1974).

All evidence suggests that disturbing a sensitive clay leads to a breakdown of the fabric, probably through rupture of connectors between assemblages. Interestingly, there seems not yet to have been published an electron photomicrograph of the fabric of a remolded quick clay.

11.4 CAUSES OF SENSITIVITY

At least six different phenomena may contribute to the development of sensitivity.

1. Metastable fabric.
2. Cementation.
3. Weathering.
4. Thixotropic hardening.
5. Leaching, ion exchange and change in monovalent/divalent cation ratio.
6. Formation or addition of dispersing agents.

Metastable fabric

For any fine-grained soil in which particles and particle groups tend to flocculate, the initial fabric after sedimentation will be open and involve some amount of edge-to-edge and edge-to-face associations. During consolidation this fabric can carry effective stress at a void ratio higher than would be possible if the particles and particle groups were arranged in an efficient, parallel array. When a clay formed in this manner is remolded, the fabric is disrupted, effective stresses are reduced because of the tendency for the volume to decrease, and the strength is less.

The importance of metastable particle arrangements in the development of sensitivity has been demonstrated for both pure clays and sand–clay mixtures. Sensitivities from 2 to 12 were measured in undrained triaxial tests on a nonthixotropic kaolinite composed of less than 2 μm particles (Houston, 1967). High initial water content, low load-increment ratio, and low rate of loading all tend to give a higher water content for a given effective stress and, therefore, higher values of sensitivity.

Figure 11.12 shows the results of consolidated–undrained triaxial tests on kaolinite–sand mixtures. It may be seen that the loss in strength due to disturbance is accompanied by a large increase in pore water pressure and decrease in effective stress to almost zero from the initial value of 2.0 kg/cm².

Cementation

Many soils contain free carbonates, iron oxide, alumina, and organic matter which may precipitate at interparticle contacts and act as cementing agents. On disturbance of the soil fabric cemented bonds are destroyed leading to a loss of strength. Four naturally cemented Canadian clays were found to have sensitivities of 45 to 780 (Sangrey, 1970).

A late glacial plastic clay from near Lilla Edit in the Gota Valley of Sweden exhibits a sensitivity of 30 to 70, and the apparent maximum preconsolidation pressure as determined by oedometer tests is much greater than the maximum past overburden pressure (Bjerrum and Wu, 1960). If a consolidation pressure greater than the apparent maximum past pressure is applied, a marked reduction in cohesion is observed. This is interpreted to result from a rupture of cemented interparticle bonds that resulted from carbonation of microfossils and organic matter and precipitation of pore water salts at particle contacts. Removal of carbonates, gypsum, and iron oxide by leaching with EDTA (a disodium salt of ethylene-diamine-tetra acetic acid) resulted in a marked reduction in the apparent preconsolidation pressure of a quick clay from Labrador (Bjerrum, 1967).

A "quasipreconsolidation" effect (Leonards and Ramiah, 1960) results if a clay remains under a constant stress for a long period. Whether the additional resistance to compression results from a true chemical cementation is debatable; however, the effect is the same and an increase in sensitivity results.

Weathering

Weathering processes cause changes in the types and relative proportions of ions in solution, which in turn can alter the flocculation–deflocculation tendencies of the soil after disturbance. Some change in the undisturbed strength is also probable; however, the major effect on sensitivity is usually through change in the remolded strength. Strengths and sensitivities may be increased or decreased depending on the nature of the changes in ionic distributions (Moum and Rosenqvist, 1957, 1961; Moum, Löken, and Torrance, 1971). The total salt content and the relative proportions of sodium, potassium, calcium, and magnesium are of greatest importance, as discussed later.

Thixotropic hardening

Thixotropy is defined as an isothermal, reversible, time-dependent process occurring under conditions of constant composition and volume, whereby a material stiffens while at rest and softens or liquifies upon remolding. The properties of a purely thixotropic material are shown in Fig. 11.13. Thixotropic hardening may account for low to medium sensitivities and for a part of the sensitivity of quick clays (Skempton and Northey, 1952).

The mechanism of thixotropic hardening is hypothesized to be as follows (Mitchell, 1960): sedimentation, remolding, and compaction of soil produce a structure compatable with conditions at that time. Once the externally applied energy of remolding or compaction is removed, however, the structure may no longer be in equilibrium with the surroundings. If the interparticle force balance is such that attraction is somewhat in excess of repulsion, there will be a tendency towards flocculation of particles and aggregates and for a reorganization of the water–cation structure to a lower energy state. Both effects, which have been demonstrated experimentally, take time because of the viscous resistance to particle and ion movement.

The effect of time after disturbance on the pressure in the pore water is particularly interesting. Several studies have shown that there is a continual decrease in pore water pressure, or increase in pore water tension, with time after compaction or remolding. Figure 11.14 and Ripple and Day (1966) show that shear of

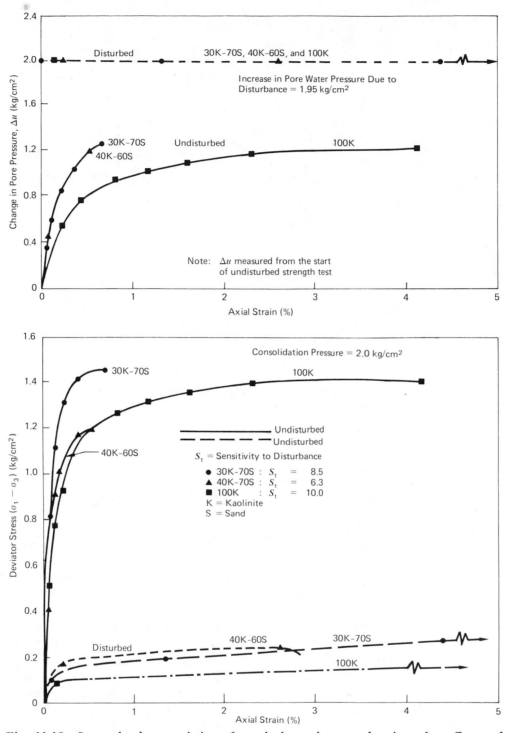

Fig. 11.12 Strength characteristics of sand–clay mixtures showing the effects of disturbance.

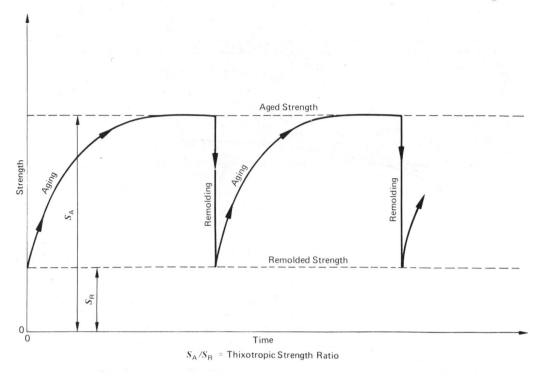

Fig. 11.13 Properties of a purely thixotropic material.

thixotropic clay pastes causes an abrupt drop in pore water tension followed by a slow regain during periods of rest. The increase in effective stress that results accounts for the observed increase in undrained strength.

Unfortunately, the extent to which thixotropic hardening may account for the sensitivity of a clay in the field is impossible to determine. Studies of thixotropy in the laboratory start with remolded and compacted clays at a specific present composition and density. The initial state of a clay deposit in nature was far different than at the present time, and the history of an undisturbed clay bears little resemblance to that of a remolded sample of the same clay which is allowed to rest at constant water content and pore fluid composition. The results of studies on samples allowed to harden starting from present composition (Skempton and Northey, 1952; Seed and Chan, 1957; Mitchell, 1960) suggest that sensitivities of up to about 8 or so may be possible due to thixotropy.*

Leaching, ion exchange, and changes in the monovalent/divalent cation ratios

The importance of salt leaching from the pores of marine clays to the development of high sensitivities

was first suggested by Rosenqvist (1946). Subsequent study has confirmed that reduction in the salinity of marine clays is an essential step in the development of a quick clay. Leaching of salt resulted after a drop in sea level or rise in land level caused the clay to be lifted above sea level so that it became exposed to a freshwater environment. The presence of percolating freshwater in silt and sand lenses is sufficient to cause removal of salt from the clay by diffusion without the requirement that the water flow through zones of intact clay, as demonstrated by laboratory experiment (Torrance, 1974).

Leaching causes little change in fabric (Kazi and Moum, 1973); however, the interparticle forces may be changed, resulting in a decrease in undisturbed strength of up to 50 percent, and such a large reduction in remolded strength that a quick clay may form. The large increase in interparticle repulsion, which is responsible for the deflocculation and dispersion of the clay on remolding, results in part from the decrease in electrolyte concentration, which causes an increase in double layer thickness. Changes in strength and the increase in sensitivity accompanying the leaching of salt from a Norwegian marine clay are shown in Fig. 11.15. The relationship between sensitivity and present salt content for several Norwegian marine clays is shown in Fig. 11.16. Confirmation of the leaching hypothesis has been obtained by means

* Sherard (1975, personal communication) indicates that thixotropic strength ratios of up to 100 have been measured in Champlain clay.

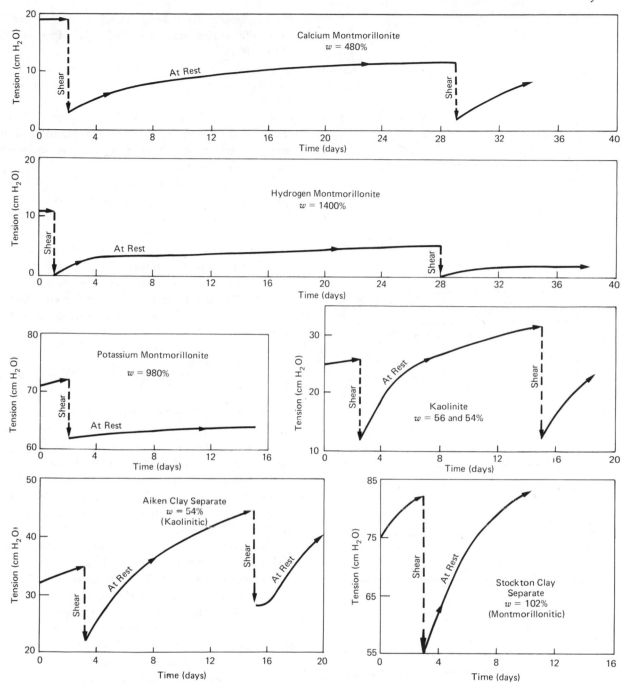

Fig. 11.14   The effect of shear on pore water tensions for various clays (After Day, 1955).

of leaching tests on artificially sedimented clays (Bjerrum and Rosenqvist, 1956. Åsrum clay sedimented in salt water (35 gm/liter) and then leached of salt exhibited an increase in sensitivity from 5 to 110. A sample sedimented in fresh water had a sensitivity of 5 to 6.

Although leaching of salt is necessary, it may not be sufficient for the development of quick clay. Studies on Swedish clays (Osterman, 1964; Söderblom, 1969) have also indicated that removal of salt does not insure development of quick clay. The salt content of Champlain (formerly termed Leda) clay in eastern Canada rarely exceeds 1 to 2 gm/liter and is usually less than 1 gm/liter. The sensitivities of dif-

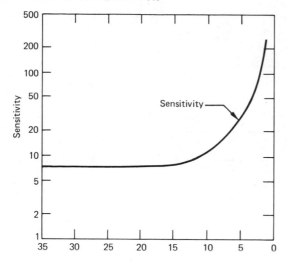

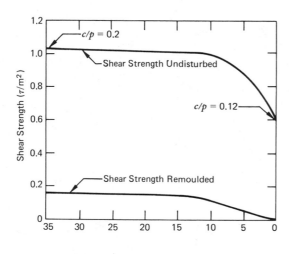

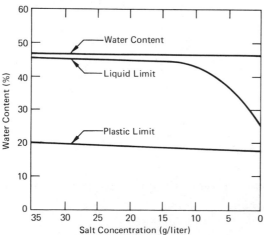

Fig. 11.15   Changes in properties of a normally consolidated marine clay when subjected to leaching by freshwater (Bjerrum, 1954).

ferent samples have been found to range from 10 to over 1000 (Eden and Crawford, 1957; Penner, 1963b; 1964, 1965).

The reason for this is that the essential condition for development of a quick clay is an increase in interparticle repulsions through expansion of the double layer. Considerations in Chapter 7 show that the type of cations and the relative amounts of mono- and divalent cation have a controlling influence on the structural status.

The electrokinetic or zeta potential in Champlain clay as determined using electro-osmosis (see Chapter 15) correlates well with sensitivity (Penner, 1965) as in Fig. 11.17. The electrokinetic potential is a measure of the double layer potential, with higher values associated with thicker double layers and higher sensitivity. For clays of low salinity (less than 1 or 2 g of salt/liter of pore water), the sensitivity correlates well with the percent of monovalent cations in the pore water

$$\frac{Na^+ + K^+}{Na^+ + K^+ + Ca^{2+} + Mg^{2+}} \times 100, \text{ all in meq/liter}$$

as in Fig. 11.17, where data for several clays are shown. Data for a Norwegian clay are shown in Fig. 11.18, where the independence of sensitivity and salt content, but dependence of sensitivity on monovalent to total cation ratio are both evident. An analysis in terms of the sodium adsorption ratio (Section 7.11) leads to a similar result (Balasubramonian and Morgenstern, 1972).

The percent monovalent cation in sea water is only about 75 on a meq/liter basis. Thus, according to the relationship in Fig. 11.17, if sea water is leached from a clay without change in the relative concentrations of Na^+, K^+, Mg^{2+}, and Ca^{2+} very high sensitivities cannot develop. Selective removal of divalent cations would appear necessary. In a quick clay, the Ca^{2+} and Mg^{2+} are complexed out of the system, possibly by organic matter (Söderblom, 1969).

Formation or addition of dispersing agents

Quick clays are frequently found in the vicinity of organic layers (Osterman, 1964; Söderblom, 1966). Organic substances are introduced at the time of sediment deposition and a variety of organic compounds are found in the pores of the clay. Some of these compounds may act as dispersing agents and lead to increased double layer repulsions. Some substances have been isolated from quick clays and peat water that

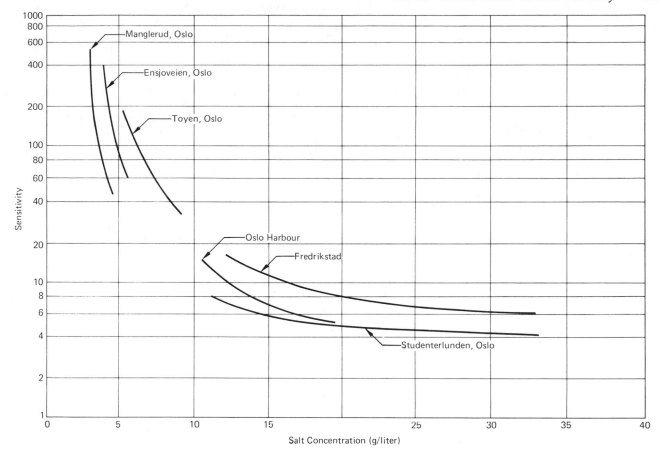

Fig. 11.16 Relationship between sensitivity and salt concentration for some Norwegian clay deposits (Bjerrum, 1954).

possess strong dispersing characteristics such as acids, polyphenols, and tannins (Söderblom, 1966). The sensitivity of leached artificially sedimented clays free of organic matter is less than that of the same clay from which organic matter has not been removed (Pusch, 1973b; Pusch and Arnold, 1969). It is not yet clear the extent to which organic substances aid the development of sensitivity as regards the relative contributions of direct dispersing agent action and through complexing of divalent cations.

Inorganic dispersing agents may also cause the formation of quick clay. In one case (Söderblom, 1966), a phosphate releasing waste dump was placed over a normal clay, and 3 years later the clay was in a quick condition.

The six causes for the development of sensitivity that have been described are summarized in Table 11.2, where an estimate of the upper limit of sensitivity is also given. Virtually all natural coherent soils are sensitive in that they lose some strength on re-

molding. Exceptions are heavily overconsolidated stiff fissured clays that gain strength because of the elimination of fissures and planes of weakness. Quick clays are formed from soft glacial marine clays only after removal of excess salt by leaching and further increase in double layer repulsions as a result of an increase in the relative proportion of monovalent cations (mainly sodium) in the pore water or as a result of the formation or leaching in of dispersing agents. These dispersing agents are possibly derived from organic compounds. More than one mechanism may contribute to the development of sensitivity in any one soil.

As a final comment, an aging effect has been reported in Swedish quick clays (Söderblom, 1969), the effect of which is to produce a decrease in undisturbed strength and an increase in remolded strength with time after sampling. There is also an increase in the concentration of Ca^{2+} and Mg^{2+} ions in the pore water with time. The causes of these effects, which

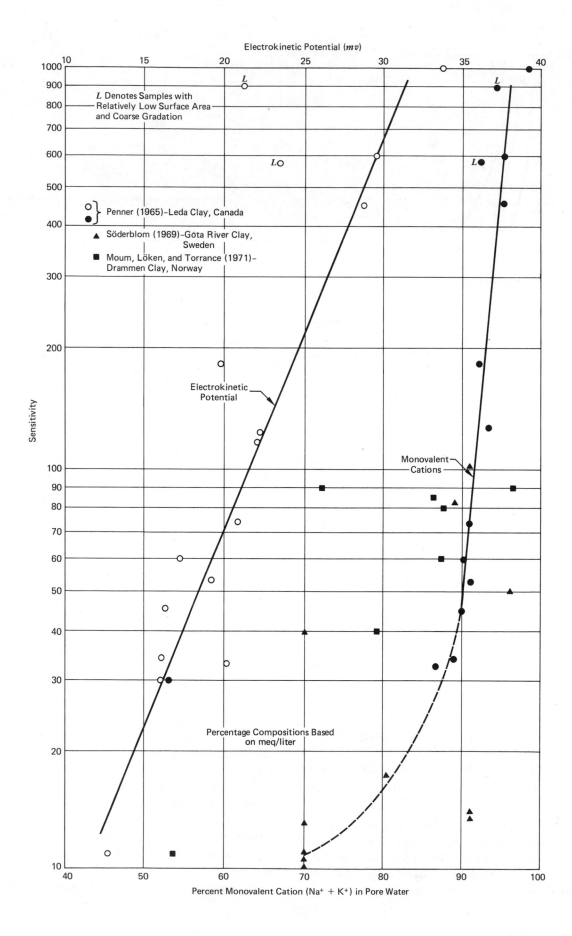

Electrokinetic Potential (*mv*)

L Denotes Samples with Relatively Low Surface Area and Coarse Gradation

○ } Penner (1965)–Leda Clay, Canada
●

▲ Söderblom (1969)–Göta River Clay, Sweden

■ Moum, Löken, and Torrance (1971)– Drammen Clay, Norway

Electrokinetic Potential

Monovalent Cations

Percentage Compositions Based on meq/liter

Sensitivity

Percent Monovalent Cation (Na$^+$ + K$^+$) in Pore Water

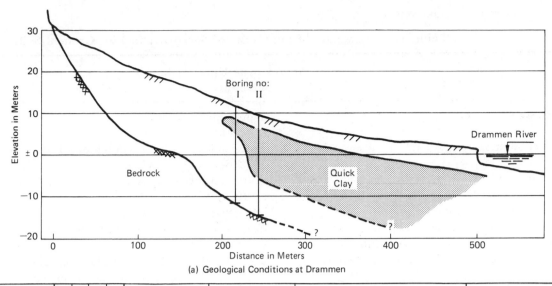

(a) Geological Conditions at Drammen

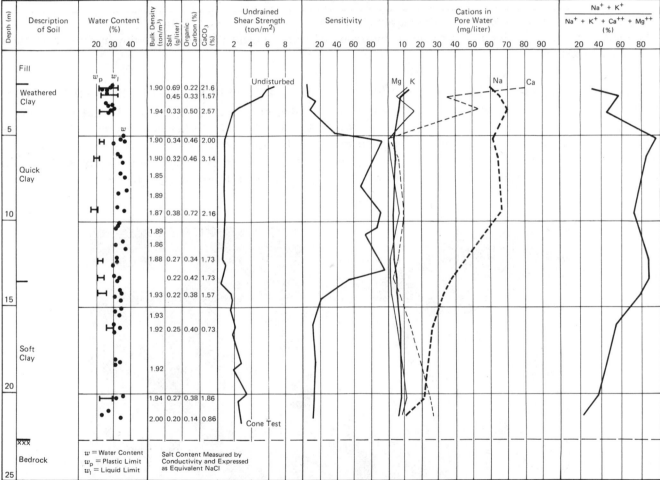

(b) Geotechnical Data for Boring II

Fig. 11.18 Sensitivity and pore water chemistry for a boring at Drammen, Norway (After Moum, Löken, and Torrance, 1971).

Fig. 11.17 Relationship between sensitivity and percent monovalent cations in low salt content clays and between sensitivity and electrokinetic potential.

Table 11.2 Summary of the Causes of Sensitivity in Fine-Grained Soils

Mechanism	Approximate Upper Limit of Sensitivity[a]	Predominant Soil Types Affected
Metastable fabric	Slightly quick (8–16)	All soils
Cementation	Extra quick (>64)	Soils containing Fe_2O_3, Al_2O_3, $CaCO_3$, free SiO_2
Weathering	Medium sensitive (2–4)	All soils
Thixotropic hardening	Very sensitive[b]	Clays
Leaching, ion exchange, and change in monovalent/divalent cation ratio	Extra quick (>64)	Glacial and postglacial marine clays
Formation or addition of dispersing agents	Extra quick (>64)	Inorganic clays containing organic compounds in solution or on particle surfaces

[a] Adjectival descriptions according to Rosenqvist (1953).
[b] Pertains to samples starting from present composition and water content. Role of thixotropy in causing sensitivity in situ is indeterminate.

may have important implications concerning the relevance of properties measured in the laboratory to field conditions, have yet to be fully understood. Possibilities include the effects of changes in temperature and pressure, and organic effects that may develop in the new environment, including bacterial action.

11.5 TUNNELING FAILURES AND EROSION

Recent research, mainly in Australia and in the U.S.A. (Sherard, Decker, and Ryker, 1972, 1975) has established that certain clays are structurally unstable, easily dispersed, and, therefore, highly erodible. The consequences following the action of water on such

Fig. 11.19 Erosion pattern in excavated slope of dispersive clay (Courtesy of J. L. Sherard).

Fig. 11.20 Erosion damage on crest of 15-ft high clay flood control dike caused by rain runoff concentrating in drying cracks, Rio Zulia, Venezuela (Courtesy, J. L. Sherard).

materials may be several, as shown in Figures 11.19, 11.20, and 11.21. Figure 11.19 shows the surface erosion pattern on an excavated slope. Erosion tunnels in a flood control dike caused by rainfall are shown in Fig. 11.20, and Fig. 11.21 shows erosion tunnels extending completely through a dam that were created by seeping reservoir water. A number of failures of this type has developed in well-constructed, low, homogenous dams, and recognition of the causes has been one of the major advances in embankment dam design and construction in recent years.

In each case the soil contained readily dispersed clay particles that went easily into suspension in flowing water. Most failures of this type have occurred in embankments, dams, and slopes composed of clays with low to medium plasticity (CL and CL–CH) that contain montmorillonite. Dispersive piping in dams has occurred either on the first reservoir filling or, less frequently, after raising the reservoir to a higher level than previously.

The initiation of failures of the type shown in Figs. 11.19 and 11.20 requires the presence of small cracks or fissures along which the erosion may initiate. Tunneling failures (Fig. 11.21) commence at the upstream face. When the reservoir is filled for the first time, settlement may accompany saturation of the soil, particularly if the soil were placed dry and not well compacted. Settlement below the phreatic line and arching above can lead to crack formation. Water moving along the crack erodes according to the dispersibility of the soil and flow velocity. This is a fundamentally

Fig. 11.21 Tunneling failure of low Australian Dam. Note three erosion tunnels entering upstream face.

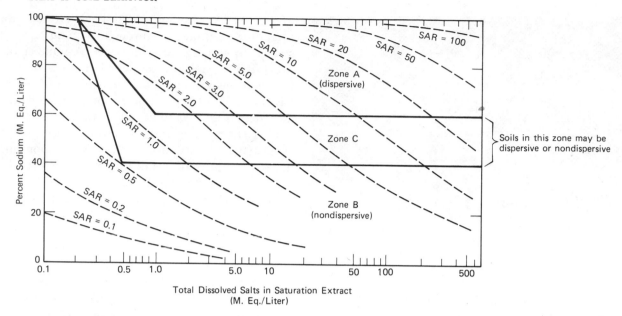

Fig. 11.22 Relationship between dispersibility (susceptibility to colloidal erosion) and dissolved pore water salts based on pinhole tests and experience with erosion in nature. From Sherard, Dunnigan, and Decker, 1976b. $SAR = \dfrac{Na}{\sqrt{0.5\,(Ca + Mg)}}$, all in m. eq./liter. *Note:* Relationship shown is valid only when eroding water is relatively pure.

different mechanism than the piping of cohesionless soils, which develops starting at the discharge face. Tunneling has been initiated in soils with a permeability as low as 1×10^{-5} cm/sec.

Visual classification, Atterberg limits and particle size analyses do not provide a basis for differentiation between dispersive clays and ordinary erosion resistant clays. Relatively simple chemical tests can be used, however, as well as a straightforward dispersion test and a recently developed pinhole test (Sherard, Dunnigan, and Decker, 1976a) for identification of dispersive clays.

The percent sodium in a saturation extract from the soil is as reliable an indicator of probable performance (Sherard et al., 1972, 1975b). Study of the behavior of many dams led to the correlation shown in Fig. 11.22, which can be used as a basis for estimation of a tendency for spontaneous dispersion. In the pinhole test (Sherard, Dunnigan, and Decker, 1976a) distilled water is allowed to flow through a 1.0-mm-diameter hole drilled through a compacted specimen. The water becomes muddy and the hole rapidly erodes in dispersive clays. For nondispersive clays the water is clear and there is no erosion.

A laboratory dispersion test can also be used to measure the dispersibility or erodibility of clayey soils

(Sherard, Decker, and Ryker, 1972). The percentage of particles finer than 5 μm is determined by hydrometer analyses on samples with and without dispersing agent in the suspension. The higher the ratio of percentage of material less than 5 μm without dispersing agent to that measured with a dispersing agent, the greater the probability of dispersion in the field. Dispersion values greater than 20 to 25 percent are indicative that dispersion in the field may be a problem, and values greater than 50 percent are nearly always indicative of soils susceptible to severe erosion damage.

A soil that is stable initially can, at a later date, become susceptible to dispersion failures. This could develop in a situation where the percolating rain, reservoir water, or ground water caused a chemical change resulting in transition from Zone B to Zones A or C, although there is no evidence to support this in the case of earth dams (Sherard et al., 1976b).

In those cases where dispersion is likely to be a problem, the addition of 2 to 3 percent hydrated lime by direct mixing during construction may be sufficient to prevent movement of clay particles. In the case of an existing dam where tunneling erosion is expected to develop, lime can be added at the upstream face to be carried inwards by the percolating water.

11.6 CATION EXCHANGE AND SLOPE STABILITY

Since the strength of clays is dependent, among other factors, on the nature of the adsorbed cations, there is reason to believe that ion exchange might impair the stability of slopes in certain cases. The possibility that a slope failure may have resulted from a change from a predominantly calcium exchange complex to one with sodium dominant was suggested by Matsuo (1957).

A number of soils from slide zones in the East Bay Hills of the San Francisco area have been studied to see whether failure may have been associated with the development of dispersive clay and reduced strength (Mitchell and Woodward, 1973). The results indicated that in most cases the soil classified as dispersion resistant; however, a few samples are classified in the unstable or transition zone of Fig. 11.22. The development of slope instability as a result of dispersion appears possible, but further study is needed.

SUGGESTIONS FOR FURTHER STUDY

Aitchison, G. E., and Wood, C. C. (1965), "Some Interactions of Compaction, Permeability, and Post-Construction Affecting the Probability of Piping Failure in Small Earth Dams," *Proc. 6th Int. Conf. Soil Mech. and Found. Eng., Montreal,* Vol. II, p. 442.

Bjerrum, L. (1954), "Geotechnical Properties of Norwegian Marine Clays," *Geotechnique,* Vol. 4, pp. 49–69.

Collins, K., and McGown, A. (1974), "The Form and Function of Microfabric Features in a Variety of Natural Soils," *Geotechnique,* Vol. 24, No. 2.

Lafeber, D., and Willoughby, D. R. (1971), "Fabric Symmetry and Mechanical Anisotropy in Natural Soils," *Proceedings of the Australia-New Zealand Conference on Geomechanics, Melbourne,* 1971, Vol. 1, pp. 165–174.

Mitchell, J. K. (1956), "The Fabric of Natural Clays and its Relation to Engineering Properties," *Proceedings of the Highway Research Board,* Vol. 35, pp. 693–713.

Sherard, J. L., Decker, R. S., and Ryker, N. L. (1972a), "Piping in Earth Dams of Dispersive Clay," *Proceedings of the ASCE Specialty Conference on the Performance of Earth and Earth-Supported Structures,* Purdue University, pp. 589–626.

Sherard, J. L., Dunnigan, L. P., and Decker, R. S. (1976b), "Identification and Nature of Dispersive Soil," *Journal of the Geotechnical Division,* ASCE., Vol. 102, No. GT 4, pp. 287–301.

CHAPTER 12

Fabric, Structure, and Property Relationships

12.1 INTRODUCTION

The variety of fabrics encountered in different soils is great, and the possible interparticle force systems associated with each are many; thus, the potential number of soil structures is almost limitless. The mechanical properties of a soil reflect directly the influences of structure to a degree that depends on the soil type, the structure type, and the particular property of interest. A knowledge of the interrelationships among these factors is useful in understanding soil behavior.

12.2 FABRIC–PROPERTY INTERRELATIONSHIPS: PRINCIPLES

Several principles can be stated that relate fabric and structure to physical properties of interest in engineering. They are listed below and used in later sections to interpret behavior.

1. Under a given consolidation pressure, a flocculated system is less dense than a deflocculated system.
2. At the same void ratio, a flocculated system with randomly oriented particles and particle groups is more rigid than a deflocculated system.
3. Beyond the precompression range, an increment of pressure causes a greater reorientation of fabric in a flocculated fabric than in a deflocculated fabric.
4. The average pore diameter and range of pore sizes is smaller in deflocculated and/or remolded soils than in flocculated and/or undisturbed soils.
5. In general, shear strains orient platy particles

and particle groups with their long axes in the direction of shear.
6. Anisotropic consolidation stresses tend to align platy particles and particle groups with their long axes in the major principle plane.
7. The equilibrium structure at the time of formation is not necessarily the same as that at a later date.
8. Stresses are usually not distributed equally among all particles and particle groups. There may be particles (termed "idle" particles by Marsal, 1973) or particle groups that are essentially stress free as a result of arching by surrounding fabric elements.
9. If, because of a change in structure, a saturated soil tends to compress, but drainage is prevented, then positive pore pressures develop, and the effective stress decreases.
10. If, because of change in structure, a saturated soil tends to expand or dilate, but drainage is prevented, then negative pore pressures develop, and the effective stress increases.

12.3 PROPERTY INTERRELATIONSHIPS IN SENSITIVE CLAYS

The geotechnical properties of normally consolidated, noncemented, sensitive clays fit a pattern that is predictable using basic concepts of the influences of effective stress and void ratio on volume change tendencies during deformation. These concepts can be used (Houston and Mitchell, 1969) to develop a consistent picture of the properties of sensitive clays in terms of sensitivity, liquidity index, and effective stress.

The liquidity index (LI) given by

$$LI = \frac{water\ content - plastic\ limit}{plasticity\ index} \quad (12.1)$$

is a useful parameter for expressing and comparing the consistencies of different clays. The considerations that follow do not apply to heavily overconsolidated, highly cemented, or residual clays, or to partly saturated or expansive clays.

General characteristics of sensitive clays

Glacial and postglacial clays of high and low sensitivity exhibit important differences in geotechnical properties, as shown by the profiles in Fig. 12.1 for a normal clay at Drammen and a quick clay at Manglerud, both in Norway. One of the most important distinguishing characteristics between these two profiles is that at Manglerud the water content is well above the liquid limit. This is characteristic of quick clays.

Plasticity and activity. Since normal clays are converted to highly sensitive or quick clays as a result

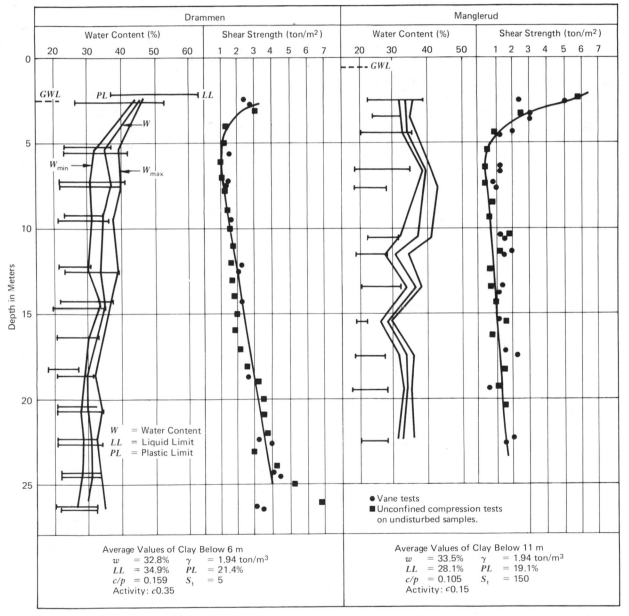

Fig. 12.1 Typical soil profiles for marine clays of high and low sensitivity (Bjerrum, 1954).

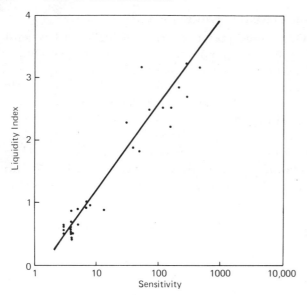

Fig. 12.2 Sensitivity as a function of liquidity index for Norwegian marine clays. Relationship was averaged from many more data points than those shown (Data from Bjerrum, 1954).

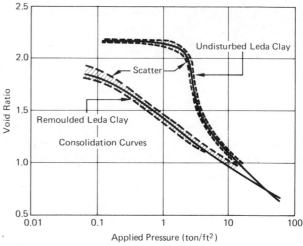

Fig. 12.3 Consolidation curves for Leda clay (After Quigley and Thompson, 1966).

of leaching, ion exchange, or the action of dispersing agents, the liquid limit, plasticity index, and activity decrease. These changes are reflected by an increase in liquidity index at constant effective stress. The liquid limit of a sensitive clay is usually less than 40 percent and rarely above 50 percent. Plastic limits are usually about 20 percent. The activity of most normal inorganic marine clays is of the order of 0.5 to 1.0; whereas, the activity of quick clays can be as low as 0.15. The sensitivity of a given clay type usually correlates uniquely with liquidity index as may be seen in Fig. 12.2 for Norwegian marine clays.

The pore pressure parameter $\bar{A}_f$.* High pore pressures are developed when sensitive clays are sheared. For some quick clays pore pressures as high as 1.5 to 2.2 times the peak deviator stress have been measured.

Undrained shear strength to consolidation pressure ratio, c/p. The *c/p* ratio decreases with increasing sensitivity, ranging from 0.5 or more for insensitive clays to less than 0.1 for quick clays. The c/p ratio increases with increasing plasticity index, Fig. 13.24.

Stress–strain relationships. In general, strain at failure decreases with increasing sensitivity. Some quick clays are quite brittle during unconfined loading and fracture at very low strains, sometimes by

axial splitting. Further working of the fractured specimen may cause it to turn into a fluid mass.

Compressibility. The compressibility of highly sensitive clays is relatively low until the consolidation stress exceeds the preconsolidation pressure. It then increases sharply as shown by Fig. 12.3 for Leda (Champlain) clay. As the void ratio reduces under higher consolidation pressures, the compressibility eventually assumes a lower value.

Property, effective stress, and water content relationships

Consolidation. Because the initial structure of a clay depends on many factors, and the volume changes under pressure are a function of structure, a clay does not have a unique consolidation curve. Consolidation from different initial conditions can yield a family of "normal consolidation" curves. Two such curves, indicated as Paths 1 and 4, are shown in Fig. 12.4. Path 1 might result after initial deposition or remolding a sample to point *E* on curve *FG*, and then reconsolidating to *D*.

The consequences of remolding a sensitive clay are either a decrease in void ratio if the original stresses are reapplied, or an increase in pore pressure and decrease in effective stress if no drainage is allowed. Thus remolding from *A* causes transfer along Path 3 to a new state at *E* if no drainage is allowed.

A specimen at *C* can be brought to *D* along Path 2, which is a conventional swelling curve. The effective stress of a specimen initially at *B* can be brought to the value at *D* by partial remolding along Path 5. Thus, there are at least three means by which a specimen with coordinates of *D* can be obtained.

---

* The pore pressure parameter $\bar{A}$ is the ratio of pore pressure increase to deviator stress increase during shear. $\bar{A}_f$ refers to $\bar{A}$ at failure.

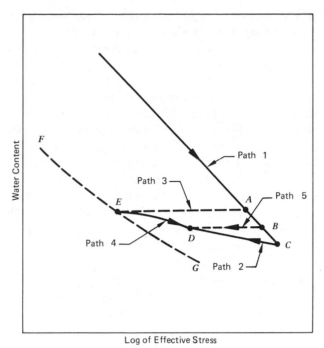

Fig. 12.4 Water content–effective stress relationships for clay consolidated from different initial conditions.

Strength of normally consolidated soil. The higher the effective stress at a given water content, the greater the undrained strength, because of increased frictional resistance. For constant effective stress, strength increases with decreasing water content, because of increased cohesive resistance. Thus, the general behavior shown in Fig. 12.5 results.

Pore pressure parameter, $\bar{A}_f$. The pore pressure at failure is controlled by the tendency of the soil to dilate or contract. The lower the water content, the greater the geometrical interference between particles and the greater the tendency for dilation at a given effective stress. Thus $\bar{A}_f$ decreases with decreasing water content at constant initial effective stress. At constant water content, dilation is easier the lower the effective stress, since less energy is required for expansion against low pressures than high. Therefore, the maximum gradient of $\bar{A}_f$ increase is as shown in Fig. 12.6.

Strain at failure. As dilation tendencies increase effective stress, thus increasing shearing resistance, the deformation required to cause failure increases with increasing dilation. Therefore, strain at failure should increase with dilation. It will decrease with increase in $\bar{A}_f$, because $\bar{A}_f$ varies inversely with dilation tendencies. Consequently, the maximum positive gradient of strain at failure should be opposite to the maximum gradient for $\bar{A}_f$.

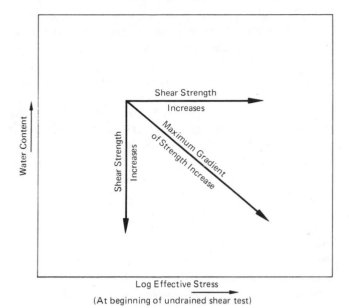

Fig. 12.5 Gradient of strength increase with water content and effective stress variation.

Sensitivity. Each point on curve *FG* in Fig. 12.4 represents a completely remolded clay. As the sensitivity of clay at any point on this curve must be unity, curve *FG* is a line of constant sensitivity, or a sensitivity contour. A saturated clay at a given water content and pore fluid composition cannot be made weaker than its thoroughly remolded strength; therefore, a water content–effective stress condition to the

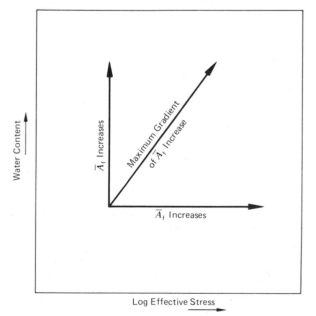

Fig. 12.6 Gradient of pore pressure parameter $\bar{A}_f$ with water content and effective stress variations.

left of *FG* is not possible. As the undisturbed strength increases with effective stress at constant water content (Fig. 12.5), the sensitivity of all points to the right of *FG* is greater than one, and the maximum gradient of sensitivity increase is generally normal to *FG*.

Example of relationships. The results of triaxial compression tests on kaolinite (Houston, 1967) serve to illustrate the above relationships. By consolidating different samples from several different initial water contents and remolding and reconsolidating in various ways, samples covering a range of initial effective stress and water content values, each reflecting a different structure, were obtained. The results of undrained triaxial tests were used to obtain values of strength, $\bar{A}_f$, strain at failure, and sensitivity. Contours based on these values are shown in Fig. 12.7. The variations in the measured values are in accord with predictions.

Sensitivity–effective stress–liquidity index relationship

General relationships between sensitivity, effective stress, and water content can be established for all clays of the type under consideration in this section, based on an appropriate normalization of the remolded strength versus water content relationship. The liquidity index provides the basis for such a relationship. A unique relationship between sensitivity, liquidity index, and effective stress exists if:

1. The LI-effective stress relationship is the same for thoroughly remolded specimens of all clays. This represents the contour for sensitivity equals one.
2. The relationship between remolded strength and liquidity index is the same for all clays.
3. At any value of liquidity index, the variation of c/p with effective consolidation pressure is the same for all clays. This fixes the undisturbed strength in terms of LI and effective stress.

For most sensitive clays, variations from these conditions are not large enough to prevent development of useful relationships. Remolded shear strength as a function of liquidity index for several clays is shown in Fig. 12.8. Several types of strength tests, for example, vane, unconfined, and triaxial compression, were used to obtain the data shown, which may account in part for the range of values.

By averaging data for several clays, the relationship between liquidity index, effective stress, and sensitivity shown in Fig. 12.9 is obtained. Although Fig.

12.9 was developed for normally consolidated clays, it can be used for moderately overconsolidated clays provided the preconsolidation pressure is used instead of the present effective stress. This is because the undrained strength depends more on the preconsolidation pressure than on the present effective stress. Some deviations from the values in Fig. 12.9 are to be expected because of the extensive averaging used in its preparation. These deviations may be greatest in the case of extra quick clays, because of the difficulty in determination of the sensitivity of such materials, due to the very low remolded strength. Nonetheless, the relationships in Figs. 12.8 and 12.9 can be used to estimate sensitivity and strength when undisturbed samples or *in situ* strength test data are not available, to estimate changes in strength and sensitivity due to change in effective stress or liquidity index, and as a guide for extrapolating a small amount of data into a larger pattern.

12.4 FABRIC AND PROPERTY ANISOTROPY

Anisotropic consolidation, shear, a transportational component during or after deposition, and method of sample preparation in remolded and compacted soils may result in anisotropic fabrics. Fabric anisotropy on a macroscale usually results in mechanical property anisotropy, and property differences in different directions may be significant. The amount of shear or compression required for development of anisotropic fabric varies for different soils, and depends on such factors as soil mineralogy, composition of pore fluid, and initial fabric. Stresses as low as 10 kN/m^2 (0.1 ton/ft^2) may be sufficient to develop intense preferred orientation in kaolinite sedimented in flocculating and deflocculating pore fluids (Morgenstern and Tchalenko, 1967a). Considerably higher stresses may be needed in other soils. Orientation development, as indicated by X-ray diffraction measurements of the (001) illite peak height, in undisturbed Champlain (Leda) clay is shown in Fig. 12.10. Examples of anisotropic fabrics in sands are shown in Figs. 8.11 and 8.12, and even assemblages of spherical particles, when allowed to fall freely, may develop anisotropic fabrics (Kallstenius and Bergau, 1961).

Some examples are presented in this section to illustrate the general nature and magnitudes of anisotropy in properties that may be associated with an homogeneous anisotropic fabric. Property anisotropy caused by stratification may be important in the field but is not considered here.

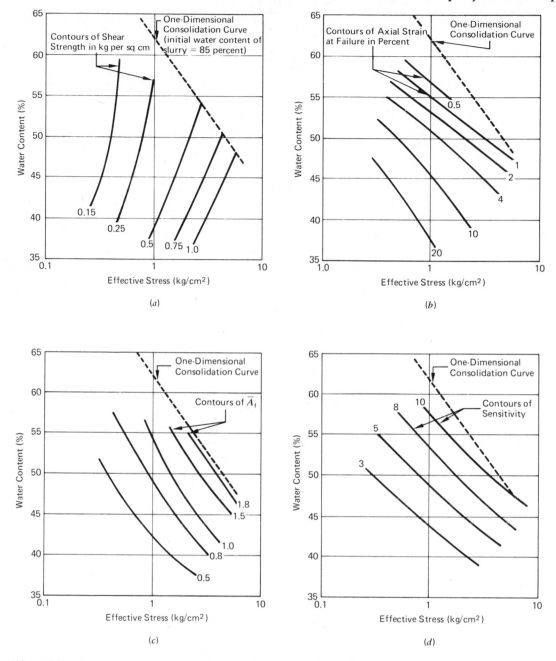

Fig. 12.7 Strength properties of normally consolidated kaolinite as a function of effective stress and water content. (*a*) Shear strength. (*b*) Strain at failure. (*c*) Pore pressure parameter $\bar{A}_f$. (*d*) Sensitivity.

Sands and silts

The strength of crushed basalt specimens both along and across the direction of preferred orientation of grains is shown in Fig. 12.11 (Mahmood and Mitchell, 1974). Preferred orientation of the somewhat elongated particles (mean particle length to width ratio = 1.64) was obtained by pouring the soil into the shear box; Fig. 8.12 shows that intense preferred orientation was obtained at moderate relative densities. It may be seen in Fig. 12.11 that at the lower densities the strength was about 40 percent higher for shear across the plane of orientation than for the shear along it. This difference decreased with increasing density, and for relative densities above 90

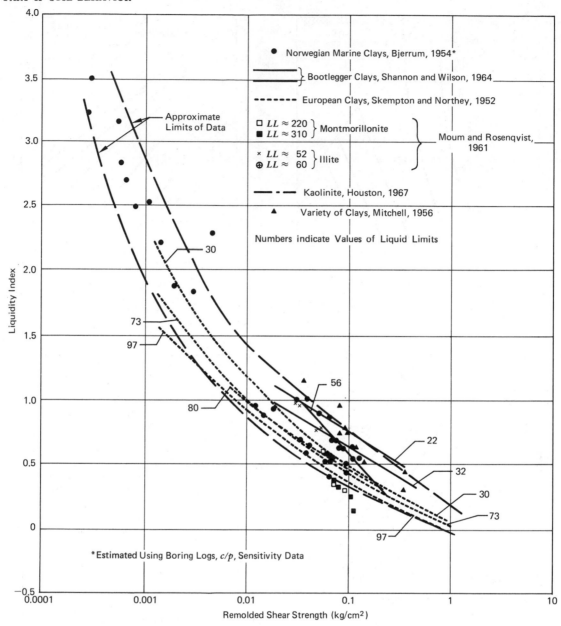

Fig. 12.8   Remolded shear strength vs. liquidity index relationships for several clays.

percent, the strengths in the two directions were the same. This was consistent with the finding that as the density increased the intensity of preferred orientation decreased.

The sample stiffness, as measured by the ratio of stress to shear displacement, at 50 percent of peak strength was about twice as high for shear across the direction of preferred orientation than parallel to it.

The results of direct shear tests on Leighten Buzzard sand, which contains rounded particles, (Arthur and Menzies, 1972) gave a maximum shear to normal stress ratio 24 percent greater for samples poured

through the side or end of the shear box than for samples prepared in the normal way. Additional tests were done on the same sand at a porosity of 34 percent using a cubical triaxial cell in which samples could be tilted to the direction of pouring. The results, shown in Fig. 12.12 in terms of effective principal strains, indicate strength variations of about 10 percent of maximum σ_1'/σ_3' for different orientations of σ_1' and σ_3' relative to the direction of pouring. The variations of peak effective stress ratio and effective stress friction angle ϕ' are shown in Fig. 12.12d.

Figures 12.12 b and c show also that the strain

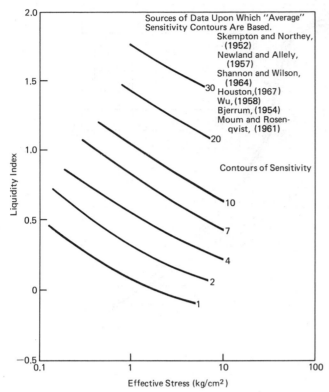

Fig. 12.9 General relationship between sensitivity, liquidity index, and effective stress.

needed to mobilize a given stress ratio varies with orientation of samples relative to direction of pouring, and that the slopes of the stress–strain curves may differ substantially for values of σ_1'/σ_3' greater than about 2. As the slope of the curve is the tangent modulus, it follows that the modulus is anisotropic as well as the strength. Fabric-caused anisotropy in modulus of granular soils may be considerably more important than fabric-induced strength anisotropy. If the results shown in Fig. 12.12 are normalized in terms of effective stress ratios and maximum shear strain, as shown in Fig. 12.13, then a common curve results.

The anisotropic distribution of contact plane normals $E(\beta')$ (Fig. 8.14) for four sands poured through water into a cylindrical mold followed by tapping the mold to obtain a desired density are shown in Fig. 8.15 (Oda, 1972a, 1972b, 1972c). Triaxial tests were made on samples of two sands (*B* and *D*) having different values of angle θ between the original horizontal plane and the direction of maximum principal stress. Sand *B* was composed mainly of elongated, flat particles, axial ratio of 1.65, and sand *D* had more nearly spherical particles with an axial ratio of 1.41.

Triaxial compression test results are shown in Figs.

12.14 and 12.15. Also shown are the distribution of contact plane normals in the initial fabric. Shear planes or shear zones were not observed before peak deviator stress was reached, but in some cases they were formed at postpeak strains. There was no preferred orientation of the failure planes relative to the original fabric. Additional tests were done using samples prepared at the same angle θ but different void ratios. The results of these tests indicate that:

1. At a given void ratio, the direction of σ_1 relative to the original horizontal plane θ has a significant effect on stress–strain, volumetric strain–axial strain relationships as well as on strength.
2. The effects are greater on a sand with the more platy, elongated grains (sand B) than on a more rounded sand (sand D).
3. The sands become less brittle as the angle θ decreases from 90 to 0 degrees.
4. With a decrease in θ from 90 to 0 degrees, the dilation decreases.
5. The stress–strain–volume change relationships for dense sand prepared at $\theta = 0$ degrees are similar to those for loose sand prepared at $\theta = 90$ degrees.
6. The secant modulus at 50 percent of peak strength (E_{50}) decreases with decreasing value of θ. The ratio of E_{50} values for $\theta = 90$ degrees and $\theta = 0$ degrees specimens is 2 to 3 for dense samples of both sands.
7. As samples of Sand *D* do not show preferred orientation of the long axes of grains, it follows that interparticle contact plane orientation has a significant influence on behavior. The major effect of changes in principal stress direction in relation to anisotropic initial particle and interparticle contact orientations is to give different volume change (dilatency) tendencies, which, in turn, account for the differences in modulus and strength. In a sand with rounded grains such as Sand *D*, rearrangement of grains can occur relatively easily at low strains. Thus, although the properties at low strains may differ for different values of θ, the fabrics at failure become approximately the same, as do the peak strengths. With the elongated, flatter particles of sand B, rearrangement of grains is more difficult and fabric differences persist to failure.

Fabric and mechanical property anisotropy is found in undisturbed sands and silts in the field. Undis-

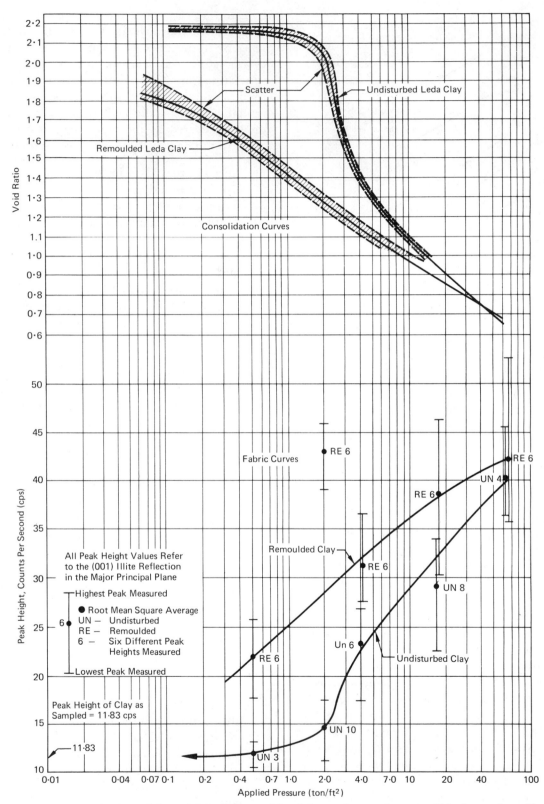

Fig. 12.10 Clay particle orientation produced by anisotropic consolidation of undisturbed and remolded samples of Leda clay (Quigley and Thompson, 1966).

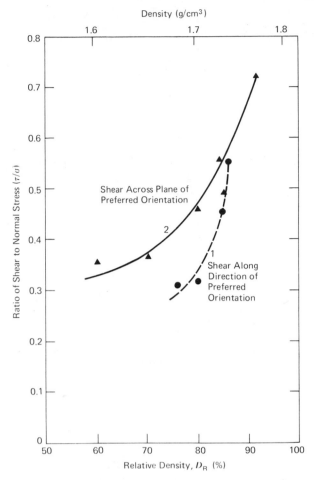

Fig. 12.11 Effect of shear direction on strength of samples of crushed basalt prepared by pouring into the shear box.

turbed samples of the same Vicksburg loess described in Section 11.2 exhibit up to 12 percent higher strength when sheared perpendicular to grain orientation (Matalucci, Abdel-Hady, and Shelton, 1970a, 1970b). Triaxial test results showed that the friction angle decreased from 34 to 31 degrees for dry loess and from 24 to 21 degrees for moist samples as the direction of σ_1 was changed from normal to orientation to an inclination of 45 degrees to the orientation direction.

Anistropic fabric in undisturbed Portsea beach sand is shown in Fig. 11.7. The effect of this anisotropy on behavior in triaxial compression was studied by testing undisturbed samples* cut as shown in Fig. 12.16.

* To handle undisturbed sand samples, Lafeber and Willoughby (1971) used a two-stage replacement of the original seawater by polyethylene glycol (Carbowax 4000). Triaxial tests were done after first heating the samples to melt the Carbowax.

Table 12.1 Effect of Sample Orientation on Secant Modulus of Undisturbed Samples of Portsea Beach Sand

Sample Axis Direction	Sample Axis Azimuth	Secant Modulus (kN/m²)	Standard Deviation (kN/m²)
Vertical		5.41×10^4	$\pm 0.27 \times 10^4$
Horizontal	Parallel to coastline	4.01×10^4	$\pm 0.24 \times 10^4$
Horizontal	30 degrees with coast-line	3.85×10^4	$\pm 0.18 \times 10^4$
Horizontal	60 degrees with coast-line	3.76×10^4	$\pm 0.23 \times 10^4$
Horizontal	Perpen-dicular to coastline	3.55×10^4	$\pm 0.53 \times 10^4$

Data from Lafeber and Willoughby (1971).

Values of mean secant modulus for samples at different orientations are given in Table 12.1. Significant differences exist for samples tested in different directions, and there is no horizontal plane of isotropy for deformation modulus.

Available information on fabric anisotropy and its effects on properties of granular soils is consistent, and the following conclusions can be drawn.

1. The development of anisotropic fabric, as measured both by particle orientations and interparticle contact plane orientations, is probable in natural deposits, compacted fills, and laboratory samples.
2. Granular soils with anisotropic fabrics exhibit anisotropic mechanical properties.
3. Strengths and deformation moduli are higher for shear directions across planes of preferred particle orientation than along them.
4. The magnitude of strength and modulus anisotropy depends on density and the extent to which particles are platy and elongated. Differences in peak strength of the order of 10 to 15 percent may be observed in the case of particles with axial ratios of 1.6 or more.
5. Differences in moduli in different directions are greater than differences in peak strength. Mod-

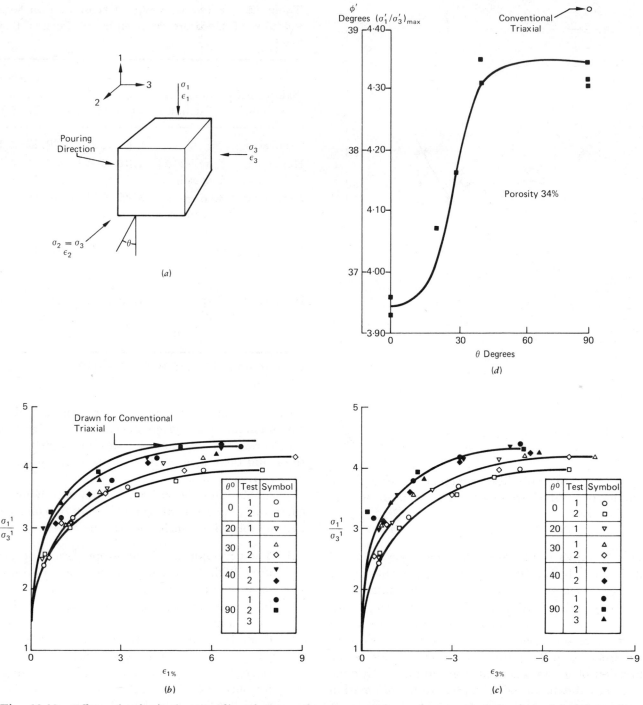

Fig. 12.12 Effect of principal stress directions on the stress–strain and strength behavior of Leighton buzzard sand (Arthur and Menzies, 1972). (a) Reference directions. (b) Stress ratios–major principal axial strain. (c) Stress ratios–minor principal lateral strain. (d) Variation in drained strength with angle of tilt.

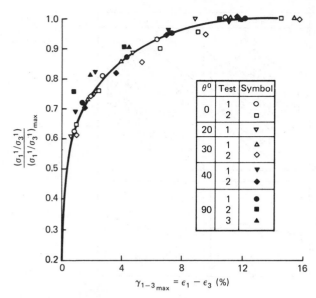

θ^0	Test	Symbol
0	1	○
	2	□
20	1	▽
30	1	▲
	2	◇
40	1	▼
	2	◆
90	1	●
	2	■
	3	▲

Fig. 12.13 Normalized stress ratios vs. maximum shear strain for Leighton Buzzard sand (Arthur and Menzies, 1972).

uli in different directions may differ by a factor of two or three.

6. The effect of fabric anisotropy on mechanical property anisotropy is primarily through its effect on volume change tendencies during deformation.

Clays

Studies of the effects of fabric anisotropy in clays have dealt mainly with strength and permeability.

Analysis (Brinch-Hansen and Gibson, 1949) has shown that an undrained strength anisotropy should result from stress anisotropy during consolidation, apart from any possible fabric anisotropy. In terms of the effective stress strength parameters c' and ϕ', the analysis leads to

$$\frac{c_u}{p} = \frac{c'}{p} \cos \phi' + \frac{1}{2}(1+K_0)\sin \phi' - \sin \phi'(2\overline{A}_f - 1)$$

$$\times \left[\left(\frac{c_u}{p}\right)^2 - \frac{c_u}{p}(1-K_0)\cos 2\left(45 + \frac{\phi'}{2} - \alpha\right) \right.$$

$$\left. + \left(\frac{1-K_0}{2}\right)^2 \right]^{1/2} \tag{12.2}$$

where c_u is undrained strength, p is vertical consolidation pressure, K_0 is the coefficient of lateral earth pressure at rest, $\overline{A}_f$ is given by $\Delta u_f / (\Delta \sigma_1 - \Delta \sigma_3)_f$, Δu_f is

the change in pore pressure at failure, and α is the inclination of the failure plane to the horizontal.

The degree of mobilization of c' and ϕ' at peak stress difference and the strain at failure in an undrained test vary with orientation of principal stresses. Data on the variation of compressive strength with orientation of the failure plane are summarized in Fig. 12.17. Strengths in vertical and horizontal directions may differ by as much as 40 percent as a result of fabric anisotropy. The differences in undrained strength in the different directions result from differences in the pore pressures developed during shear (Duncan and Seed, 1966; Bishop, 1966). The effective stress strength parameters are independent of sample orientation. The drained strength is independent of shear stress orientation relative to fabric orientation, as demonstrated by tests on kaolin (Duncan and Seed, 1966; Morgenstern and Tchalenko, 1967b). Stress paths (Lambe and Whitman, 1969) for two samples from a clay with anisotropic fabric but isotropic initial stresses appear as shown schematically in Fig. 12.18.

The facts that both the effective stress strength parameters and the drained strength are independent of fabric anisotropy, but that pore pressures in undrained shear are strongly influenced by anisotropy suggest that the effect of fabric anisotropy on strength is the same for both sands and clays. Changes in stress orientation relative to fabric orientation influence volume change tendencies. This, in turn, influences the dilatancy contribution to strength of sands and the volume changes in drained deformation and the pore pressures in undrained shear of clays.

In most analyses of seepage problems, an anisotropic permeability, with a higher hydraulic conductivity in the horizontal direction than in the vertical direction, is either measured or assumed. The difference is often a direct result of stratification in natural deposits. Anisotropic permeability may result also from preferred orientation of elongated or platy particles.

Ratios of horizontal to vertical permeability from less than one to more than seven were measured for undisturbed samples of several different clays (Mitchell, 1956). These ratios correlated reasonably well with preferred orientation of the clay as determined using the petrographic microscope and thin sections. Ratios of 1.3 to 1.7 were measured for kaolinite consolidated one-dimensionally in the range from 4 to 256 atm. Ratios for illite and Boston blue clay consolidated to a maximum pressure of over 200 atm

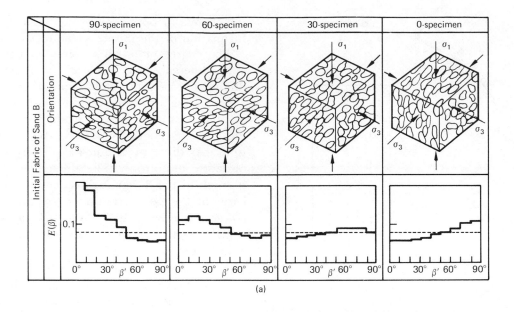

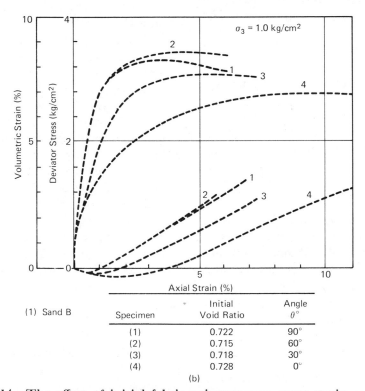

Specimen	Initial Void Ratio	Angle $\theta°$
(1)	0.722	90°
(2)	0.715	60°
(3)	0.718	30°
(4)	0.728	0°

(1) Sand B

(b)

Fig. 12.14 The effect of initial fabric anisotropy on stress–strain and volume change behavior of sand B—elongated, flat particles (Oda, 1972a). (a) Initial fabric. (b) Volumetric strain and stress–strain behavior.

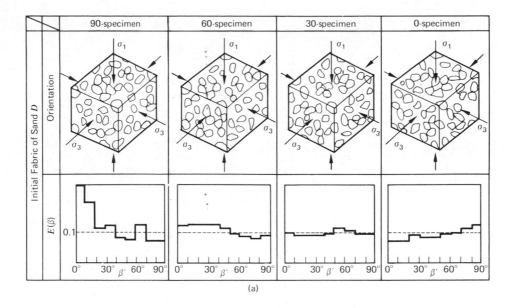

(a)

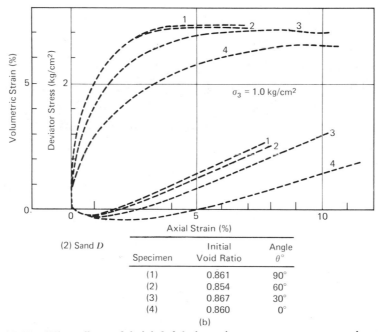

(2) Sand D		
Specimen	Initial Void Ratio	Angle $\theta°$
(1)	0.861	90°
(2)	0.854	60°
(3)	0.867	30°
(4)	0.860	0°

(b)

Fig. 12.15 The effect of initial fabric anisotropy on stress–strain and volume change behavior of sand D—rounded, more spherical particles than sand B (Oda, 1972a). (a) Initial fabric. (b) Volumetric strain and stress–strain behavior.

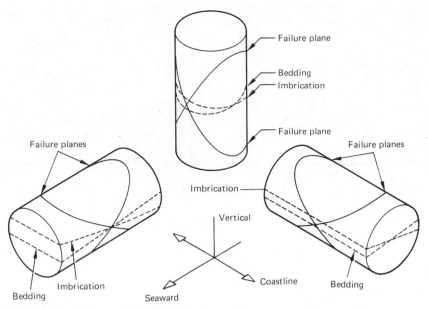

Fig. 12.16 Orientations of triaxial cylinders of Portsea Beach sand in relation to *in situ* conditions (Lafeber and Willoughby, 1971).

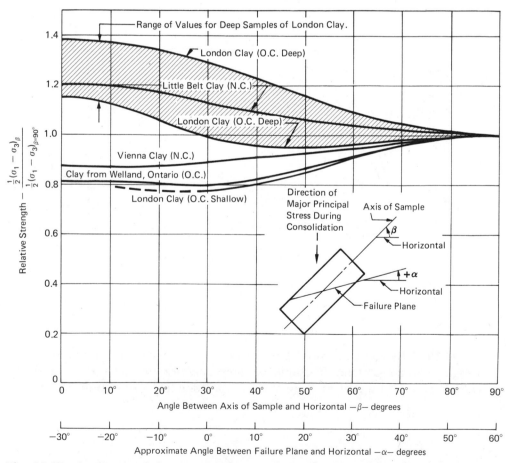

Fig. 12.17 Summary of data concerning the variation of compressive strength with orientation of the failure plane (Duncan and Seed, 1966).

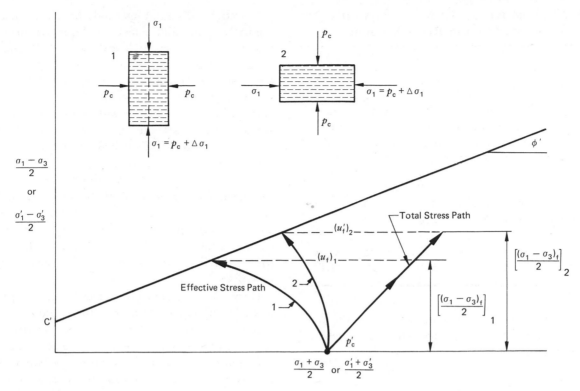

Fig. 12.18 Stress paths in triaxial compression for differently oriented samples from a clay with anisotropic fabric.

ranged from 0.9 to 4.0 (Olsen, 1962). A permeability ratio of approximately 2 was measured for kaolinite over a range of void ratios corresponding to consolidation pressures up to 4 atm (Morgenstern and Tchalenko, 1967a). If all particles were oriented parallel to the horizontal, then according to analysis of a tortuous flow path model (Olsen, 1962), permeability ratios of up to 20 for kaolinite and up to 100 for other clay types should be measured.

12.5 FABRIC, STRUCTURE, AND VOLUME CHANGE

Shrinkage, swell, collapse, and compression are influenced by the fabric and structure of the soil. In the absence of grain crushing, which may be important in granular soils under very high pressures, collapse, shrinkage, and compression involve shear at interparticle contacts. Resistance to shear depends both on the arrangement of particles and particle groups and on the forces holding them in place. Swelling depends strongly on physico-chemical interactions between particles, but fabric also plays a role.

Aspects of volume change that are uniquely related to fabric are discussed in this section; volume change behavior in general is treated in Chapter 13.

Shrinkage

The drying shrinkage of fine-grained soils depends on particle movements as a result of pore water tensions developed by capillary menisci. If two samples of a given clay are at the same initial water content, but have different fabrics, the one that is the more deflocculated and dispersed shrinks most. This is because the average pore sizes are smaller, thus allowing greater capillary stresses, and because of easier relative movements of particles and particle groups.

An illustration of such differences is provided by the data in Table 12.2, where dry void ratios of several undisturbed and remolded clays are listed. In each case, the clay was dried from its natural water content either undisturbed or after thorough remolding. The substantially lower dry void ratios for the remolded samples indicate greater shrinkage than in the undisturbed samples.

Structure anisotropy on a macroscale may be reflected by anisotropic shrinkage. For preferred orientation of platy particles parallel to the horizontal direction, vertical shrinkage is greater than lateral shrinkage on drying. As an example, the vertical shrinkage of Seven Sisters clay is three times greater than the horizontal shrinkage (Warkentin and Bozo-

Table 12.2 Void Ratios of Several Clays After Drying in the Undisturbed and Remolded States

Clay	Natural Water Content (%)	Sensitivity	Dry Void Ratio Undisturbed	Dry Void Ratio Remolded
Boston blue	35.6	6.8	0.69	0.50
Boston blue	37.5	5.8	0.75	0.53
Fore River, Maine	41.5	4.5	0.65	0.46
Goose Bay, Labrador	29.0	2.0	0.60	0.55
Chicago	39.7	3.4	0.65	0.55
Beauharnois, Quebec	61.3	5.5	0.76	0.70
St. Lawrence	53.6	5.4	0.79	0.66

zuk, 1961), and the thermal conductivity in the horizontal direction is 1.6 to 1.7 times that in the vertical direction (Penner, 1963b).

Collapse

Collapse as a result of wetting under constant total stress is an apparent contradiction to the principle of effective stress. The addition of water increases the pore water pressure and reduces the effective stress; hence, expansion might be expected. The apparent anomaly of volume decrease under decreased effective stress results because of the application of continuum concepts to a phenomenon that is controlled by particulate behavior. Collapse requires:

1. An open, partly unstable, partly saturated fabric.
2. A high enough total stress that the structure is metastable.
3. A strong enough clay binder or other cementing agent to stabilize the structure when dry.

When water is added to a collapsing soil in which the silt and sand grains are stabilized by clay coatings or buttresses, the effective stress in the clay is reduced, the clay swells, becomes weaker, and contacts fail in shear. Thus, compatability with the principle of effective stress is maintained on a microscale.

Compression

Sands. Figures 12.14 and 12.15 and other data (Mahmood, 1973) show that the compressibilities of two samples of sand at the same void ratio but with different initial fabrics can be different. Different volume change tendencies for samples at the same density

prepared by different methods have also manifested themselves in differences in liquefaction behavior under undrained cyclic loading (Ladd, 1974).

Clays. Consolidation curves obtained by oedometer tests on undisturbed and remolded Leda (Champlain) clay, illite, and kaolinite are shown in Fig. 12.19. Liquidity index is used as an ordinate, and the sensitivity curves from Fig. 12.9 are superimposed.

Curve A is for undisturbed Leda clay at an initial water content corresponding to a liquidity index of 1.82. Because the sensitivity contours were developed for normally consolidated clays, they cannot be used to estimate sensitivity for stresses less than the preconsolidation pressure. After the preconsolidation stress has been exceeded the curve cuts sharply across the sensitivity contours, indicating a large decrease in sensitivity.

Curve B is for kaolinite remolded at a liquidity index of 2.06. The early part of the consolidation curve is not shown in Fig. 12.19. Immediately after remolding at a high water content the effective stress is very low, and the sensitivity is equal to one. Curve B shows that consolidation results in an increase in sensitivity to a maximum of about 15 to 18, at an effective consolidation pressure of about 0.2 kg/cm². At this point the interparticle and interaggregate shear stresses caused by the applied compressive stress begin to exceed the bond strengths, the degree of structural metastability decreases, and the sensitivity decreases.

Curve D is for kaolinite remolded at a liquidity index of 0.98. It differs considerably from Curve B for liquidity index values less than 0.98. This is consistent with the results of other studies, for example, Morgenstern and Tchalenko (1967a), which show that the compression behavior, and therefore also the structure, are different for a given clay remolded at different water contents. Significantly lower sensitivity is developed in the kaolinite of Curve D than in that of Curve A. These observations suggest that both the concentration of clay in suspension and the rate of sediment accumulation may be important for clay deposits.

Curve E is for a well-graded illitic clay remolded at a liquidity index of 1.36. The consolidation curve indicates a low sensitivity at all consolidation pressures. Results of strength tests showed that the actual sensitivity ranged from 1.0 to 2.6.

Curve C is for Leda clay remolded at a liquidity index of 1.82. The sensitivity increases from 1 to about 8 with reconsolidation, indicating development of metastability after remolding. The sensitivity de-

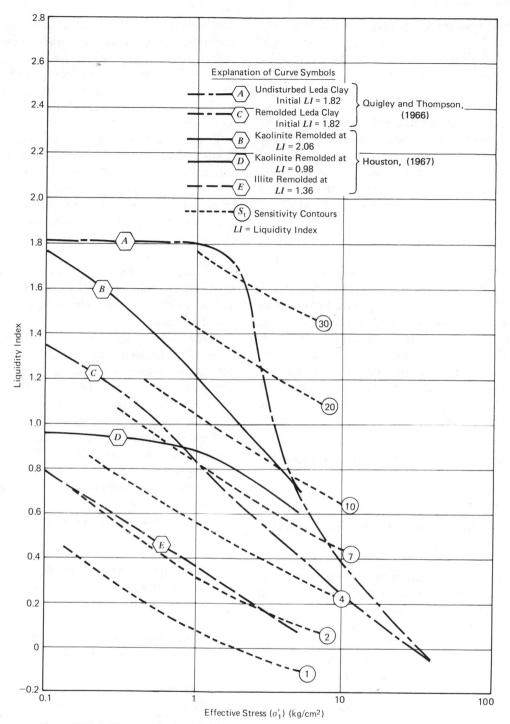

Fig. 12.19 Change in sensitivity with consolidation for various clays.

239

creases at high pressures as convergence with Curve *A* is approached.

All of these findings are consistent with principles 1, 2, 3, and 7 stated in Section 12.2.

Swelling

The swelling of fine-grained soils as a result of reduction in effective stress caused by unloading and/or addition of water is related to the structure. For example, an expansive soil compacted dry of optimum water content can swell more than when compacted wet of optimum to the same density (Seed and Chan, 1959). This difference cannot be accounted for in terms of differences in initial water content and, therefore, must be ascribed to differences in structure.

A "swell sensitivity" has been observed in some clays wherein the swell index for the remolded clay is higher than that for the same clay undisturbed. The increased swelling of the disturbed material can result from both the rupture of interparticle bonds which inhibit swelling in the undisturbed state and differences in fabric. Old, unweathered, overconsolidated clays may be particularly swell sensitive. Values of swell sensitivity as high as 20 were measured in one case (Schmertmann, 1969).

12.6 STRESS-DEFORMATION AND STRENGTH BEHAVIOR

Fabric changes during shear

Shear deformations break down particle and aggregate assemblages. If slip planes develop, platy and elongated particles may align with their long axes in the direction of slip. Fabrics associated with slip planes and well-defined, narrow shear zones have been studied using thin sections and the electron microscope (Morgenstern and Tchalenko, 1967*b*, 1967*c*; Tchalenko, 1968; McKyes and Yong, 1971). Figure 8.26 shows a shear zone, whose width is 40 to 80 μm, that formed in a triaxial test specimen of compacted kaolin. Fabric changes accompanying failure wherein no clearly defined shear zone has developed have not been studied directly.

The development of direct shear-induced fabric as samples deformed to residual strength has been studied by Morgenstern and Tchalenko (1967*b*) and Tchalenko (1968). A shear-induced fabric is generally composed of regions of homogeneous fabric separated by discontinuities. No discontinuities appear before peak strength is reached although there is some particle rotation in the direction of motion. The

microfabric units that appear at or near the peak in the shear stress versus shear displacement curve indicate the central part of the shear zone to be in simple shear. Although near perfect preferred orientation develops during yield after peak strength is reached, local deviations prevent failure at residual strength at the outset.

Virtually perfect parallel alignment of particles is needed before the strength reduces to the residual value. By then the basal planes of the platy clay particles are approximately normal to the major principal stress. Zones of these particles are enclosed between two highly oriented bands of particles on opposite sides of the shear plane. The dominant mechanism of deformation in the displacement shear zone is basal plane slip (Fig. 12.20), and the overall thickness of the shear zone is usually less than 50 μm (Tchalenko, 1968). Residual strength is treated in more detail in Section 14.8.

Shear zones from five clays[*] in which slides had developed showed (Morgenstern and Tchalenko, 1967*c*) continuous shear planes (displacement shears), usually between 10 and 100 μm thick, located close to, or at the boundaries of the shear zone. The development of discontinuous shear planes (Reidel shears) and preferred orientation in the zone between the shear planes depended on the clay composition and the magnitude of the displacement. The shear zones were several millimeters thick, although in the case of one clay it was several centimeters wide.

The deformation of sands, gravels, and rockfills is influenced by the initial fabric, and the deformation process causes changes in the fabric. Under conditions of high confining pressure, particle crushing may have an important effect on both the structure and the properties (Lee and Farhoomand, 1967; Marsal, 1973), as discussed further in Chapter 13.

Fabric changes associated with the sliding and rolling of grains during triaxial compression have been determined (Oda, 1972*b*) using a uniform sand with rounded to sub-rounded grains with sizes in the range of 0.84 to 1.19 mm and a mean axial ratio of 1.45. Samples were prepared to a specified void ratio (0.640), using two methods (tapping the side of the mold and tamping), and tested in triaxial compression, using a delayed setting water–resin solution as the pore fluid. Different samples were tested to successively higher strains, the resin was allowed to harden, and thin sections were prepared.

---

[*] One of them is shown in Fig. 8.23.

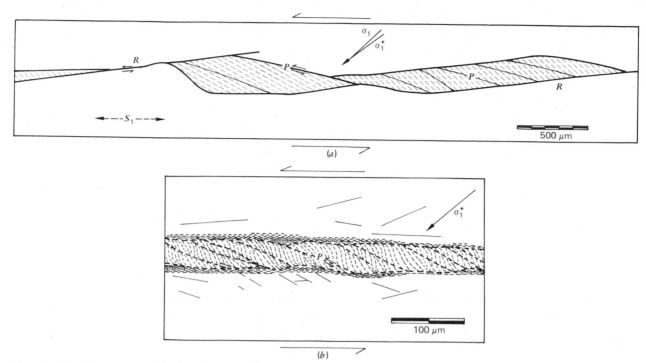

Fig. 12.20 Shear zone fabrics at the residual strength state. (From Tchalenko, 1968b). (*a*) Structures formed in the direct shear test. *P*, thrust shears; *R*, Riedel shears; S_1, initial fabric attitude; σ_1, σ_1^*, major principal stress directions at peak and residual strengths respectively; hatchings, particle orientation in the compression texture; white areas, particles in the initial fabric attitude. (*b*) Principal displacement shears in the direct shear test. *P*, thrust shears; σ_1^*, major principal stress direction at residual strength; hatchings, particle orientation in the compression texture; white areas, particles in the initial fabric attitude.

Differences in the initial fabrics were sufficiently great, even though the same void ratio was obtained in each case, to give the markedly different stress–strain and volumetric strain curves shown in Fig. 12.21. A statistical analysis of changes in particle orientation with increase in axial strain up to failure showed:

1. For samples prepared by tapping, the initial fabric tended toward some preferred orientation of long axes parallel to the horizontal plane and the intensity of orientation increased slightly during deformation.

2. For samples prepared by tamping, there was very weak preferred orientation in the vertical direction initially, but this disappeared with deformation.

Shear planes or zones did not appear until after peak stress had been reached. On the other hand, the distribution of normals to the interparticle contact planes $E(\beta')$ did change with shear, as may be seen in Fig. 12.22. This figure shows different initial distribu-

tions for samples prepared by the two methods and a concentration of contact plane normals within 50 degrees of the vertical as deformation progresses. Thus, the fabric tended toward greater anisotropy in each case in terms of contact plane orientations. After peak stress was reached little further change in $E(\beta')$ was observed.

By considering the projection of interparticle contact areas on the three orthogonal planes, Oda calculated a Fabric Index from $E(\beta')$. If S_X, S_Y, and S_Z are summations of the projected area of contact surfaces on the YZ, ZX, and XY planes, the fabric index is given by S_Z/S_X ($= S_Z/S_Y$ for axisymmetric fabric and deformation). This index can be related to $(\sigma_1/\sigma_3)_{max}$ and to dilatancy rate as defined by the rate of volume change with strain.

Effect of structure on stiffness and elastic and plastic deformations

The sensitivity behavior of quick clays illustrates well the principle that flocculated open fabrics are more rigid, but more unstable than deflocculated

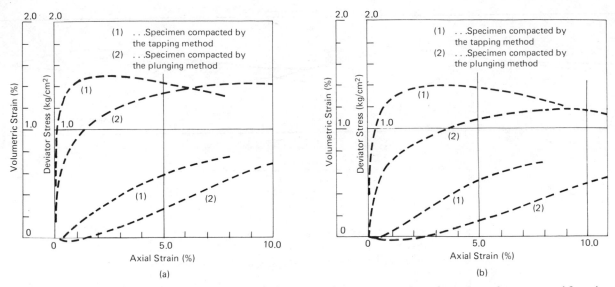

Fig. 12.21 Stress–strain and volumetric strain curves for two samples of sand at the same void ratio, but prepared by different methods. (Oda, 1972b). (*a*) Stress, strain, and volumetric strain of specimens saturated with water. (*b*) Stress, strain, and volumetric strain of specimens saturated with water-resin solution.

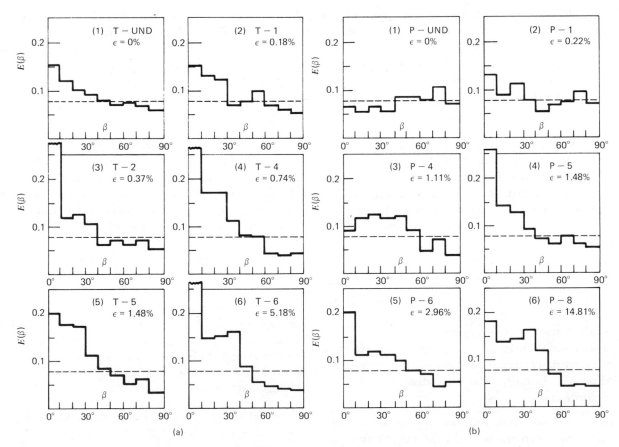

Fig. 12.22 Distribution of interparticle contact normals as a function of axial strain for sand samples prepared in two ways (Oda, 1972b). (*a*) Specimens prepared by tapping. (*b*) Specimens prepared by tamping.

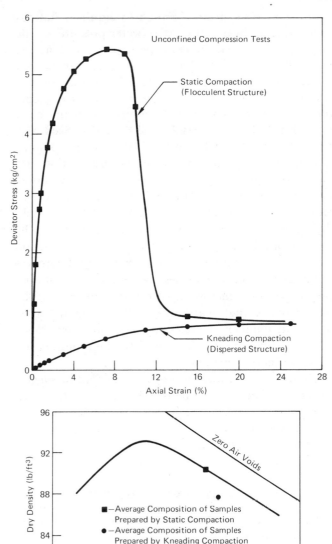

Fig. 12.23 Stress—strain behavior of kaolinite compacted using two methods.

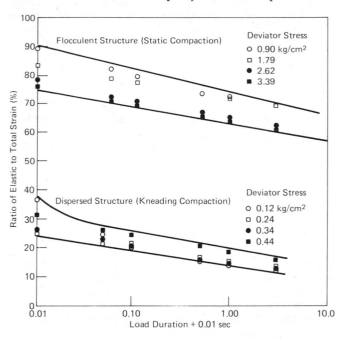

Fig. 12.24 Ratio of recoverable to total strain for samples of kaolinite with different structures.

fabrics. Similar behavior may be observed in compacted soils, and the results of a series of tests on structure–sensitive kaolinite are illustrative of differences in stress–deformation behavior that may be observed (Mitchell and McConnell, 1965). Compaction conditions and stress–strain curves for samples of kaolinite compacted using kneading and static methods are shown in Fig. 12.23. The high shear strain associated with compaction wet of optimum is known (Seed and Chan, 1959) to break down flocculated structures. Marked differences in peak strengths and

stiffnesses as a result of the different compaction methods may be seen.

Figure 12.24 shows that the recoverable deformation of specimens with flocculent structures ranges between 60 and 90 percent of the total deformation. The recovery of samples with dispersed structures is only of the order of 15 to 30 percent of the total deformation. This illustrates the much greater ability of the braced-box type of fabric that remains after static compaction to withstand stress without permanent deformations than is possible with the broken down fabric associated with kneading compaction. Values of recoverable elastic strain as a function of stress intensity are shown in Fig. 12.25, and again the substantial difference between the two structures is emphasized. The nonlinear character of the elastic response may also be noted.

Structure, effective stresses, and strength

The effective stress strength parameters c' and ϕ' are isotropic properties, with anisotropy in undrained strength explainable in terms of anisotropy of pore pressure parameter $\overline{A}_f$. The undrained strength loss associated with remolding an undisturbed clay can also be accounted for in terms of differences in effective stress, provided part of the undisturbed strength does not result from cementation. Remolding breaks

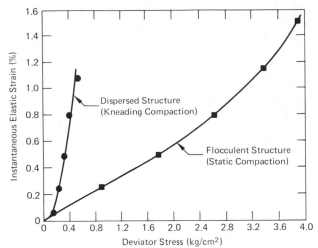

Fig. 12.25 Variations of instantaneous elastic strain with deviator stress for two different structures.

down the structure and causes a transfer of effective stress to the pore water.

Figure 12.26 is an example, and shows the results of incremental loading triaxial tests on two undisturbed and remolded samples of San Francisco bay mud. In these tests, the undisturbed sample was first brought to equilibrium under an isotropic consolidation pressure of 0.8 kg/cm². After undrained loading to failure, the triaxial cell was disassembled, and the sample was remolded in place. The apparatus was reassembled, and pore pressure was measured. Thus, the effective stress at the start of compression of the remolded clay was known. Stress–strain and pore pressure–strain curves for two samples, tested undisturbed and remolded, are shown in Fig. 12.26, and stress paths for Test 1 are shown in Fig. 12.27.

Differences in strength that result from fabric differences caused by differences in compaction method or from thixotropic hardening can be explained in the same way. Thus, in the absence of chemical or mineralogical changes, different strengths in two samples of the same soils at the same void ratio that result from different structures can be accounted for in terms of differences in effective stress.

Sand fabric and liquefaction

Much study of liquefaction in sands has been made in recent years because of the important role of liquefaction in soil failures during earthquakes. Liquefaction is taken here to mean* the transformation of a

* There have been ambiguities in the use of the term liquefaction. These are discussd by Youd (1973) and are not dealt with here.

granular material from a solid to a liquified state as a consequence of increased pore water pressure. Cyclic loading due to earthquakes is perhaps the commonest cause of liquefaction. The resistance to liquefaction in any case depends on characteristics of the sand including gradation, particle size, and particle shape; relative density; confining pressure; and principal stress ratio (Lee and Seed, 1967a, 1967b; Seed and Lee, 1966; Seed and Peacock, 1971). It is now known (Ladd, 1974) that method of sample preparation also has a very significant effect on the resistance to liquefaction determined in the laboratory. Since differences in sample preparation have been shown to yield differences in fabric, it follows that fabric should be an important factor influencing the liquefaction potential of a sand because of the resulting differences in volume change tendencies. The importance of sample preparation method is illustrated by Fig. 12.28, which shows the relationship between cyclic stress level and number of cycles to cause liquefaction for samples of three sands prepared to the same density by two different methods.

Findings such as those shown in Fig. 12.28 suggest that more than relative density is required to characterize a given sand for liquefaction analyses. In addition, the important question arises as to how best to fabricate samples in the laboratory to duplicate the fabric in the field.

12.7 FABRIC AND PERMEABILITY

Of the properties of importance in the analysis of geotechnical problems in fine-grained soils, none is more influenced by fabric than the hydraulic conductivity or permeability. Fabric anisotropy can cause substantial permeability anisotropy. Remolding an undisturbed clay reduced the permeability of several clays by a factor of from 1 to 4, with an average of about 2 (Mitchell, 1956).

Theoretical equations for prediction of permeability (Chapter 15) based on capillary tube models for flow through porous media indicate that the flow velocity should depend on the square of the pore radius, and the flow rate on the fourth power of the radius. Thus, fabrics with a high proportion of large pore sizes are much more pervious than those with small pores. The reduction in permeability associated with remolding a soft, sensitive clay can be explained in terms of the breakdown of a flocculated open fabric and the destruction of large pores.

The profound influence of compaction water content on the permeability of a silty clay is shown in

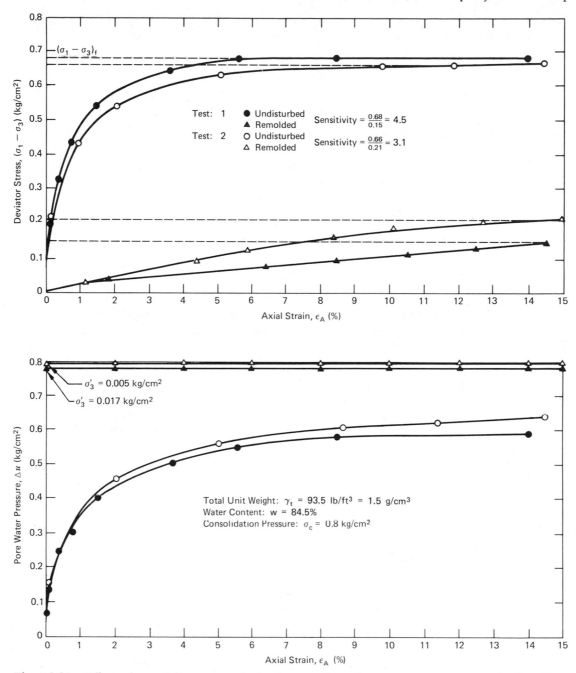

Fig. 12.26   Effect of remolding on undrained strength and pore water pressures in San Francisco Bay mud.

Fig. 12.29. This is typical of the variation of permeability with water content in compacted fine-grained soils (Mitchell, Hooper, and Campanella, 1965). For the case shown, all samples were compacted to the same density. For samples compacted using the same compactive effort, curves such as those in Fig. 12.30 are typical. For compaction dry of optimum clay particles and aggregates are flocculated, the resistance to rearrangement during compaction is high, and a fabric with comparatively large pores is formed. For higher water contents, the particle groups are weaker, and fabrics with smaller average pore sizes are formed. Considerably lower values of permeability are obtained wet of optimum in the case of kneading com-

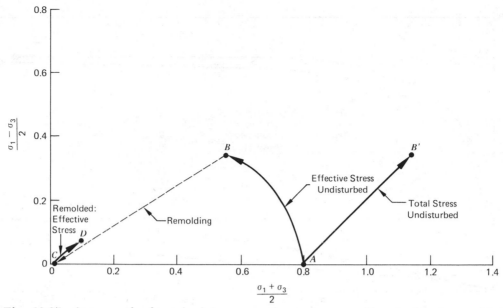

Fig. 12.27 Stress paths for triaxial compression tests on undisturbed and remolded samples of San Francisco Bay mud.

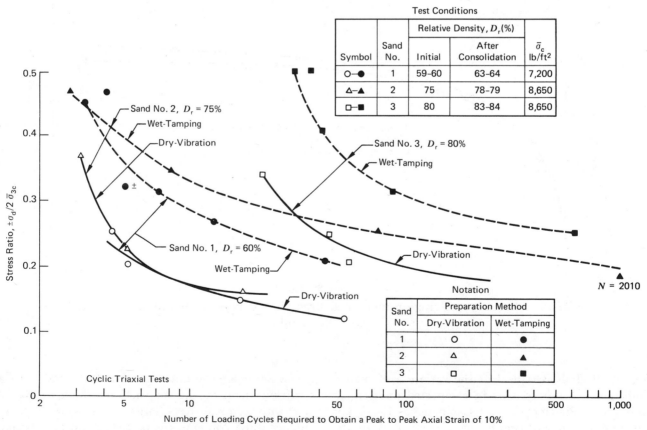

Fig. 12.28 The influence of sand sample preparation method on liquefaction behavior (Ladd, 1974).

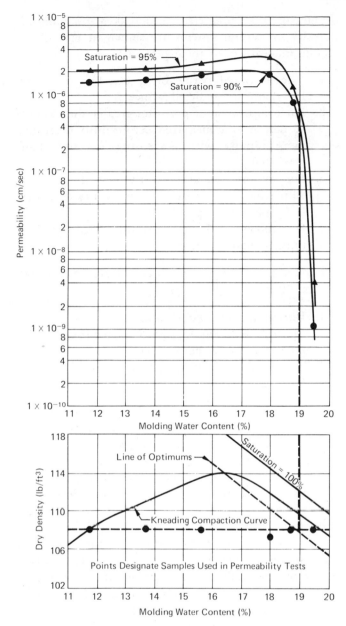

Fig. 12.29 Permeability as a function of molding water content for samples of silty clay prepared to constant density by kneading compaction.

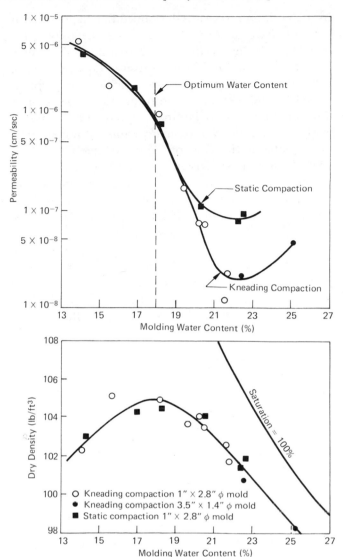

Fig. 12.30 Influence of the method of compaction on the permeability of silty clay.

paction than by static compaction, because the high shear strains induced in the kneading compaction method break down flocculated fabric units.

Under conditions of constant saturation and no fabric or other changes during flow Darcy's law is valid for the range of hydraulic gradients usually encountered. According to Darcy's law, the flow rate q is given by

$$q = kiA \qquad (12.4)$$

where k is the hydraulic conductivity, i is the gradient, and A is the area of cross section normal to flow. The conductivity k with units of LT^{-1} is related to the "absolute permeability" K with units of L^2 by

$$K = k\left(\frac{\eta}{\gamma}\right) \qquad (12.5)$$

where η and γ are the viscosity and unit weight of the pore fluid, respectively.

The well-known Kozeny–Carman equation is widely used for the prediction of K in saturated porous media:

$$K = \frac{1}{k_0 T^2 S_0{}^2} \frac{n^3}{(1-n)^2} \qquad (12.6)$$

In this equation, which is developed in Chapter 15, k is the pore shape factor (≈ 2.5), T is the tortuosity factor ($\approx \sqrt{2}$), S_0 is the specific surface area per unit volume of particles, and n is the porosity. Equation (12.6) is based on the assumptions that particles are approximately equidimensional, uniform, and larger than 1 μm and that flow is laminar. It has been found to work well in the case of many sands; however, it is inadequate in the case of clays (Michaels and Lin, 1954; Lambe, 1954; Olsen, 1962; and others).

The reasons for the failure of the Kozeny–Carman equation in clays provide an insight into the fabric of fine-grained soils and its relation to flocculation, dispersion, consolidation, and permeability (Olsen, 1962). The results of consolidation–permeability tests, in which the permeability is measured at a number of porosities during compression and rebound, can be compared with values predicted by equation (12.6); a typical result is shown in Fig. 12.31 for illite. Ratios

of measured to predicted values as a function of porosity for several systems are shown in Fig. 12.32. Significant features of these plots are:

1. The measured permeability can be greater or less than the predicted value.
2. For compression at porosites greater than about 40 percent, the measured permeability decreases more rapidly with decreasing porosity than predicted.
3. For compression at porosities less than about 40 percent, the measured permeability decreases less rapidly than predicted.
4. For rebound, the permeability increases less rapidly than predicted.

Explanations for these discrepancies might be (Olsen, 1962):

1. Deviations from Darcy's law.
2. Electrokinetic coupling (see Chapter 15).
3. High viscosity of the pore water.
4. Tortuous flow paths.
5. Unequal pore sizes.

It may be shown that the possible effects of (1) and (2) are insignificant, and the discrepancies between predicted and measured flow rates cannot be accounted for in terms of (3) and (4).

To estimate the consequences of unequal pore sizes, the cluster model shown in Fig. 12.33 was developed (Olsen, 1962). Clusters are assumed equidimensional and of the same size. Only flow through the intercluster pores is considered, which is reasonable because of the dependence of flow rate on the fourth power of pore radius. The number of particles per cluster is N, the intracluster void ratio is e_c, and the intercluster void ratio e_p equals the total void ratio e_T less the cluster void ratio e_c.

The following relationship can be derived for the ratio of estimated flow rate for the cluster model q_{CM}, to flow rate predicted by the Kozeny–Carman equation q_{KC}:

$$\frac{q_{CM}}{q_{KC}} = N^{2/3} \frac{[1 - e_C/e_T]^3}{[1 - e_C]^{4/3}} \qquad (12.7)$$

To apply equation (12.7), an assumption must be made for the variations of e_C with e_T that occur with compression and rebound. One relationship, shown in Fig. 12.34, is based on considerations of the relative compressibilities of individual clusters and cluster as-

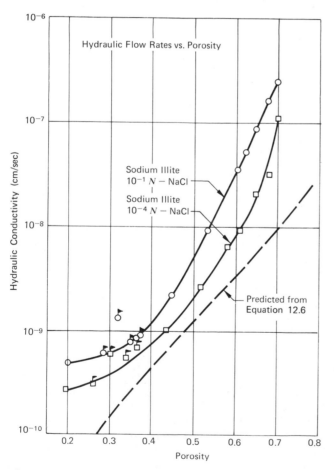

Fig. 12.31 Hydraulic flow rates as a function of porosity for illite (Olsen, 1962).

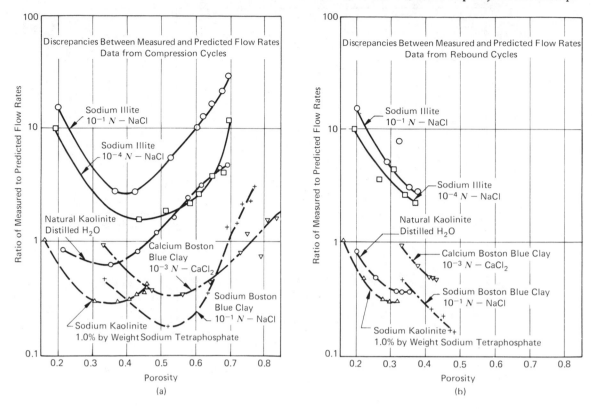

Fig. 12.32 Discrepancies between measured and predicted permeabilities (Olsen, 1962). (a) Compression cycles. (b) Rebound cycles.

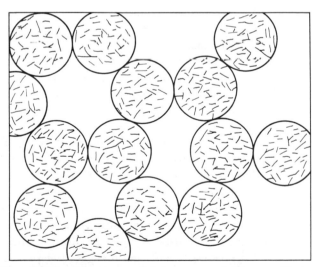

Fig. 12.33 Cluster model for permeability prediction (After Olsen, 1962).

semblages. The compressibility of individual clusters is small at high total void ratios, and compression develops at the expense of intercluster pores. After

compression to a state where the intercluster pore space is comparable to that in a system of closely packed spheres, the clusters themselves begin to compress. Further decreases in porosity involve decreases in both e_C and e_p.

Predictions using equation (12.7), and the relationships in Fig. 12.34 are shown in Fig. 12.35 for three values of N. It may be seen that the agreement between Figs. 12.35 and 12.32 is remarkably close, both quantitatively and qualitatively. Computed values of the initial cluster void ratios range from 0.4 to 1.1, the number of particles per cluster from 3 to 2000, and the initial cluster diameters are from 0.12 to 1.06 μm (Olsen, 1962). In each case, the number of particles per cluster decreases in the direction of increasing dispersion according to the pore water chemistry. Essentially the same values of cluster parameters were deduced also from electrical conductivity data.

The existence of unequal pore sizes is shown clearly by observations of fabric using both the optical and the electron microscopes (Morgenstern and Tcha-

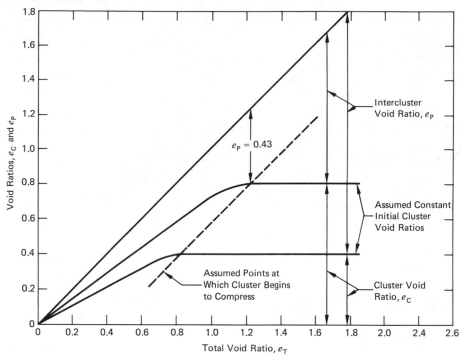

Fig. 12.34   Assumed relationship between the total, cluster, and intracluster void ratios (Olsen, 1962).

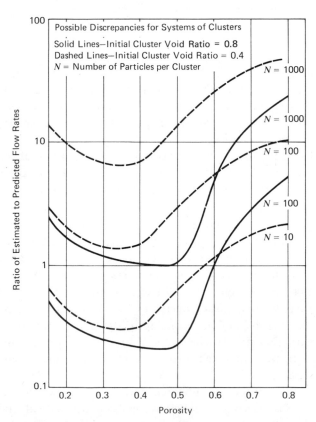

Fig. 12.35 Permeability discrepancies according to the cluster model (Olsen, 1962).

lenko, 1967a; Collins and McGown, 1974). Fabric changes during compression develop as follows:

1. At high porosities, compression results mainly from decreases in the size of intercluster or interaggregate pores.
2. For consolidation at low porosities and for rebound volume changes occur mainly in the clusters themselves with relatively small changes in the intercluster pores.

Fabric changes as a result of the clogging and unclogging of pores during permeation may cause apparent deviations from Darcy's law. Figure 12.36 shows a nonlinearity between flow velocity and hydraulic gradient for an undisturbed soft clay from Skå-Edeby in Sweden. In addition to the nonlinearity shown in the figure, the flow rate varied with time and flow direction under a given gradient, again suggesting particle migration. Electron photomicrographs of the undisturbed clay fabric and others like it have shown the presence of loosely held mineral and organic particles in the large pores. These particles probably can be dislodged easily and account for the gradient-dependence of permeability in such clays. Similar effects have been observed in compacted fine-grained soils (Mitchell and Younger, 1967).

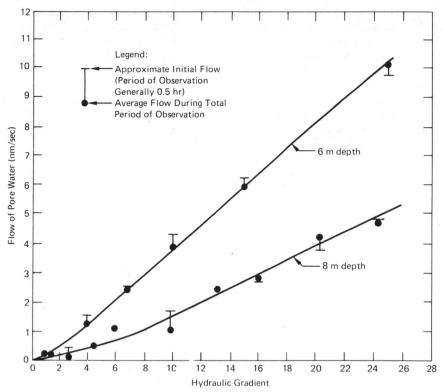

Fig. 12.36   Dependence of flow velocity on hydraulic gradient, undisturbed soft clay from Skå Edeby, Sweden (Hansbo, 1973).

12.8 PRACTICAL APPLICATIONS

The main conclusion to be drawn from the considerations of this chapter is that the geotechnical properties of any given soil are dependent on the structure to such an extent that analyses based on properties determined from the same material but with a different structure may be totally in error. This has been recognized for many years in the case of clays but only recently has the importance of sand fabric become fully appreciated.

The fact that the effects of fabric differences on stress–strain and strength behavior can be accounted for in terms of differences in volume change (drained conditions) and pore pressure changes (undrained conditions) provides insight into soil behavior and unifies effective stress and total stress approaches to soil mechanics. The pattern of properties in sensitive clays, developed in Section 12.3, can be used to estimate values before samples are available and to extrapolate to possible future changed conditions.

Anisotropy is likely to be the rule rather than the exception; estimates of its potential effects on deformation and strength behavior and on hydraulic conductivity can be inferred from some of the data presented.

The principles stated in Section 12.2 can be used as a basis for assessing both the properties at any time and the probable consequences of a change in the state of a soil due to consolidation, shear, or other effect.

SUGGESTIONS FOR FURTHER STUDY

Duncan, J. M., and Seed, H. B. (1966), "Anisotropy and Stress Reorientation in Clay," *Journal of the Soil Mechanics and Foundations Division*, A.S.C.E., Vol. 92, No. SM5, pp. 21–50.

Houston, W. N., and Mitchell, J. K. (1969), "Property Interrelationships in Sensitive Clay," *Journal of the Soil Mechanics and Foundations Division*, A.S.C.E., Vol. 95, no. SM4, pp. 1037–1062.

Lambe, T. W. (1958), "The Structure of Compacted Clay," *Journal of the Soil Mechanics and Foundations Division*, A.S.C.E., Vol. 84, No. SM2, p. 34.

Mitchell, J. K. (1960), "Fundamental Aspects of Thixotropy in Soils," *Journal of the Soil Mechanics and Foundations Division*, A.S.C.E., Vol. 86, No. SM3, pp. 19–52.

Morgenstern, N. R., and Tchalenko, J. S. (1967), "Microscopic Structure in Kaolin Subjected to Direct Shear," *Geotechnique,* Vol. 17, No. 4, pp. 309–328.

Oda, M. (1972a), "Initial Fabrics and Their Relations to Mechanical Properties of Granular Material," *Soils and Foundations,* Vol. 12, No. 1, pp. 17–37.

Oda, M. (1972b), "The Mechanism of Fabric Changes During Compressional Deformation of Sand," *Soils and Foundations,* Vol. 12, No. 2, pp. 1–18.

Olsen, H. W. (1962), "Hydraulic Flow Through Saturated Clays," *Proceedings of the Ninth National Conference on Clays and Clay Minerals,* pp. 131–161.

Seed, H. B., and Chan, C. K. (1959), "Structure and Strength Characteristics of Compacted Clays," *Journal of the Soil Mechanics and Foundations Division,* A.S.C.E., Vol. 85, No. SM5, pp. 87-128.

CHAPTER 13

Volume Change Behavior

13.1 INTRODUCTION

Volume changes in soils are important because of their consequences in terms of settlement due to compression and heave due to expansion. In addition, changes in volume lead to changes in strength and deformation properties that in turn influence stability. Changes in volume may result from changes in pressure, temperature, and chemical environment. Of these, the effects of pressure are generally the most important and have been the most studied.

In this chapter, all three factors contributing to volume change are discussed, and their relative importance is considered. Emphasis is on consolidation and swelling. Shrinkage can be considered a special case of consolidation, wherein the consolidation pressure is derived from capillary menisci and the surface tension of water.

Familiarity of the reader with the phenomenological aspects of compression and swelling as ordinarily treated in soil mechanics is assumed, to include pressure–void ratio relationships as described by the coefficient of compressibility ($a_v = - de/dp$), the compressibility ($m_v = - a_v/(1 + e)$), the compression index ($C_c = - de/d \log p$), the swelling index ($C_s = - de/d \log p$ on unloading), and the maximum preconsolidation pressure (p_c).

13.2 GENERAL RELATIONSHIPS BETWEEN SOIL TYPE, PRESSURE, AND VOID RATIO

Soils ordinarily have void ratios in the range of about 0.5 to 4.0 as shown in Fig. 13.1. Although the range of pressures of interest in most problems (to a few tens of atmospheres at most) is relatively small on

a geological scale,* the void ratios encountered encompass virtually the full range from fresh sediments to shale. A complex of mechanical, physico-chemical, and chemical changes accompany and influence the densification process. In general, the void ratio–pressure relationship is related to the grain size and plasticity in the manner shown by Fig. 13.1b.

Particle size is probably the most significant single factor influencing both the void ratio at any pressure and the effects that physico-chemical and mechanical factors have on the process of consolidation (Meade, 1964). A similar statement can be made relative to swelling. Particle size is a direct manifestation of mineralogical composition, with increasing colloidal activity and expansiveness associated with decreasing particle sizes.

Values of compression index for the clay minerals range from less than 0.2 for kaolinite to as high as 17 (Bolt, 1956) for specially prepared sodium montmorillonite under low pressures, although values less than 2 are more usual. The compression index for most natural clays is usually less than 1.0 with a value less than 0.5 in most cases. The swell index is less than the compression index, usually by a substantial amount, as a result of particle rearrangement during compression. After one or more cycles of recompression and unloading the compression and swelling indices may become equal. Swell index values for three clay minerals, muscovite, and sand are listed in Table 13.1. For undisturbed natural soils the swell index values are usually less than 0.1 for nonexpan-

---

* Rieke and Chilingarian (1974) present a comprehensive treatise on the consolidation of argillaceous sediments to depths of several kilometers beneath the earth's surface.

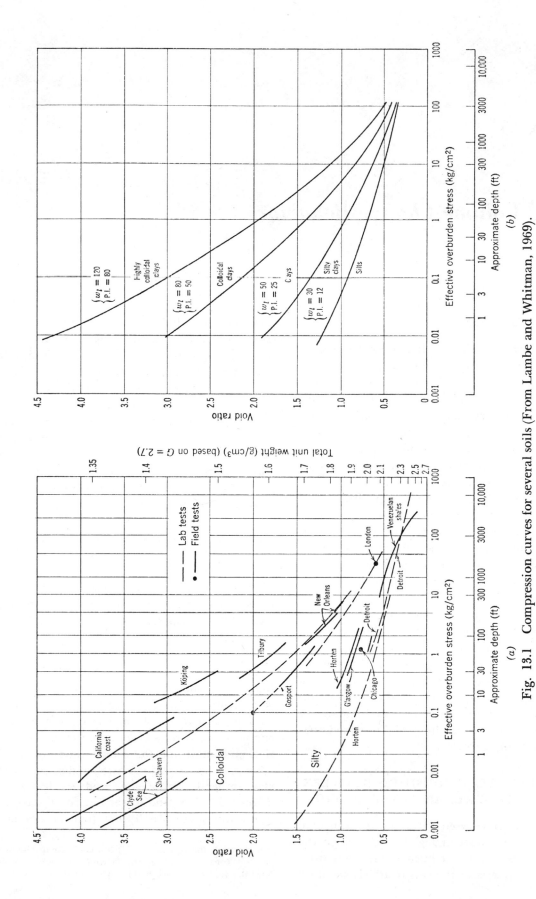

Fig. 13.1 Compression curves for several soils (From Lambe and Whitman, 1969).

254

Table 13.1 Swelling Index Values for Several Minerals

Mineral (1)	Pore Fluid, Adsorbed Cations, Electrolyte Concentration, in Gram Equivalent Weights per Liter (2)	Void Ratio at Effective Consolidation Pressure of 100 psf (3)	Swelling Index (4)
Kaolinite	Water, sodium, 1	0.95	0.08
	Water, sodium, 1×10^{-4}	1.05	0.08
	Water, calcium, 1	0.94	0.07
	Water, calcium, 1×10^{-4}	0.98	0.07
	Ethyl alcohol	1.10	0.06
	Carbon tetrachloride	1.10	0.05
	Dry air	1.36	0.04
Illite	Water, sodium, 1	1.77	0.37
	Water, sodium, 1×10^{-3}	2.50	0.65
	Water, calcium, 1	1.51	0.28
	Water, calcium, 1×10^{-3}	1.59	0.31
	Ethyl alcohol	1.48	0.19
	Carbon tetrachloride	1.14	0.04
	Dry air	1.46	0.04
Smectite	Water, sodium, 1×10^{-1}	5.40	1.53
	Water, sodium, 5×10^{-4}	11.15	3.60
	Water, calcium, 1	1.84	0.26
	Water, calcium, 1×10^{-3}	2.18	0.34
	Ethyl alcohol	1.49	0.10
	Carbon tetrachloride	1.21	0.03
Muscovite	Water	2.19	0.42
	Carbon tetrachloride	1.98	0.35
	Dry air	2.29	0.41
Sand			0.01 to 0.03

From Olson and Mesri (1970).

sive materials to more than 0.2 in the case of expansive soils.

The compressibility of dense sands and gravels is far less than that of normally consolidated clays; nonetheless, volume changes under high pressures may be substantial in granular materials. Compressibility data for several sand, gravel, and rockfill materials are shown in Fig. 13.2. It may be seen that at a pressure of 700 kN/m² (100 psi) a compression of 3 percent is common, and a value as high as 6.5 percent has been observed. The compacted shells of a rockfill dam are sometimes more compressible than the compacted clay core.

13.3 FACTORS CONTROLLING RESISTANCE TO VOLUME CHANGE

The amount of compression or swell in any case depends on both compositional and environmental factors (Section 9.1), and meaningful quantitative predictions are possible only if undisturbed samples or *in situ* tests are used for evaluation of needed parameters. The following factors are important in determining the resistance to volume change.

Physical interactions

Physical interactions include bending, sliding, rolling, and crushing of soil particles. Physical interactions are more important than physico-chemical interactions at high pressures and low void ratios.

Physico-chemical interactions

These interactions depend on particle surface forces and are responsible for double layer interactions, surface and ion hydration, and interparticle attractive forces. Physico-chemical interactions may be more important than physical interactions at low pressures and high void ratios.

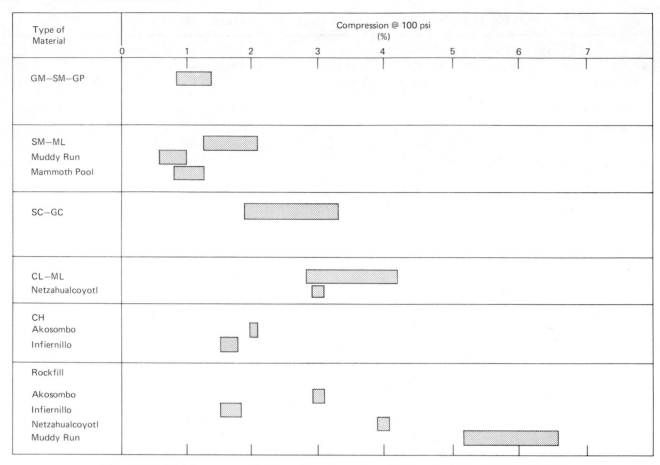

Fig. 13.2   Field compressibility of earth and rockfill materials (Wilson, 1973).

Chemical and organic environment

Chemical precipitates may serve as cementing agents between particles. Organic materials influence surface forces and water adsorption properties. "Salt heave" (Blaser and Scherer, 1969; Blaser and Arulanandan, 1973), which results from temperature related crystallization of sodium sulfate, may be an important swelling mechanism in some areas.

Mineralogical detail

Small differences in certain characteristics of the expansive clay minerals can have a major effect on the swelling of a soil.

Fabric and structure

Compacted, expansive soils with flocculent structures are likely to be more expansive than with dispersed structures. Figure 13.3 is an example. At pressures less than the preconsolidation pressure, a soil with a flocculent structure is less compressible than the same soil with a dispersed structure. The reverse

is true for pressures greater than the preconsolidation pressure.

Stress history

An overconsolidated soil is less compressible but more expansive than the same material at the same void ratio but normally consolidated. Figure 13.4 is an illustration of this. If anisotropic stress systems have been applied to a soil in the past, then anisotropic compression and swelling characteristics usually develop.

Temperature

An increase in temperature usually causes a decrease in volume for a fully drained soil. If drainage is prevented, an increase in temperature causes a decrease in effective stress.

Pore water chemistry

In the case of a soil containing expansive clay minerals, any change in the pore solution chemistry or surrounding water chemistry that tends to depress

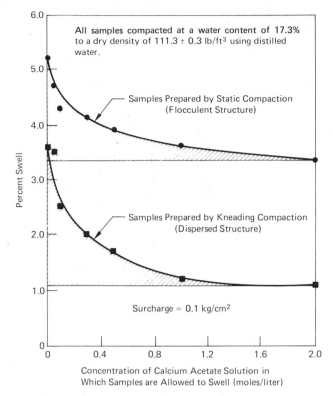

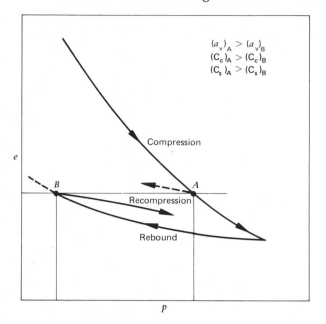

Fig. 13.3 Effect of structure and electrolyte concentration of absorbed solution on swell of compacted clay (Seed, Mitchell, and Chan, 1962).

Fig. 13.4 Comparison of compressibility and swell characteristics for normally consolidated and over-consolidated soil.

double layers leads to a reduction in swell or swell pressure. Figure 13.3 shows an example of this in terms of the electrolye concentration of the solution imbibed by a clay during swelling. For soils containing nonexpansive clay minerals, the pore water chemistry has relatively little effect on the compression behavior after the initial fabric has formed and the structure has stabilized under a moderate effective stress. In the case of the leaching of a normally consolidated marine clay at high water content, however, the change in interparticle forces may be sufficient to cause a small reduction in volume (Kazi and Moum, 1973; Torrance, 1974).

Stress path

The amount of compression or swelling associated with a given change in stress usually depends on the path followed. Unloading from stress *A* to stress *B* in one step can give considerably different behavior than if unloading is done in stages. An example is shown in Fig. 13.5. Similarly, differences in load increment magnitude and duration influence the equilibrium pressure–void ratio relationships for clays as discussed in Section 13.8.

13.4 PHYSICAL INTERACTIONS IN VOLUME CHANGE

Physical interactions between particles that are important during soil compression include particle bending, particle sliding, particle rolling, and particle crushing. In general, the coarser the gradation, the more important physical interactions relative to physico-chemical interactions are. The resistance to particle sliding depends on interparticle friction. Frictional resistance between particles is discussed in Chapter 14.

Particle bending is important in the case of platy particles. The presence of even small amounts of mica in coarse grained soils can greatly increase the compressibility as shown in Fig. 3.1. Mixtures of a dense sand having rounded grains with mica flakes can duplicate the form of the compression and swelling curves of clays, as shown in Fig. 13.6. A sample of Chattahoochie River sand with a mica content of 5 percent is twice as compressible as one with no mica (Moore, 1971). On the other hand, a well-graded soil may be little affected in terms of compressibility by the addition of mica.

Cross-linking adds rigidity to the clay fabric. Particles and particle groups act as struts whose resistance depends both on their bending resistance and on the

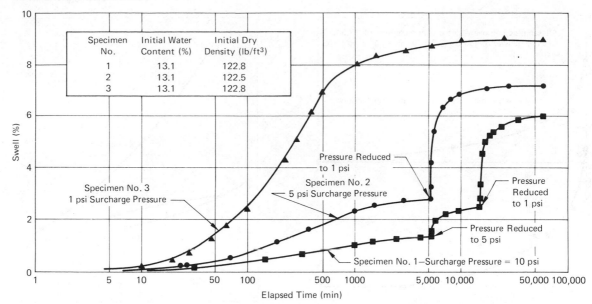

Fig.   13.5   Effect of unloading stress path on swelling (Seed, Mitchell, and Chan, 1962).

strength of the junctions at their ends. According to van Olphen (1963), cross-linking is important even in "pure clay" systems, where the confining pressure is sometimes interpreted as balanced entirely by interparticle repulsion (van Olphen, 1963).

The importance of grain crushing increases with particle size and load magnitude. Particle breakage is a progressive process that starts at relatively low stress levels, because of the wide dispersion of the magnitudes of interparticle contact forces. The number of contacts per particle depends on gradation and density, and the average contact force increases greatly with particle size, as summarized in Table 13.2. Statistical analyses of the probable frequency distribution

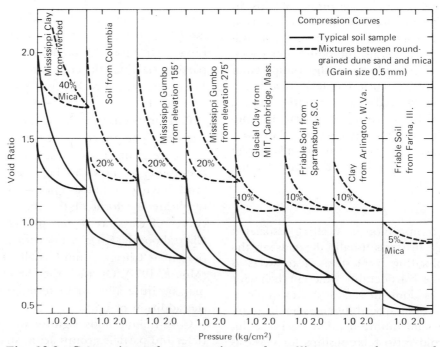

Fig.   13.6   Comparison of compression and swelling curves for several clays and sand–mica mixtures (Terzaghi, 1931).

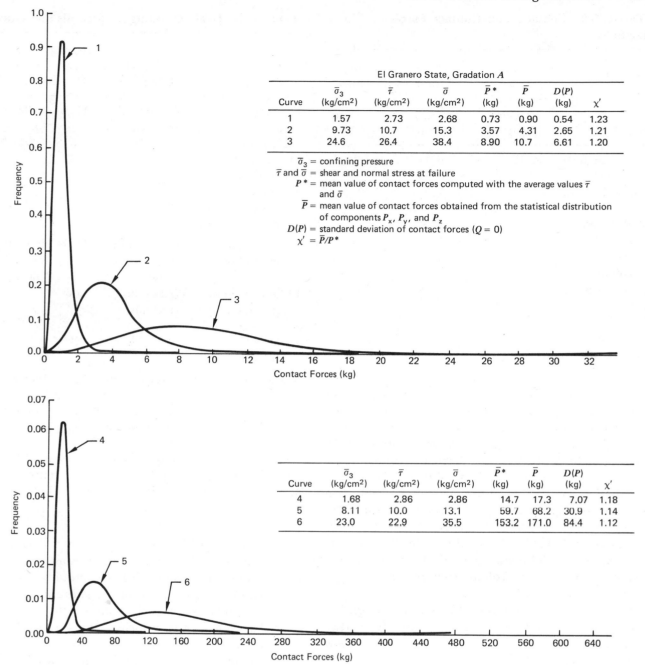

Fig. 13.7 Frequency curves of contact forces for El Granero slate on the failure plane (Dense Packing) (From Marsal, 1973).

of contact forces (Marsal, 1973) show large deviations from the mean. Examples for two gradations of a slate and several confining pressures are shown in Fig. 13.7.

Some unstressed "idle particles" occupy the voids between larger particles or particle arches. The percentage of idle particles depends on gradation, fabric,

void ratio, and stress level. With idle particles the effective void ratio, from the standpoint of resistance to deformation, is greater than the measured void ratio.

In soils containing idle particles, particulate mechanics approaches to behavior that depend on such

Table 13.2   Contacts and Contact Forces in Granular Soils

Soil Type	Grain Contacts/Particle (Range)	Grain Contacts/Particle (Mean)	Average Contact Force for $\sigma' = 1$ atm (N)
Loose Uniform Gravel	4–10	6.1	
Dense Uniform Gravel	4–13	7.7	
Well-Graded Gravel, 0.8 mm $< d <$ 200 mm	5–1912	5.9	
Medium Sand			10^{-2}
Gravel			10
Rockfill, $\bar{d} = 0.7$ m			10^4

Table 13.3   Grain Crushing in Rockfills and Gravels

Samples	Grain Size distribution	Crushing Strength of Grains	Particle Breakage $B_q q_i$
El infiernillo silicified conglomerate Pinzandaran sand and gravel San Francisco basalt (gradations 1 and 2)	Well-graded rockfills and gravels	High	0.02–0.10 for $5 \leq \sigma_{1f} \leq 80\,\text{kg/cm}^2$
El infiernillo diorite	Somewhat uniform rockfills	High	0.10–0.20 for $5 \leq \sigma_{1f} \leq 80\,\text{kg/cm}^2$
El Granero slate (gradation A) Mica granitic-gneiss (gradation X)	Well-graded rockfills	Low	
Mica granitic-gneiss (gradation Y)	Uniform rockfill produced by blasting metamorphic rocks ($C_u < 5$)	Low	Increases with σ_{1f}: maximum value = 0.30

Note that B_q, grain breakage; q_i, initial concentration of solids; σ_{1f}, major principal stress at failure. Marsal (1973).

quantities as average number of particles per unit area or per unit volume, average number of contacts per particle, and so on lose some of their relevance.

The resistance to grain crushing or breakage depends on the strength of the particles, which in turn depends on mineralogy and the influences of fissures, weathering, and voids. Failure may be by compression shear, or in a split tensile mode. The amount of grain crushing to be expected for rockfills and gravels is summarized in Table 13.3. In this table, B_q is the proportion of the solid phase by weight that will undergo breakage, and q_i is the volume concentration of solids ($V_s/V = 1/(1 + e)$).

Studies of grain crushing in sands and gravels under isotropic and anisotropic triaxial stresses (Lee and Farhoomand, 1967) to more than 200 atm show the following:

1. Coarse granular soils compress more and show more particle breakage than fine granular soils. A comparison of gradation curves before and after compression is shown in Fig. 13.8. Compressibility data are shown in Fig. 13.9.

2. Soils with angular particles compress more and show more particle crushing than soils with rounded particles (Fig. 13.9a).

3. Uniform soils compress and crush more than well-graded soils with the same maximum grain size.

4. Under a given load, compression and crushing continue indefinitely at a decreasing rate.

5. Volume change during compression depends only on the major principal stress and is independent of the principal stress ratio.

6. The higher the principal stress ratio ($K_c = \sigma_{1c}/\sigma_{3c}$) during consolidation, the greater the amount of grain crushing, as shown in Fig. 13.10.

13.5   OSMOTIC PRESSURE CONCEPT OF VOLUME CHANGE

The adsorption of cations by clays and the formation of double layers is responsible for long range repulsive forces between particles (Chapters 7 and 10). Quantitative prediction of these forces is possible in some cases, and the extent to which they can account for the swelling and compression behavior of clays is known.

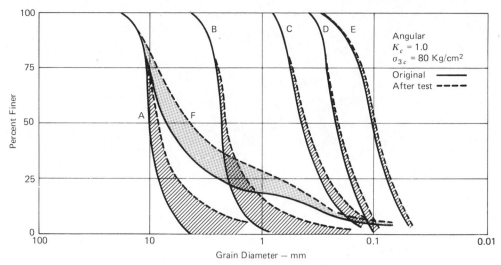

Fig. 13.8   Comparison of crushing of soils with different initial grain sizes (From Lee and Farhoomand, 1967).

Calculation of interparticle repulsions due to interacting double layers may be done in more than one way; however, the osmotic pressure concept is convenient. By this approach, the pressure that must be applied to a system to prevent movement of water either in or out is determined as a function of solution concentration. Solution concentrations can be related to particle spacings, and theoretical relationships between void ratio or water content and pressure can be developed.

The concept of osmotic pressure is illustrated by Fig. 13.11. The two sides of the cell in Fig. 13.11a are separated by a semipermeable membrane, through which solvent (water) may pass, but solute (salt) cannot. Because the salt concentration in solution is greater on the left side of the membrane than on the right side, the free energy or chemical potential of the water on the left is less than on the right.* Because solute cannot pass to the right to equalize concentrations because of the presence of the membrane, solvent passes into the chamber on the left.

---

* Formal treatment of the concepts stated here and derivation of equation (13.1) are given in standard texts on chemical thermodynamics.

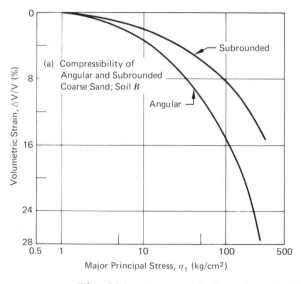

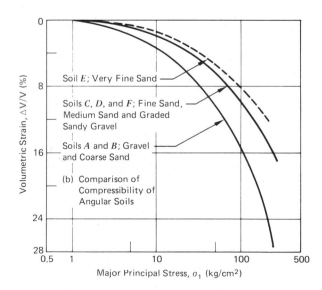

Fig. 13.9   Compressibility of sands and gravels (From Lee and Farhoomand, 1967).

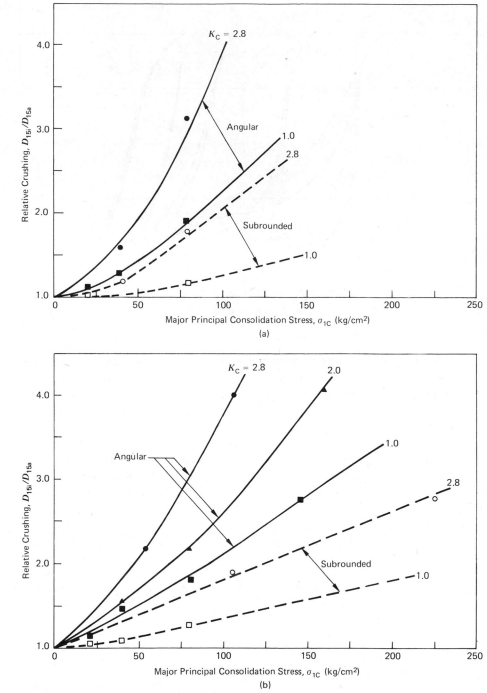

Fig. 13.10 Relative crushing of coarse soils at different stress conditions (From Lee and Farhoomand, 1967). $K_c = \sigma_{1c}/\sigma_{3c}$; D_{15i} = 15 percent size before test; D_{15a} = 15 percent size after test. (a) Soil A–gravel. (b) Soil B–coarse sand.

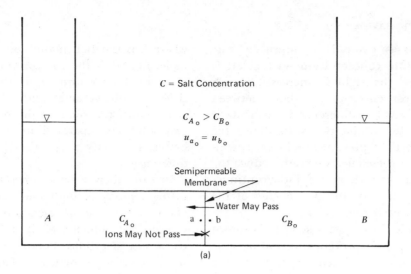

(a)

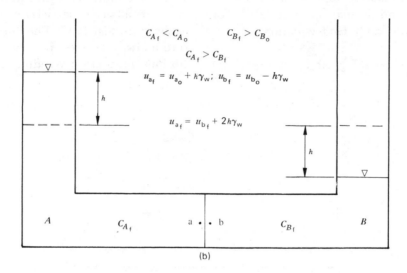

(b)

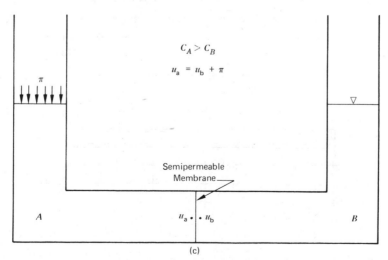

(c)

Fig. 13.11 Osmotic pressure. (*a*) Initial condition: no equilibrium. (*b*) Final condition: equilibrium. (*c*) Osmotic pressure equilibrium.

The effect of this is twofold as shown by Fig. 13.11b. First, the solute concentration on the left is reduced and that on the right is increased, which serves to reduce the concentration imbalance between the two chambers. Second, a difference in hydrostatic pressure develops between the two sides. Since the free energy varies directly as pressure and inversely with concentration, both of these effects serve to reduce the imbalance between the two chambers. Flow continues through the membrane until the free energy of the water is the same on each side.

It would be possible, in a system such as is shown by Fig. 13.11a to completely prevent flow through the membrane by applying a sufficient pressure to the solution in the left chamber, as shown by Fig. 13.11c. The pressure needed to exactly stop flow is termed the osmotic pressure π, and it may be calculated, for dilute solutions, by the van't Hoff equation

$$\pi = kT \sum (n_{iA} - n_{iB}) = RT \sum (c_{iA} - c_{iB}) \quad (13.1)$$

where k is the Boltzmann constant (gas constant per molecule), R is the gas constant per mole, T is the absolute temperature, n_i is the concentration (particles per unit volume), and c_i is the molar concentration. Thus, the osmotic pressure difference between two solutions separated by a semipermeable membrane is directly proportional to the concentration difference.

In a soil there is no semipermeable membrane separating regions of high and low concentration. The effect of a restrictive membrane is introduced, however, by the influence of the negative clay surfaces on the adsorbed cations. Because of the attraction of adsorbed cations to particle surfaces, the cations are not free to diffuse, and concentration differences responsible for osmotic pressures develop whenever double layers or adjacent particles overlap. The situation is as shown in Fig. 13.12. The difference in osmotic pressure midway between particles and that in the equilibrium solution surrounding the clay is the interpar-

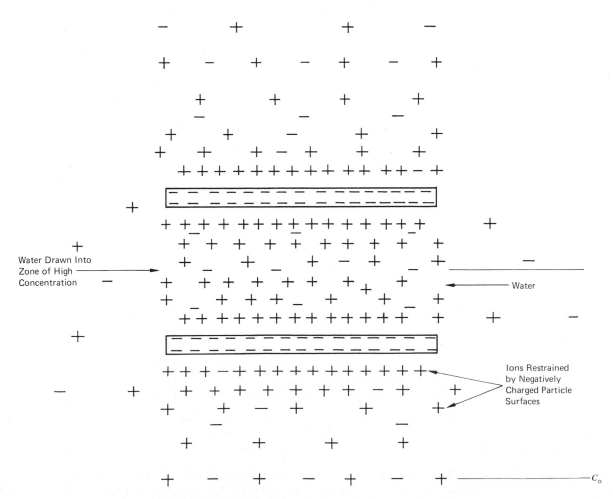

Fig. 13.12 Mechanism of osmotic pressure in clay.

ticle repulsive pressure or swelling pressure P_s. It may be expressed in terms of midplane potentials (see Section 7.6); according to

$$P_s = p = 2n_0 kT \, (\cosh u - 1) \quad (7.30)$$

where n_0 is the concentration (molecules per unit volume) in the external solution, and u is the midplane potential function.

The relationships are more convenient in terms of midplane cation and equilibrium solution concentrations c_c and c_0 (Bolt, 1956). Equation (13.1) becomes

$$P_s = \pi = RT \sum (c_{ic} - c_{io}) \quad (13.2)$$

For single cation and anion species of the same valence

$$P_s = RT(c_c + c_a - c_0{}^+ - c_0{}^-) \quad (13.3)$$

where c_a is the midplane anion concentration and $c_0{}^+$ and $c_0{}^-$ are the equilibrium solution concentrations of cations and anions. It may be shown that at equilibrium in dilute solutions

$$c_c \cdot c_a = c_0{}^+ \cdot c_0{}^- = c_0{}^2 \quad (13.4)$$

because $c_0{}^+ = c_0{}^-$. Thus, equation (13.3) becomes

$$P_s = RTc_0 \left(\frac{c_c}{c_0} + \frac{c_0}{c_c} - 2 \right) \quad (13.5)$$

Equation (7.26) gives midplane concentration in terms of particle spacing $2d$ and equilibrium solution concentration. This equation, which assumes parallel flat plates, may be written in terms of void ratio for a saturated clay. The water content w, in terms of volume of water per unit weight of soil solids, divided by the specific surface of soil solids A_s gives the average thickness of water layer, which is half the particle spacing or d. Thus,

$$d = \frac{w}{\gamma_w A_s} \quad (13.6)$$

For a saturated soil the void ratio is related to the water content by

$$e = G_s w \quad (13.7)$$

where G_s is the specific gravity of soil solids, for A_s measured in square centimeters per gram. Noting that $\gamma_w \approx 1.0 \text{ g/cm}^3$, equation (13.6) becomes

$$d = \frac{e}{G_s A_s} \quad (13.8)$$

and equation (7.26) can be written

$$v\sqrt{\beta c_0}\left(x_0 + \frac{e}{G_s A_s} \right) = 2\left(\frac{c_0}{c_c} \right)^{1/2}$$
$$\times \int_{\phi=0}^{\pi/2} \frac{d\phi}{[1 - (c_0/c_c)^2 \sin^2 \phi]^{1/2}} \quad (13.9)$$

Combinations of (P_s/RTc_0) and $v(\beta c_0)^{1/2} (x_0 + e/G_s A_s)$ that satisfy equations (13.5) and (13.9) are given in Table 13.4. These values may be used to calculate theoretical curves of void ratio vs pressure for consolidation and swelling. For any value of log $[P_s/(RTc_0)]$ the swelling pressure may be calculated. The void ratio can be computed from the corresponding value of $v(\beta c_0)^{1/2} (x_0 + e/G_s A_s)$. For a given soil P_s depends completely on c_c and c_0. Those factors that cause c_c to be large relative to c_0; for example, low c_0, low valence of cation, high pH, and large ions, cause high interparticle repulsions, high swelling pressures, and large physico-chemical resistance to compression.

Table 13.4 Relation Between the Distance Variable Expressed as a Function of the Void Ratio and the Swelling Pressure of Pure Clay Systems

$v(\beta c_0)^{1/2}$ $(x_0 + e/G_s A_s)$	log $P_s/$ RTc_0	$v(\beta c_0)^{1/2}$ $(x_0 + e/G_s A_s)$	log $P_s/$ RTc_0
0.050	3.596	0.997	0.909
0.067	3.346	1.188	0.717
0.100	2.993	1.419	0.505
0.200	2.389	1.762	0.212
0.300	2.032	2.076	−1.954
0.400	1.776	2.362	−1.699
0.500	1.573	2.716	−1.427
0.600	1.405	3.09	−1.111
0.700	1.258	3.57	−2.699
0.801	1.130	4.35	−2.045
0.902	1.012		

Note: v is the cation valence; β is $8\pi F/1000 \, DRT \approx 10^{15}$ cm/mmole for water at normal T; c_0 is the concentration in bulk solution (mmole/cm^3); $x_0 = 4/v\beta T \approx 1/v$ Å for illite, $2/v$ Å for kaolinite, and $4/v$ Å for montmorillonite; e is the void ratio; G_s is the unit weight of solids (g/cm^3); A_s is the specific surface area of clay; P_s is the swelling pressure; R is the gas constant; and T is the absolute temperature. Bolt (1956).

The preceding relationships were developed for soils containing a single electrolyte. Approximate equations for mixed cation systems that cover most of the moisture suction or overburden pressure ranges of interest in soil mechanics and soil science are available (Collis-George and Bozeman, 1970). They are suitable for

$$\frac{|\tau| \times 4 \times 10^{-5}}{\sum c_o} \geq 20 \qquad (13.10)$$

where $|\tau|$ is the swelling pressure or matrix suction (see Section 10.5) measured in centimeters of water.

Since the sum of the applied constraint $|\tau|$ in concentration units and the external solution concentration must equal the midplane concentration, the pressure or suction is given by

$$|\tau| = \frac{\sum c_m - \sum c_o}{4 \times 10^{-5}} \qquad (13.11)$$

For homovalent and di-cation/mono-anion systems, $\sum c_m$ is found from

$$v(\beta)^{1/2}\left(\frac{e}{G_s A_s}\right) = \frac{\pi}{\sqrt{\sum c_m}} - \frac{2}{(\frac{1}{4}\beta\Gamma^2 + \sum c_m)^{1/2}} \qquad (13.12)$$

where $\beta \approx 1.0 \times 10^{15}$ cm/mmole at 20°C and $\Gamma =$ double layer charge (meq/cm²). For dilute concentrations in the external solution equations (13.11 and 13.12) reduce to

$$|\tau| = 0.25 \times 10^5 \frac{\pi^2}{v^2 \; \beta \; (e/G_s A_s)^2} \qquad (13.13)$$

For mixed cation heterovalent systems, $\sum c_m$ is given by

$$v(\beta)^{1/2}\left(\frac{e}{G_s A_s}\right)$$
$$= \frac{\pi}{(\sum c_m)^{1/2}} \frac{-\cos^{-1}\{1/a[1-(\sum c_m/(\frac{1}{4}\beta\Gamma^2 + \sum c_m))^{1/2}]\}^{1/2}}{c_m} \qquad (13.14)$$

$$a = \qquad (13.15)$$
$$\frac{2c'_m - (c_m^+ + c_m^{++}) + [4c'_m c_m^{++} + (c_m^+ + c_m^{++})^2]^{1/2}}{2c'_m}$$

where c'_m is the midplane anion concentration. Since evaluation of equation (13.15) requires knowledge of the midplane concentrations of the different ions separately, the application of equation (13.14) is not as straightforward as is the case for equations (13.12) and (13.13).

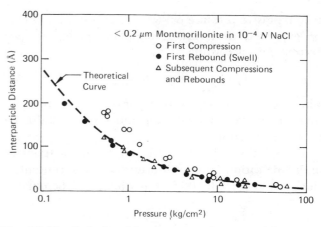

Fig. 13.13 Relationship between interparticle spacing and pressure for montmorillonite (From Warkentin, Bolt, and Miller, 1957).

13.6 APPLICATION OF OSMOTIC PRESSURE CONCEPTS FOR DESCRIPTION OF VOLUME CHANGE BEHAVIOR OF SOILS

A reasonably clear picture of the extent to which the osmotic pressure concept can account for the compression and swelling behavior of fine-grained soils is now available.

Homoionic cation systems

Early testing of the applicability of the osmotic pressure theory was done using "pure clays" consisting of specially prepared very fine-grained samples. For example, Fig. 13.13 shows good agreement between theoretical and experimental values of interparticle spacing and pressure for a less than 0.2 μm fraction of montmorillonite in 10^{-4} N NaCl.

Theoretical and experimental compression curves for sodium and calcium montmorillonite in 10^{-3} M electrolyte solutions are compared in Fig. 13.14. Agreement is fairly good as regards the influence of cation valence. The fact that compression curves are above decompression and recompression curves is attributed to cross-linking and nonparallel particle arrangements that are either eliminated or made constant after the first cycle. Similar behavior is evident in Fig. 13.13.

Compression curves for sodium illite with particles finer than 0.2 μm are shown in Fig. 13.15. The effect of electrolyte concentrations is as predicted. The experimental curves are substantially above the theoretical curves, the reason being perhaps the presence of a "dead" volume of liquid resulting from terraced particle surfaces (Bolt, 1956).

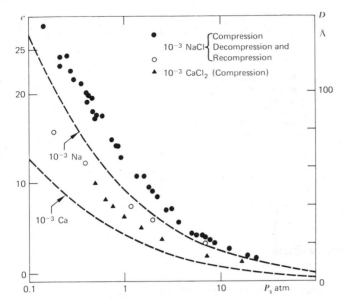

Fig. 13.14 Compression curves of Na-montmorillonite and Ca-montmorillonite, fraction −0.2 μm, in equilibrium with 10^{-3} M NaCl and $CaCl_2$, respectively. The dashed lines represent the theoretical curves for $A_s = 800$ m²/g (From Bolt, 1956).

Theoretical and experimental relationships between water content, pressure, and electrolyte concentration are shown in Fig. 13.16 for homoionic Na samples and in Fig. 13.17 for homoionic Ca samples of Wyoming bentonite, vermiculite, and kaolinite, where the particles are finer than 0.2 μm (El Swaify and Henderson, 1967). In each case, the influence of increasing electrolyte concentration is to decrease the water content at a given pressure in accordance with theory. The experimental values of water content were higher than theoretical values, again reflecting the effects of dead zones.

Agreement between theory and experiment has not been good for clays containing larger particle sizes. The coarse fraction (0.2 to 2.0 μm) of two bentonites gave swelling pressures less than predicted; whereas, the fine fraction (< 0.2 μm) gave values close to theoretical, even though the charge densities of the two fractions were the same (Kidder and Reed, 1972). The behavior of three size fractions of sodium illite is shown in Fig. 13.18. Even though the discrepancies between theory and experiment are fairly large for the < 0.2 μm fraction, the experimental curves are in the predicted relative positions, Fig. 13.18a. For samples containing coarser particles, Fig. 13.18b and

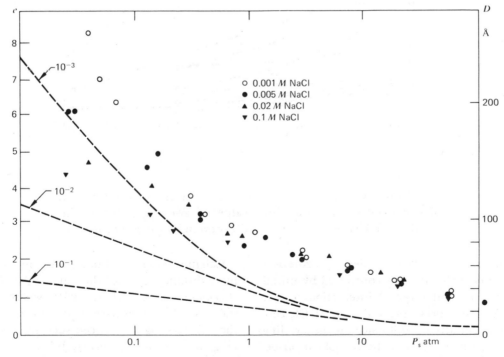

Fig. 13.15 Compression curves of Na-illite, fraction −0.2 μm in equilibrium with different concentrations of NaCl, given as molarity. The dashed curves represent the theoretical curves for $A_s = 120$ m²/g (From Bolt, 1956).

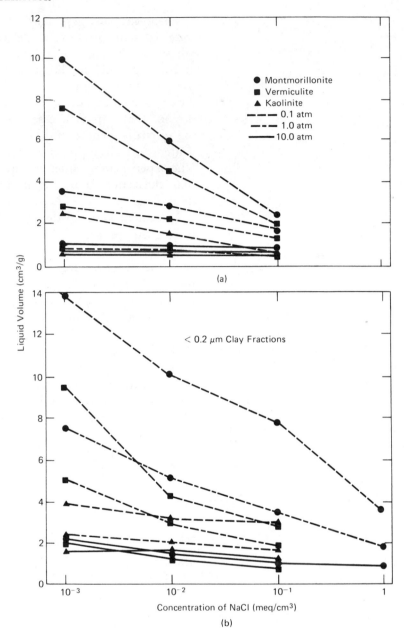

Fig. 13.16 Water retention curves for Na clays as a function of NaCl concentration at three values of swelling pressure (El-Swaify and Henderson, 1967). (a) Theoretical. (b) Experimental.

13.18c, the curves are in reverse order to theoretical prediction, because behavior was controlled by initial particle orientations and physical interactions rather than osmotic repulsive pressures. CaCl₂ or MgCl₂ concentration has essentially no influence on the swelling of a 2 μm fraction of illite, and the consolidation behavior is influenced only as the changes in electrolyte concentration change the initial structure (Olson and Mitronovas, 1962).

Calcium montmorillonites do not swell to interplate distances greater than about 9 Å (Norrish, 1954; Blackmore and Miller, 1961) where the particles stabilize. The following equation may be used for the half spacing d' between such plate groups (Shainberg, Bresler, and Klausner, 1971)

$$d' = \frac{NV}{WA_s} - \frac{d(N-1)}{W} \qquad (13.16)$$

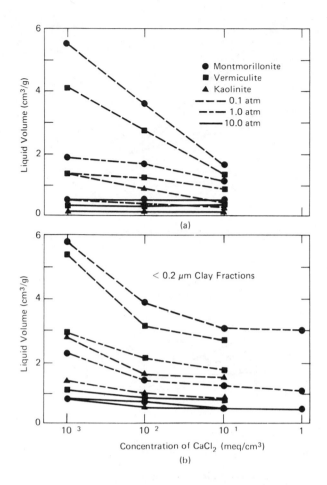

Fig. 13.17 Water retention curves for Ca clays as a function of CaCl$_2$ concentration at three values of swelling pressure (El-Swaify and Henderson, 1967). (a) Theoretical. (b) Experimental.

where N is the number of particles per plate group (4 to 9), v is the volume of water, w is the dry weight of clay, A_s is the specific area of clay, and d is the half spacing between platelets (4.5 Å). A theoretical swelling curve computed using equation (13.16) to relate water content to spacing was in good agreement with the experimental curve for a calcium montmorillonite when it was assumed that $N = 3$ at low pressures and $N = 5$ at high pressures.

Factors other than clay particle size may also contribute to failure of the theory in natural soils. These include deviations from the assumed parallelism between clay plates, cross-linking, the effects of other long and short range repulsions and attractions, for example, van der Waal's forces, that are neglected, and the effects of impurities such as organic matter.

Mixed cation systems

Most natural soils contain mixtures of sodium, potassium, calcium, and magnesium in their adsorbed cation complex. As shown in Chapter 7 and Section 13.5, modifications are required of the double layer and osmotic pressure equations that hold for homoionic systems. The extent to which the resulting equations may be suitable depends on the structural status of the clay as well as on the particle size. In addition, the equations for mixed cation systems are derived on the assumption that ions of all species are distributed uniformly over the clay surfaces in proportion to the amounts present. There is evidence, however, that in some cases sodium and calcium ions separate into distinct regions. This is termed "demixing" (Glaeser and Mering, 1954; McNeal, Norwell, and Coleman, 1966; McNeal, 1970; and Fink, Nakayama, and McNeal, 1971).

A demixed ion model for interlayer swelling has been developed (McNeal, 1970), and observed behavior was good for most cases examined (5 out of 6) for values of exchangeable sodium percentage (ESP) less than about 50 percent. Based on X-ray determinations of interplate spacings in montmorillonite, it appears (Fink, Nakayama, and McNeal, 1971) that for

ESP > 50%	Random mixing of Na$^+$ and Ca^{2+}, unlimited swelling between all plates on water addition
10% < ESP < 50%	Demixing on interlayer exchange sites, with progressively more sets of plates collapsing to a 20 Å repeat spacing with decrease in ESP
ESP < 10 to 15%	Interlayer exchange complex is predominantly Ca-saturated; Na ions on external planar and edge sites

13.7 IMPORTANCE OF MINERALOGICAL DETAIL IN SOIL EXPANSION

Most expansive soils contain montmorillonite or vermiculite. Details of mineral structure and the presence of interlayer materials may have significant effects on the swelling properties.

Crystal lattice configuration effects

Lattice charge has a great influence on swelling of different minerals, and greatest expansion is observed

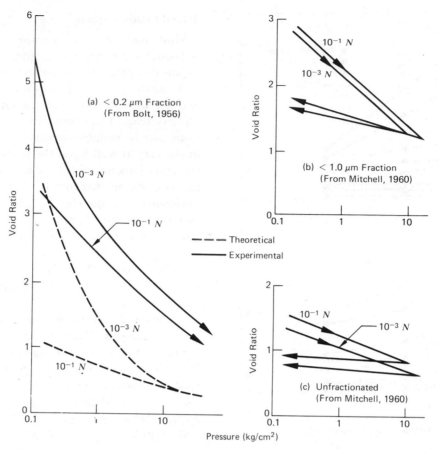

Fig. 13.18   Influence of NaCl concentration and particle size on compression and swelling behavior of Fithian illite.

for charge deficiencies of about one per unit cell as indicated in Table 13.5. However, within the range of charge deficiencies where swell is observed, there is no consistant relationship between charge, as measured by the cation exchange capacity, and the amount of swell (Foster, 1953, 1955). There is, however, inverse correlation between free swell and the *b*-dimension of the montmorillonite crystal lattice (Davidtz and Low, 1970) as shown in Fig. 13.19. Differences in the two curves probably relate to differences in sample preparation or method for evaluation of free swell used in the two investigations from which the data shown were obtained.

The differences in *b* dimension, which could be caused by differences in isomorphous substitution, are presumed to result in alterations of the adsorbed water structure, which in turn give a different free energy and swelling. Furthermore, as water content increases, so also does the *b* dimension, as shown in Fig. 6.6. The swelling ceases when the *b*-dimension reaches 9.0 Å.

Table 13.5   Influence of Lattice Charge on Expansion

Mineral	Negative Charge Per Unit Cell	Tendency to Expand
Margarite	4	None
Muscovite		
Biotite	2	Only with drastic chemical treatment, if at all
Paragonite		
Hydrous mica and illite	>1.2	
Vermiculite	1.4–0.9	Expanding
Montmorillonite		
Beidellite	1.0–0.6	Readily expanding
Nontronite		
Hectorite		
Pyrophyllite	0	None

From Brindley and MacEwen (1953).

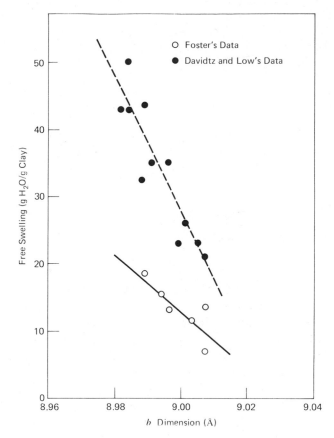

Fig. 13.19 Relation between free swelling and *b*-dimension for Na-montmorillonites (Davidtz and Low, 1970).

Thus, the potential swell or swell pressure of a clay may depend on the difference between the *b*-dimension and 9.0 Å. Insufficient measurements have been made, however, to demonstrate this for clay minerals other than montmorillonite or cations other than sodium. Nonetheless, an alternative theory to the osmotic pressure theory based on the interactions between water and mineral structures may be possible.

Hydroxy interlayering

The occurrence, formation, and properties of hydroxy–cation interlayers (Fe–OH, Al–OH, Mg–OH) have been studied (Rich, 1968) regarding their effects on physical properties of expansive clays. Some aspects of interlayering between the basic sheets in the expansive clay minerals are:

1. Optimum conditions for interlayer formation are
   a. A supply of Al^{3+} ions.
   b. Moderately acid pH ($\sim$ 5).
   c. Low oxygen content.
   d. Frequent wetting and drying.
2. Hydroxy-Aluminum is the principal interlayer material in acid soils, but Fe–OH layers may be present.
3. $Mg(OH)_2$ is probably the principal interlayer component in alkaline soils.
4. Randomly distributed islands of interlayer material bind adjacent layers together. The degree of interlayering in soils is usually small (10 to 20 percent), but this is enough to fix the basal spacings of montmorillonite and vermiculite at 14 Å.
5. Reduction of cation exchange capacity by interlayer formation.
6. Increased swelling as a result of removal of interlayer material.

Figure 13.20 shows that swelling is reduced at any electrolyte content as a result of treatment by hydroxides of iron and aluminum. The salt concentration at which dispersion develops is lowered. The presence of hydroxides around aggregates and particles can reduce swell at high electrolyte concentrations. Under high swell pressure conditions, which can develop at lower electrolyte concentrations, the coatings can be broken leading to aggregate breakdown. Removal of free amorphous silica resulted in an increase of free swell in several sodium montmorillonites, as shown in Table 13.6.

Table 13.6 Effect of Silica Removal on Free Swell of Some Sodium Montmorillonites

	Swelling (g water/g clay)	
Sample	Untreated Na-Saturated Clay	Free SiO_2 Removed by Boiling in 0.5 N NaOH
Utah, before drying	5.58	7.75
No. 4, before drying	10.63	11.88
No. 4, after drying	11.70	15.32
No. 6, before drying	6.44	12.45
No. 6, after drying	6.72	16.84
No. 2, before drying[a]	7.86	7.79
Sm. Miss., before drying[a]	8.67	8.01

[a] Naturally occurring clay does not contain free SiO_2.
Barshad (1973).

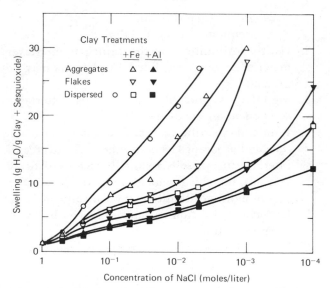

Fig. 13.20   Swelling of oriented flakes of montmorillonite in NaCl solutions (El Rayah and Rowell, 1973).

13.8 PRECONSOLIDATION PRESSURE AND SECONDARY COMPRESSION

When the consolidation characteristics of an undisturbed clay sample are determined, the behavior is ordinarily expressed in the form of Fig. 13.21, where void ratio is shown as a function of effective consolidation pressure p'. The maximum past consolidation stress p_c is determined (commonly by means of the Casagrande, 1936, construction) and compared with the present overburden effective stress p_o. The three possible relationships between p_c and p_o are

1. $p_c < p_o$—underconsolidated clay. The clay has not yet reached equilibrium under the present overburden.
2. $p_c = p_o$—normally consolidated clay. The clay is in void ratio equilibrium with the present overburden effective stress.
3. $p_c > p_o$—overconsolidated clay. The clay has been consolidated, or behaves as if it had been consolidated, under an effective stress greater than the present overburden effective stress.

Accurate knowledge of p_c is needed for settlement analysis and also as an aid for interpretation of geological history and for prediction of possible future changes. In the majority of cases, p_c is greater than p_o. Overconsolidation by dessication and by unloading due to erosion or by a rise in the groundwater table have been widely recognized as among the principal causes for this.

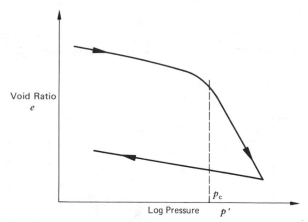

Fig. 13.21   Typical consolidation curve.

More recently, weathering and cementation have been identified as important causes of an apparent preconsolidation. Changes in ion concentrations and pH, leaching, oxidation, precipitation, mineralogy, and other consequences of chemical weathering can cause substantial changes in strength and compressibility. A weathered clay may not show a pronounced p_c value (Bjerrum, 1973), and the $e - \log p$ curve is likely to show a gentle curvature representing a reduced compressibility compared to the unweathered clay. Cemented clays have particles held together with chemically derived cementing bonds of a nature different than those responsible for cohesion and friction in noncemented clays.

A condition where the preconsolidation pressure p_c is greater than p_o may also develop in soft clays without chemical change or unloading. When such a clay is subjected to a constant consolidation pressure, the resulting deformations may be divided into three parts.

1. *Immediate compression* due to compression of gas in partly saturated clays and shear deformation at constant volume.
2. *Primary consolidation*, the rate of which is controlled by the rate of dissipation of excess pore water pressure.
3. *Secondary or delayed compression* that involves a time-dependent adjustment of the soil structure and can be considered a creep–type phenomenon (Chapter 14). Secondary compression has been recognized for many years as an important contribution to total compression; its influence on the observed value of p_c has been clarified more recently.

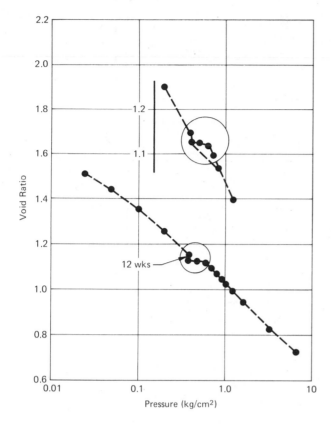

Fig. 13.22 The development of a quasi-preconsolidation pressure (Leonards and Ramiah, 1960).

The concept of a "quasi-preconsolidation" pressure (Leonards and Ramiah, 1960) is shown in Fig. 13.22. As a result of sustained loading at the point indicated, the compression is greater than developed under previously applied load increments where less time was allowed for secondary compression. Upon resumption of consolidation under increased stress, the clay behaves as if preconsolidated to a somewhat higher stress.

The equilibrium void ratio versus vertical effective stress relationships for a normally consolidated clay are shown in Fig. 13.23. The different curves correspond to different times for consolidation under each effective stress. A young normally consolidated clay can just sustain the applied effective stress. Secondary compression leads to a reduction in void ratio, development of a more stable arrangement of soil particles, greater strength, and reduced compressibility. The rate of decrease of void ratio is roughly proportional to logarithm of time.

A result of this delayed compression is development of a reserve resistance against further consolidation,

whereby more load may be carried in addition to the overburden stress without significant volume change. Thus, after a long time under a constant stress the compression behavior will be similar to that indicated by the "aged normally consolidated clay" in Fig. 13.23.

The value of p_c, for a given time of loading, increases in proportion to p_o (Bjerrum, 1972, 1973). Thus, in a homogeneous clay deposit, p_c/p_o is constant with depth. Since the p_c/p_o ratio depends on the amount of secondary compression and, for a given time of loading, the amount of secondary compression increases with plasticity index, the value of p_c/p_o increases with plasticity. Figure 13.24 shows the relationship between p_c/p_o and Plasticity Index for normally consolidated late glacial and postglacial clays that have consolidated for several thousand years.

The behavior of clays that exhibit both a p_c effect and overconsolidation due to load removal is shown in Fig. 13.25. Original consolidation and aging was under effective overburden pressure p_1. Reduction of the overburden pressure to p_o by erosion then occurred, giving an overconsolidation ratio p_1/p_o. A delayed swelling may develop, if cohesive bonds are of a viscous character. When such a clay is reloaded, compression will be small until pressure p_c is reached, at which point a break in the curve is observed and virgin compression resumes.

The p_c effect may be significant in terms of the additional stress that can be applied without development of large compressions. Procedures for settlement prediction in soft clays exhibiting a p_c effect, to include immediate consolidation and secondary or delayed settlements, are available (Bjerrum, 1972, 1973). Correct determination of the value of p_c is of major importance in making these predictions. Bjerrum (1973) suggests that after placing the sample in the consolidometer it be loaded in two or three steps to p_o. Small load increments, of the order of $(p_c - p_o)/3$, should be used until p_c is exceeded. Then load increments of 50 to 100 percent of the previous load should be used.

The determination of the delayed or secondary compression depends on knowledge of the positions of the curves corresponding to different times in plots of the form of Figs. 13.23 and 13.25. Their location cannot be obtained by laboratory tests within reasonable time periods, but an estimate may be made in the following way. The laboratory test data enable construction of a 24 hr (or other time of loading, de-

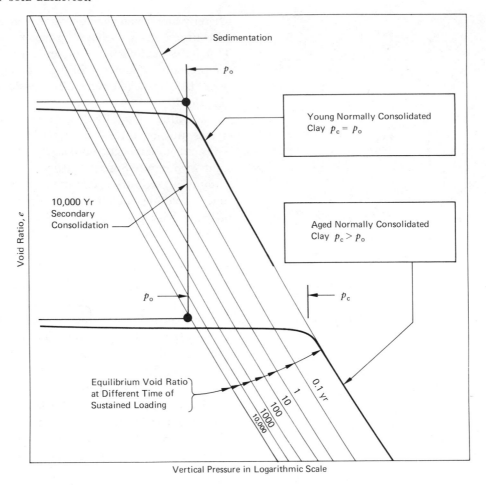

Fig. 13.23 Geological history and compressibility of normally consolidated
clays (From Bjerrum, 1972).

pendent on load increment duration) curve. A parallel second curve drawn through (e_o, p_o) gives a curve corresponding to the loading age of the deposit. The time between a 24-hr curve and a 10,000-year curve will represent about six logarithmic cycles. About 50 percent of the delayed compression will occur in the first year and about 80 percent after 100 years.

The procedures suggested by Bjerrum for estimation of settlements are based on the assumption of separate primary consolidation and secondary compression. A solution for the rate of consolidation combining both a hydrodynamic time lag and creep of the soil skeleton was developed by Garlanger (1972).

Classification of soft clays

Based on age and geological history, soft clays can be classified according to the scheme in Table 13.7.

This table is useful in anticipating the behavior of different deposits.

13.9 TEMPERATURE–VOLUME RELATIONSHIPS

Temperature variations can cause significant changes in the volume and effective stress of saturated soils. Figure 13.26 shows the percentage of original pore water volume drained from a saturated specimen of illite subjected to a temperature increase from 66 to 140°F followed by cooling to 66°F while maintained under isotropic effective stress of 2.0 atm. The temperature changes were in the sequence defined by the numbered points. Figure 13.27 shows the variation in effective stress σ_3' under the same temperature variations but with drainage prevented. Changes such

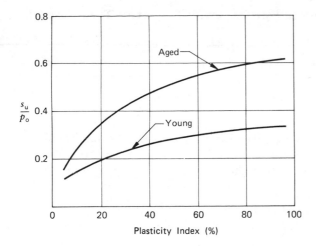

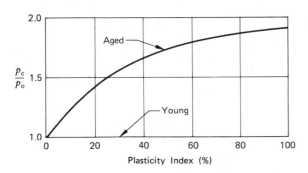

Fig. 13.24 Typical values of s_u/p_o and p_c/p_o observed in normally consolidated late glacial and post glacial clays (From Bjerrum, 1972).

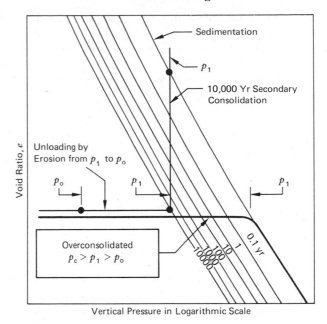

Fig. 13.25 Geological history and compressibility of an overconsolidated clay (Bjerrum, 1972).

where α_s is the thermal coefficient of cubical expansion of mineral solids and V_s is the volume of solids.

If a saturated soil is free to drain due to a change in temperature while under constant effective stress, the volume of water drained is

$$(\Delta V_{DR})_{\Delta T} = (\Delta V_w)_{\Delta T} + (\Delta V_s)_{\Delta T} - (\Delta V_m)_{\Delta T} \quad (13.19)$$

as those shown in Figs. 13.26 and 13.27 may have important consequences, both in the laboratory and in the field.

Theoretical analysis

Drained conditions. Thermal expansion of the mineral solids and pore water and thermally-induced changes in soil structure lead to volume changes due to temperature variations. For a temperature change ΔT, the volume change of the pore water is

$$(\Delta V_w)_{\Delta T} = \alpha_w V_w \Delta T \quad (13.17)$$

where α_w is the thermal expansion coefficient of soil water and V_w is the pore water volume. The change in volume of mineral solids is

$$(\Delta V_s)_{\Delta T} = \alpha_s V_s \Delta T \quad (13.18)$$

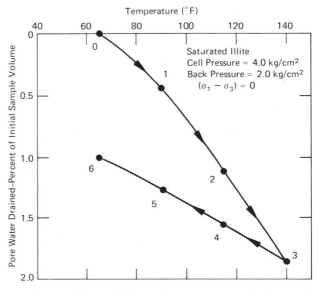

Fig. 13.26 Effect of temperature variations on volume change under drained conditions.

Table 13.7   Classification of Some Frequently Encountered Soft Clays with Respect to Their Strength and Compressibility

Classification	Water Content	Shear Strength	Compressibility
Weathered clays in upper crust			
Frost treated dried-out clay	$w \approx w_p$	Very stiff, fissured open cracks	—
Dried-out clay	$w \approx w_p$	Very stiff fissured	Low compressibility
Weathered clay	$w_p < w < w_L$	Shear strength decreases with depth	Low compressibility; curved e-log p curve
Unweathered clays			
Young normally consolidated clays	$w \approx w_L$	s_u/p_0 constant with depth	$p_c \approx p_0$
Aged normally consolidated clays	$w \approx w_L$	s_u/p_0 constant with depth	p_c/p_0 constant with depth
Over consolidated aged clays	$w_p < w < w_L$	s_u/p_0 constant with depth	$p_c \approx p_1\left(\dfrac{p_c}{p_0}\right)I_p$
Young normally consolidated quick clay	$w_L < w$	s_u/p_0 constant with depth	$p_c \approx p_0$
Aged normally consolidated quick clay	$w_L < w$	s_u/p_0 constant with depth	p_c/p_0 constant with depth

Bjerrum (1972).

in which $(\Delta V_m)_{\Delta T}$ is the change in total volume due to ΔT, with volume increases considered positive.

For a soil mass with grains in contact, and assuming the same coefficient of thermal expansion for all the soil minerals, the soil grains and the soil mass undergo the same volumetric strain, $\alpha_s \, \Delta T$. In addition, the change in temperature may induce a change in interparticle forces, cohesion and/or frictional resistance that necessitates some reorientations or movements of soil grains to permit the soil structure to

carry the same effective stress. If the volume change due to this effect is $(\Delta V_{ST})_{\Delta T}$, then

$$(\Delta V_m)_{\Delta T} = \alpha_s V_m \Delta T + (\Delta V_{ST})_{\Delta T} \qquad (13.20)$$

and

$$(\Delta V_{DR})_{\Delta T} = \alpha_w V_w \Delta T + \alpha_s V_s \Delta T$$
$$- (\alpha_s V_m \Delta T + (\Delta V_{ST})_{\Delta T}) \quad (13.21)$$

Undrained conditions. The governing criterion for undrained conditions is that the sum of the sepa-

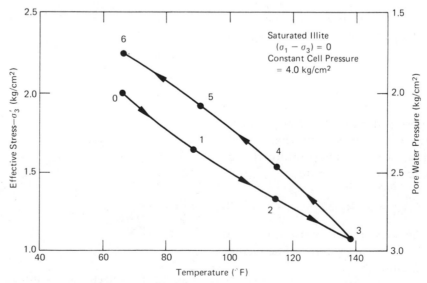

Fig. 13.27   Effect of temperature variations on effective stress (undrained conditions).

rate volume changes of the soil constituents due to both temperature and pressure changes must equal the sum of the volume changes of the total soil mass due to both temperature and pressure changes; that is,

$$(\Delta V_w)_{\Delta T} + (\Delta V_s)_{\Delta T} + (\Delta V_w)_{\Delta P} + (\Delta V_s)_{\Delta P}$$
$$= (\Delta V_m)_{\Delta T} + (\Delta V_m)_{\Delta P} \quad (13.22)$$

where the subscripts ΔT and ΔP refer to temperature and pressure changes, respectively. If m_w, m_s, and m_s' refer to the compressibility of water, the compressibility of mineral solids under an all round pressure, and the compressibility of mineral solids under concentrated loadings, respectively, then

$$(\Delta V_w)_{\Delta P} = m_w V_w \Delta u \quad (13.23)$$

$$(\Delta V_s)_{\Delta P} = m_s V_s \Delta u + m_s' V_s \Delta \sigma' \quad (13.24)$$

where Δu is the change in pore water pressure and $\Delta \sigma'$ is the change in effective stress. The term $m_s' V_s \Delta \sigma'$ is the change in volume of mineral solids due to a change in effective stress, which manifests itself by changes in forces at interparticle contacts. Also

$$(\Delta V_m)_{\Delta P} = m_v V_m \Delta \sigma' \quad (13.25)$$

where m_v is the compressibility of the soil structure.

From equations (13.17), (13.18), (13.23), (13.24), and (13.25), equation (13.22) becomes

$$\alpha_w V_w \Delta T + \alpha_s V_s \Delta T - (\Delta V_m)_{\Delta T}$$
$$= m_v V_m \Delta \sigma' - m_w V_w \Delta u - V_s(m_s \Delta u + m_s' \Delta \sigma') \quad (13.26)$$

For constant total stress during a temperature change

$$\Delta \sigma' = -\Delta u \quad (13.27)$$

Thus, equation (13.26) becomes

$$\alpha_w \Delta T + \alpha_s V_s \Delta T - (\Delta V_m)_{\Delta T}$$
$$= m_v V_m \Delta \sigma' - m_w V_w \Delta u - \Delta u V_s(m_s - m_s') \quad (13.28)$$

Since both m_s and m_s' are not likely to differ significantly and both are much less than m_v and m_w, little error results from assuming $m_s - m_s' = 0$, so equation (13.28) can be written

$$\alpha_w V_w \Delta T + \alpha_s V_s \Delta T - (\Delta V_m)_{\Delta T} = -m_v V_m \Delta u - m_w V_w \Delta u \quad (13.29)$$

The left side of equation (13.29) is equal to $(\Delta V_{DR})_{\Delta T}$, and the right side is an equivalent volume change caused entirely by a change in pore pressure. Because

$$V_m = V_w + V_s \quad (13.30)$$

equation (13.29) may be written, after substitution for $(\Delta V_m)_{\Delta T}$ by equation (13.20),

$$\alpha_w V_w \Delta T - \alpha_s V_w \Delta T - (\Delta V_{ST})_{\Delta T}$$
$$= -m_v V_m \Delta u - m_w V_w \Delta u \quad (13.31)$$

Equation (13.31) can be rearranged to give the pore pressure change accompanying a temperature change:

$$\Delta u = \frac{n \Delta T(\alpha_s - \alpha_w) + (\Delta V_{ST})_{\Delta T}/V_m}{m_v + n m_v}$$
$$= \frac{n \Delta T(\alpha_s - \alpha_w) + \alpha_{ST} \Delta T}{m_v + n m_w} \quad (13.32)$$

in which the porosity $n = V_w/V_m$ and α_{ST} is the physico-chemical coefficient of structural volume change defined by

$$\alpha_{ST} = \frac{(\Delta V_{ST}) \Delta T / V_m}{\Delta T}$$

Thus, the factors controlling pore pressure changes are the magnitude of ΔT, porosity, the difference between thermal expansion coefficients for soil grains and water, the volumetric strain due to physico-chemical effects, and the compressibility of the soil structure. For most soils (but not rocks) $m_v \gg n m_w$, so

$$\Delta u = \frac{n \Delta T(\alpha_s - \alpha_w) + \alpha_{ST} \Delta T}{m_v} \quad (13.33)$$

In the application of the above equations, consistency in algebraic signs is required. Both α_s and α_w are positive and correspond to a volumetric increase with increasing temperature. The compressibilities m_v and m_w are negative, because an increase in pressure causes a decrease in volume, and α_{ST} is negative if an increase in temperature causes a decrease in volume of the soil structure.

Volume change behavior

Permanent volume decreases may develop when the temperature of a normally consolidated clay is increased, as shown by Fig. 13.28. Temperature changes in the order indicated were carried out on a sample of saturated, remolded illite after initial consolidation to an effective stress of 2.0 kg/cm². Water drained from the sample during an increase in temperature and was absorbed during temperature decreases. The shape of the curves is similar to normal consolidation curves for volume changes caused by changes in applied stresses.

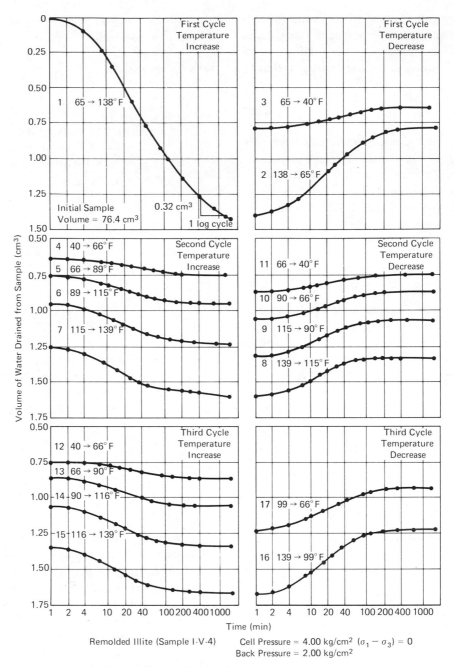

Fig. 13.28 Relationship between volume of water drained and time during temperature changes at constant stress.

When the temperature is increased, two effects occur. If the increase is rapid, a significant positive pore pressure develops, due to a greater volumetric expansion of the pore water than of the mineral solids. The lower the permeability of the soil, the longer the time required for its dissipation. Dissipation of this pressure can account for the parts of the curves in Fig. 13.28 that resemble primary consolidation.

The second effect results because an increase in temperature causes a decrease in the shearing strength of individual interparticle contacts (Chapter 14). As a consequence, there is a partial collapse of the soil structure and a decrease in void ratio until a sufficient number of additional bonds are formed to enable the soil to carry the stress at the higher temperature. This effect is analogous to secondary compression under a stress increase.

When the temperature is decreased, differences in the volumetric shrinkage of the soil grains and water cause a tension in the pore water, which in turn causes soil to absorb water, as shown by the temperature decrease curves in Fig. 13.28. No secondary volume change effect is observed, because the temperature decrease causes a strengthening of the soil structure, and no further structural adjustment is needed to carry the effective stress. On subsequent temperature increases, the secondary effect is negligible, because the structure has already been strengthened in previous cycles.

The final height changes and volumes of water drained associated with each temperature change shown in Fig. 13.28 are plotted as a function of temperature in Fig. 13.29, and clay structure volume

changes are shown in Fig. 13.30. There is similarity in the form of these plots and conventional compression curves involving virgin compression, unloading, and reloading. Again, the effect of an increase in temperature is analogous to an increase in pressure. The slope of the curves in Fig. 13.30 gives the coefficient of thermal expansion for the soil structure, α_{ST}, defined previously, as

$$\alpha_{ST} = \frac{\Delta V_{ST}/V_m}{\Delta T} \qquad (13.34)$$

For the cases shown it has a value of about $-0.5 \times 10^{-4} °C^{-1}$.

The effect of temperature on the compression of a clay depends on the pressure range. Figure 13.31 shows void ratio as a function of logarithm of pres-

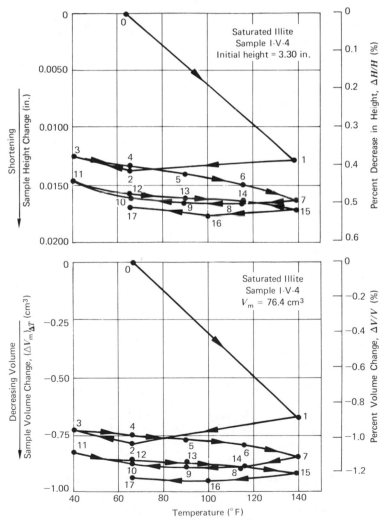

Fig. 13.29 Effect of temperature variations on height and volume change.

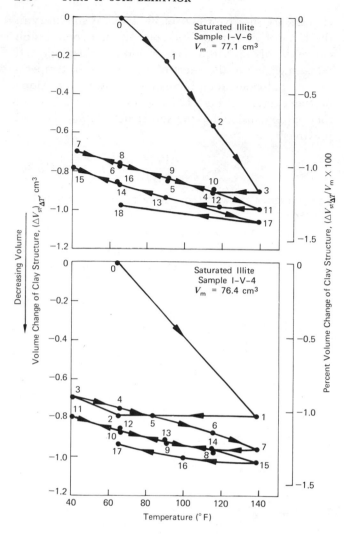

Fig. 13.30 Clay structure volume changes due to physico-chemical phenomena.

sure for illite at two temperatures (Plum and Esrig, 1969). Initial consolidation was at a temperature of 24°C under a pressure of 1.7 psi. At pressures less than 30 psi, compressibilities were essentially the same at the two temperatures.

Data for another series of tests (Campanella and Mitchell, 1968) are shown in Fig. 13.32 for consolidation of illite at three temperatures. Equal compression indices were measured for pressures greater than 2 kg/cm². As consolidation was started from the same initial water content for all three specimens, the clay at higher temperatures must have been more compressible at lower pressures to account for the void ratio differences observed at 2.0 kg/cm².

Observations of the type shown in Figs. 13.31 and

13.32 indicate that the weaker structure at low stresses caused by higher temperatures causes consolidation to a lower void ratio in order to carry the stress. Adjustment of the structure to compensate for the temperature effects is complete by the time an effective stress of the order of 2.0 atm is reached, and the weakening effect of the higher temperature is compensated by the strengthening effect of the lower void ratio.

The effect of heating followed by cooling at two stages in a consolidation test is shown in Fig. 13.33. The effect is remarkably similar to the p_c-effect caused by aging under sustained stress, discussed in Section 13.8. Thus, a normally consolidated clay in nature that had been previously subjected to a higher temperature could exhibit an apparent preconsolidation. Laboratory samples that have been heated and recooled can give an incorrect evaluation of the maximum preconsolidation pressure.

Pore pressure behavior

Pore pressure changes accompanying temperature changes under undrained conditions can be predicted reasonably well using equation (13.33). The most important factors are the thermal expansion of the pore water, the compressibility of the soil structure, and the initial effective stress. The value of compressibility m_v that is appropriate depends on the rebound and recompression characteristics of the soil. When the temperature increases, pore pressures increase, and effective stresses decrease, which is a condition analogous to unloading. When the temperature decreases, pore pressures decrease, and effective stresses increase. Since the previous temperature history caused permanent volume decreases at the higher temperature, the condition is analogous to recompression. Thus, the appropriate value of m_v is that based on the slope of the rebound or recompression curves, both of which are approximately the same.

$$(m_v)_R = \frac{\Delta V_m/V_m}{\Delta \sigma'} = \frac{0.435}{(1 + e_0)} \frac{C_S}{\sigma'} \quad (13.35)$$

where C_S is the swelling index, e_0 is the initial void ratio, and σ' is the effective stress at which $(m_v)_R$ is to be evaluated.

A pore pressure–temperature parameter F may be defined as the change in pore pressure per unit change in temperature per unit effective stress, or alternatively, the change in unit effective stress per unit change in temperature; that is,

$$F = \frac{\Delta u/\Delta T}{\sigma'} = -\frac{\Delta \sigma'/\sigma'}{\Delta T} = \frac{e_0[(\alpha_s - \alpha_w) + \alpha_{ST}/n]}{0.435 C_S} \quad (13.36)$$

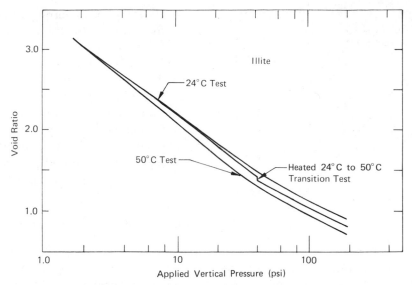

Fig. 13.31 Influence of temperature on consolidation of illite (After Plum and Esrig, 1968).

Some values of F are summarized in Table 13.8. The values listed for σ' are averages for the indicated temperature changes.

The influence of effective stress on change in pore pressure can be seen from the data for Vicksburg buckshot clay and for the saturated sandstone. The greater change in pore pressure for a given ΔT for the case of a higher effective stress is predicted by the theory. Also, the much lower compressibility of the sandstone is responsible for a much higher temperature sensitivity of pore water pressure and effective stress than for the buckshot clay.

The parameter F is approximately the same for different clays (Table 13.8). Knowledge of F values allows determination of laboratory temperature con-

Table 13.8 Temperature-Induced Pore Pressure Changes under Undrained Conditions

Soil Type	σ' (kN/m²)	ΔT (°C)	Δu (kN/m²)	F $(\Delta u/\Delta T)$ σ' (°C⁻¹)
Illite (Grundite)	200	21.1–43.4	+58	0.013
San Francisco Bay mud	150	21.1–43.4	+50	0.015
Weald Clay[a]	710	25.0–29.0	+51	0.018
Kaolinite	200	21.1–43.4	+78	0.017
Vicksburg	100	20.0–36.0	+28	0.017
buckshot clay[b]	650	20.0–36.0	+190	0.018
Saturated	250	5.3–15.0	+190	0.079
sandstone (porous stone)	580	5.3–15.0	+520	0.092

[a] From Henkel and Sowa (1963).
[b] From Ladd (1961) Fig. VIII-6.

trol to assure accurate pore pressure measurements in undrained tests. For example, if it were desired to keep pore pressure fluctuations to within ± 5 kN/m² (±.05 kg/cm²) for one of the clays in Table 13.8, the required temperature control would be about 0.5°C for a sample at an effective stress of 500 kN/m².

13.10 PRACTICAL APPLICATIONS

In addition to providing insight into factors influencing volume changes in soils, the information in

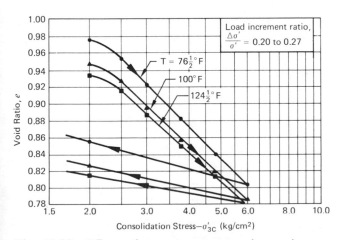

Fig. 13.32 Effect of temperature on isotropic consolidation behavior of saturated illite.

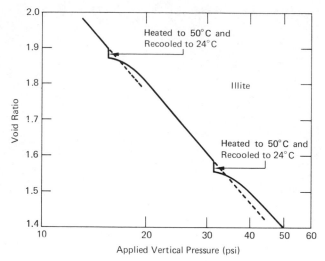

Fig. 13.33 Effect of heating and cooling on void ratio vs. pressure relationship of illite (Plum and Esrig, 1969).

this chapter can be applied in a variety of ways. Some of them are listed here.

1. Considerations from Section 13.2 can be used to make preliminary estimates of compression and swell index values.
2. The considerations of grain crushing (Section 13.4) are important in any case where coarse-grained soils are to be subjected to high stresses.
3. The osmotic pressure theory of swelling is useful to estimate the effects of changes in environment.
4. The information in Section 13.7 suggests that examination of subtle differences in mineralogical detail may be necessary if differences in the swelling behavior of apparently similar soils are to be explained.
5. Delayed compression and the preconsolidation effect (Section 13.8) may have important consequences on the compression behavior of soft

clays *in situ*. Thorough study of geological history and careful laboratory testing are indicated if useful deformation and settlement analyses are to be made in such materials.

6. The considerations in Section 13.9 on temperature effects point up the need for temperature controls in soil testing. They also provide a basis for analyses of the possible consequences of temperature change that might be caused by power plants, underground power cables, liquified natural gas storage, and pipelines.

SUGGESTIONS FOR FURTHER STUDY

Bjerrum, L. (1973), "Problems of Soil Mechanics and Construction on Soft Clays and Structurally Unstable Soils," General Report, Session 4, *Proceedings of the Eighth International Conference on Soil Mechanics and Foundation Engineering*, Vol. 3, pp. 111–159.

Bolt, G. H. (1956), "Physico-Chemical Analysis of the Compressibility of Pure Clay," *Geotechnique*, Vol. 6. No. 2, pp. 86–93.

Campanella, R. G., and Mitchell, J. K. (1968), "Influence of Temperature Variations on Soil Behavior," *Journal of the Soil Mechanics and Foundations Division*, A.S.C.E., Vol. 94, No. SM3, pp. 709–734.

Highway Research Board (1969), "Effects of Temperature and Heat on the Engineering Behavior of Soil," Special Report 103.

Meade, R. H. (1964) "Removal of Water and Rearrangement of Particles During The Compaction of Clayey Sediments—Review," U. S. Geological Survey Professional Paper 497-B, U. S. Government Printing Office.

Mesri, G., and Olson, R. E. (1971), "Consolidation Characteristics of Montmorillonite," *Geotechnique*, Vol. 21, No. 4, pp. 341–352.

Rich, E. I. (1968), "Hydroxy Interlayers in Expansible Layer Silicates," *Clays and Clay Minerals*, Vol. 16, pp. 15–30.

Rieke, H. H. III, and Chilingarian, G. V. (1974), *Compaction of Argillaceous Sediments*, Elsevier, New York.

CHAPTER 14

Strength and Deformation Behavior

14.1 INTRODUCTION

All problems of stability, bearing capacity, and *in situ* stress depend in some manner on soil strength. The stress–strain and stress–strain–time behavior of soils under subfailure stress conditions is also of interest in many problems. Most of the relationships that are used for the characterization of strength and stress–deformation properties of soils are empirical, based on phenomenological description of soil behavior. The Mohr-Coulomb theory is by far the most widely used. It states that

$$\tau_{ff} = c + \sigma_{ff} \tan \phi \qquad (14.1a)$$

$$\tau_{ff} = c' + \sigma_{ff}' \tan \phi' \qquad (14.1b)$$

where τ_{ff} is shear stress at failure on the failure plane, c is a cohesion intercept, σ_{ff} is a normal stress on the failure plane, and ϕ is a friction angle. This relationship recognizes that neither shear stress nor normal stress alone cause failure in a soil but that failure develops under critical combinations of shear and normal stress. Equation (14.1a) is applicable for σ_{ff} defined as a total stress, and c and ϕ are termed total stress parameters. Equation (14.1b) applies for σ_{ff}', defined as an effective stress, and c' and ϕ' are effective stress parameters.

In reality, the shearing resistance of a soil depends on many factors, and a complete equation might be of the form

$$\text{Shearing resistance} = F(e, \phi, C, \sigma', c', H, T, \epsilon, \dot{\epsilon}, S) \qquad (14.2)$$

in which e is the void ratio, C is the composition, H is the stress history, T is the temperature, ϵ is the strain, $\dot{\epsilon}$ is the strain rate, and S is the structure. All of the parameters in equation (14.2) may not be independent, however, and their quantitative functional forms are not all known. As a consequence, values of c and ϕ are determined using specified test type (for example, direct shear, triaxial compression, triaxial extension, simple shear), drainage conditions, rate of loading, range of confining pressures, and stress history. As a result, a variety of types of "friction angle," "cohesion," and so on, have been defined, including parameters for total stress, effective stress, drained, undrained, peak strength, and residual strength. The values of c and ϕ applicable in practice depend on such factors as whether the problem is one of loading or unloading, whether short or long term stability is of interest, and stress orientation.

Emphasis in this chapter is on the factors controlling the strength and stress–deformation behavior of soils, rather than on the strength parameters themselves, which are treated in classical soil mechanics texts. Following a review of general characteristics of the strength, stress–strain, and stress–strain–time behavior of soils, fundamentals of bonding, effective stress and strength, friction, particulate behavior, cohesion, and time dependencies are treated in some detail.

14.2 GENERAL CHARACTERISTICS

1. The basic factor responsible for the strength of a soil is frictional resistance between soil particles in contact. The magnitude of this resistance depends, for a given soil type, on the effective stress, which in turn is controlled by applied stresses, physicochemical forces of interaction, and the volume change tendencies of the soil. The strength of cohesionless soils is influenced most by relative density, effective minor principal stress (σ_3'), and test type (for

example plane strain versus triaxial compression). The strength of saturated clay is influenced most by drainage conditions, disturbance (which is manifested by a change in effective stress and a loss of cementation), overconsolidation ratio, and creep effects. The peak strength of a clay may be considerably greater than the strength after very large strain or shear displacement.

2. In the absence of chemical cementation between grains, the strength of sand and normally consolidated clay depends directly on effective stress,

$$\tau_{ff} = \sigma_{ff}' \tan \phi' \qquad (14.3)$$

where the primes designate effective stresses.

3. The peak value of ϕ' for clays decreases with increasing plasticity index and activity, as may be seen in Fig. 14.1.

4. The residual friction angle decreases with increasing plasticity as shown in Fig. 14.2. The residual strength is the shear strength along a well-defined failure surface at large displacements. It is independent of stress history and original structure. For a given set of testing conditions it depends only on composition and effective stress.

5. The post-peak shearing displacement required to cause a reduction in friction angle to the residual value varies with soil type, normal stress on the shear plane, and test conditions. For example, for shale mylonite in contact with smooth steel or other polished, hard surfaces, a shearing displacement of only 1 or 2 mm is sufficient to give residual strength (Deere, 1974, personal communication). For soil against soil, however, a slip along the shear plane of many centimeters may be required; as shown by Fig. 14.3.

6. Failure envelopes for peak and residual values of friction angle are curved in the manner shown in Fig. 14.4. This behavior in sands can be caused by dilatancy effects and, at higher confining pressures, by grain crushing. Curved failure envelopes are also observed for some clays. Two examples of the stress-dependency of the residual friction angle are shown in Fig. 14.5. Possible causes of this behavior are discussed in Section 14.8. Not all clays exhibit this stress-dependency; Bishop et al. (1971) found the residual strength of blue London clay to be independent of normal stress.

7. Overconsolidated clays may have a higher strength at a given effective stress than nor-

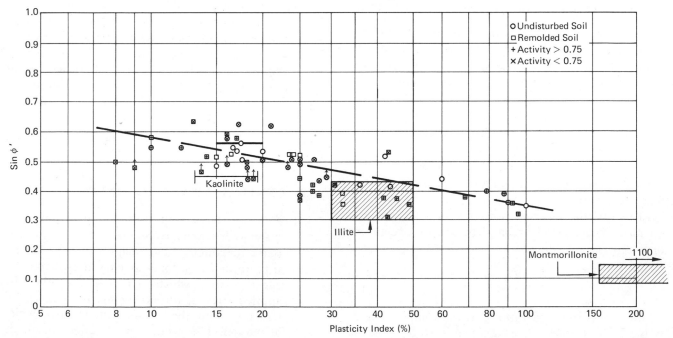

Fig. 14.1 Relationship between sin ϕ and plasticity index for normally consolidated soils (Kenney, 1959). (Data for pure clays from Olson, 1974.)

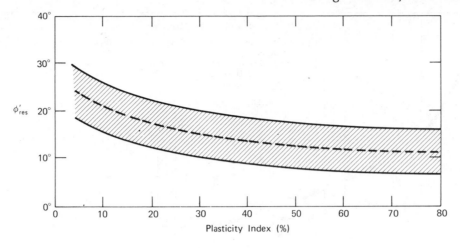

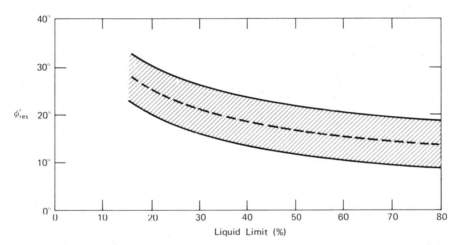

Fig. 14.2 Relationship between residual friction angle and plasticity (Compiled by Deere, 1974).

mally consolidated clays, as shown in Fig. 14.6. The difference between the two strength envelopes at any value of effective stress depends on type of clay, drainage conditions during shear, and amount of overconsolidation.

8. The envelope for overconsolidated clay shown in Fig. 14.6 reflects the influences of both stress history and differences in water content. For comparisons at the same water content but different effective stress, as for points A and A' in Fig. 14.6 the Hvorslev strength parameters c_e and ϕ_e are obtained (Hvorslev, 1937, 1960). The parameter ϕ_e has been assumed to be independent of water content and effective normal stress; whereas, c_e depends only on water content.

The Hvorslev parameters have been termed "true friction angle" and "true cohesion," and considered by some to reflect the mechanism of shear strength in terms of interparticle forces (c_e) and friction (ϕ_e). Such an interpretation is controversial, however, because as shown in Chapter 12, two samples at the same water content but different effective stresses must have different structures. This will manifest itself by differences in volume change in the case of drained tests, and by differences in pore pressures in undrained tests. In addition, the procedures for evaluation of c_e and ϕ_e fail to take the curvature of the failure envelope into account. Present evidence is that a true cohesion

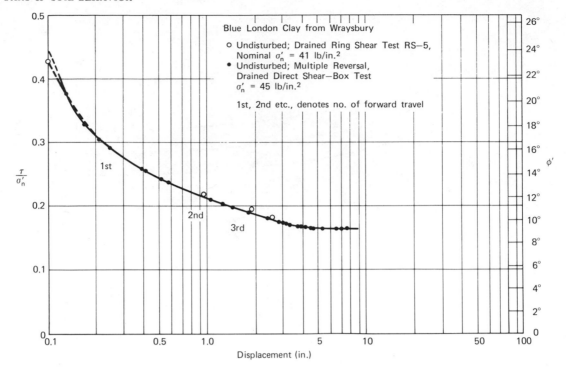

Fig. 14.3 Development of residual strength with increasing displacement (Bishop et al, 1971).

at zero effective stress is negligible in the absence of chemical bonding between particles caused by cementation.

9. In the absence of chemical cementation, the differences in the strengths of two samples of the same soil at the same void ratio but with different fabrics are accountable in terms of different effective stresses (Chapter 12).

10. Undrained strength anisotropy may result from both stress anisotropy and fabric anisotropy. In clays, anisotropy decreases with increasing plasticity (Bjerrum, 1973).

11. Undrained strength in triaxial compression may differ from the strength in triaxial exten-

sion; however, the influence of type of triaxial test on the effective stress parameters c' and ϕ' is relatively small.

12. At failure, dense sands and heavily overconsolidated clays have a greater volume after drained shear or a higher effective stress after undrained shear than at the start of deformation. At failure, loose sands and normally consolidated to moderately overconsolidated clays (OCR up to about 4) have a smaller volume after drained shear or a lower effective stress after undrained shear than initially.

13. Effective stress friction angles measured in plane strain are typically about 10 percent greater than those determined by triaxial compression.

14. For a saturated soil a change in temperature causes either a change in void ratio or a change in effective stress (or a combination of both) as demonstrated in Chapter 13. Thus, a change in temperature can cause a strength increase or a strength decrease, depending on the circumstances, as illustrated by Figs. 14.7 and 14.8. For the tests on kaolinite shown in Fig. 14.7, all samples were prepared by isotropic triaxial consolidation at 75°F. Then, with no further drainage allowed, temperatures were increased

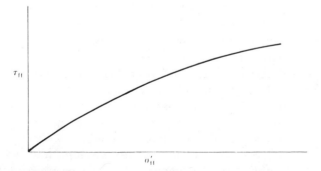

Fig. 14.4 Curvature of Mohr strength envelope.

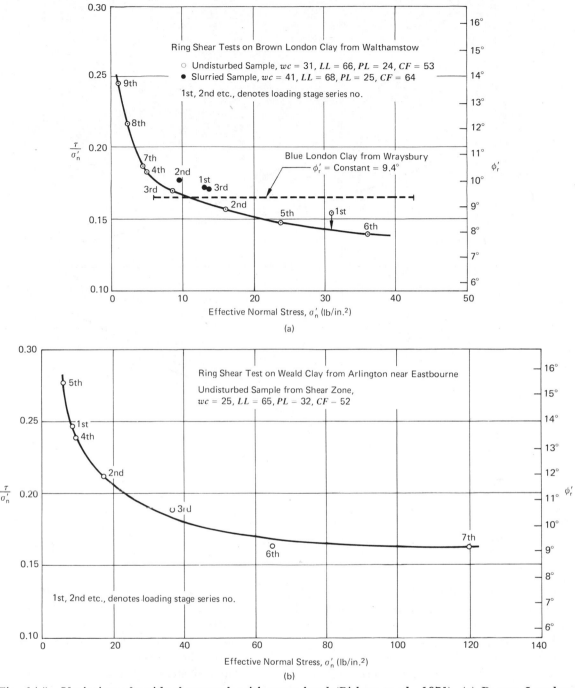

Fig. 14.5 Variation of residual strength with stress level (Bishop et al., 1971). (*a*) Brown London clay. (*b*) Weald clay.

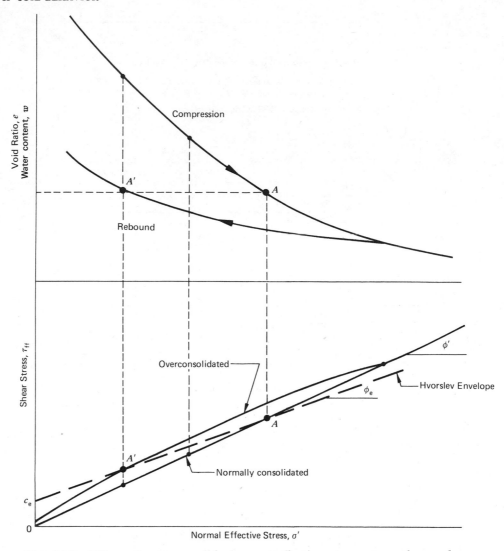

Fig. 14.6   Effect of overconsolidation on effective stress strength envelope.

to the values indicated, and the samples were tested in unconfined compression. Substantial reductions in strength accompanied the increases in temperature.

In Fig. 14.8, T_c represents the temperature at consolidation and T_s the temperature of shear for consolidated undrained direct shear tests on a highly plastic alluvial clay. The higher the consolidation temperature, the greater the shear strength at any given test temperature because of the greater decrease in void ratio at the higher consolidation temperatures.* For a

given consolidation temperature, however, the strength decreases in a regular manner with increasing test temperature. Data such as these have established that for given initial conditions the undrained strength of a saturated clay may decrease by about 10 percent for a temperature increase from 0 to 40°C.

Stress–strain behavior

1. Stress–strain behavior may range from very brittle for some quick clays, cemented soils, heavily overconsolidated clays, and dense sands to ductile for insensitive and remolded clays and loose sands, as illustrated by Fig. 14.9.

2. An increase in confining pressure causes an increase in the deformation modulus of cohesion-

---

* For all tests, $T_s \lesssim T_c$ to prevent further consolidation under a higher temperature with the result that the strength would be the same as if the soil had been consolidated under the higher temperature.

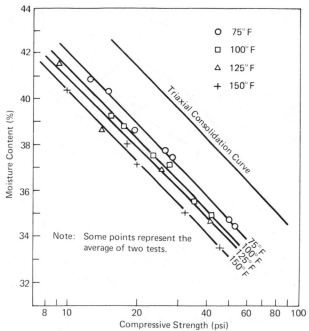

Fig. 14.7 Effect of temperature on the undrained strength of kaolinite in unconfined compression (Sherif and Burrous, 1969).

Young's modulus E_i can be expressed by a relationship adapted from Janbu (1963)

$$E_i = K p_a \left(\frac{\sigma_3}{p_a} \right)^n \qquad (14.4)$$

where K is a dimensionless modulus number that varies from about 300 to 2000, n is an exponent usually in the range of 0.3 to 0.6, and p_a is a units constant equal to atmospheric pressure in the same units as σ_3. Values of K and n for a variety of soils are summarized by Wong and Duncan (1974). For saturated clays the initial tangent modulus for undrained loading is typically in the range of 50 to 1000 times the undrained shear strength.

3. Shear deformations of saturated soils are accompanied by volume changes when drainage is permitted or changes in pore water pressure and effective stress when drainage is prevented. The general nature of this behavior is shown in Fig. 14.11. The magnitudes of the volume or pore pressure changes, in any case, depend on interactions between fabric and stress state and the ease with which shear deformations can develop without overall changes in volume or transfer of normal stress from the soil structure to the pore water.

4. Fabric anisotropy may have a greater influence

less soils as well as an increase in strength, as shown by Fig. 14.10. An increase in the preshear consolidation pressure has a similar effect in clays. For cohesionless soils the initial tangent

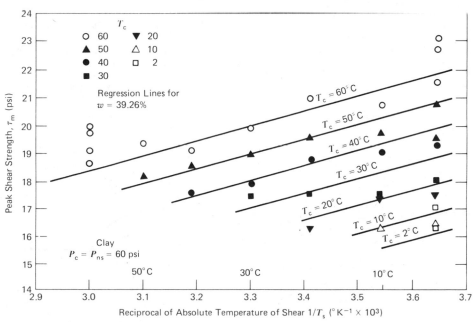

Fig. 14.8 Effect of consolidation and test temperatures on the strength of alluvial clay in direct shear (Noble and Demirel, 1969).

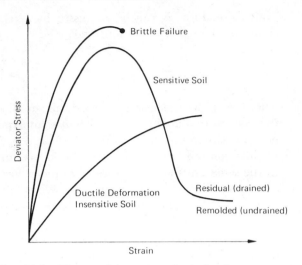

Fig. 14.9   Types of stress–strain behavior.

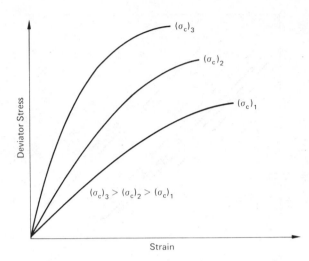

Fig. 14.10   Effect of confining pressure on the consolidated–drained, stress–strain behavior of soils.

on the stress–strain behavior in different directions than on the strength in different directions as shown in Chapter 12.

5. For a constant value of total minor principal stress the magnitudes of the pore pressures de-

veloped in undrained loading may depend more on the strain than on the stress (Lo, 1969a, 1969b).

6. An increase in temperature causes a decrease in modulus, that is, a softening of the soil. Imme-

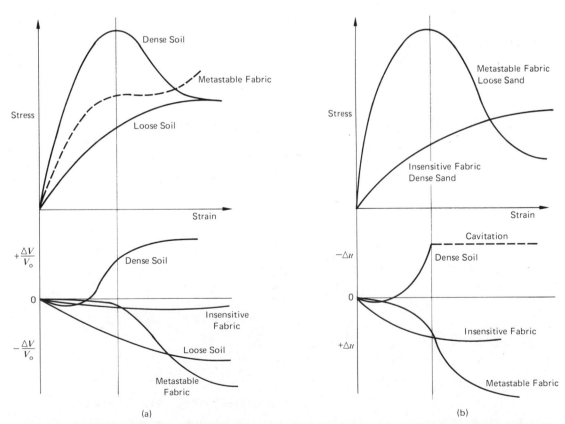

Fig. 14.11   Volume and pore pressure changes during shear. (a) Drained conditions. (b) Undrained conditions.

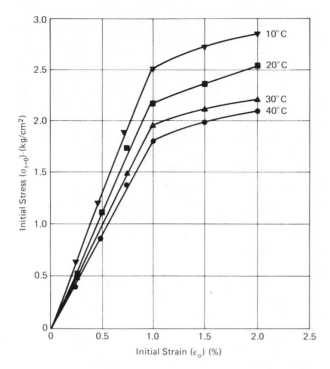

Fig. 14.12 Effect of temperature on the stiffness of Osaka clay in undrained triaxial compression. (After Murayama, 1969).

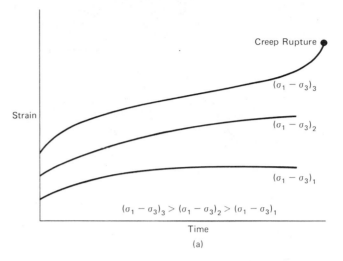

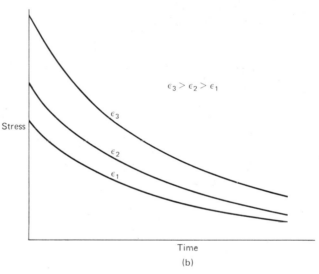

Fig. 14.13 Creep and stress relaxation. (*a*) Creep under constant stress. (*b*) Stress relaxation under constant strain.

diate strain as a function of stress is shown in Fig. 14.12 for samples of Osaka clay tested in undrained triaxial compression at different temperatures as an illustration of this.

Stress–strain–time behavior

1. Soils exhibit both creep* and stress relaxation, Fig. 14.13. The magnitude of these effects increases with the plasticity, activity, and water content of the soil, and it is influenced by whether conditions are drained or undrained, but the form of the behavior is essentially the same for all soils.

2. Some soils may fail under a sustained creep stress significantly less (as much as 50 percent) than the peak stress measured in an undrained test where a sample is loaded to failure in a few minutes or hours (Casagrande and Wilson, 1951; Hirst and Mitchell, 1968). This is termed "creep

rupture." The time to failure increases logarithmically with decrease in the stress level for stresses greater than some limiting value below which failure does not develop (Murayama and Shibata, 1961, 1964; Singh and Mitchell, 1969). Saturated soft sensitive clays under undrained conditions and heavily overconsolidated clays under drained conditions are most susceptible to strength loss during creep.

3. Deformation under sustained stress ordinarily produces an increase in stiffness to the action of subsequent stress increase, as shown schematically in Fig. 14.14. The effect is analogous to the preconsolidation effect due to secondary

---

* The term "creep" is used herein to refer to time-dependent shear strains and/or volumetric strains that develop at a rate controlled by the viscous resistance of the soil structure. "Secondary compression" refers to the specific case of volumetric strain that follows primary consolidation, and is controlled in rate by the viscous resistance of the soil skeleton as opposed to hydrodynamic lag.

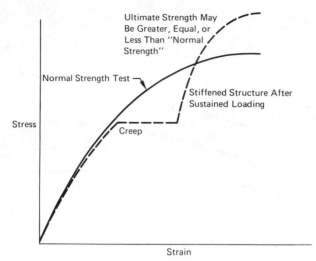

Fig. 14.14 Effect of sustained loading on stress–strain and strength behavior.

compression discussed in Section 13.8; however, it may develop under undrained as well as drained conditions.

4. Pore pressures may increase, decrease, or remain constant during undrained creep.

5. Undrained strength increases with increase in rate of strain (see Fig. 14.15). The magnitude of the effect increases with increasing plasticity index, but is typically of the order of 5 to 10 percent for each order of magnitude increase in strain rate.

14.3 SOIL DEFORMATION AS A RATE PROCESS

Deformation and shear failure of soils involves time-dependent rearrangement of matter. As such,

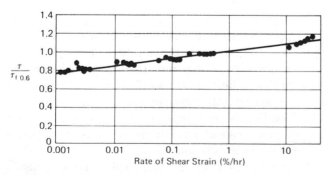

Fig. 14.15 Effect of strain rate on strength for undrained triaxial compression tests on plastic clay from Drammen, Norway. Strength is normalized to that at a strain rate of 0.6 percent/hr (Bjerrum, 1972).

these phenomena are amenable for study as "rate processes" through application of the theory of absolute reaction rates (Glasstone, Laidler, and Eyring, 1941). This theory is helpful in providing both insights into the fundamental nature of soil strength and functional forms for the influences of certain variables on soil behavior.

Detailed development of the theory, which is based on statistical mechanics, may be found in Eyring (1936), Glasstone et al. (1941), and elsewhere in the physical chemistry literature. Adaptations to the study of soil behavior are given by Abdel-Hady and Herrin (1966), Andersland and Douglas (1970), Christensen and Wu (1964), Mitchell (1964), Mitchell, Campanella, and Singh (1968), Mitchell, Singh, and Campanella (1969), Murayama and Shibata (1958, 1961, 1964), Noble and Demirel (1969) and Wu, Resindez, and Neukirchner (1966).

The concept of activation

The basis of rate process theory is that atoms, molecules, and/or particles participating in a time-dependent flow or deformation process, termed "flow units," are constrained from movement relative to each other by virtue of energy barriers separating adjacent equilibrium positions, as depicted schematically by Fig. 14.16. The displacement of flow units to new positions requires the introduction of an *activation energy*, ΔF of sufficient magnitude to surmount the barrier. The potential energy of a flow unit may be the same following the activation process, or higher or lower than it was initially. These conditions are shown by analogy with the rotation of three blocks in Fig. 14.17. In each case an energy barrier must be crossed. The assumption of a steady-state condition is implicit in most applications to soils as regards the at-rest barrier height between successive equilibrium positions.

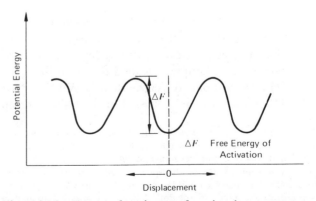

Fig. 14.16 Energy barriers and activation energy.

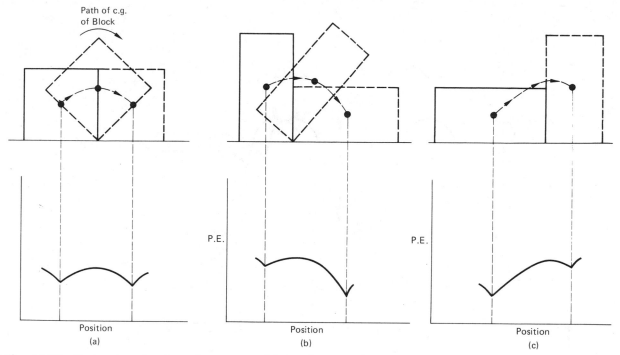

Fig. 14.17 Examples of activated processes. (*a*) Steady state. (*b*) Increased stability. (*c*) decreased stability.

The value of activation energy depends on the material and the type of process. For example, values of ΔF for viscous flow of water, chemical reactions, and solid-state diffusion of atoms in silicates are 3 to 4, 10 to 100, and 25 to 40 kcal/mole of flow units, respectively.

Activation frequency

Energy to enable a flow unit to cross a barrier may be acquired from thermal energy and various applied potentials. For a material at rest, the potential energy–displacement relationship is as shown by curve *A* in Fig. 14.18. From statistical mechanics it is known that the average thermal energy per flow unit is kT, where k is Boltzmann's constant (1.38×10^{-16} erg °K^{-1}) and T is the absolute temperature (°K). Even in a material at rest, however, thermal vibrations occur at a frequency given by kT/h where h is Planck's constant (6.624×10^{-27} erg sec^{-1}). As a result the actual thermal energies are divided among the flow units according to a Boltzmann distribution.

It may be shown that the probability of a given unit becoming activated, or the proportion of units that are activated during any one oscillation is given by

$$p(\Delta F) = e^{-\Delta F/N\,kT} \qquad (14.5)$$

where N is Avogadro's number (6.02×10^{23}). Nk is equal to R, the universal gas constant (1.98 cal

K^{-1}mole^{-1}). The frequency of activation ν then is

$$\nu = \frac{kT}{h} \exp\left(\frac{-\Delta F}{NkT}\right) \qquad (14.6)$$

In the absence of directional potentials, barriers are crossed with equal frequency in all directions, and no consequences of the periodic activation are observed.* If however, a directed potential, such as a shear stress, is applied, then the barrier heights become distorted as shown by curve *B* in Fig. 14.18. If f represents the force acting on a flow unit, then the barrier height is reduced an amount ($f\lambda/2$) in the direction of the force and raised a like amount in the opposite direction, where λ represents the distance between successive equilibrium positions.† Minimums in the energy curve are displaced a distance δ from their original positions, representing an elastic distortion of the material structure.

The reduced barrier height in the direction of force f increases the activation frequency in that direction to

$$\underset{\rightarrow}{\nu} = \frac{kT}{h} \exp\left[-\frac{(\Delta F/N - f\lambda/2)}{kT}\right] \qquad (14.7)$$

---

* Unless the temperature is sufficiently high that softening, melting, or evaporation ensue.
† Work ($f\lambda/2$) done by the force f as the flow unit drops from the peak of the energy barrier to a new equilibrium position is assumed to be given up in heat.

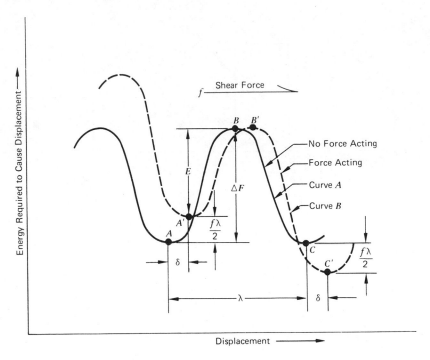

Fig. 14.18   The effect of a shear force on energy barriers.

and in the opposite direction, the increased barrier height decreases the activation frequency to

$$\underset{\leftarrow}{\nu} = \frac{kT}{h} \exp\left[- \frac{(\Delta F/N + f\lambda/2)}{kT} \right] \quad (14.8)$$

The net frequency of activation in the direction of the force then becomes

$$\underset{\rightarrow}{\nu} - \underset{\leftarrow}{\nu} = 2 \frac{kT}{h} \exp\left(- \frac{\Delta F}{RT} \right) \sinh\left(\frac{f\lambda}{2kT} \right) \quad (14.9)$$

Strain rate equation

Of the total number of activated flow units at any instant, some may fall back into their original positions. For each unit that is successful in crossing the barrier, there will be a displacement λ'. The component of λ' in a given direction times the number of successful jumps per unit time gives the rate of movement per unit time. If this rate of movement is expressed on a per unit length basis, then the strain rate $\dot{\epsilon}$ is obtained.

Let $X = F$ (proportion of successful barrier crossings and λ') such that

$$\dot{\epsilon} = X(\underset{\rightarrow}{\nu} - \underset{\leftarrow}{\nu}) \quad (14.10)$$

Then from (14.9)

$$\dot{\epsilon} = 2X \frac{kT}{h} \exp\left(- \frac{\Delta F}{RT} \right) \sinh\left(\frac{f\lambda}{2kT} \right) \quad (14.11)$$

The parameter X may be both time and structure dependent.

If $(f\lambda/2kT) < 1$, then $\sinh (f\lambda/2kT) \approx (f\lambda/2kT)$, and rate is directly proportional to f. This is the case for ordinary Newtonian fluid flow and diffusion where

$$\frac{dx}{dt} = \frac{1}{\eta} \tau \quad (14.12)$$

in which dx/dt is the flow rate, η is viscosity, and τ is shear stress. For most soil deformation problems, however, $(f\lambda/2kT) > 1$ (Mitchell, Campanella, and Singh, 1968), so

$$\sinh\left(\frac{f\lambda}{2kT} \right) \approx \frac{1}{2} \exp\left(\frac{f\lambda}{2kT} \right) \quad (14.13)$$

and

$$\dot{\epsilon} = X \frac{kT}{h} \exp\left(- \frac{\Delta F}{RT} \right) \exp\left(\frac{f\lambda N}{2RT} \right) \quad (14.14)$$

Equation (14.14) is valid except for very small stress intensities, where the exponential approximation of the hyperbolic sine is not justified. Equations (14.11) and (14.14) or comparable forms have been used to obtain dashpot coefficients for rheological models, to obtain functional forms for the influences of different factors on strength and deformation rate, and to study deformation mechanisms in soils.

Soil deformation as a rate process

Although there appears to be no rigorous proof of the correctness of the detailed statistical mechanics formulations, even for simple chemical reactions, the real behavior of many systems has been substantially in accord with rate process theory. It has been possible to test different parts of equation (14.14) separately (Mitchell, Campanella, and Singh, 1968). It was found that the temperature dependence of creep rate, the stress dependence of creep rate, and the stress dependence of the experimental activation energy (equation (14.16)) were in accord with predictions. These results do not prove the correctness of the theory; they do, however, support the concept that soil deformation is a thermally activated process.

The Arrhenius equation

Equation (14.14) may be written

$$\dot{\epsilon} = X \frac{kT}{h} \exp\left(-\frac{E}{RT}\right) \qquad (14.15)$$

where

$$E = \Delta F - \frac{f\lambda N}{2} \qquad (14.16)$$

is termed the experimental activation energy. For all conditions constant except T, assuming $X(kT/h) \approx$ constant,

$$\dot{\epsilon} = A \exp\left(-\frac{E}{RT}\right) \qquad (14.17)$$

Equation (14.17) is the same as the well-known empirical equation proposed by Arrhenius around 1900 to describe the temperature-dependence of chemical reaction rates. It has been shown suitable also for characterization of the temperature-dependence of processes such as creep, stress relaxation, secondary compression, thixotropic strength gain, diffusion, and fluid flow.

14.4 BONDING, EFFECTIVE STRESSES, AND STRENGTH

Deformation parameters from creep test data

If the shear stress on a material is τ, and it is distributed uniformly among S flow units per unit area, then

$$f = \frac{\tau}{S} \qquad (14.18)$$

Displacement of a flow unit requires that interatomic or intermolecular forces be overcome so that it can be moved. Let it be assumed that the number of flow units and the number of bonds are equal.

If D represents the deviator stress under triaxial stress conditions, the value of f on the plane of maximum shear stress is

$$f = \frac{D}{2S} \qquad (14.19)$$

so equation (14.14) becomes

$$\dot{\epsilon} = X \frac{kT}{h} \exp\left(-\frac{\Delta F}{RT}\right) \exp\left(\frac{D\lambda}{4SkT}\right) \qquad (14.20)$$

Determination of activation energy. From equation (14.15)

$$\frac{\partial \ln (\dot{\epsilon}/T)}{\partial (1/T)} = -\frac{E}{R} \qquad (14.21)$$

provided strain rates are considered under conditions of unchanged soil structure. Thus, the value of E can be determined from the slope of a plot of $\ln (\dot{\epsilon}/T)$ versus $(1/T)$. Procedures for evaluation of strain rates at different temperatures but the same structure are given by Mitchell, Campanella, and Singh (1968) and Mitchell, Singh, and Campanella (1969).

Determination of number of bonds. For stresses large enough to justify approximation of the hyperbolic sine function by a simple exponential in the creep rate equation and small enough to avoid tertiary creep, the logarithm of strain rate varies directly as the deviator stress. For this case, equation (14.20) may be written

$$\dot{\epsilon} = K(t) \exp (\alpha D) \qquad (14.22)$$

where

$$K(t) = X \frac{kT}{h} \exp\left(-\frac{\Delta F}{RT}\right) \qquad (14.23)$$

and

$$\alpha = \frac{\lambda}{4SkT} \qquad (14.24)$$

Parameter α is a constant for a given value of effective consolidation pressure and is given by the slope of the relationship between log strain rate and stress. It is evaluated using strain rates at the same time after start of creep for tests at several stress intensi-

Table 14.1 Activation Energies For Creep of Several Materials

Material	Activation Energy (kcal/mole)[a]	Reference
(1) Remolded illite, saturated, water contents of 30 to 43%	25 to 40	Mitchell, Singh, and Campanella (1969)
(2) Dried illite: samples air-dried from saturation, then evacuated	37	Mitchell, Singh, and Campanella (1969)
(3) San Francisco Bay mud, undisturbed	25 to 32	Mitchell, Singh, and Campanella (1969)
(4) Dry Sacramento River sand	~25	Mitchell, Singh, and Campanella (1969)
(5) Water	4 to 5	Glasstone, Laidler, and Eyring (1941)
(6) Plastics	7 to 14	Ree and Eyring (1958)
(7) Montmorillonite–water paste, dilute	20 to 26	Ripple and Day (1966)
(8) Soil asphalt	27	Abdel-Hady and Herrin (1966)
(9) Lake clay, undisturbed and remolded	23 to 27	Christensen and Wu (1964)
(10) Osaka clay, overconsolidated	29 to 32	Murayama and Shibata (1961)
(11) Concrete	54	Polivka and Best (1960)
(12) Metals	50+	Finnie and Heller (1959)
(13) Frozen Soils	94	Andersland and Akili (1967)
(14) Sault Ste. Marie clay, suspensions, discontinuous structures	Same as water	Andersland and Douglas (1970)
(15) Sault Ste. Marie clay, Li+, Na+, K+ forms, in H_2O and CCl_4, consolidated	28	Andersland and Douglas (1970)

[a] The first four values are experimental activation energies, E. Whether the remainder are values of ΔF or E is not always clear in the references cited.

ties. With α known, λ/S is calculated as a measure of the number of interparticle bonds.*

Activation energies for creep

Activation energies for the creep of several soils and other materials are given in Table 14.1. The free energy of activation for creep of soils is in the range of 20 to 45 kcal/mole. Four features of the values for soils in Table 14.1 are significant:

1. The activation energies are relatively high, much higher than for viscous flow of water, for example.
2. Variations in water content (including complete drying), ionic form, consolidation pressure, void ratio, and pore fluid have no significant effect.
3. The values for sand and clay are about the same.
4. Clays in suspension with insufficient solids to form a continuous structure deform with an activation energy equal to that of water.

* A procedure for evaluation of α from the results of a test at a succession of stress levels on a single sample is given by Mitchell, Singh, and Campanella (1969).

Number of interparticle bonds

Specific evaluation of S requires knowledge of λ, the separation distance between successive equilibrium positions in the interparticle contact structure. A value of 2.8×10^{-10}m (2.8 Å) has been assumed for λ for evaluation of the values of S presented below. This distance is the same as that separating atomic valleys in the surface of a silicate mineral. It is assumed that deformation involves displacement of oxygen atoms along contacting particle surfaces, as well as periodic rupture of bonds at interparticle contacts. Figure 14.19 shows this interpretation for λ schematically. If the above assumption for λ is not correct, calculated values of S will still be in the correct relative proportion so long as λ remains constant during deformation.

Normally consolidated clay. Results of creep tests at different stress intensities for different consolidation pressures enable computation of S as a function of consolidation pressure. Figure 14.20 shows the results obtained for undisturbed San Francisco Bay mud. The open point is for remolded Bay mud. This specimen was normally consolidated to 4.0 kg/cm², re-

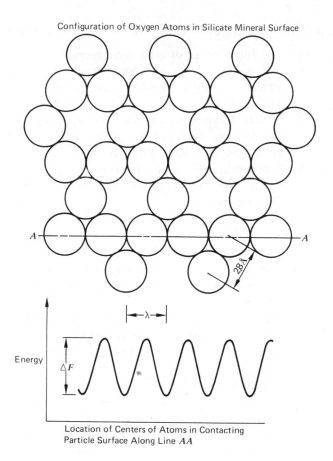

Configuration of Oxygen Atoms in Silicate Mineral Surface

Energy

Location of Centers of Atoms in Contacting
Particle Surface Along Line *AA*

Fig. 14.19 Interpretation of λ in terms of silicate mineral surface structure.

bounded to 0.5 kg/cm², and then remolded at constant water content. The effective consolidation pressure dropped to 0.25 kg/cm² as a result of remolding. It may be seen that the drop in effective stress was accompanied by a decrease in the number of interparticle bonds.

Tests on remolded illite gave comparable results. Figure 14.21 shows a continuous inverse relationship between the number of bonds and water content over a range of water contents from more than 40 percent to air dried and vacuum-desiccated clay. The dried material had a water content of 1 percent on the usual oven-dry basis. The very large number of interparticle bonds developed by drying is responsible for the high dry strength of clay.

Overconsolidated clay. Samples of undisturbed San Francisco Bay mud were prepared to overconsolidation ratios of 1, 2, 4, and 8 following the stress paths in the upper part of Fig. 14.22. Point *d′* refers to a specimen remolded at a water content of 52.3 percent. The undrained compressive strength as a function of consolidation pressure is shown in the middle section of Fig. 14.23, and the number of bonds, deduced from creep tests, is shown in the lower part of the figure. The effect of the overconsolidation is to increase the number of interparticle bonds over the values for normally consolidated clay. Some bonds formed during consolidation are retained after removal of much of the consolidation pressure.

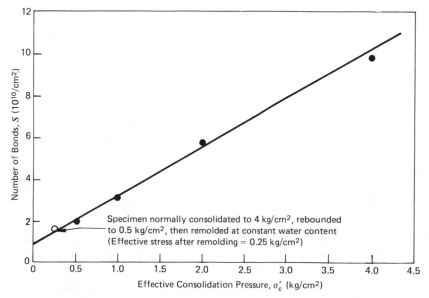

Specimen normally consolidated to 4 kg/cm², rebounded to 0.5 kg/cm², then remolded at constant water content (Effective stress after remolding = 0.25 kg/cm²)

Fig. 14.20 Number of interparticle bonds as a function of consolidation pressure for normally consolidated San Francisco Bay mud.

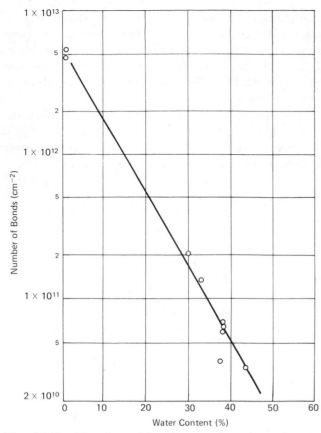

Fig. 14.21 Number of bonds as a function of water content for illite.

Data from Fig. 14.22 are replotted in Fig. 14.23 in the form of compressive strength versus number of bonds. Figure 14.23 shows that strength depends only on the number of bonds and is independent of whether the clay is undisturbed, remolded, normally consolidated, or overconsolidated.

Dry sand. Figure 14.24 shows that creep tests on oven-dried sand yields results of the same type as for clay, suggesting that the strength generating and creep-controlling mechanisms may be similar for both types of material.

Composite strength–bonding relationship. Data for S and strength of bay mud, illite, and sand are combined in Fig. 14.25. The same proportionality exists for all the materials, which may at first seem somewhat surprising, but is in reality to be expected based on a reasonable model for bonding, effective stress, and strength.

Significance of activation energy and bond number values

The following aspects of activation energies and numbers of interparticle bonds are helpful in the understanding of deformation and strength behavior.*

1. The high values of activation energy (30 to 45 kcal/mole) in soils in comparison with other materials suggest breaking of strong bonds during shear.
2. Similar creep behavior for wet and dry clay and for dry sand indicate deformation is not controlled by viscous flow of water.
3. Comparable values of activation energy for wet and dry soil indicate that water is not responsible for bonding.
4. Comparable values of activation energy for clay and sand support the concept that interparticle bond strengths are the same for both types of material. This is supported also by the uniqueness of the strength vs. number of bonds relationship for all soils.
5. The activation energy and presumably, therefore, the bonding type are independent of consolidation pressure, void ratio, and water content.
6. The number of bonds is directly proportional to effective consolidation pressure for normally consolidated clays.
7. Overconsolidation leads to more bonds than for a normally consolidated clay at the same effective consolidation pressure.
8. Strength depends only on the number of bonds.
9. Remolding causes a decrease in the effective consolidation pressure which means also a decrease in the number of bonds.
10. There are about 100 times as many bonds in dry clay as in wet clay.

It may be possible to account for these results in more than one way; however, a reasonably consistant interpretation is as follows. ΔF represents the energy to activate a mole of flow units. This may involve rupture of a single interatomic or intermolecular bond or the simultaneous rupture of several such bonds. Shear of dilute montmorillonite–water pastes involves breaking single bonds (Ripple and Day, 1966). In the viscous flow of water, the activation energy is approximately that for a single hydrogen bond rupture per flow unit displacement, even though each water molecule may form simultaneously up to four hydrogen bonds with its neighbors. If the single bond rupture interpretation is also correct for soils, then con-

* Some of these statements are not valid for chemically cemented soils.

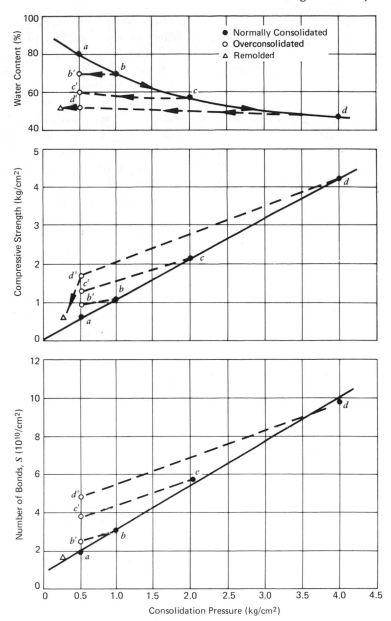

Fig. 14.22 Consolidation pressure, strength, and bond numbers for San Francisco Bay mud.

sistency in equation (14.11) requires that shear force f pertain to the force per bond. On this basis, parameter S indicates the number of single bonds per unit area. In the event activation of a flow unit requires simultaneous rupture of n bonds, then S represents $1/n$th of the total bonds in the system.

That the activation energy for deformation of soil is well into the chemical reaction range (10 to 100 kcal/mole) does not prove that bonding is of the primary valence type, because simultaneous rupture

of several weaker bonds could yield values of the magnitude observed. On the other hand, the facts that the activation energy is much greater than for flow of water, that it is the same for wet and dry soils, and that it is essentially the same for different adsorbed cations and pore fluids (Andersland and Douglas, 1970) do suggest that bonding is through solid interparticle contacts.

Activation energy values of 30 to 40 kcal/mole are compatible with those for solid-state diffusion of oxy-

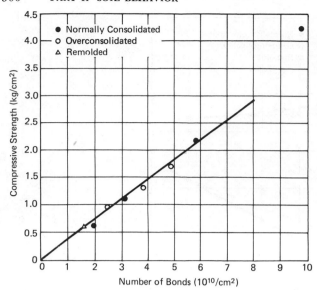

Fig. 14.23 Strength as a function of number of bonds for San Francisco Bay mud.

gen in silicate minerals. This supports the concept (Rosenqvist, 1965)* that creep movements result from slow diffusion of oxygen ions in and around interparticle contacts. The important soil minerals, both sand and clay, are silicates, and their surface layers

* Lecture, University of California, Berkeley, December 1965.

consist of oxygen atoms held together by silicons. Water in some form is adsorbed onto these surfaces. The water structure consists of oxygen held together by hydrogen. It is not too different from that of the silicate layer in minerals. Thus, a distinct boundary between particle surface and water may not be discernible, as suggested by both Rosenqvist and Trollope (1964). Under these conditions a more or less continuous solid structure containing water molecules that propogates through interparticle contacts might be visualized, as opposed to direct contact between clean mineral surfaces.

An individual flow unit can be an atom, a group of atoms or molecules, or a particle. The preceding arguments are based on the interpretation that individual atoms are the flow units. This is consistent with both the relative and actual values of S that have been determined for different soils. Furthermore, Andersland and Douglas (1970) used a formulation of the rate equation that enabled calculation of flow unit volume from their test data. A value of about 1.7 Å^3 was obtained, which is of the same order as that of individual atoms.

If particles were assumed to be the flow units, not only would it be difficult to visualize their thermal vibrations, but then S would relate to the number of interparticle contacts. It is then difficult to conceive

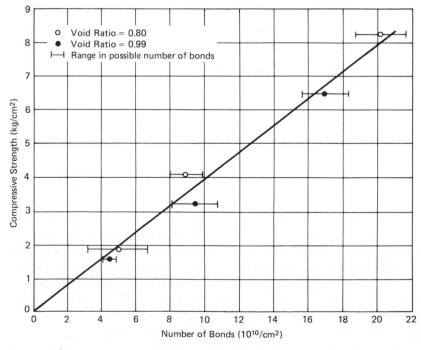

Fig. 14.24 Strength as a function of number of bonds for dry Antioch River sand.

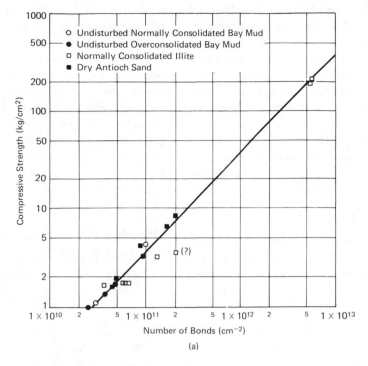

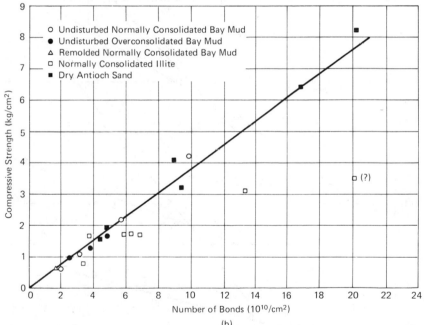

Fig. 14.25 Compressive strength as a function of number of bonds. (a) Logarithmic scale to include values for dry clay. (b) Natural scale.

how simply drying a clay could give a hundredfold increase in the number of interparticle contacts, as would have to be the case based on Fig. 14.21. A more plausible interpretation is that drying, while causing some increase in the number of interparticle contacts during shrinkage, causes mainly an increase in the number of bonds per contact because of the increased effective stress.

At any value of effective stress, the value of S is about the same for both sand and clay. The number of interparticle contacts should be vastly different, however. For equal numbers of contacts per particle, the number per unit volume should vary inversely with the cube of particle size. Thus the number for clay particles of 1 μm average size would be some nine orders of magnitude greater than for sand of 1 mm average particle size. Each contact between sand

particles would involve many bonds; in a clay, the much greater number of contacts would mean fewer bonds per contact.

The required interparticle contact area to develop bonds of the number indicated in Fig. 14.26 is very small. For example, for a compressive strength of 3 kg/cm² there are 8×10^{10} bonds/cm² of shear surface required. Oxygen atoms on the surface of a silicate mineral have a diameter of 2.8×10^{-8} cm. Allowing an area 3×10^{-8} cm on a side for each oxygen gives 9×10^{-16} cm² per bonded oxygen for a total contact area of $9 \times 10^{-16} \times 8 \times 10^{10} = 7.2 \times 10^{-5}$ cm²/cm² of soil cross section.

Hypothesis for bonding, effective stress, and strength

Normal effective stresses and shear stresses can be transmitted only at interparticle contacts in most

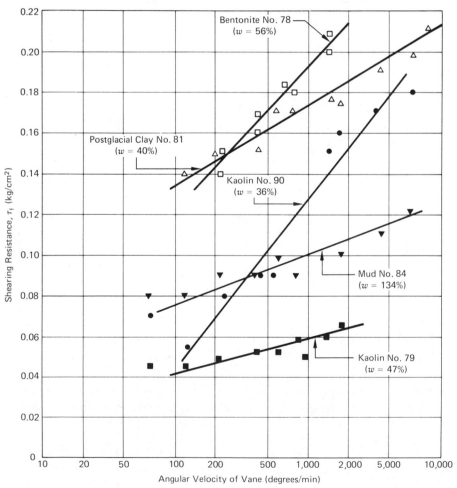

Fig. 14.26 Effect of rate of shear on shearing resistance of remolded clays as determined by the Laboratory Vane apparatus (Prepared from data of Karlsson, 1963).

soils.* The predominant effects of long range physico-chemical forces of interaction are to control the initial soil fabric and to alter the forces transmitted at contact points from what they would be due to applied stresses alone.

Interparticle contacts are effectively solid,† and it is likely that both adsorbed water and cations in the contact zone participate in the structure. An interparticle contact can contain many bonds that may be strong, approaching the primary valence type. The number of bonds at any contact depends on the compressive force transmitted at the contact, and the Terzaghi–Bowden and Tabor adhesion theory of friction, developed in Section 14.6, can account for strength. The macroscopic strength is directly proportional to the number of bonds.

For normally consolidated soils the number of bonds is directly proportional to the effective stress. As a result of particle rearrangements and contacts formed during virgin compression, an overconsolidated soil at a given effective stress has a greater number of bonds than a normally consolidated soil. This accounts for higher peak strength as a result of overconsolidation. This effect is more pronounced in clays than in sands, because the larger, bulky sand grains tend to recover their original shapes when unloaded, thus rupturing most of the excess of interparticle bonds over those needed to resist the present stress.** The strength of interparticle contacts can vary over a wide range, depending on the number of bonds per contact.

The unique relationship between strength and number of bonds for all soils, as indicated by Fig. 14.25, reflects the fact that the minerals comprising most soils are silicates and all have similar surface structures.

In the absence of chemical cementation, interparticle bonds may form in response to interparticle contact forces generated by either applied stresses, physico-chemical forces of interaction or both. Any bonds existing in the absence of applied effective stress, that is, $\sigma' = 0$, are responsible for true cohe-

---

* A pure sodium montmorillonite may be an exception, since a part of the normal stress can be carried by physico-chemical forces of interaction. Specifically, the true effective stress may be less than the apparent effective stress by $(R-A)$ as discussed in Section 10.8.
† Microstructural study by Chattopadhyay (1972) using the scanning electron microscope supports the existence of solid interparticle contacts between clay minerals.
** There may be some bond formation in sands due to cold welding under high contact pressures between particles and to physico-chemical changes at contacting asperities (Lee, 1975).

sion. There should be no difference between friction and cohesion in terms of the shearing process. Complete failure in shear involves simultaneous rupture or slipping of all bonds along the shear plane.

14.5 SHEARING RESISTANCE AS A RATE PROCESS

The general creep rate equation (14.20) becomes, if the maximum shear stress τ is substituted for the deviator stress D,

$$\tau = \frac{\sigma_1 - \sigma_3}{2}$$

$$\dot{\epsilon} = X \frac{kT}{h} \exp\left(-\frac{\Delta F}{RT}\right) \exp\left(\frac{\tau\lambda}{2SkT}\right) \quad (14.25)$$

Taking logarithms of both sides of (14.25) gives

$$\ln \dot{\epsilon} = \ln\left(X\frac{kT}{h}\right) - \frac{\Delta F}{RT} + \frac{\tau\lambda}{2SkT} \quad (14.26)$$

By assuming, $X\,kT/h$ is a constant and equal to B (Mitchell, 1964),

$$\tau = \frac{2S}{\lambda N}\Delta F + \frac{2SkT}{\lambda}\ln\left(\frac{\dot{\epsilon}}{B}\right) \quad (14.27)$$

From the relationships in Section 14.4, the following relationship between bonds per unit area and effective stress is suggested

$$S = a + b\sigma_f' \quad (14.28)$$

where a and b are constants and σ_f' is the effective stress on the shear plane. Thus, equation (14.27) becomes

$$\tau = \frac{2a\Delta F}{\lambda N} + \frac{2akT}{\lambda}\ln\frac{\dot{\epsilon}}{B}$$
$$+ \left(\frac{2b}{\lambda N}\Delta F + \frac{2bkT}{\lambda}\ln\left(\frac{\dot{\epsilon}}{B}\right)\right)\sigma_f' \quad (14.29)$$

Equation (14.29) is of the same form as the Coulomb equation for strength

$$\tau = c + \sigma_f' \tan \phi \quad (14.30)$$

By analogy,

$$c = \frac{2a\Delta F}{\lambda N} + \frac{2akT}{\lambda}\ln\left(\frac{\dot{\epsilon}}{B}\right) \quad (14.31)$$

and

$$\tan \phi = \frac{2b}{\lambda N}\Delta F + \frac{2bkT}{\lambda}\ln\left(\frac{\dot{\epsilon}}{B}\right) \quad (14.32)$$

These relationships state that both cohesion and friction depend on the number of bonds times the bond strength, as reflected by the activation energy, and that the values of c and ϕ, in any case, should depend on the rate of deformation and the temperature.

Strain rate effects

All other factors being equal, the shearing resistance should increase linearly with the logarithm of the rate of strain. Figure 14.15 shows this to be the case for Drammen clay. Data for several clays are shown in Fig. 14.26, where shearing resistance as a function of the speed of vane rotation in a vane shear test is plotted. Analysis of the relationship between shearing stress and rate of vane rotation w (Karlsson, 1963) showed that $\Delta\tau/\Delta \log w$ decreases with an increase in water content. This follows directly from equation (14.29) because

$$\frac{d\tau}{d \ln (\dot{\epsilon}/B)} = \frac{2akT}{\lambda} + \frac{2bkT}{\lambda} \sigma_f' = \frac{2kT}{\lambda} (a + b\sigma_f')$$

(14.33)

that is, $d\tau/d \ln (\dot{\epsilon}/B)$ is proportional to the number of bonds, which decreases with increasing water content. This interpretation of the data in Figs. 14.15 and 41.26 assumes that the effective stress was unaffected by changes in the strain rate, an assumption that may not be strictly true in all cases.

Effect of temperature

Assumptions for parameters (Mitchell, 1964) shows that the term $(\dot{\epsilon}/B)$ is less than one. Thus, the quantity in $(\dot{\epsilon}/B)$ in equation (14.29) is negative, and an increase in temperature should give a decrease in strength, all other factors being constant. That this is the case is demonstrated by Fig. 14.27, which shows deviator stress as a function of temperature for samples of San Francisco Bay mud compared under conditions of equal mean effective stress and structure. Other examples of the influence of temperature on strength are shown in Figs. 14.7 and 14.8.

Fundamental shear strength properties—An alternate view

Mechanisms that differ somewhat from those advanced herein can be postulated. For example, Bjerrum (1973) suggested that a mineral–mineral contact forms if the contact force is high, as would be the case between large particles. For such contacts, the number of interatomic bonds is a function of the effective stress, and the effective friction angle de-

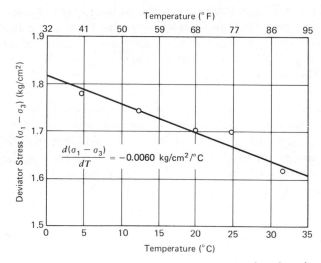

Fig. 14.27   Influence of temperature on the shearing resistance of San Francisco Bay mud. Comparison is for samples at equal mean effective stress and at the same structure.

pends on the strength of the atomic bonds between these larger particles. On unloading there is recovery of elastic distortion, and the bonds are broken.

Where the contact stresses are small, as between clay particles, the contact is nonmineral and consists of merged water films. The area of these contacts is proportional to the load. Since this area does not depend on elastic distortion, it is presumed to remain after the load is removed, thus, accounting for true cohesion. c_e is given by

$$c_e = Kp_e$$

(14.34)

where p_e is the equivalent consolidation pressure (Hvorslev, 1937, 1960). The effective cohesion and friction develop separately with strain, with cohesion peaking at very low strain and friction mobilized gradually.* The gradual mobilization of friction is attributed by Bjerrum to particle rotation without sliding in the early stages of deformation.

Although this hypothesis of mineral–mineral frictional contacts and merged water films at cohesive contacts appears to account for a number of observa-

---

* Data have been obtained that can be interpreted to support this concept (Schmertmann and Osterberg, 1960; Schmertmann 1963; Campanella and Gupta, 1969). The values of c_e and ϕ_e are obtained from plots of the type shown in the lower part of Fig. 14.6 prepared for several values of strain. Because of possible curvature in the effective stress failure envelope and different fabrics in the normally consolidated and overconsolidated samples, the correctness of this interpretation is uncertain in clays free of chemical cementation.

tions, it does not accord with the findings of Section 14.4 that suggest solid contacts between particles of all sizes in both wet and dry soils.

Interparticle creep is presumed by Bjerrum to result from thermal vibrations of adsorbed water atoms. Failure of each contact point occurs after some critical displacement, and the shear stress is transferred to more stable frictional contact points. In the event that the transfer results in an overloading of the frictional contacts, failure results. Creep, transfer of cohesion to friction, stiffening of the soil structure, and creep rupture are explained in this way. However, the overall similarity of all contacts and comparable activation energy for wet and dry clay cannot be accounted for.

The following explanation can account for the observations equally well. The interparticle contacts of a soil subjected to a change in stress conditions pick up both normal and shear forces. Slow creep displacements of particles in contact occur whenever the ratio of contact shear to normal force is greater than some minimum value, which is probably a small fraction of the contact strength. Interparticle movement continues until the shear force at the contact can be relieved by transfer as a *normal force* through other contacts that form as a result of the movements. Creep or secondary compression would continue until each particle had rotated and translated such that there is a preferential orientation of interparticle contact plane normals in the direction of shear so that the contact shear to normal force ratio is low.

Considerations of effective stress and numbers of bonds would require that, for a macroscopic stress state where $(\sigma_1'/\sigma_3') > 1$, there be more bonds in a plane normal to the σ_1' direction than in a plane normal to the σ_3' direction. In the event the contact force transfers could no longer be accommodated after the process had continued for some time, then forces thrown onto previously stable contacts would cause their rupture, progressive failure would ensue, and creep rupture would follow.

14.6 FRICTION BETWEEN SOLID SURFACES

The friction angle used in equations such as (14.1) and (14.3) contains contributions from several sources, including sliding of grains in contact, resistance to volume change (dilatancy), grain rearrangement, and grain crushing. The true friction, contributed by interparticle resistance to sliding, can account for half or more of the peak strength and nearly all of the residual strength. The importance of the contribu-

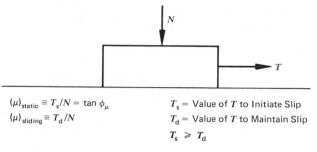

$(\mu)_{static} \equiv T_s/N = \tan \phi_\mu$ $\qquad T_s$ = Value of T to Initiate Slip

$(\mu)_{sliding} \equiv T_d/N$ $\qquad T_d$ = Value of T to Maintain Slip

$$T_s \geqslant T_d$$

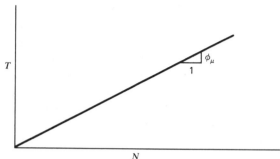

Fig. 14.28 Coefficient of friction for surfaces in contact.

tions from dilatancy, grain rearrangement, and grain crushing depend on particle size, particle shape, void ratio, and confining pressure. The true friction coefficient is shown in Fig. 14.28 and represented by

$$\mu = \frac{T}{N} = \tan \phi_\mu \qquad (14.33)$$

where ϕ_μ, the true intergrain friction angle,* is a compositional property that is determined by the type of soil minerals.

Basic laws of friction

Two "laws" of friction have been recognized, beginning with Leonardo da Vinci in about 1500. They were restated by Amontons in 1699 and are frequently referred to as "Amontons' laws." They are:

1. The frictional force is directly proportional to the normal force, as in Fig. 14.28 and equation (14.33).

2. The frictional resistance between two bodies is independent of the size of the bodies. In Fig. 14.28, the value of T is the same for a given value of N, regardless of the size of the sliding block.

* ϕ_μ should not be confused with the Hvorslev friction parameter ϕ_e. ϕ_e is supposed to represent the total frictional contribution to strength. ϕ_μ represents true sliding friction between grains.

Although these principles of frictional resistance have been known for many years, explanations for them were long in coming. It was at one time believed that interlocking between irregular surfaces could account for the behavior. On this basis, μ would be determined by the tangent of the average inclination of surface irregularities to the sliding plane. This cannot be the case, however, because such an explanation would require that μ decrease as surfaces became smoother and be zero for perfectly smooth surfaces. In fact, the coefficient of friction is essentially constant over a range of surface roughnesses. Hardy (1936) suggested instead that static friction originates from cohesive forces between contacting surfaces. He observed that the actual area of contact is very small, because of surface irregularities, and thus, the cohesive forces per unit area must be large.

The foundation for the present understanding of the mechanism of friction between surfaces in contact was laid by Terzaghi (1920). He hypothesized that the normal load N, acting between two bodies in contact causes yielding at asperities (local "hills" on the surface), where the actual interbody solid contact develops. The actual contact area A_c is given by

$$A_c = \frac{N}{\sigma_y} \tag{14.34}$$

where σ_y is the yield strength of the material. The shearing strength of the material in the yielded zone is assumed to have a value τ_m. The maximum shearing force T that can be withstood by the contact is then

$$T = A_c \tau_m \tag{14.35}$$

the coefficient of friction is given by T/N,

$$\mu = T/N = \frac{A_c \tau_m}{A_c \sigma_y} = \frac{\tau_m}{\sigma_y} \tag{14.36}$$

This concept of frictional resistance was subsequently proposed by Bowden and Tabor (1950, 1964): The Terzaghi–Bowden and Tabor hypothesis, commonly referred to as the adhesion theory of friction, forms the basis of almost all modern studies of friction. Two characteristics of surfaces play a key role in the adhesion theory of friction: roughness and surface adsorption.

Surface roughness

The surfaces of almost all solids are rough on a molecular scale. Accurate measurements of profiles of

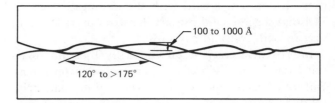

Fig. 14.29   Contact between two smooth surfaces.

polished surfaces show a succession of asperities and depressions, ranging generally from 100 to over 1000-Å in height. The slopes of the asperities are rather flat, with individual angles ranging from about 120 to 170 degrees as shown in Fig. 14.29. The average slope of asperities on metal surfaces is an included angle of 150 degrees; on rough quartz it may be over 175 degrees (Bromwell, 1965, 1966). When two surfaces are brought together, contact is established at the asperities, and the actual contact area is a small fraction of the total surface area.

Quartz surfaces polished to mirror smoothness may consist of peaks and valleys with an average height of about 5000 Å. Rougher quartz surfaces may have asperities about ten times higher (Lambe and Whitman, 1969). Even these surfaces are probably smoother than most soil particles composed of bulky minerals. The actual surface texture of sand particles depends on geologic history as well as mineralogy (Fig. 4.13).

Among the smoothest surfaces obtainable are the cleavage faces of mica flakes. Even in mica, however, there is some waviness due to rotation of tetrahedra in the silica layer, and surfaces contain steps ranging in height from 10 to 1000 Å, representing different numbers of unit layers in the particle.

Therefore, large areas of solid contact between grains are not probable in soils, solid to solid contact is through asperities, and corresponding contact stresses are high. Although the molecular structure and composition in the region of contacting asperities are not known, the composition and structure determine the magnitude of τ_m in equation (14.36).

Surface adsorption

Because of unsatisfied force fields at the surfaces of solids, the surface structure may differ from that in the interior, and material may be adsorbed from adjacent phases. "Clean" surfaces, prepared by fracture of a solid or by evacuation at high temperature are rapidly contaminated when exposed to normal atmospheric conditions.

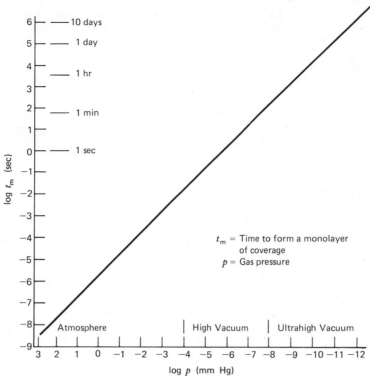

Fig. 14.30 Monolayer formation time as a function of pressure.

According to the kinetic theory of gases, the time for adsorption of a monolayer t_m is given by

$$t_m = \frac{1}{\sigma SZ} \qquad (14.37)$$

where σ is the area occupied per molecule, S is the fraction of molecules striking the surface that stick to it, and Z is the number of molecules per second striking a square centimeter of surface. For a value of S equal to 1.0, which is reasonable for a high energy surface, the relationship between t_m and gas pressure is shown in Fig. 14.30. The conclusions to be drawn from this are that adsorbed layers are present on the surfaces of soil particles in the terrestrial environment,* and contacts through asperities involve adsorbed material, unless it is extruded under the high pressure, as well as the solid.

Adhesion theory of friction

The basis for the adhesion theory of friction is in equation (14.35); namely, the tangential force that

* Conditions may be different on the Moon, where ultrahigh vacuum conditions exist. It may be that the hard vacuum on the Moon gives rise to cleaner surfaces that can account in part for the higher cohesion observed in lunar soils than in terrestrial soils of comparable gradation. In the absence of suitable adsorbate, clean surfaces can reduce their surface energy by cohering with like surfaces.

causes sliding depends on the solid contact area and the shear strength of the contact. Plastic and/or elastic deformations may control the contact area at asperities.

Plastic junctions. If asperities yield and deform plastically, then the contact area is proportional to the normal load on the asperity as shown in equation (14.34). Because surfaces are not clean, but contain adsorbed films, actual solid contact may develop only over a fraction δ of the contact area as shown in Fig. 14.31. If the contaminant film strength is τ_c, the strength of the contact will be

$$T = A_c[\delta\tau_m + (1 - \delta)\,\tau_c] \qquad (14.38)$$

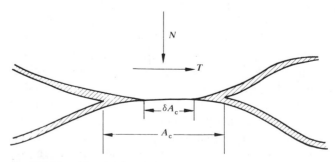

Fig. 14.31 Plastic junction between asperities with adsorbed surface films.

Equation (14.38) cannot be applied in practice because δ and τ_c are unknown. It may, however, provide a basis for explanation of the fact that measured values of friction angle for bulky minerals, such as quartz and feldspar, are greater than the values obtained for the clay minerals and the platy minerals, such as mica, even though the surface structure is essentially the same for all the silicate minerals. The small particle size of clays means that the load per particle, for a given effective stress, will be small relative to that in silts and sands composed of bulky minerals. The surfaces of platy silt and sand size particles are smoother than those of bulky mineral particles. The asperities, caused by waviness of the surface, are more regular but not as high as those for the bulky minerals.

Thus, it can be postulated that for a given number of contacts per particle the load per asperity decreases with decreasing particle size and, for particles of the same size, is less for platy minerals than for bulky minerals. Because δ should increase as the normal load per asperity increases, and it seems reasonable that the adsorbed film strength is less than the strength of the solid material ($\tau_c < \tau_m$), it follows that (ϕ_μ) small and platy particles is less than (ϕ_μ) large and bulky particles.

In the event that two platy particles are in face-to-face contact and the surface waviness is insufficient to cause direct solid–solid contact, shear will be through the adsorbed films, and the effective value of δ will be zero, again giving a lower value of ϕ_μ.

The behavior of plastic junctions is actually somewhat more complicated than outlined above. Under combined compression and shear stresses, deformation follows the von Mises–Henky criterion, which is, for two dimensions,

$$\sigma^2 + 3\tau^2 = \sigma_y{}^2 \qquad (14.39)$$

For asperities loaded initially to $\sigma = \sigma_y$, the application of a shear stress requires that σ become less than σ_y. The only way that this can happen is for the contact area to increase. A continued increase in τ causes a continued increase in contact area. The phenomenon is called junction growth and is responsible for cold welding in some materials (Bowden and Tabor, 1964). If the shear strength of the junctions equals that of the bulk solid, then gross seizure occurs. For the case where the ratio of junction strength to bulk material strength is less than 0.9 the amount of junction growth is small. This is the probable situation in soils.

Elastic junctions. For a perfectly elastic material the contact area is not defined in terms of plastic yield. For two spheres in contact, application of the Hertz theory leads to

$$d = (\delta NR)^{1/3} \qquad (14.40)$$

where d is the diameter of a plane circular area of contact; δ is a function of geometry, Poisson's ratio, and Young's modulus*; and R is the sphere radius. The contact area is

$$A_c = \frac{\pi}{4} (\delta NR)^{2/3} \qquad (14.41)$$

If the shear strength of the contact is τ_i, then

$$T = \tau_i A_c$$

and

$$\mu = \frac{T}{N} = \tau_i \frac{\pi}{4} (\delta R)^{2/3} N^{-1/3} \qquad (14.42)$$

According to this analysis, the friction coefficient for the case of two elastic asperities in contact should decrease with increasing load, which is in contradiction to Amonton's first law. Nonetheless, the adhesion theory should still be applicable to explain the strength of the junction, with the frictional force proportional to the area of real contact.

If it is assumed that the number of contacting asperities in a soil mass is independent of particle size and effective stress, then the influences of particle size and effective stress on the frictional resistance of a soil with asperities deforming elastically may be analyzed. For uniform spheres arranged in a regular packing, the gross area covered by one sphere along a potential plane of sliding is ($4R^2$). The load per contacting asperity, assuming one asperity per contact, is

$$N = 4R^2\sigma' \qquad (14.43)$$

where σ' is the normal effective stress across the plane.

The area per contact becomes

$$A_c = \frac{\pi}{4} (4\delta R^3\sigma')^{2/3} \qquad (14.44)$$

and the total contact area per unit gross area is

$$(A_c)_T = \frac{1}{4R^2}\left(\frac{\pi}{4} R^2\right)(4\delta\sigma')^{2/3} = \frac{\pi}{16} (4\delta\sigma')^{2/3} \quad (14.45)$$

---

* For a sphere in contact with a plane surface
$$\delta = 12(1 - \nu^2)/E$$

The total shearing resistance of $\mu\sigma'$ is equal to the contact area times τ_i, so

$$\mu = \frac{\pi}{16}(4\delta\sigma')^{2/3}\frac{\tau_i}{\sigma'} = \tau_i K(\sigma')^{-1/3} \quad (14.46)$$

where $K = \pi(4\delta)^{2/3}/16$. On this basis, the coefficient of friction should decrease with increasing σ', but it should be independent of sphere radius or particle size.

Data are available to both support and contradict these predictions. A 50-fold variation in the normal load on assemblages of quartz particles in contact with a quartz block was found to have no effect on frictional resistance (Rowe, 1962). The residual friction angle of quartz, feldspar, and calcite is independent of normal stress (Kenney, 1967) as shown in Fig. 14.32.

On the other hand, a decreasing frictional resistance with increasing normal load up to some limiting value of normal stress is evident for mica and the clay minerals in Fig. 14.32, for several clays and clay shales (Bishop, 1971), for diamond (Bowden and Tabor, 1964), and for solid lubricants such as graphite and molybdenum disulphide (Campbell, 1969). Additional data for clay minerals and analysis by Chattopadhyay (1972) show that this decrease in friction varies as $(\sigma')^{-1/3}$ as predicted by equation (14.46) up to a normal stress of the order of 200 kN/m² (30 psi).

There are at least two possible explanations for the normal stress independence of the frictional resistance of quartz, feldspar and calcite:

1. As the load per particle increases, the number of asperities in contact increases proportionally, and the deformation of each asperity remains essentially constant. In this case, the assumption of one asperity per contact used in the development of equation (14.46) is no longer valid.

2. As the load per asperity increases the value of δ in equation (14.38) increases, reflecting a greater proportion of solid contact relative to adsorbed film contact. Thus, the average strength per contact increases more than proportionally with the load while the area increases less than proportionally: the net result is an essentially constant frictional resistance.

Quartz is a hard, brittle material that can deform both elastically and plastically. A normal pressure of 11 kN/mm² (1,500,000 psi) is required to produce plastic deformation, and brittle failure usually occurs before plastic deformation. Plastic deformations are evidently confined to small, highly confined asperities, and elastic deformations control at least part of the behavior (Bromwell, 1965). Either of the above two explanations might be applicable depending on details of surface texture on a microscale and characteristics of the adsorbed films.

With the exception of some data for quartz (Rowe, 1962), there appears little information concerning possible variations of the true friction angle with particle size. Rowe found that the value of ϕ_μ for assemblages of quartz particles on a flat quartz surface decreased from 31 degrees for coarse silt to 22 degrees for coarse sand. This is an apparent contradiction with the independence of particle size predicted by equation (14.46). On the other hand, the assumption of an asperity per contact may not have been valid for all particle sizes, and additionally, particle surface textures on a microscale could have been size depend-

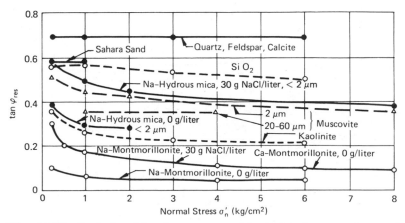

Fig. 14.32 Relationship between residual friction angle and normal stress for different minerals (Kenney, 1967).

ent. Furthermore, the test procedure could have enabled different amounts of particle rearrangement and rolling.

Sliding friction

The frictional resistance once sliding has been initiated may be equal to or less than (it cannot be greater than) the resistance that had to be overcome to start movement. In other words, the coefficient of sliding friction may be less than the coefficient of static friction. A higher value of static than sliding friction is accounted for in terms of time-dependent bond formations at asperity junctions. "Stick-slip" motion, wherein μ varies more or less erratically as two surfaces in contact are displaced, appears common to all friction measurements of minerals involving single contacts (Procter and Barton, 1974). Stick-slip is not observed during shear of assemblages of large numbers of particles, because the slip of individual contacts would be masked within the behavior of the mass as a whole.

14.7 FRICTIONAL BEHAVIOR OF MINERALS

Few studies have been made that enable evaluation of the true coefficient of friction μ and friction angle ϕ_μ between soil minerals without some influence of effects due to particle rearrangement, volume changes,

surface preparation, and so on. This is due mainly to lack of reliable techniques for direct measurement of the sliding resistance between two small soil particles and for controlling particle surface conditions.

Nonclay minerals

Values of true friction angle ϕ_μ for several nonclay minerals are summarized in Table 14.2. The type of test and test conditions are listed in each case.

A pronounced antilubricating effect of water is evident for polished surfaces of the nonsheet minerals quartz, feldspar, and calcite. This apparently results from a disruptive effect of water on adsorbed films that may have served as a lubricant for dry surfaces. Evidence to support this is shown in Fig. 14.33, where it may be seen that the presence of water had no effect on the frictional resistance of quartz surfaces that had been chemically cleaned prior to measurement of the friction coefficient. The samples tested by Horn and Deere (1962) in Table 14.2 had not been chemically cleaned.

An apparent antilubrication effect due to water might also arise as a result of attack of the silica surface (quartz and feldspar) or carbonate surface (calcite) and the formation of silica and carbonate cement at the interparticle contacts. Although such effects probably played little, if any part in the behavior

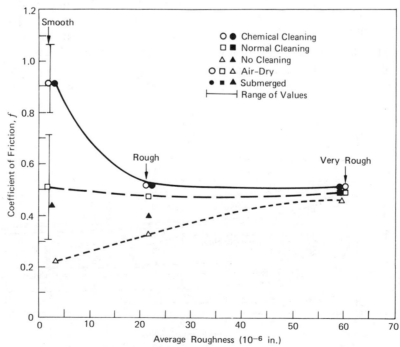

Fig. 14.33 Friction of quartz (Data from Bromwell, 1966 and Dickey, 1966).

Table 14.2 Values of Friction Angle (ϕ_μ) Between Mineral Surfaces

Mineral	Type of Test	Conditions	ϕ_μ (degrees)	Comments	Reference
Quartz	Block over particle set in mortar	Dry	6	Dried over $CaCl_2$ before testing	Tschebotarioff and Welch (1963)
		Moist	24.5		
		Water-saturated	24.5		
Quartz	Three fixed particles over block	Water-saturated	21.7	Normal load per particle increasing from 1 g to 100 g	Hafiz (1950)
Quartz	Block on block	Dry	7.4	Polished surfaces	Horn and Deere (1962)
		Water-saturated	24.2		
Quartz	Particles on polished block	Water-saturated	22–31	ϕ decreasing with increasing particle size	Rowe (1962)
Quartz	Block on block	Variable	0–45	Depends on roughness and cleanliness	Bromwell (1966)
Quartz	Particle–particle	Saturated	26	Single point contact	Procter and Barton (1974)
	Particle–plane	Saturated	22.2		
	Particle–plane	Dry	17.4		
Feldspar	Block on block	Dry	6.8	Polished surfaces	Horn and Deere (1962)
		Water-saturated	37.6		
Feldspar	Free particles on flat surface	Water-saturated	37	25–500 sieve	Lee (1966)
Feldspar	Particle–plane	Saturated	28.9	Single point contact	Procter and Barton (1974)
Calcite	Block on block	Dry	8.0	Polished surfaces	Horn and Deere (1962)
		Water-saturated	34.2		
Muscovite	Along cleavage faces	Dry	23.3	Oven-dry	Horn and Deere (1962)
		Dry	16.7	Air-equilibrated	
		Saturated	13.0		
Phlogopite	Along cleavage faces	Dry	17.2	Oven-dry	Horn and Deere (1962)
		Dry	14.0	Air-equilibrated	
		Saturated	8.5		
Biotite	Along cleavage faces	Dry	17.2	Oven-dry	Horn and Deere (1962)
		Dry	14.6	Air-equilibrated	
		Saturated	7.4		
Chlorite	Along cleavage faces	Dry	27.9	Oven-dry	Horn and Deere (1962)
		Dry	19.3	Air-equilibrated	
		Saturated	12.4		

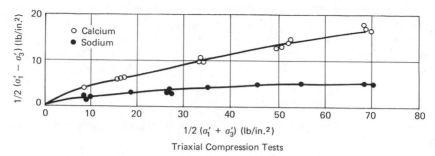

Fig. 14.34   Effective stress failure diagrams for calcium and sodium montmorillonite (Mesri and Olson, 1970).

shown in Table 14.2 and Fig. 14.33, they might be important in nature in the formation of cemented sands.*

As surface roughness increases, the apparent anti-lubricating effect of water decreases (Horn and Deere, 1962; Bromwell, 1966; Dickey, 1966). Figure 14.33 shows this for quartz surfaces that had not been cleaned. Chemically cleaned quartz surfaces, which gave the same value of friction coefficient when both dry and wet, exhibited a loss in frictional resistance with increasing surface roughness. Evidently, in the case of contaminated surfaces, increased roughness makes it easier for asperities to break through surface films, resulting in an increase in δ, equation (14.38). The decrease in friction with increased roughness for chemically cleaned surfaces is not readily explainable. One possibility is that the cleaning process used was not effective on the rough surfaces. For soils in nature, the surfaces of bulky mineral particles are most probably rough relative to the scale in Fig. 14.33, and they will not be chemically clean, thus values of $\mu = 0.5$ and $\phi_\mu = 26$ degrees are reasonable for quartz, both wet and dry.

Table 14.2 shows that water apparently acts as a lubricant in the case of the sheet minerals. The explanation for this is that in air the adsorbed film is thin, surface ions are not fully hydrated, and thus the adsorbed layer is not easily disrupted. Horn and Deere (1962) provided evidence for this by their observation that the surfaces of sheet minerals were scratched when tested in air. When the surfaces of the layer silicates are wetted, the mobility of the surface films is increased because of the increased thickness and because of greater surface ion hydration and dissocia-

tion. Thus, the values of ϕ_μ listed in Table 14.2 for saturated conditions (7 to 13°) are probably appropriate for sheet mineral particles in soils.

From the data in Table 14.2 and Fig. 14.33, it is clear that attempts to relate the strength of a soil mass to interparticle surface friction are likely to be successful only if representative conditions of surface roughness and cleanliness are used. Results for polished, chemically clean surfaces may be misleading.

Clay minerals

Directly measured values of ϕ_μ for the clay minerals are not available. Because the structure of these minerals is similar to that of the layer silicates discussed above, approximately the same values would be anticipated, and the range of residual friction angles measured for highly plastic clays and clay minerals (see, e.g., Fig. 14.2); Kenney (1967), Bishop et al. (1967), and Chattapadhyay (1972) support this.

In highly active colloidal pure clays, such as montmorillonite, very low friction angles may be measured. Residual values of friction angle as low as 4 degrees for sodium montmorillonite are indicated by the data in Fig. 14.32. The effective stress failure envelopes for calcium and sodium montmorillonite are different, as shown by Fig. 14.34, and the friction angle is stress dependent (Mesri and Olson, 1970). For each material the effective stress failure envelope is the same in drained and undrained triaxial compression tests and unaffected by electrolyte concentration over the range investigated, 0.001 N to 0.1 N. The water content at any effective consolidation pressure is independent of electrolyte concentration for calcium montmorillonite, but varies in the manner shown in Fig. 14.35 for sodium montmorillonite.

The observed consolidation behavior is consistent with that described in Chapter 13, and reflects the fact that interlayer expansion in calcium montmorillonite is restricted to a c-axis spacing of 19 Å, leading to

* Sand deposits exhibiting characteristics of cementation have been encountered in some areas. A so-called "seasoning" effect has been reported for some sandfills placed under water, wherein the strength increases with time. Precipitation of cements derived either from dissolved material in the pore water or from particle surfaces could be responsible.

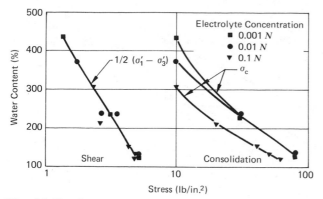

Fig. 14.35 Stress–water content curves for sodium montmorillonite (Mesri and Olson, 1970).

domains or aggregates of several unit layers. The intersheet spacing of sodium montmorillonite is sensitive to double layer repulsions, which in turn are dependent on electrolyte concentration. It is significant that the influence of system chemistry on the behavior of sodium montmorillonite is to change the water content at any consolidation pressure but not the strength at a given apparent effective stress ($\sigma - u$). This suggests that the strength generating mechanism is independent of system chemistry.

The platelets of sodium montmorillonite behave as thin films held apart by high repulsive forces (osmotic pressure) that carry the effective stress. For this case, if it is assumed that there is essentially no intergranular contact, equation (10.28) becomes

$$\sigma_i' = \sigma + A - u_0 - R = 0 \qquad (14.47)$$

Since ($\sigma - u_o$) is the conventionally defined effective stress σ', and assuming negligible long range attractions, equation (14.47) becomes

$$\sigma' = R \qquad (14.48)$$

This accounts for the increase in consolidation pressure required to decrease the water content, while at the same time there is little increase in shear strength, since the shearing strength of water and solutions is essentially independent of hydrostatic pressure. The small friction angle that is observed for the sodium montmorillonite at low effective stresses can be ascribed primarily to the few interparticle contacts that resist particle rearrangement. Resistance from this source evidently approaches constant value at the higher effective stresses as evidenced by the nearly horizontal failure envelope at values of average effective stress greater than about 50 psi (Fig. 14.34). The viscous resistance of the pore fluid can be presumed

to contribute a small portion of the strength at all effective stresses.

The aggregation of clay plates in the calcium montmorillonite leads to particle groups that behave more like equidimensional particles. More physical interference and more intergrain contact are probable than in the sodium montmorillonite, since the water content range for the strength data shown in Fig. 14.34 was only 50 to about 97 percent for the calcium montmorillonite, whereas it was about 125 to 450 percent for the sodium montmorillonite. At a consolidation pressure of about 5 atm, the slope of the failure envelope for the calcium montmorillonite was about 10 degrees, which is in the middle of the ranges of the values found by Horn and Deere (1962) for nonclay sheet minerals (Table 14.2) and of the residual friction angle corresponding to the highest values of plasticity index in Fig. 14.2.

It is probable that the frictional behavior of most clays in nature is more like that of the nonclay layer silicates and calcium montmorillonite than that of sodium montmorillonite. The presence of some amount of sodium montmorillonite may be indicated for cases where the residual friction angle is less than 10 degrees, however.

14.8 RESIDUAL STRENGTH

For a given set of stress conditions, a soil will, if deformed under drained conditions and given sufficient time to consolidate or swell, arrive ultimately at the same final water content (saturated clays) or void ratio (sands), regardless of the initial state of the soil. The strength of the soil is referred to as the "ultimate" or "residual" strength, and the friction angle corresponding to the strength is termed the residual friction angle ϕ_r'. The relationships between stress–strain, volume change–strain, and different friction angles are shown in Fig. 14.36. At the residual state, the strength–determining factors listed in equation (14.2) are reduced to the effective stress, composition, and friction angle for a given test type and strain rate.

The peak strength of normally consolidated clay exceeds the ultimate value because of rupture of cemented bonds, particle reorientations, and other factors contributing to sensitivity. The shear of a heavily overconsolidated clay leads to a rupture of cemented bonds and swelling that cause strength loss beyond the peak.

The behavior of dense sand is similar to that of overconsolidated clays. Loose sands behave somewhat

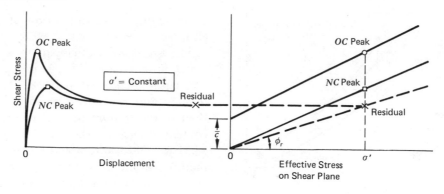

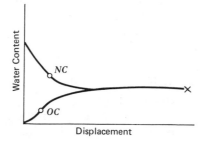

Fig. 14.36 Shear characteristics of clays under fully drained conditions.

like normally consolidated clays, except the peak and ultimate strengths of a loose sand are the same. Differences between the peak and residual strengths of an initially dense sand are attributable to differences in void ratio, dilatancy effects, and fabric.

The residual condition, termed the "critical state" by Roscoe, Schofield, and Wroth (1958), represents deformation under conditions of constant volume and constant fabric. Residual friction angles bear a more direct relationship to true friction between grains than those determined at peak strength.

That values of residual friction angle for clays are related to composition is suggested by Fig. 14.2 where ϕ'_r versus PI and ϕ'_r versus liquid limit are shown. This is confirmed by Fig. 14.32 where values of tan ϕ'_r for several minerals are shown as a function of effective normal stress on the failure plane.

The residual strength of the massive nonclay minerals is not much different than the peak strength if interlocking and dilatancy are accounted for, and quartz, feldspar, and calcite all have the same value of $\phi'_r = 35$ (Fig. 14.32). Slip at contacting asperities is the dominant deformation mechanism. That ϕ'_r for quartz exceeds ϕ_μ (Table 14.2) is to be expected because ϕ'_r includes contributions for particle rearrangement and rolling; whereas, ϕ_μ considers only slip at contacts. The values of ϕ'_r and ϕ_μ for water saturated feldspar and calcite are about the same, however.

Evidently particle rearrangements are relatively unimportant in these minerals at the residual state.

Basal plane slip is the dominant deformation mechanism at large strain in the clay minerals and other layer silicates. Compression textures with basal planes approximately perpendicular to the normal load are formed in the shear zone (Fig. 12.20), and most of the deformation takes place in this zone as well as in the highly oriented zones that enclose it. The behavior of the layer silicates and solid lubricants, such as graphite and molybdenum disulphide (MoS_2), is similar.

For this mode of deformation it would be expected that the type and number of bonds along the cleavage planes would be important. This is borne out by the relationship between interlayer bonding and ϕ'_r listed in Table 14.3 for all the materials except attapulgite. The high residual strength of attapulgite results because it occurs in aggregates of intermeshed crystallites and the crystal structure gives a stair-step mode of cleavage instead of basal cleavage. Hence, attapulgite behaves more like a massive mineral than a clay mineral (Chattopadhyay, 1972).

The values of residual strength shown in Fig. 14.32 were determined by shearing samples in direct shear back and forth through a displacement of 2 to 2.5 mm each side of center until minimum values were obtained. In some cases, a total displacement of several inches in the same direction may be required

Table 14.3 Bonding Along Cleavage Planes, Cleavage Mode, and Residual Strength

Mineral	Mode of Cleavage	Bonding Along Cleavage Planes	ϕ'_r	Particle Shape
Quartz	No definite cleavage		35 degrees	Bulky
Attapulgite	Along (110) plane	Si–O–Si, weak	30 degrees	Fibrous and needle-shaped
Mica	Good basal (001)	Secondary valence (0.5 to 5 kcal/mole) + K− linkages	17 to 24 degrees	Sheet
Kaolinite	Basal (001)	Secondary valence (0.5 to 5 kcal/mole) + H-bonds (5–10 kcal/mole)	12 degrees	Platy
Illite	Basal (001)	Secondary valence (0.5 to 5 kcal/mole) + K linkages	10.2 degrees	Platy
Montmorillonite	Excellent basal (001)	Secondary valence (0.5 to 5 kcal/mole) + exchangeable ion linkages	4 to 10 degrees	Platy–filmy
Talc	Basal (001)	Secondary valence (0.5 to 5 kcal/mole)	6 degrees	Platy
Graphite	Basal (001)	van der Waal's	3 to 6 degrees	Sheet
MoS₂	Basal (001)	Weak interlayer	2 degrees	Sheet

Adapted from Chattopadhyay (1972).

(Fig. 14.3) before minimum residual strength can be obtained. A series of small back and forth displacements may not be equivalent to a test with a displacement of the same total magnitude always in the same direction.* Thus, some of the values in Fig. 14.32 may be high, particularly those for the layer silicates.

Figure 14.32 and additional data by Chattopadhyay (1972) show stress dependency of residual friction angle of the clay minerals, and Fig. 14.5 shows similar stress-dependency for brown London clay and Weald clay. In Section 14.2, it is noted that residual strength can be obtained after only 1 or 2 mm of displacement for clay materials in contact with hard surfaces, such as steel; whereas, a displacement of several centimeters may be required for clay against clay. Fabric studies of kaolin subjected to direct shear show that platy particles must be in perfect parallel orientation before minimum ultimate strength can be obtained (Morgenstern and Tchalenko, 1967). These findings suggest that just a few particles projecting into the plane of shear are sufficient to provide enough interlocking to account for an increment of several degrees in the friction angle.

One explanation for the stress-dependency shown by Figs. 14.5 and 14.32 could be that under low nor-

mal stresses less work is required to shear the clay in the absence of perfect orientation of clay plates in the shear zone than would be required to develop parallelism during shear. Under higher normal stresses, however, least work is expended if perfect particle reorientations develop during shear displacement. If a hard, smooth surface is on the opposite side of the shear plane, less displacement is needed to effect the orientation of clay plates because during shear:

1. There will not be particles projecting across the shear plane and catching on each other.
2. The hard, smooth surface provides a fixed plane to guide the alignment of particles.

An alternative explanation of the stress-dependency of ϕ'_r for clays can be based on the elastic junction theory developed in Section 14.6. On this basis, the area of real contact between sliding surfaces increases less than proportionally with increase in normal effective stress; and according to equation (14.46), $\tan \phi'_r$ should vary as $(\sigma_n')^{-1/3}$. Data for several clays are shown in Fig. 14.37, which shows agreement with this theory in the low pressure range (0 to 30 psi), but at higher stresses, σ'_r is independent of stress indicating that the solid contact area varies as (load)$^{1.0}$.

Both hypotheses appear tenable and evidence is not available to favor one over the other. Nonetheless, it is clear from what is now known that determinations

---

* Bishop et al. (1971) developed a new ring shear device that can be used to shear specimens through large displacements always in the same direction.

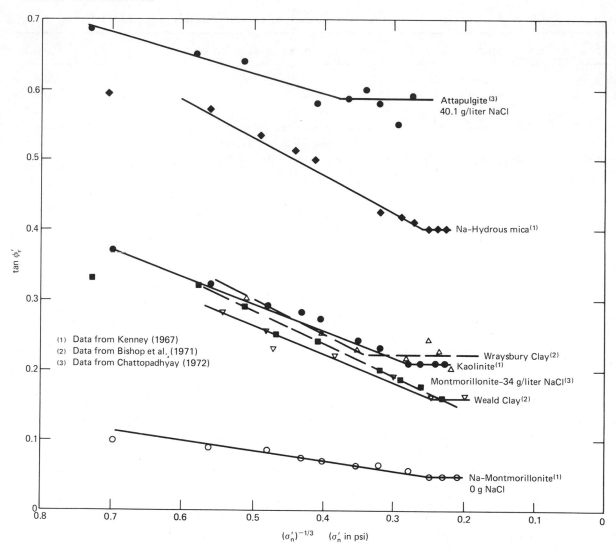

Fig. 14.37   Residual friction angle as a function of normal effective stress on the shear plane raised to the minus 1/3 power (Replotted from data in Chattopadhyay, 1972).

of values of residual strength to be used for analysis of particular problems should be made under stress conditions approximating those in the field. Displacements in one direction should be provided for of a magnitude that have or will develop in the field.

14.9   THE STRENGTH OF GRANULAR SOIL

The strength of a cohesionless soil cannot be accounted for entirely in terms of intergrain sliding friction. For example, peak values of friction angle for quartz sands range from about 30° to more than 50°; whereas, ϕ_μ is only 26°. The peak friction angle ϕ_m can be represented as the sum of three contributions (Rowe, 1962): ϕ_μ, the frictional resistance at con-

tacts; particle rearrangements; and dilation, as shown by Fig. 14.38. In this figure, ϕ_m is the peak value, ϕ_μ is the true friction or sliding between grains, ϕ_r is the friction angle corrected for the work of dilation, and ϕ_{cv} is for deformation at constant volume.

At the lowest porosities, the peak strength is reached before significant interparticle movement can develop; thus, the work of rearrangement is small. Failure requires volume expansion against the confining stress, giving the large dilation contribution shown. If the confining stresses are very high, there will be less dilation but more grain crushing to accommodate the shear deformations. At higher porosities some grain rearrangement develops prior to failure as particles roll and slide along planes inclined at various angles.

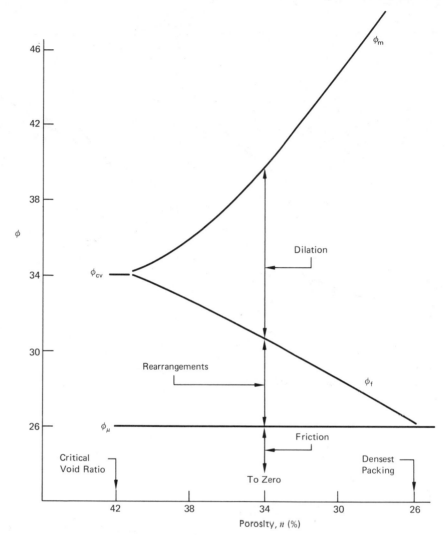

Fig. 14.38 Components of shear strength in granular soils (After Rowe, 1962).

The critical void ratio shown in Fig. 14.38 represents a condition where failure occurs at constant volume. In this case, no work is required to produce dilation, and ϕ_m is composed only of ϕ_μ and particle rearrangements. Figure 14.38 provides a graphic illustration of the mechanism of ϕ_m mobilization and shows clearly the components of strength at different porosities.

Theoretical analyses of ideal packings of uniform spheres and rods have been made to explain quantitatively the difference between $\phi_m = \phi_{cv}$ and ϕ_μ for the loosest state and between ϕ_m and ϕ_μ for the densest state. Some of the results are summarized in Fig. 14.39. In the loosest state, the porosity of uniform spheres is 42 percent; in the densest state, it is 26 percent. The basis for some of the relationships in Fig. 14.39 is as follows.

It was assumed by Rowe (1962) that in the densest state all deformation requires sliding at contacts and there is no rolling of grains. The resulting equations are

$$\left(\frac{\sigma_1'}{\sigma_3'}\right)_{max} = \tan^2\left(45 + \frac{\phi_m}{2}\right) = \tan\alpha\,\tan\left(\phi_\mu + \beta\right) \tag{14.49}$$

$$\left(\frac{\sigma_1'}{\sigma_3'}\right)_{max} = \left(1 + \frac{dv}{d\epsilon}\right)\tan^2\left(45 + \frac{\phi_\mu}{2}\right) \tag{14.50}$$

where β is the inclination of the plane of sliding relative to the direction of the major principal stress, $\tan\alpha$ represents the ratio of the numbers of interparticle contacts sliding in the major and minor principal planes, respectively, and the rate of change of volume with strain $dv/d\epsilon$, is termed the dilatancy rate.

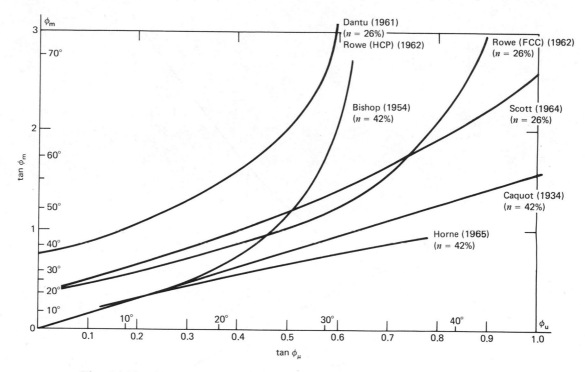

Fig. 14.39   Theoretical curves for relating ϕ_m and ϕ_μ (Bromwell, 1966).

For face centered cubic packing (FCC), $\alpha = 60°$ and $\beta = 45°$. For hexagonal close packing (HCP), $\alpha = 71.6°$ $\beta = 54.7°$, and the same curve is obtained as derived by Dantu (1961); see Fig. 14.39. Later work (Rowe, 1973) shows that equations (14.49) and (14.50) can also be obtained for an assembly of irregular particles in contact.

In the analysis leading to equations (14.49) and (14.50) the close-packed planes of spheres are assumed fixed in orientation relative to the σ_1' and σ_3' directions. Analysis based on shear and normal forces at sphere contacts and with the external stress system free to reorient itself relative to the packing geometry so as to minimize the work leads to (Scott, 1964)

$$\tan \phi_m = \frac{\sqrt{3} + 4\sqrt{2} \tan \phi_\mu}{2(\sqrt{6} - \tan \phi_\mu)} \qquad (14.51)$$

The assumption of plane strain deformation of uniform spheres in loose packing leads to (Bishop, 1954)

$$\sin \phi_m = \frac{3}{2} \tan \phi_\mu \qquad (14.52)$$

If sliding is assumed to occur simultaneously at sphere-to-sphere contacts inclined in all tangential directions on a spherical surface, and normal and tangential forces are integrated over the surface of a semisphere (Caquot, 1934), the following equation is obtained

$$\tan \phi_m = \frac{\pi}{2} \tan \phi_\mu \qquad (14.53)$$

The principal stress ratio and dilatancy of a sand can be related to a "fabric index," S_Z/S_X or S_Z/S_Y (Oda, 1972a, 1972b). The meaning of fabric index is shown by Fig. 14.40. The area ΔS_i of any grain contact can be projected on the YZ, ZX, and XY planes as shown. Equations are presented by Oda to calculate S_X, S_Y, and S_Z, the sum of the projected areas on each orthogonal plane resulting from m contacts. The ratios S_Z/S_X and S_Z/S_Y, which should be equal for a cross-anisotropic material such as a triaxial specimen can be evaluated from fabric measurements using thin sections.

Test results show that both the principal stress ratio, σ_1'/σ_3', and the dilatancy rate, $(-dv/d\epsilon_1)$, where v is volumetric strain and ϵ_1 is major principal strain, vary directly as S_Z/S_X. This ratio changes continuously during deformation and reaches a maximum at peak (σ_1'/σ_3'). The equations developed by Oda based on the results of fabric measurements are similar in form to those derived theoretically by Rowe (1962) and Horne (1969).

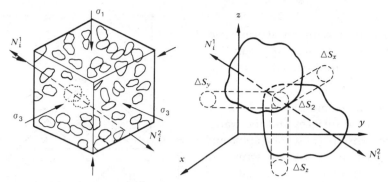

Fig. 14.40 Basis for determination of fabric index for a sand (Oda, 1972a).

14.10 COHESION

True cohesion at any effective stress is shear strength in excess of the strength mobilized by that effective stress times the tangent of the effective stress friction angle ϕ'. The existence of a tensile or shear strength in the absence of any effective stress in the soil skeleton or on the failure plane can be taken as evidence of a true cohesion. Evaluation of the magnitude of true cohesion is difficult because projection of the failure envelope back to $\sigma' = 0$ is uncertain unless tests are done at very low effective stresses. Interpretation of the Hvorslev parameters ϕ_e and c_e (Fig. 14.6) is complicated by the facts that there may be substantial differences in fabric between the samples tested and that ϕ_e may be stress-dependent. Tensile tests cannot readily be made on most soils, and there is no convenient way to run a triaxial compression test while maintaining the effective stress equal to zero on the potential failure plane.

Nonetheless, there is evidence to indicate that a true cohesion may develop in soils, especially clays, and the importance of cohesion to the total strength increases with increased clay content and activity, decreased water content, increased overconsolidation ratio, an increased amount of cementing material in the soil, and increased time for bond formation.

True cohesion

Several sources of cohesion have been identified. A true, stress-independent cohesive strength may result from

1. *Cementation.* Chemical bonding between particles can develop through cementation by carbonate, silica, alumina, iron oxide and organic compounds. Cementing materials may be derived from the soil minerals themselves as a result of various weathering processes, or they may be taken from solution. An analysis of the strength of cemented bonds is given by Ingles (1962) and is summarized in Section 10.3 and equations (10.2) to (10.8). Cohesive strengths of as much as several hundred kN/m^2 (several tens of psi) may develop due to cementation.

2. *Electrostatic and electromagnetic attractions.* Electrostatic and electromagnetic attractions are discussed in Sections 7.7 and 10.3. Electrostatic attractions may become significant (greater than $7 kN/m^2$ or 1 psi) for separation distances less than 25 Å. Electromagnetic attractions or van der Waals forces can be a source of tensile strength only between closely spaced particles of very small size ($< 1 \mu m$). Attractions of these types can be a source of cohesion in both normally consolidated and overconsolidated clays.

3. *Primary valence bonding and adhesion.* According to the Hvorslev concept, a normally consolidated clay can possess a true cohesion, even through the effective stress failure envelope passes through the origin of the Mohr diagram. The amount of cohesion is proportional to the effective consolidation pressure. When a normally consolidated clay is unloaded, thus becoming overconsolidated, the strength may not decrease in proportion to the effective stress reduction, but a part of the strength is retained as shown in Fig. 14.6. This indicates that at least some of the interparticle junctions in clays are plastic rather than elastic. The exact nature of the bonding that causes the particles to remain together after the normal stress is removed is not known. Perhaps an essentially solid junction with primary valence bonding, wherein mineral surface atoms and adsorbed water atoms participate, is formed.

Apparent cohesion

An apparent cohesion may develop in at least two ways. This type of cohesion is termed apparent because it does not depend on interparticle cementations or bonding.

1. *Capillary stresses.* The combination of water attraction to soil particle surfaces and the water property of surface tension causes an apparent attraction between particles in a partly saturated soil. Equation (10.9) can be used to estimate the magnitude of tensile strength that can be developed in a soil due to capillary stresses. Strength developed by capillary stresses is not a true cohesion; it is really a strength generated by frictional resistance, because the negative pore water pressure is the source of an equal and opposite effective stress.

2. *Apparent mechanical forces.* Particle geometry and packing can cause an apparent cohesion in a system with no physical or chemical attractions between particles. Interlocking rough surfaces as shown in Fig. 14.41 illustrate this. A resistance to shear displacement in the direction shown develops even in the absence of a normal stress on the macroscopic surfaces of sliding during shear.

Summary

Several contributions to cohesion are summarized in Fig. 14.42, in terms of the potential tensile strengths for each mechanism as a function of particle size. For all the mechanisms except chemical cementation, cohesion is a consequence of normal stresses between particles generated by internal attractive forces. The mechanism of shear resistance should be the same as

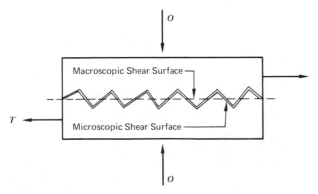

Fig. 14.41 Apparent cohesion due to mechanical forces.

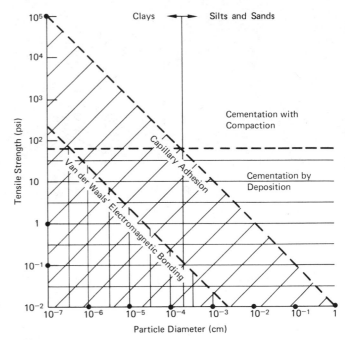

Fig. 14.42 Potential contributions of various bonding mechanisms to soil strength (Ingles, 1962).

if the contact normal stress were derived from effective stresses carried by the soil. It is convenient, therefore, to think of cohesion (except for cementation and apparent mechanical forces) as due to interparticle friction derived from interparticle attractions; whereas, the friction term in the Mohr–Coulomb equation is developed by interparticle friction caused by applied stresses. Essentially the same concept was suggested by Taylor (1948), where cohesion was attributed to an "intrinsic pressure." A similar hypothesis was advanced by Trollope (1960) who attributed shear strength to the Terzaghi–Bowden and Tabor adhesion theory, with both applied stresses and interparticle forces contributing to the effective stress which developed the frictional resistance. Present evidence indicates that cohesion due to interparticle attractive forces is in almost all cases quite small; whereas, that attributable to chemical cementation can be significant.

14.11 CREEP AND STRESS RELAXATION— GENERAL CONSIDERATIONS

Time-dependent deformations and stress relaxation in soils (Fig. 14.13) are important in a variety of geotechnical problems where long term behavior is of concern. The time-dependent responses of soils may assume a variety of forms, depending on such factors as soil type, soil structure, stress history, drainage con-

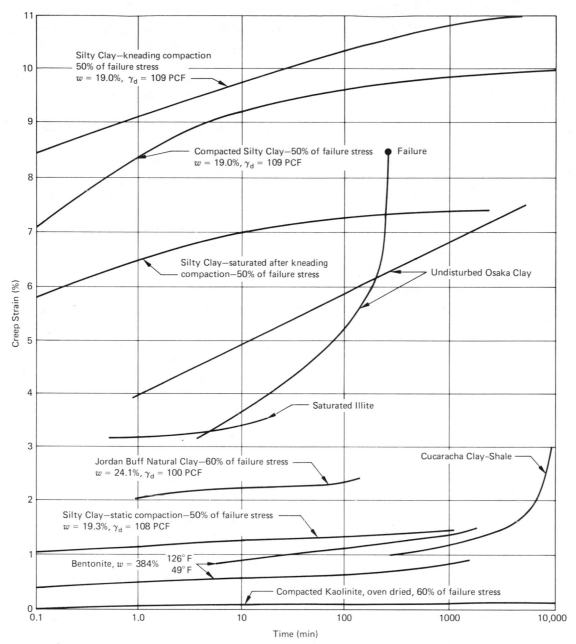

Fig. 14.43 Sustained stress creep curves illustrating different forms of behavior.

ditions, type of stress system, and other factors. Figure 14.43, for example, illustrates the wide variety of creep curves that may be exhibited depending on soil type and test conditions. In spite of this apparently random behavior, time-dependent deformations and stress relaxations are now known to follow logical and often predictable patterns.

General characteristics

Some of the aspects of the creep and stress relaxation behavior of soils are noted in Section 14.2. In many cases application of a stress leads first to a period of transient creep, during which the strain rate decreases continuously with time, followed by creep at constant rate for some period (Finnie and Heller, 1959). For materials susceptible to creep rupture the steady-state period is followed by an acceleration in creep rate leading to failure. These three stages, shown in Fig. 14.44, are often termed primary, secondary and tertiary.

Although for some soils the strain versus time curve can be approximated as steady state for a part of the

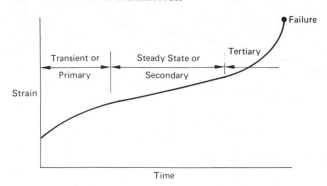

Fig. 14.44 Stages of creep.

time span as shown for example by Fig. 14.45, a true steady state does not really exist. On the other hand, a characteristic relationship between strain rate and time does exist for most soils, as illustrated by Fig. 14.46 for drained triaxial creep of London clay (Bishop, 1966) and Fig. 14.47 for undrained compression creep of Osaka clay (Murayama and Shibata, 1958). The same general pattern has been observed (Singh and Mitchell, 1968) for undisturbed and remolded clay, wet clay and dry clay, normally consolidated and overconsolidated soil, and sand. At any stress level,* the logarithm of the strain rate decreases linearly with the logarithm of time. The slope of this relationship is essentially independent of the creep stress, and increases in stress shift the line vertically upwards. The onset of failure is signalled by a reversal in the slope of the relationship, as shown by the topmost curve in Fig. 14.47.

* The stress level is the stress as a percentage of the normal strength before creep.

The relationship between creep strain and the logarithm of time may be linear, concave upward, or concave downward (Fig. 14.43). A linear relationship is often assumed as an engineering approximation because of its simplicity in analysis. There is no fundamental "law" of behavior to dictate one form or another.

A schematic representation of the influence of creep stress intensity on creep rate at some given time after stress application is shown in Fig. 14.48. At low stresses, creep rates are small and of little practical importance. The curve shape is compatible with the hyperbolic sine function predicted by rate process theory as shown in equation (14.11). In the midrange of stresses, a nearly linear relationship is found between logarithm of strain rate and stress, also as predicted by equation (14.11) for the case where the argument of the hyperbolic sine is greater than 1. At stresses approaching the strength of the material, the strain rates become very large and signal the onset of failure. An example for undrained creep tests on remolded illite is shown in Fig. 14.49 for the midrange of stresses. Results of drained tests on London clay are shown in Fig. 14.50. Undrained test results for undisturbed San Francisco Bay mud, also for the midrange of stresses are given in Fig. 14.51.

Although stress relaxation has been less studied than creep, it appears that equally regular patterns of behavior are observed (Murayama and Shibata, 1964; Murayama, 1969; Lacerda and Houston, 1973).

Effect of composition

In general, the higher the clay content and the more active the clay, the more important are stress

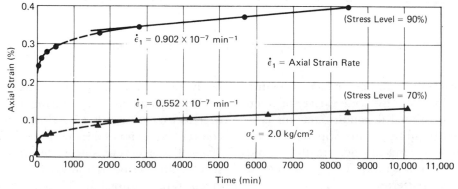

Fig. 14.45 Variation of axial strain with time during creep of a kaolinite–sand mixture (40 percent kaolinite, 60 percent sand) (Paduana, 1965).

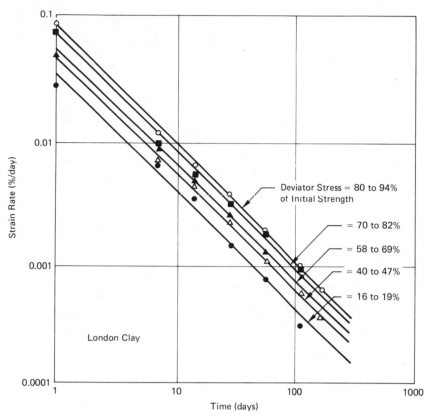

Fig. 14.46   Strain rate vs. time relationships during drained creep of London clay (Data from Bishop, 1966).

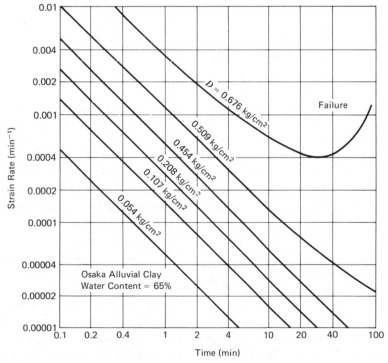

Fig. 14.47   Strain rate vs. time relationships during undrained creep of Osaka alluvial clay (After Murayama and Shibata, 1958).

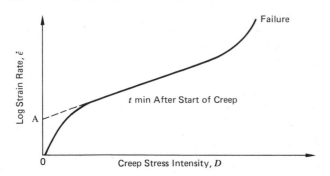

Fig. 14.48 Influence of creep stress intensity on creep rate.

relaxation and creep, as illustrated by Figs. 9.18 and 9.19, where creep rates, approximated by steady-state values are related to clay type, clay content, and plasticity. Time-dependent deformations are more important at high water contents than at low. For stresses greater than the preconsolidation pressure, creep effects, including secondary compression, are greater for metastable, sensitive fabrics than for insensitive, remolded fabrics. Although the magnitude of creep strains and strain rates may be small in a sand or dry soil, the form of the behavior conforms with the patterns shown above. This is consistent with the concepts developed in Section 14.4 that the creep mechanisms are the same in all cases.

Pore pressures and volume change

Rapid application of a stress or a strain invariably results in rapid development of pore water pressures in a saturated soil. For a constant total minor principal stress, the magnitude of the pore pressure depends on the volume change tendencies of the soil when subjected to shear distortions. These tendencies are in turn controlled by the void ratio, structure, and effective stress, as discussed in Chapters 12 and 13. Following the initial large change in pore pressure, pore pressures may either dissipate, with accompanying volume change, if drainage is allowed, or change slowly during creep or stress relaxation, if drainage is prevented. An example showing pore pressure increase with time for consolidated undrained creep tests on illite at several stress intensities is shown in Fig. 14.52. Figure 14.53 shows a slow decrease in pore pressure during the sustained loading of kaolinite.

Changes in effective stress caused by changes in pore pressure during undrained creep loading and changes in water content brought about by changes in volume during drained creep can lead to strength

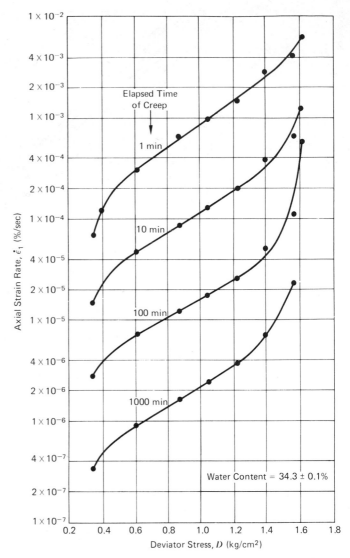

Fig. 14.49 Variation of strain rate with deviator stress for undrained creep of remolded illite.

changes. Since soil strength in terms of effective stresses does not change during creep, in the absence of chemical effects, considerations of pore pressure and volume change can account for this behavior.

A further aspect of the pore pressure behavior of clays under undrained loading complicates the interpretation of pore pressure measurements in creep tests on consolidated-undrained specimens. Figure 14.54 shows pore pressures during undrained creep of undisturbed San Francisco Bay mud. In each instance, the sample was allowed to consolidate under an effective confining pressure of 1.0 kg/cm² for 1800 min prior to the cessation of drainage and the start of the creep test. This consolidation period was greater than

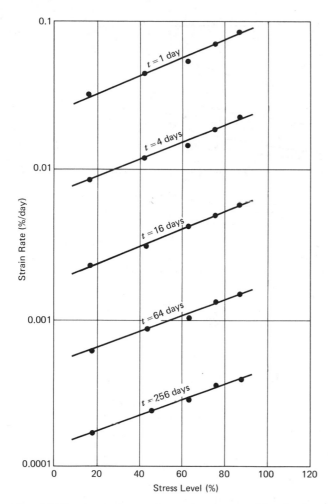

Fig. 14.50 Variation of strain rate with deviator stress for drained creep of London clay (Bishop, 1966).

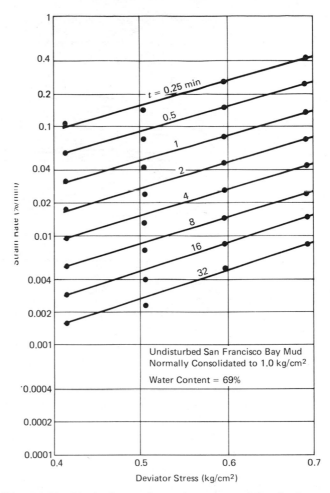

Fig. 14.51 Variation of strain rate with deviator stress for undrained creep of normally consolidated San Francisco Bay mud.

that required for 100 percent primary consolidation. The curve marked 0 percent stress level refers to a specimen maintained undrained but not subjected to a deviator stress. This curve indicates that each of the other tests was influenced by a pore pressure due not to the creep stress, but to a residual effect from the previous consolidation history. Specifically, it relates to the time allowed for secondary compression, as may be seen from Fig. 14.55, which shows pore pressure as a function of time after specimens were allowed to undergo different periods of secondary compression.*

Behavior of this type is further evidence that structure equilibrium under a given effective stress can be

obtained only after long periods of secondary compression. It is consistent with the picture of compressibility of normally consolidated clays developed by Bjerrum (1972, 1973) and discussed in Section 13.8.

Effects of temperature

An increase in temperature increases pore pressure, decreases effective stress, and weakens the soil structure. Creep rates ordinarily increase and the stresses corresponding to specific values of strain decrease at higher temperatures. These effects are illustrated by the data shown in Figures 14.56 and 14.57.

Effect of stress system

The vast majority of studies of time-dependent deformation and stress relaxation in soils, apart from studies of secondary compression in oedometer tests,

---

* Similar data were obtained by P. Kaldveer (1964) and presented in an unpublished research report, University of California, Berkeley.

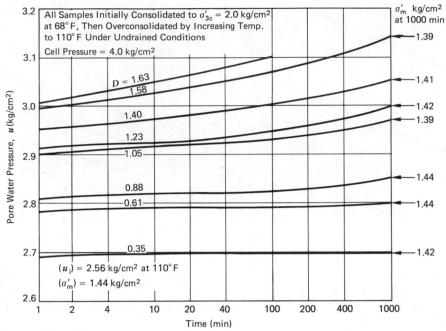

Fig. 14.52 Effect of deviator stress on deformation and pore pressure behavior during undrained creep of illite.

have been done using triaxial compression tests on samples consolidated isotropically. Most soils in nature have been subjected to an anisotropic stress history, with K_0 ranging from 0.5 or less for normally consolidated clays to values approaching 3.0 for heavily overconsolidated clays.* In addition, deformation

conditions in the field approximate more closely to plane strain than to triaxial compression in many problems.

Investigations of the effect of type of test and anisotropy on creep have been initiated only very recently. Undisturbed Haney clay, a gray silty clay with a sensitivity in the range 6 to 10, from British Columbia was tested in triaxial compression and plane strain (Campanella and Vaid, 1974). Samples were

* Values of K_o as a function of plasticity index and overconsolidation ratio are given by Brooker and Ireland (1965) and by Lambe and Whitman (1969).

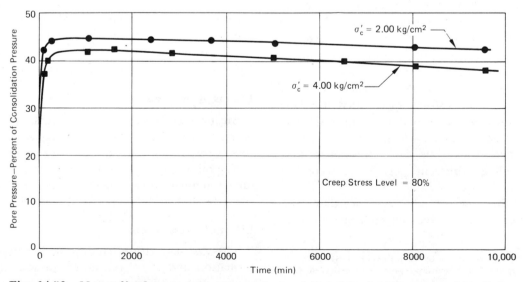

Fig. 14.53 Normalized pore pressure vs. time relationship during creep of kaolinite.

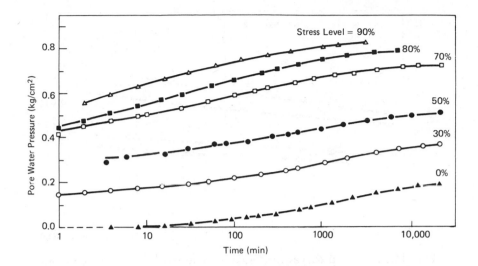

Fig. 14.54 Excess pore pressure development during undrained creep of San Francisco Bay mud after consolidation at 1.0 kg/cm² for 1800 min. (Holzer, Hoeg, and Arulanandan, 1973).

normally consolidated both isotropically and under K_0 conditions to the same vertical effective stress. Samples consolidated isotropically were tested in triaxial compression. K_0 consolidation was used for both K_0 triaxial and plane strain tests. Figure 14.58 shows that the creep behavior was influenced significantly by the pre-creep stress history. Therefore, meaningful estimates of creep behavior for use in practice require duplication in the laboratory of the stress conditions in the field.

14.12 STRESS–STRAIN–TIME FUNCTIONS AND RHEOLOGICAL MODELS

The uniqueness and simplicity of the relationships between strain rate, stress and time shown previously make possible the use of simple expressions for characterization of creep. Alternatively, rheological models can be developed in an effort to duplicate the stress–strain–time response of a soil in terms of various arrangements of springs, dashpots and sliders. Phe-

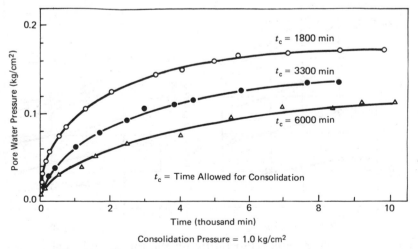

Consolidation Pressure = 1.0 kg/cm²

Fig. 14.55 Excess pore pressure development following cessation of drainage after secondary compression for different time periods (Holzer, Hoeg, and Arulanandan, 1973).

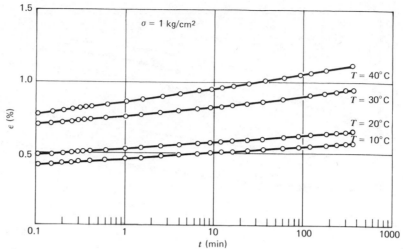

Fig. 14.56　Creep curves for Osaka clay tested at different temperatures—undrained triaxial compression (Murayama, 1969).

nomenological relationships developed by both methods are empirical curve-fitting techniques that do not necessarily imply anything about the mechanisms underlying the deformation process. Since such relationships are useful in practice and also as a basis for organization of data for different soils, some examples are given here.

A general stress–strain–time function

Relationships between strain rate $\dot{\epsilon}$ and time t of the type shown in Figs. 14.46 and 14.47 can be expressed by

$$\ln \frac{\dot{\epsilon}}{\dot{\epsilon}(t_1,D)} = -m \ln \left(\frac{t}{t_1} \right) \tag{14.54}$$

or

$$\ln \dot{\epsilon} = \ln \dot{\epsilon}(t_1,D) - m \ln \left(\frac{t}{t_1} \right) \tag{14.55}$$

where $\dot{\epsilon}(t_1,D)$ is the strain rate at unit time and is a function of stress intensity D, m is the absolute value of the slope of the straight line on the log strain rate vs. log time plot, and t_1 is a reference unit, for example, 1 min. Values of m generally fall in the range

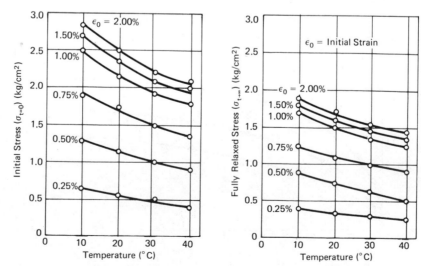

Fig. 14.57　Influence of temperature on the initial and final stresses in stress relaxation tests on Osaka clay—undrained triaxial compression (Murayama, 1969).

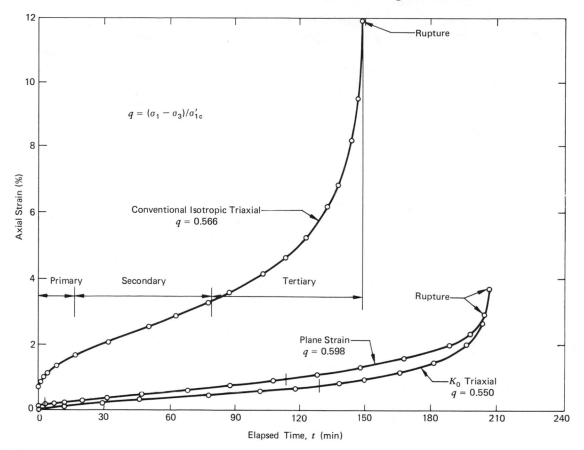

Fig. 14.58 Creep curves for isotropically and K_o-consolidated samples of undisturbed Haney clay tested in triaxial and plane strain compression (Campanella and Vaid, 1974).

of 0.7 to 1.3. For the development shown here, the stress intensity D is taken as the deviator stress ($\sigma_1 - \sigma_3$). A shear stress or stress level could also be used.

The same data plotted in the form of Figs. 14.49, 14.50, and 14.51 can be expressed by

$$\ln\left(\frac{\dot{\epsilon}}{\dot{\epsilon}(t,D_0)}\right) = \alpha D \qquad (14.56)$$

or

$$\ln \dot{\epsilon} = \ln \dot{\epsilon}(t,D_0) + \alpha D \qquad (14.57)$$

in which $\dot{\epsilon}(t,D_0)$ is a fictitious value of strain rate at $D = 0$, a function of time after start of creep; and α is the slope of the linear part of the log strain rate versus stress plot. From (14.55) and (14.57)

$$\ln \dot{\epsilon}(t_1,D) - m \ln\left(\frac{t}{t_1}\right) = \ln \dot{\epsilon}(t,D_0) + \alpha D \quad (14.58)$$

For $D = 0$,

$$\ln \dot{\epsilon}(t,D_0) = \ln \dot{\epsilon}(t_1,D_0) - m \ln\left(\frac{t}{t_1}\right) \quad (14.59)$$

in which $\dot{\epsilon}(t_1,D_0)$ is the value of strain rate obtained by projecting the straight-line portion of the relationship between log strain rate and deviator stress at unit time to a value of $D = 0$. Designation of this quantity by A and substitution of (14.59) into (14.57) gives

$$\ln \dot{\epsilon} = \ln A + \alpha D - m \ln\left(\frac{t}{t_1}\right) \qquad (14.60)$$

which may be written

$$\dot{\epsilon} = Ae^{\alpha D}\left(\frac{t_1}{t}\right)^m \qquad (14.61)$$

This simple three parameter relationship has been found suitable for the description of the creep rate behavior of a wide variety of soils. The parameter A is shown in Fig. 14.48. Since it reflects an order of magnitude for the creep rate under a given set of conditions, it is in a sense a soil property.

A minimum of two creep tests are needed to establish the values of A, α, and m for a soil. If identical specimens are tested using different creep stress intensities, a plot of log strain rate vs. log time yields the value of m, and a plot of log strain rate vs. stress for different values of time can be used to find A and α from the intercept at unit time and the slope, respectively.

The parameter α has units of reciprocal stress. If stress is expressed as the ratio of creep stress to strength at the beginning of creep D/D_{max}, then the dimensionless quantity αD_{max} should be used. For a given soil type, values of αD_{max} do not vary greatly for different water contents, as the change in α with change in water content is compensated by a change in D. Thus, the strain rate vs. time behavior for any stress at any water content can be predicted from the results of creep tests at any other water content, provided the strength variation with water content is known. Since normal strength tests are considerably simpler and less time consuming than creep tests, the uniqueness of the quantity αD_{max} can be useful, because the results of a limited number of tests can be used to predict behavior over a range of conditions. A further generalization of equation (14.61) then is

$$\dot{\epsilon} = Ae^{\bar{\alpha}\bar{D}}\left(\frac{t_1}{t}\right)^m \qquad (14.62)$$

where

$$\bar{\alpha} = \alpha D_{max} \quad \text{and} \quad \bar{D} = \frac{D}{D_{max}}.$$

A general relationship between strain ϵ and time can be obtained by integration of equation (14.61). Two solutions are obtained, depending on the value of m. If $\epsilon = \epsilon_1$ at $t = 1$ and $t_1 = 1$, then

$$\epsilon = \epsilon_1 + \frac{A}{1-m}e^{\alpha D}(t^{1-m} - 1) \qquad (m \neq 1) \quad (14.63)$$

and

$$\epsilon = \epsilon_1 + Ae^{\alpha D}\ln t \qquad (m = 1, t = 1) \quad (14.64)$$

Creep curves corresponding to these relationships are as shown in Fig. 14.59. It may be seen that these curves embrace the variety of creep curves shown in Fig. 14.43. Equations (14.63) and (14.64) have been shown to fit actual creep data well (Singh and Mitchell, 1968).

Although stress relaxation has not been studied as extensively as has creep, most evidence suggests that the decay of stress is essentially linear with logarithm of time. Equation (14.62) takes the following form when stress relaxation is started after deformation under constant rate of strain (Lacerda and Houston, 1973):

$$\frac{D}{D_0} = \frac{\bar{D}}{\bar{D}_0} = 1 - s\log\frac{t}{t_0} \text{ for } t > t_0 \qquad (14.65)$$

where s is the slope of the stress relaxation curve, and the subscripts 0 refer to conditions at the start of stress relaxation.

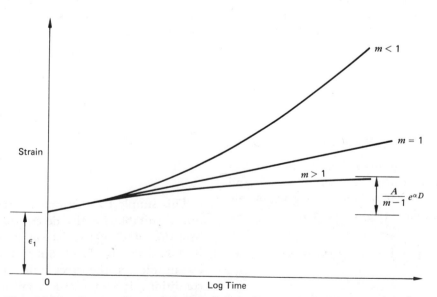

Fig. 14.59　Creep curves predicted by the general stress–strain–time function, equations (14.63) and (14.64).

Also

$$s = \frac{\Phi}{D} \qquad (14.66)$$

where

$$\Phi = \frac{2.3(1 - m)}{\bar{\alpha}} \qquad (14.67)$$

The validity of this equation has been established for $m < 1.0$. Pore pressures decrease slightly during undrained stress relaxation (Murayama and Shibata, 1964; Lacerda and Houston, 1973). This is in accord with the hypothesis of Lo (1969a, 1969b) and others that pore pressures depend almost exclusively on strain.

Stresses may not begin to relax immediately after the strain rate is reduced to zero. The time t_0 between the time strain rate is reduced to zero and the time relaxation begins is a variable that depends on soil type and strain rate. The effect of prior strain rate on stress relaxation is shown schematically by Fig. 14.60. This figure shows that the greater the initial rate of strain used to bring a soil to a given deformation, the more quickly stress relaxation begins. This is a direct reflection of the relative differences in equilibrium soil structures during and after deformation. Data showing t_0 as a function of $\dot{\epsilon}$ are shown in Fig. 14.61 for several soils. These curves can be described empirically by

$$t_0 = \frac{h_0}{\dot{\epsilon}} \qquad (14.68)$$

where h_0 is the strain rate to give a "delay time" of $t_0 = 1$ min before stresses begin to relax. The data

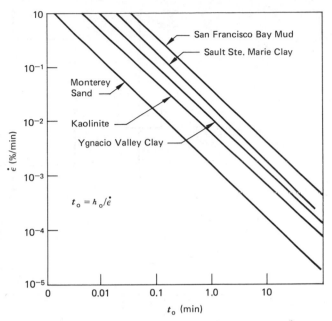

Fig. 14.61 Influence of prior strain rate on time to start of stress relaxation (Lacerda and Houston, 1973).

presented by Lacerda and Houston show that the values of Φ and h_0 increase with increasing plasticity of the soil, and they suggest the following equations

$$\Phi = \frac{PI}{4.4PI + 280} + 0.022 \quad \text{(PI in percent)} \quad (14.69)$$

$$h_0 = 10^{(0.044PI - 3.82)} \quad \text{(h in percent per minute)} \qquad (14.70)$$

In contrast with the large effect of anisotropic consolidation on creep (Campanella and Vaid, 1974), no measurable deviation between the stress relaxation of isotropically and anisotropically consolidated samples has been observed (Lacerda and Houston, 1973).

Rheological models

Different rheological models have been proposed for mathematical description of the stress–strain–time behavior of soils, for example, Geuze and Tan (1953), Schiffman (1959), Murayama and Shibata (1958, 1961, 1964), Christensen and Wu (1964), and Abdel-Hady and Herrin (1966). Linear springs, linear and nonlinear dashpots, and sliders are combined so as to provide a reasonable approximation of behavior for certain soils and loading conditions. Some of these models as well as three basic models (Maxwell, Kelvin or Voight, and Bingham models) proposed for use in characterization of material behavior are shown in Fig. 14.62.

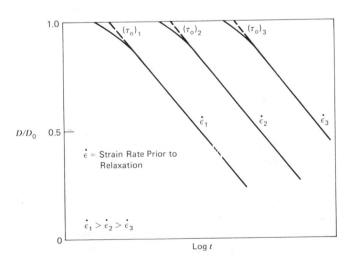

Fig. 14.60 Influence of prior strain rate on stress relaxation.

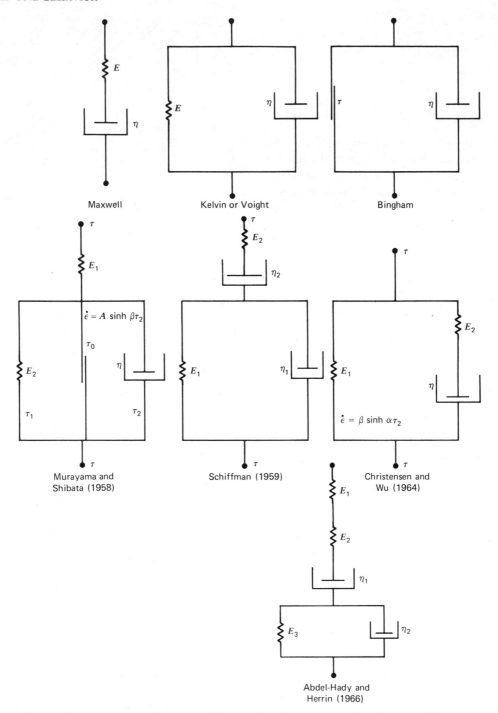

Fig. 14.62   Some rheological models proposed for characterization of the stress–strain–time behavior of soils.

In most rheological models the springs are assumed to be linear. In the Murayama and Shibata, Christensen and Wu, and Abdel-Hady and Herrin models the dashpots are nonlinear with stress-flow rate behavior governed by the functional forms dictated by rate process theory.

Rheological models are useful conceptually to aid in recognition of elastic and plastic components of deformation. They are helpful for visualization by analogy of viscous flow that accompanies time-dependent change of structure to a more stable state. Mathematical relationships can be developed in a straightforward manner for the description of creep, stress relaxation, steady-state deformation, and so on in terms of the model constants. In most cases, these relationships are complex and necessitate the evaluation of several parameters that may not be valid for different stress intensities. None has been proposed that yields relationships as simple as those given by equations (14.61), (14.63), and (14.64), which depend on only three parameters and the strain at one known time.

A model consisting of Voight and Bingham elements in series can account well for the creep behavior of several soils (Komamura and Huang, 1974). To account for initial elastic deformations, a spring is also needed, giving the model shown in Fig. 14.63. The complete strain–time equation for this model is

$$\epsilon = \frac{\sigma}{E_1} + \frac{\sigma - \sigma_0}{\eta_1} t + \frac{\sigma}{E_2} (1 - e^{E_2 t / \eta_2}) \quad (\sigma > \sigma_0)$$
(14.71)

For low stress intensities the second term on the right drops out and creep ultimately ceases. For values of $\sigma > \sigma_0$, creep continues indefinitely. The model coefficients vary with water content, and for sufficiently high values, σ_0, E_1, and E_2 approach zero so that the model predicts essentially viscous behavior.

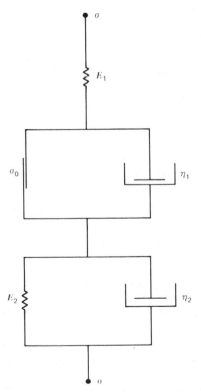

Fig. 14.63 Modified Komamura–Huang model.

14.13 CREEP RUPTURE

In Section 14.2 it was noted that the strength of a soil may decrease under the sustained action of a creep stress and that the stress vs. strain curve was changed as a result of creep (Fig. 14.14). In some cases strength increases have also been observed. Changes in strength may be as much as 50 percent or more of the strength measured in normal undrained tests prior to creep.

Causes of strength loss during creep

Loss of strength during creep appears particularly important in the case of soft clays deformed under undrained conditions and heavily overconsolidated clays tested drained. Both of these conditions are pertinent to certain types of engineering problems; the former in connection with stability of soft clays immediately after construction, and the later in connection with problems of long-term stability.

The loss of strength as a result of creep may be explained in terms of the following principles of behavior:

1. If a significant portion of the strength of a soil is due to cementation and creep deformations lead to failure of cemented bonds, then strength will be lost.
2. In the absence of chemical or mineralogical changes, strength depends on effective stresses at failure. If creep causes changes in effective stress, then strength changes will also occur.
3. In almost all soils, shear causes changes in pore pressure in undrained deformation and changes in volume and water content in drained deformation.
4. Water content changes cause strength changes.

For both strength gain and strength loss, the strength after creep plots on the same effective stress failure envelope as the strength prior to creep (Hirst and Mitchell, 1968; Campanella and Vaid, 1972, 1974). Figure 14.64 shows failure envelopes for undrained tests on illite–sand mixtures. These results show that the strength after creep was either higher or lower than the normal strength depending both on compositional factors and creep stress level. Although it is not yet possible to quantify the influences of these factors directly, it is clear that the effective stress failure envelope is a unifying factor for all the data for a given mixture. Changes in undrained strength during creep are due to changes in pore pressure, unless cementation bonds are broken caus-

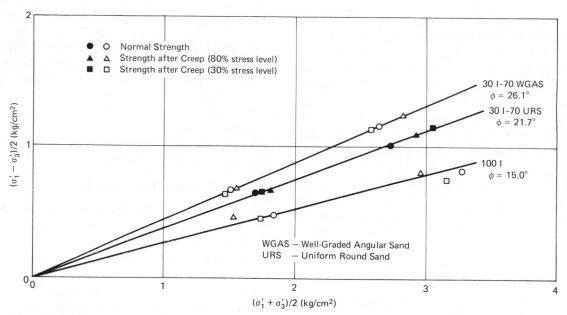

Fig. 14.64   Influence of type of sand on effective stress strength characteristics of illite–sand mixtures. WGAS, Well-graded angular sand and URS, Uniform round sand.

ing a change to a different effective stress failure envelope.

Strength loss as a result of the undrained creep of saturated samples of heavily overconsolidated clay and clay shales reported by Casagrande and Wilson (1951), Goldstein and Ter-Stepanian (1957), and Vialov and Skibitsky (1957) may be accounted for in the following way. The application of shear stresses causes a tendency for dilation and the development of negative pore pressures. These negative pore pressures do not develop uniformly but tend to concentrate along planes where higher shear strains develop. With time during sustained loading, migration of water into zones of high negative pore pressure occurs leading to softening and strength decrease relative to the strength in "normal" undrained tests.

In drained creep deformation, if positive pore pressures are generated by creep stress, then the soil consolidates and the strength increases with time. For overconsolidated clays, however, the dissipation of the negative pore pressures induced by the strains associated with the application of the creep stress leads to a higher water content and lower strength after creep than determined by normal undrained strength tests prior to creep. Stress paths for the two conditions are as shown in Fig. 14.65. In this figure, AB represents the effective stress path, and AC represents the total stress path in a conventional consolidated–undrained (CU) test. The negative pore pressure at failure is CB. If a creep stress DE is applied to the same clay, a neg-

ative pore pressure EF will be induced. This negative pore pressure dissipates during creep so at the end of the creep period the effective stress will be as represented by point E. Shear starting from these conditions leads to strength G which is less than the original value at B.

Time to failure

For soils susceptible to loss of strength, the time to failure is dependent on the stress level. Behavior in this respect is quite similar to that of other materials in that the time to failure is a negative exponential function of the stress, for stresses greater than some critical limiting value. For stresses less than this value,* no failure develops even after very long times. Figure 14.66 shows the relationship between deviator stress, normalized to the pretest major principal effective stress, and time to failure for Haney clay.

There are certain principles relating to the probability of creep rupture and time to failure:

1. Values of the parameter m less than 1.0 in equations (14.61) through (14.63) are indicative of a high potential strength loss during creep and eventual rupture (Singh and Mitchell, 1969).
2. The minimum strain rate $\dot{\epsilon}_{min}$ prior to the onset of creep rupture decreases, and the time to failure increases for a given soil as the stress in-

---

* Termed the "upper yield" by Murayama and Shibata (1958, 1962, 1964).

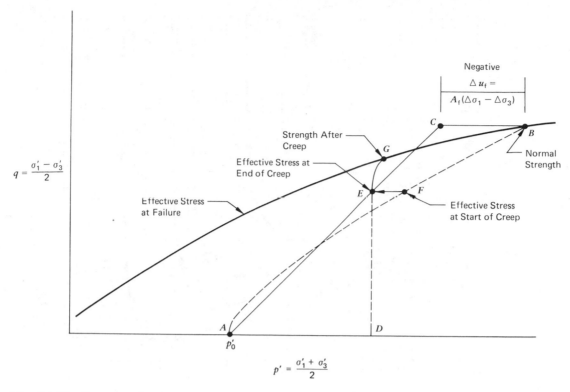

Fig. 14.65 Stress paths for normal undrained shear and drained creep of heavily overconsolidated clay.

tensity decreases, as shown in Fig. 14.67 for Haney clay. This relationship is unique, as may be seen by Fig. 14.68, which shows that

$$t_f = \frac{C}{\dot{\epsilon}_{min}} \qquad (14.72)$$

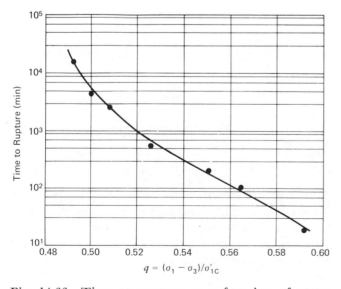

Fig. 14.66 Time to rupture as a function of creep stress for Haney clay (Campanella and Vaid, 1972).

3. The strain at failure is a constant independent of stress level, as shown in Fig. 14.69. The failure strain is taken as the strain corresponding to minimum strain rate. For the case of undrained creep rupture this is consistent with the concept that pore pressure development is uniquely related to strain and independent of the rate at which it accumulates (Lo, 1969a, 1969b).

Values of the constant C in equation (14.72) for several clays, accurate to about ± 0.2 log cycles, are listed in Table 14.4.

The relationship expressed by equation (14.72) is a direct consequence of the fact that the strain at the point of minimum strain rate is a constant independent of stress or strain rate. The general stress–strain rate–time function, equation (14.61) describes the strain rate–time behavior until $\dot{\epsilon}_{min}$ is reached. For $t_1 = 1$ and $\epsilon = 0$ at $t = 0$, the corresponding strain–time equation is

$$\epsilon = \frac{A}{1-m} e^{\alpha D} t^{(1-m)} \qquad (14.73)$$

By setting $\epsilon = 0$ at $t = 0$, the assumption is made that that there is no time-independent or instantaneous de-

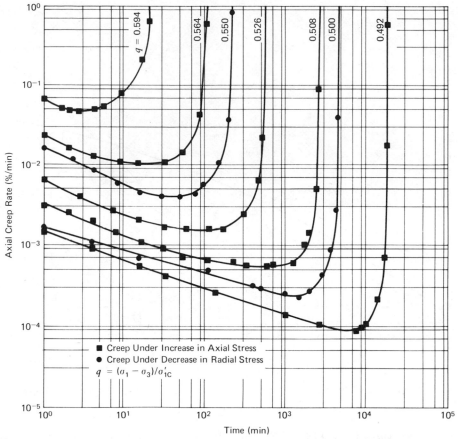

Fig. 14.67 Creep rate behavior of K_0-consolidated undisturbed Haney clay under axially symmetric loading (Campanella and Vaid, 1972)

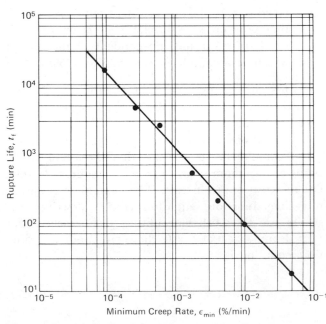

Fig. 14.68 Relationship between time to failure and minimum creep rate (Campanella and Vaid, 1972).

formation. Substitution for $Ae^{\alpha D}$ in (14.73) gives

$$\epsilon = \frac{1}{1-m} \dot{\epsilon} t^m t^{(1-m)} \qquad (14.74)$$

which at the point of minimum strain rate becomes

$$\epsilon_f = \text{constant} = \frac{1}{1-m} \dot{\epsilon}_{\min} t_f = \frac{C}{1-m} \qquad (14.75)$$

Thus, the constant in equation (14.44) should be defined by

$$C = (1-m) \epsilon_f \qquad (14.76)$$

Values of ϵ_f for Haney clay tested in three ways are shown in Fig. 14.69 and values of C and m are in Table 14.4. These quantities may be used to make the comparison shown in Table 14.5. The agreement between predicted and measured values of C is reasonable.

Predictions of the time to failure under a given stress may be made in the following way. Strain at

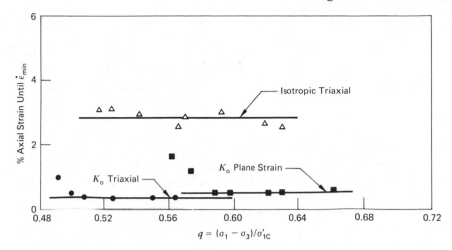

Fig. 14.69 Axial strain at minimum strain rate as a function of creep stress for undisturbed Haney clay (Campanella and Vaid, 1974).

Table 14.4 Creep Rupture Parameters for Several Clays

Soil	Test Type[a]	Creep Rate Parameter, m	$C = (\dot{\epsilon}_{min} t_f)$ (∓ 0.2 log cycles)
Undisturbed Haney Clay, N.C.[b]	ICU	0.7	1.2
Undisturbed Haney Clay, N.C.[b]	ACU	0.4±	0.2
Undisturbed Haney Clay, N.C.[b]	ACU-PS	0.5	0.3
Undisturbed Seattle Clay, O.C.[c]	ICU	0.5	0.6
Undisturbed Tonegawa Loam[c]	U	0.8	1.6
Undisturbed Redwood City Clay, N.C.[c]	ICU	0.75	2.8
Undisturbed Bangkok Mud[c]	ICU	0.70	1.4
Undisturbed Osaka Clay[c]		1.0	0.07

[a] ICU, isotropic consolidated, undrained triaxial; ACU, K_0 consolidated, undrained triaxial; ACU-PS, K_0 consolidated, plane strain; and U, compression test.
[b] Data from Campanella and Vaid (1974).
[c] Data from Singh and Mitchell (1969).

Table 14.5 Predicted and Measured Values of C for Haney Clay

Test Condition	Creep Rate Parameter m (from Table 14.4)	ϵ_f (from Fig. 14.69)	C Predicted by equation (14.76)	C Measured
ICU[a]	0.7	2.8	0.84	1.2
ACU[b]	0.4	0.3	0.18	0.20
ACU-PS[c]	0.5	0.5	0.25	0.30

[a] Isotropic Consolidated, Undrained Triaxial.
[b] Anisotropic, Consolidated, Undrained triaxial.
[c] Anisotropic Consolidated, Undrained, Plane Strain.
Data from Campanella and Vaid (1974).

failure can be determined by either a creep rupture test or by a constant rate of strain test, as strain at failure is independent of test type. If the latter is used, then the rate of strain must be slow enough for pore pressure equalization or drainage, depending on the conditions of interest, and in both cases the stress history and stress system should simulate that in the field. Parameter m can be established from a creep test, and then $C = \dot{\epsilon}_{min} t_f$ can be computed using equation (14.76). Values of A and α are established from creep tests at two stress intensities. Then, for $t_1 = 1$,

$$C = \dot{\epsilon}_{min} t_f = A e^{\alpha D} t_f^{1-m} \qquad (14.77)$$

and corresponding values of D and t_f may be calculated. From (14.77)

$$\ln t_f = \frac{1}{(1-m)}\left[\ln\left(\frac{C}{A}\right) - \alpha D\right] \quad (14.78)$$

14.14 CONCLUSION

Limit or equilibrium analyses as made, for example, in studies of bearing capacity, lateral pressure, and slope stability, depend on accurate representation of soil strength. The stresses and deformations in soil masses under subfailure loading conditions depend on stress–strain and stress–strain–time properties. The contents of this chapter have included both an analysis of how strength, stress-deformation, and stress–strain–time behavior are influenced by compositional and environmental factors and the development of relationships for analysis of such things as the influences of strain rate and temperature on strength, creep, stress relaxation, and time to failure under a sustained stress. Although no equations have been developed that can, at this stage, be considered general constitutive relations applicable to all situations, the factors responsible for and influencing strength in any case have been identified and analyzed.

Rate process theory has proven a particularly fruitful approach to the study of deformation phenomena in soils. From an analysis of the influences of stress and temperature on deformation rates it has been possible to relate strength directly to the number of interparticle bonds per unit area, which are themselves of the same strength for different soils and under different conditions. Functional forms for the influences of strain rate and temperature on shearing resistance are predicted by rate process theory that can be used in practice.

True friction due to interparticle sliding accounts for from half to almost all the strength of most uncemented soils. The remainder is contributed through work of dilatancy, particle rearrangements, particle crushing, and true cohesion. Frictional resistance is developed by adhesion between contacting asperities on opposing surfaces. Values of true friction angle (ϕ_μ) range from a low of less than 4 degrees for sodium montmorillonite to more than 30 degrees for feldspar and calcite.

The residual friction angle depends on mineralogical composition and effective stress. The value of residual friction angle for a clay may decrease by several degrees for increases in effective stress on the shear surface from 0 to 200-to-350 kN/m² (0 to 30-to-50

psi). The shear displacement in one direction required to develop residual strength may be several centimeters. These factors must be taken into account when analyzing stability problems.

In the absence of cementation true cohesion resulting from electrostatic and electromagnetic interparticle forces is usually small. Such cohesion as does exist from interparticle forces can be treated as an interparticle friction derived from internal attractive forces as opposed to externally applied compressive stresses.

Time-dependent deformations and stress relaxations in soils follow predictable patterns that are essentially the same for all soil types. The magnitude of the effects increase with water content and plasticity. A simple three parameter stress–strain–time equation (14.61) can be used to characterize the creep behavior of most soils, and a related equation (14.65) appears suitable for stress relaxation.

Soft clays deformed under undrained conditions and heavily overconsolidated clays under drained conditions can fail under sustained stresses as low as 50 percent of the strength in a conventional undrained test. The strength changes can be accounted for in terms of changes in effective stress and water content. The time to failure under a sustained creep stress decreases with increasing stress and may be predicted using procedures given in Section 14.13.

Because of the great diversity of soil types and the range of environmental conditions under which they may be found, evaluations of strength, its characterization for analyses, and prediction of future behavior will continue as major consideration in any project.

SUGGESTIONS FOR FURTHER STUDY

Bishop, A. W. (1966), "The Strength of Soils as Engineering Materials," *Geotechnique*, Vol. 16, No. 2, pp. 91–128.

Bjerrum, L. (1973), "Problems of Soil Mechanics and Construction on Soft Clays and Structurally Unstable Soils," General Report, Session 4, *Proceedings of the Eighth International Conference on Soil Mechanics and Foundation Engineering*, Vol. 3, pp. 111–159.

Bowden, F. P., and Tabor, D. (1964), *The Friction and Lubrication of Solids*, Part II, Oxford University Press, London.

Finnie, I., and Heller, W. (1959), *Creep of Engineering Materials*, McGraw-Hill, New York.

Glasstone, S., Laidler, K., and Eyring, H. (1941), *The Theory of Rate Processes*, McGraw-Hill, New York.

Hvorslev, M. J. (1960), "Physical Components of the Shear Strength of Saturated Clays," *Proceedings of the*

A.S.C.E. Research Conference on the Strength of Cohesive Soils, pp. 169–273.

Ingles, O. G. (1962), "Bonding Forces in Soils: Part 3. A Theory of Tensile Strength for Stabilized and Naturally Coherent Soils," *Proceedings of the First Conference, Australian Road Research Board,* Vol. 1, pp. 1025–1047.

Kenny, T. C. (1967), "The Influence of Mineralogical Composition of the Residual Strength of Natural Soils," *Proceedings of the Oslo Geotechnical Conference on the Shear Strength Properties of Natural Soils and Rocks,* Vol. I, pp. 123–129.

Ladd, C. C. (1971), "Strength Parameters and Stress-Strain Behavior of Saturated Clays," M.I.T. Special Summer Program on Soft Ground Construction, Department of Civil Engineering Research Report R 71-23, Soils Publication 278.

Mitchell, J. K., Campanella, R. G., and Singh, A. (1968), "Soil Creep as a Rate Process," *Journal of the Soil Mechanics and Foundations Division, A.S.C.E.,* Vol. 94, No. SM1, pp. 231–253.

Singh, A., and Mitchell, J. K. (1968), "General Stress–Strain–Time Function for Soils," *Journal of the Soil Mechanics and Foundations Division, A.S.C.E.,* Vol. 94, No. SM1, pp. 21–46.

Skempton, A. W. (1964), "Long Term Stability of Clay Slopes," *Geotechnique,* Vol. 14, p. 77.

CHAPTER 15

Conduction Phenomena

15.1 INTRODUCTION

Many geotechnical problems involve the flow of matter (water, ions, suspended solids) and energy (heat, electricity) through soils. Predictions of the rates of these flows, their changes with time, and changes in the properties and composition of both the permeated soil and the flowing material are the subject of this chapter.

Water flow has been extensively studied, because of its important bearing on problems of seepage, consolidation and stability, which form a major part of soil and rock engineering analysis and design. As a result, much is known about the hydraulic conductivity, or permeability, of earth materials.

Chemical, thermal, and electrical flows in soils have been given greatly increased attention in recent years. Chemical transport through earth materials is important in connection with groundwater pollution, waste disposal and storage, corrosion, leaching phenomena, osmotic effects in clay layers, and soil stabilization. Heat movement must be considered relative to frost action, construction in permafrost areas, insulation, underground storage, thermal pollution, temporary ground stabilization by freezing, permanent ground stabilization by heating, underground transmission of electricity, and other problems. Electrical flows are important to the transport of water and ground stabilization by electro-osmosis, insulation, and subsurface investigations.

In addition to the above four direct flow types, each driven by its own potential gradient, *coupled flows* can occur in soils. A coupled flow is a flow of one type, such as hydraulic, driven by a potential gradient of another, such as electrical. Some coupled flows, including chemical osmosis, electro-osmosis, thermal osmosis, and ion and heat transfer by flowing water, may be important in practice.

The scope of this chapter includes a review of the laws controlling the several types of flows, an analysis of the factors controlling the magnitude of the hydraulic conductivity, water movement and consolidation of soils by electro-osmosis, chemico-osmotic effects in soils, and frost action in soils, including the mechanism of frost heave and depth of frost penetration.

15.2 FLOW LAWS AND RELATIONSHIPS

Each of the four types of flows or fluxes (J_i) is linearly related to its corresponding driving force (X_i) all other factors being constant, according to

$$J_i = L_{ii}X_i \qquad (15.1)$$

where L_{ii} is the conductivity coefficient for flow. When written specifically for a particular flow, equation (15.1) becomes

Water flow	$q_h = k_h i_h$	Darcy's law	(15.2)
Heat flow	$q_t = k_t i_t$	Fourier's law	(15.3)
Electrical flow	$I = \sigma i_e$	Ohm's law	(15.4)
Chemical flow	$J_D = D i_c$	Fick's law	(15.5)

The terms in equations (15.2) through (15.5) are identified in Table 15.1, which shows analogs between the various flow types. As long as fluxes and forces are linearly related, the mathematical treatment of each flow type is the same, and solutions for flow of one type may be used for problems of another type provided there is a correspondence of boundary conditions and the needed property values are properly determined. Two well-known practical illustrations of this are the correspondence between the Terzaghi theory of consolidation and one-dimensional transient heat flow and the use of electrical analogies for study of seepage problems.

Table 15.1 Conduction Analogies

	Fluid	Heat	Electrical	Ionic (Chemical)
1. Potential	Total head h (cm)	Temperature T (°C)	Voltage V (volts)	Chemical potential μ but concentration usually used c (moles/cm^3)
2. Storage	Fluid volume W (cm^3/cm^3)	Thermal energy u (cal/cm^3)	Charge Q (coulomb)	(moles)
3. Conductivity	Hydraulic conductivity (cm/sec) (Permeability) k_h	Thermal conductivity k_t (cal/°C/cm/sec)	Electrical conductivity σ (coulomb/sec/volt)	Diffusion coefficient D (cm^2/sec)
4. Flow	q_h(cm^3/sec)	q_t(cal/sec)	Current I (amp)	J_D (moles/sec)
5. Gradient	$i_h = -\dfrac{\partial h}{\partial x}$ (cm/cm)	$i_t = -\dfrac{\partial T}{\partial x}$ (°C/cm)	$i_e = -\dfrac{\partial V}{\partial x}$ (v/cm)	$i_c = -\dfrac{dc}{dx}$ (moles cm^{-4})
6. Conduction (one-dimensional)	Darcy's law $q_h = -k_h\dfrac{\partial h}{\partial x}A$	Fouriers' law $q_t = -k_t\dfrac{\partial T}{\partial x}A$	Ohm's law $I = -\sigma\dfrac{\partial V}{\partial x}A = \dfrac{V}{R}$	Fick's law $J_D = -D\dfrac{dc}{dx}A$
7. Capacitance	Coefficient of volume change $M = \dfrac{dW}{dh}\dfrac{\gamma_w a_v}{1+e} = \dfrac{k_h}{c_v}$	Volumetric heat C (cal/°C/cm^3) $C = \dfrac{dQ}{dT}$	Capacitance C (farads = coulomb/V)	
8. Continuity	$\dfrac{\partial W}{\partial t} + \nabla q = 0$	$\dfrac{\partial u}{\partial t} + \nabla q_t = 0$	$\dfrac{\partial Q}{\partial t} + \nabla I = 0$	$\dfrac{\partial(\text{storage})}{\partial t} + \nabla J_D = 0$
9. Continuity (steady state, isotropic and homogeneous)	$\nabla^2 q_h = 0$	$\nabla^2 q_t = 0$	$\nabla^2 I = 0$	$\nabla^2 J_D = 0$
10. Diffusion (one-dimensional)	$\dfrac{\partial h}{\partial t} = \dfrac{k_h}{M}\dfrac{\partial^2 h}{\partial x^2}$ $(k/M = c_v)$	$\dfrac{\partial T}{\partial t} = \dfrac{k_t}{C}\dfrac{\partial^2 T}{\partial x^2}$ $(k/C = a)$	$\dfrac{\partial V}{\partial t} = \dfrac{\sigma}{C}\dfrac{\partial^2 V}{\partial x^2}$	$\dfrac{\partial c}{\partial t} = \text{coefficient}\,\dfrac{\partial^2 c}{\partial x^2}$

Coupled flows

The general phenomenon of a gradient of one type causing a flow of another type is expressed by

$$J_i = L_{ij}X_j \tag{15.6}$$

The L_{ij} are called "coupling coefficients." They are properties that may or may not be of significant magnitude in any given material. The types of coupled flow that can occur are listed in Table 15.2, along with the terms used to describe them. For the case of fluid flow through a soil under the combined action of hydraulic, thermal, electrical, and chemical gradients, X_H, X_T, X_E, X_C, respectively, equation (15.6) becomes

$$J_H = L_{HH}X_H + L_{HT}X_T + L_{HE}X_E + L_{HC}X_C \tag{15.7}$$

where L_{HH} is the coefficient for hydraulic flow under a hydraulic gradient, L_{HT} is the coefficient for hydraulic flow under a thermal gradient, L_{HE} is the coefficient for hydraulic flow under an electrical gradient, and L_{HC} is the coefficient for hydraulic flow under a chemical gradient. Similar relationships can be written for each of the other types of flow.

The importance of all the possible types of coupling in soil–water–electrolyte systems is not completely known. In general, however, the significance of the different contributions is about as follows.

Fluid flow. Coupled contributions to fluid flow in soils depend to a large extent on the soil type, the water content or saturation, and the presence of dissolved salts in the pore fluid. Water movement in cohesionless soils is dominated by the flow under a

Table 15.2   Coupled and Direct Flow Phenomena

Flow J	Hydraulic Head	Gradient X Temperature	Electrical	Chemical (concentration)
Fluid	Hydraulic conduction *Darcy's law*	Thermo-osmosis	Electro-osmosis	Normal osmosis
Heat	Isothermal heat transfer	Thermal conduction *Fourier's law*	Peltier effect	Dufour effect
Current	Streaming current	Thermo-electricity	Electric conduction *Ohm's law*	Diffusion and membrane potentials
Ion	Streaming current	Soret effect—thermal diffusion of electrolyte	Electro-phoresis	Diffusion *Fick's law*

hydraulic gradient. Contributions to the total due to coupling are negligible. Thermo-osmosis appears of little importance in saturated soils of any type. Electro-osmosis can cause large flows in fine-grained soils relative to those induced by a hydraulic gradient; so much so that electro-osmosis can be a practical means for dewatering. Normal osmosis under chemical concentration gradients may be important in highly active clays at low void ratios. The lower the hydraulic conductivity in any case, the greater the relative importance of electro-osmosis and normal osmosis.

To illustrate this latter point, liquid movement through kaolinite under hydraulic, electric, and osmotic gradients may be considered. For combined flow under all three gradients equation (15.7) becomes

$$q_h = -k_h \frac{\Delta H}{L} A + k_c \frac{\log (C_B/C_A)}{L} A - k_e \frac{\Delta E}{L} A$$

$$(15.8)$$

where k_h, k_c, and k_e are the hydraulic, osmotic and electro-osmotic conductivities, ΔH is the hydraulic head difference, ΔE is the voltage difference, and C_A and C_B are the salt concentrations on opposite sides of a clay layer of thickness L. In the absence of electrical and chemical gradients the hydraulic flow rate is

$$q_h = -k_h \frac{\Delta H}{L} A \qquad (\Delta C, \Delta E = 0) \quad (15.9)$$

For a chemical gradient acting alone the osmotic flow is

$$q_{he} = k_c \frac{\log (C_B/C_A)}{L} A \qquad (\Delta H, \Delta E = 0) \quad (15.10)$$

For an electrical gradient acting alone it is

$$q_{he} = -k_e \frac{\Delta E}{L} A \qquad (\Delta H, \Delta C = 0) \quad (15.11)$$

The ratio of osmotic to hydraulic flows in a given system would be

$$\frac{q_{he}}{q_h} = -\left(\frac{k_c}{k_h}\right) \frac{\log (C_B/C_A)}{\Delta H} \qquad (\Delta E = 0) \quad (15.12)$$

and the ratio of electro-osmotic to hydraulic flows is

$$\frac{q_{he}}{q_h} = \left(\frac{k_e}{k_h}\right) \frac{\Delta E}{\Delta H} \qquad (\Delta C = 0) \quad (15.13)$$

Values of (k_c/k_h) and (k_e/k_h) for kaolinite provide a conservative estimate of the importance of osmotic and electro-osmotic flows, as coupling effects in kaolinite are smaller than in the more active clays. Measurements of these quantities have been made by Olsen (1972), and the results are summarized in Fig. 15.1.

The ratio (k_c/k_h) in Fig. 15.1b indicates the hydraulic head difference in centimeters of water required across a kaolinite layer to give a flow rate equal to the osmotic flow caused by a ten-fold difference in salt concentration on opposite sides of the layer. The ratio (k_e/k_h) gives the hydraulic head difference required to balance that caused by 1 V difference between electrical potentials on opposite sides of the layer. It may be seen that during consolidation the hydraulic conductivity decreases dramatically. The ratios $(k_c/k_h$ and $k_e/k_h)$, however, increase significantly, indicating that the relative importance of osmotic and electro-osmotic flows to the total flow increase. The data are presented in Fig. 15.1b as a function of consolidation pressure; however, the changes in (k_c/k_h) and (k_e/k_h) are really a result of the decrease in void ratio that accompany the increase in pressure as may be seen from Fig. 15.1c.

In systems containing confined clay layers acted on by chemical and/or electrical gradients, Darcy's law

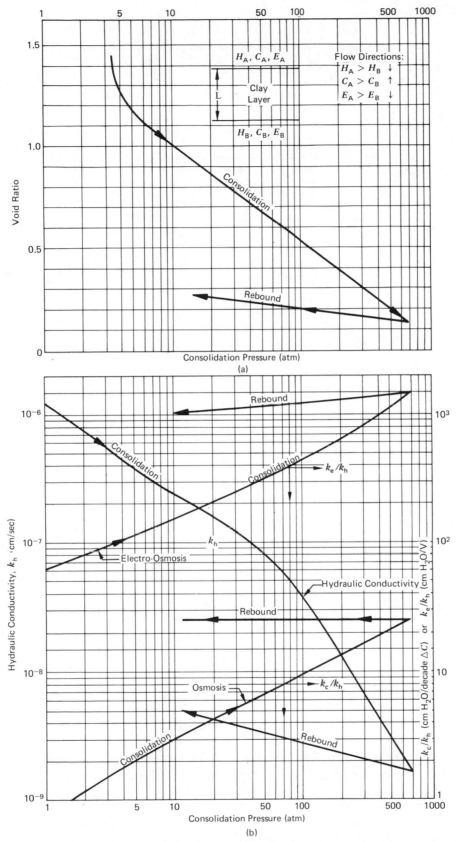

Fig. 15.1 Hydraulic, osmotic, and electro-osmotic conductivities of kaolinite (Data from Olsen, 1969, 1972). (*a*) Consolidation curve. (*b*) Conductivity values. (*c*) Conductivities as a function of void ratio.

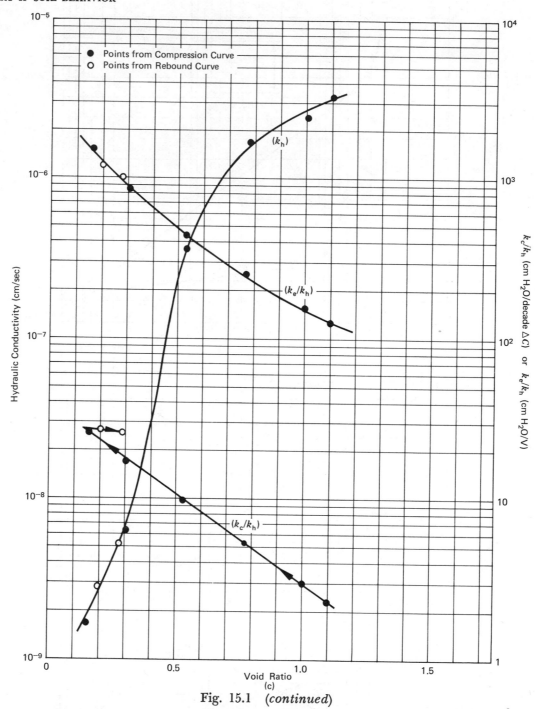

Fig. 15.1 (*continued*)

(equation (15.9)) may be an insufficient basis for prediction of hydraulic flow rates, particularly if the clay is highly plastic and at a very low void ratio. Such conditions can be found in deeply buried clays and clay shales. For very compressible active clays the ratios (k_c/k_h) and (k_e/k_h) may be sufficiently high to be useful for consolidation by electrical and chemical means (see Sections 15.7 and 15.9).

Heat flow. Heat transfer due to moving groundwater can be significant under conditions of high flow rate (permeable soil) and large temperature gradient. It may play an important role in thermal pollution related to power plants. It is known to affect seriously the feasibility of using frozen soil barriers as a temporary soil stabilization measure. The bulk of the heat transfer in soils under static water conditions is by straight thermal conduction, and the Peltier and Dufour effects are probably of little consequence.

Current flow. Electrical conduction is the main electrical transport mechanism in soils. The significance of thermo-electricity and diffusion and membrane potential flows (current flow due to a chemical gradient) is not known. A streaming current caused by ion transport under hydraulic gradient may be of importance in electrokinetic phenomena (Section 15.4).

Ion flow. Streaming currents caused by high groundwater flow between regions of different salt composition can be of concern in many problems of groundwater hydrology, water pollution, waste disposal, and so on. Electrophoresis may be important in clay suspensions, but not in dense or intact clays in nature. Capillary rise above the water table can act as an "ion pump" that feeds dissolved salts continuously to the top of the capillary fringe as water evaporates there. This results in a concentration of salts at the top and development of an osmotic gradient that attracts even more water to the top of the column (Gray, 1969).

Diffusion rates of ions in saturated soils are not precisely known, although measurements under controlled conditions have yielded values of diffusion coefficient of about 10^{-5} cm^2/sec. Diffusion can certainly be important over geologic time, and it may lead to significant ionic transport into or through a soil in time periods of more immediate concern.

Irreversible thermodynamics

Several interactions occur when a soil–water–electrolyte system is subjected to gradients of different types. The thermodynamics of irreversible processes, also termed nonequilibrium thermodynamics and thermodynamics of the steady state, is used extensively for the study of problems of this type. The formalism of irreversible thermodynamics allows for treatment of flows in terms of constituents, that is, cations, anions, and water molecules. As a result, account can be taken of diffusion and electrical phenomena in addition to liquid movement as a whole.

The formulation of irreversible thermodynamics is based on postulates relating to the rate of entropy generation, energy dissipation, and reciprocity of flow coefficients. Treatises by de Groote (1952), Katchalsky and Curran (1967), Kirkwood (1954), and others set forth the theory in detail. The relationship stated by equation (15.6) is a direct result of the theory. Furthermore, according to the Onsager reciprocity theorem (Onsager, 1931)

$$L_{ij} = L_{ji} \qquad (15.14)$$

Thus, behavior of one type may sometimes be predicted from knowledge of behavior of another type.

A major problem in the quantitative description of coupled flows is in the proper representation of coefficients and gradients. Irreversible thermodynamics provides a basis for derivation of the proper equations. The relationships are valid only if forces and fluxes are linearly related, that is, L_{ij} is a constant. This must be established in any case; it is not guaranteed by the theory. The results of several sets of careful experiments have established the applicability of the theory to a range of clay–water–electrolyte systems (Abd-El Aziz and Taylor, 1964; Gray and Mitchell, 1967; Letey and Kemper, 1969; and Olsen, 1962, 1969, 1972). Banin and Low (1971) were unable to confirm reciprocal coupling coefficients for Na-Wyoming bentonite, however, either because the theory was not applicable or the experiments were inadequate to test it satisfactorily. Thus, some question remains concerning the use of irreversible thermodynamics in soils.

In those soil systems where linearity between flows and gradients and reciprocity of coupling coefficients have been measured, extreme care was required in the preservation of constancy of composition and fabric during measurements. In many situations a truly steady state does not persist; seepage forces cause changes in effective stress and, in some instances, migration of particles and electrochemical effects can result from passage of an electrical current. Ion exchange and adsorption phenomena may accompany the flow of both organic and inorganic solutes into a fine-grained soil.

15.3 HYDRAULIC CONDUCTIVITY

Darcy's law* (Darcy, 1856) states that there exists a direct proportionality between flow velocity v_h or flow rate q_h and hydraulic gradient i_h

$$v_h = k_h i_h \qquad (15.15)$$

$$q_h = k_h i_h A \qquad (15.16)$$

---

* This so-called "law" was established empirically by Darcy based on the results of some flow tests through sands. Its general validity for the description of hydraulic flow through sand has been verified by many subsequent studies.

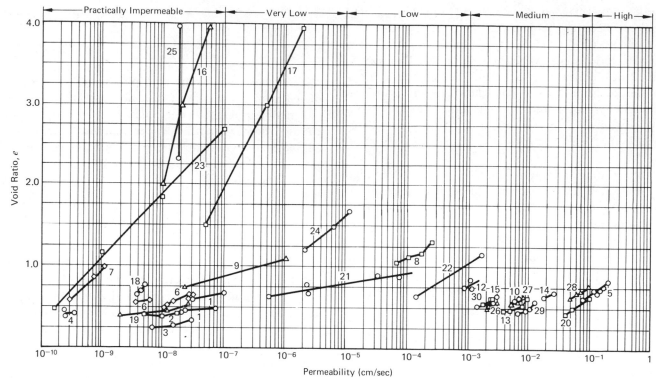

Fig. 15.2 Hydraulic conductivity values for several soils. (Lambe and Whitman, 1969). Soil identification code: 1, Compacted caliche; 2, Compacted caliche; 3, Silty sand; 4, Sandy clay; 5, Beach sand; 6, Compacted Boston blue clay; 7, Vicksburg buckshot clay; 8, Sandy clay; 9, Silt—Boston; 10, Ottawa sand; 11, Sand—Gaspee Point; 12, Sand—Franklin Falls; 13, Sand—Scituate; 14, Sand—Plum Island; 15, Sand—Fort Peck; 16, Silt—Boston; 17, Silt—Boston; 18, Loess; 19, Lean clay; 20, Sand—Union Falls; 21, Silt—North Carolina; 22, Sand from dike; 23, Sodium—Boston blue clay; 24, Calcium kaolinite; 25, Sodium montmorillonite; 26–30, Sand (dam filter).

where A is the area of cross section normal to the direction of flow. The constant k_h, usually termed the coefficient of permeability by geotechnical engineers and the hydraulic conductivity by others concerned with flow through porous media,† is a basic property of the material. Virtually all steady-state and transient flow analyses in soils are based on Darcy's law. In many instances, more attention is directed at the analysis than at the value of k. This is unfortunate, since no other property of importance in geotechnical problems is likely to exhibit the great range of values (up to ten orders of magnitude) among coarse and very fine-grained soils or show the variability in a given deposit as does the hydraulic conductivity. A given soil may exhibit manyfold variations in hydraulic conductivity as a result of changes in fabric, void ratio, and water content.

† The term hydraulic conductivity is used primarily herein to provide a distinction from the absolute permeability, discussed below, by analogy with thermal and electrical conductivity, and because it is becoming more widely used in geotechnical work.

Figure 15.2 shows values of hydraulic conductivity as a function of void ratio for several soils. Different units for hydraulic conductivity are often used by different groups or agencies, for example, centimeters per second by geotechnical engineers; feet per year by groundwater hydrologists, and Darcys by petroleum technologists. Figure 15.3 can be used to convert from one system to another.

Theoretical equations for permeability

For soils finer than coarse gravel flow is laminar. Equations are available to relate the hydraulic conductivity or permeability to properties of the soil and permeant. The usual starting point for derivation of these relationships is Poiseiulle's law for flow through a round capillary, which gives the average flow velocity v_{ave}, according to

$$v_{ave} = \frac{\gamma_p R^2}{8\mu} i_h \qquad (15.17)$$

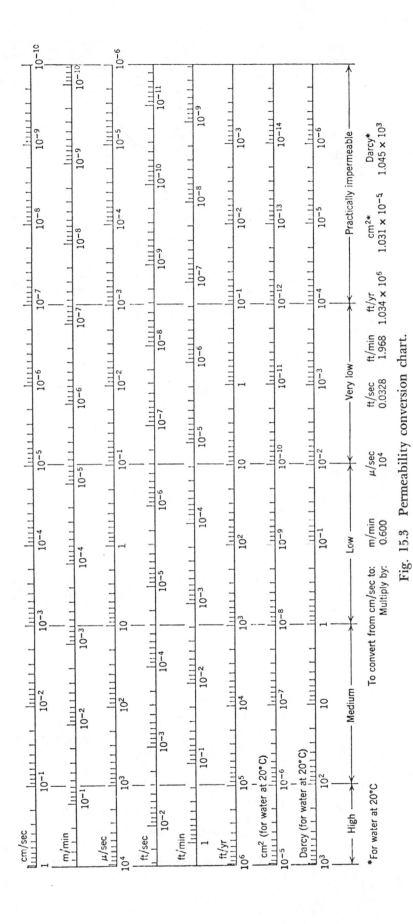

Fig. 15.3 Permeability conversion chart.

where μ is viscosity, R is tube radius, and γ_p is unit weight of the permeant. Because the flow channels in a soil are of various shapes and sizes, a characteristic dimension is needed to describe average size. The hydraulic radius

$$R_H = \frac{\text{flow channel cross section area}}{\text{wetted perimeter}}$$

is useful.

For a circular tube flowing full

$$R_H = \frac{\pi R^2}{2\pi R} = \frac{R}{2}$$

so Poiseiulle's equation becomes

$$q_{circle} = \frac{1}{2}\frac{\gamma_p}{\mu} R_H^2 \, i_h \, a \qquad (15.18)$$

where a is the cross-sectional area of the tube. For other shapes of cross section, an equation of the same form will hold, differing only in the value of a shape coefficient C_s, so that

$$q = C_s \frac{\gamma_p R_H^2}{\mu} \, i_h \, a \qquad (15.19)$$

For a bundle of parallel tubes of constant but irregular cross section contributing to a total cross sectional area, A (solids plus voids), the area of flow passages A_f filled with water is

$$A_f = SnA \qquad (15.20)$$

where S is the degree of saturation and n is the porosity, as shown in Fig. 15.4. For this condition, the hydraulic radius is given by

$$R_H = \frac{A_f}{P} + \frac{A_f L}{PL} = \frac{\text{volume available for flow}}{\text{wetted area}}$$

$$= \frac{V_{water}}{S_0} \qquad (15.21)$$

where P is the wetted perimeter, L is the length of flow channel in the direction of flow, and S_0 is the specific surface per unit volume of particles. For void ratio e and volume of solids V_S, the volume of water is

$$V_w = eV_s S \qquad (15.22)$$

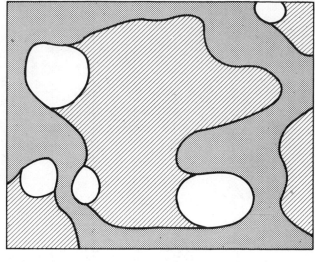

☐ Air Bubble	Total Cross Section = A
▨ Solids	Degree of Saturation = S
▨ Water	Porosity = n
	Area of Water = SnA

Fig. 15.4   Cross section in plane normal to direction of seepage.

Thus, equation (15.19) becomes

$$q = C_s\left(\frac{\gamma_p}{\mu}\right) R_H^2 \, Sni_h A = C_s\left(\frac{\gamma_p}{\mu}\right) R_H^2 S\left(\frac{e}{1+e}\right) i_h A \qquad (15.23)$$

and substitution for R_H gives

$$q = C_s V_s^2 \left(\frac{1}{S_0^2}\right)\left(\frac{\gamma_p}{\mu}\right) S^3 \left(\frac{e^3}{1+e}\right) i_h A \qquad (15.24)$$

By analogy with Darcy's law,

$$k_h = C_s V_s^2 \left(\frac{\gamma_p}{\mu}\right) \frac{1}{S_0^2}\left(\frac{e^3}{1+e}\right) S^3 \qquad (15.25)$$

For the case of full saturation, $V_s = 1$, and $C_s = 1/k_0 T^2$ where k_0 is a pore shape factor and T is a tortuosity factor, equation (15.25) becomes

$$K = k_h \left(\frac{\mu}{\gamma_p}\right) = \frac{1}{k_0 T^2 S_0^2}\left(\frac{e^3}{1+e}\right) \qquad (15.26)$$

This is the well-known Kozeny–Carman equation for the permeability of porous media (Kozeny, 1927; Carman, 1956). In principle, the effects of permeant characteristics are accounted for by the μ/γ_p term.

The pore shape factor k_0 has a value of about 2.5, and the tortuosity factor has a value of about $\sqrt{2}$ in porous media of approximately uniform pore sizes. Whereas the hydraulic conductivity k_h has units of velocity (LT^{-1}), the absolute permeability K has units of area (L^2).

Although the Kozeny–Carman equation works well for the description of permeability in uniformly graded sands and some silts, serious discrepancies are found in clays. Reasons for these discrepancies are discussed in some detail in Section 12.7. The major factor responsible for failure of the equation in fine-grained soils is that the fabrics of such materials do not contain uniform pore sizes (Olsen, 1962). Particles are grouped in clusters or aggregates that result in large intercluster pores and small intracluster pores.

Provided comparisons are made using samples of the same fabric, the influence of permeant on hydraulic conductivity is accounted for by the (γ_p/μ) term. If a clay is molded in different permeants, then fabrics may be quite different, and the permeability will deviate greatly from the predicted value (Michaels and Lin, 1954).

If in equation (15.25) C_s is taken to be a composite shape factor and V_s^2/S_0^2 is interpreted as a representative grain size D_s, then

$$k_h = CD_s^2 \left(\frac{\gamma_w}{\mu}\right)\frac{e^3}{1 + e} S^3 \qquad (15.27)$$

Like the Kozeny–Carman equation, equation (15.27) describes the behavior of cohesionless soils reasonably well, but is inadequate for clays. For a uniform sand with bulky particles and a given permeant, equations (15.25) and (15.27) predict that k_h should vary directly with $e^3/(1 + e)$ and D_s^2, and experimental observations support this.

Despite the inability of equations such as (15.26) and (15.27) to predict the permeability accurately in many cases, they do reflect the overwhelming importance of pore size. Flow velocity depends on the square of pore radius, and, hence, the flow rate depends on radius to the fourth power. The specific surface term in the Kozeny–Carman equation and the representative grain size term in equation (15.27) are both measures of pore size. The hydraulic conductivity depends more on the small particles in a soil than on the large. A small percentage of fines can clog the pores of an otherwise coarse material and result in manyfold lower hydraulic conductivity.

Equation (15.27) predicts that the hydraulic conductivity should vary directly with the cube of the degree of saturation, and experimental data support this, even in the case of fine-grained soils, as may be seen in Fig. 15.5. The values shown in Fig. 15.5 were obtained under conditions of constant sample volume.

The validity of Darcy's law

As early as 1898, instances in which hydraulic flow velocity increased more than proportionally with gradient were cited (King, 1898). The absence of water flow at finite hydraulic gradients in ceramic filters of 0.1 μm average pore diameter was reported by Derjaguin and Krylov (1944), and Oakes (1960) found no detectable flow through a 30-cm long suspension of 6 percent Wyoming bentonite subjected to a 50-cm head of water. Experiments by Miller and Low (1963) led to the conclusion that there was a threshold gradient for flow in sodium montmorillonite.

Flow rates through clay-bearing sandstones were found to increase more than directly with gradient up to gradients of 170 (von Englehardt and Tunn, 1955). Deviations from Darcy's law in pure and natural clays up to gradients of 900 were measured by Lutz and Kemper (1959). Apparent deviations from Darcy's law for flow in an undisturbed soft clay are shown in Fig. 12.36 (Hansbo, 1960, 1973).

As a result of these types of observations, a number of proposals (Hansbo, 1960; Swartzendruber, 1962a, 1962b, 1963) have been made for alternative relationships to Darcy's law to describe the relationship between flow velocity and gradient. Others (Florin, 1951; Roza and Kotov, 1958; Hansbo, 1960; Barden and Berry, 1965; Mitchell and Younger, 1967; Schmidt and Westmann, 1973) have studied the implications of a threshold gradient and nonlinearity on seepage and consolidation.

Two concepts have been used to account for the nonlinearity between flow velocity and gradient: non-Newtonian water flow properties and particle migrations that cause blocking and unblocking of flow passages. The apparent existence of a threshold gradient below which flow was not detected has been attributed to a quasicrystalline water structure. It is now known that many of the effects interpreted as due to unusual water properties can be ascribed to undetected experimental errors arising from contamination of measuring systems (Olsen, 1965), local consolidation and swelling effects caused by varying effective stresses in the direction of flow, and bacterial growth

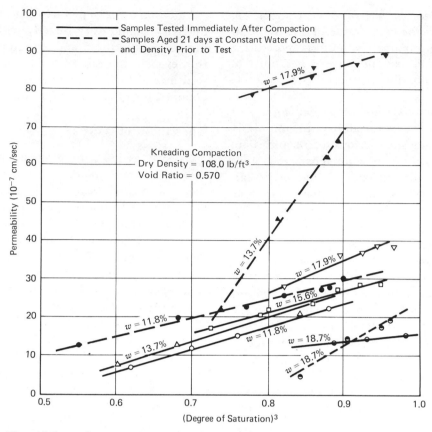

Fig. 15.5   Influence of degree of saturation on hydraulic conductivity
of compacted clay.

(Gupta and Swartzendruber, 1962). Further careful measurements have failed to confirm the existence of a threshold gradient in clays (Olsen, 1965, 1969; Gray and Mitchell, 1967; Mitchell and Younger, 1967; Miller, Overman, and Peverly, 1969; Chan and Kenney, 1973). Darcy's law was obeyed exactly in the studies of Olsen (1965, 1969), Gray and Mitchell (1967), and Chan and Kenney (1973). Thus, it seems unlikely that unusual water properties can lead to non-Darcy flow behavior in clays.

On the other hand, particle migrations leading to void plugging and unplugging, electrokinetic effects, and chemical concentration gradient effects may cause apparent deviations from Darcy's law. The behavior shown in Fig. 12.36 was attributed to particle migrations under the action of seepage forces. Analysis of interparticle bond strengths in relation to the magnitudes of seepage forces leads to the conclusion that only particles not participating in the load-carrying skeleton of a soil mass are likely to be moved under moderate values of hydraulic gradient. Soils with

open, flocculated fabrics and soils with a relatively low content of clay appear particularly susceptible to the movement of fine particles during permeation. Tests on saturated samples of compacted silty clay (Mitchell and Younger, 1967) indicated that particle migrations, as reflected by changes in pore pressure distribution with time in the direction of flow, were sensitive to compacted density and water content. The largest effects were found in samples compacted at the lowest water contents and densities. Variations in the fines content in a single coarse grading can influence the gradient-flow rate response significantly (Younger and Lim, 1969).

Internal swelling and dispersion of clay particles during permeation can be responsible for changes in flow rate and apparent non-Darcy behavior (Hardcastle and Mitchell, 1974). From tests on illite–silt mixtures, it was learned that the hydraulic conductivity depends on clay content, sedimentation procedure, compression rate, and electrolyte concentration. Subsequent behavior is quite sensitive to the type and

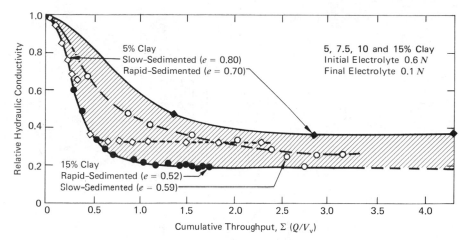

Fig. 15.6 Reduction in hydraulic conductivity as a result of internal swelling.

concentration of electrolyte used for permeation and the total throughput volume of permeant. Figure 15.6 illustrates changes in relative hydraulic conductivity that occurred while the electrolyte concentration was changed from 0.6 to 0.1 N NaCl. The cumulative throughput shown is the ratio of total flow volume at any time to sample pore volume. The hydraulic conductivities for these materials ranged from more than 1×10^{-5} to less than 1×10^{-7} cm/sec.

Practical implications. All other factors held constant, Darcy's law is valid. The only evidence to support the existence of non-Newtonian water properties or threshold gradients that cannot be accounted for by alternative explanations is that of Miller and Low (1963). On the other hand, particle migrations and fabric changes may result from seepage forces and internal swelling and dispersion brought about by changes in pore solution chemistry. Unless these effects can be shown to be of little importance, measurements should be made under conditions as closely approximating those in the field as possible.

Gradients used in laboratory determinations of hydraulic conductivity and consolidation behavior are higher than is the case for the field conditions under study. Figure 15.7 illustrates the variation of hydraulic gradient i with time factor T during consolidation according to the Terzaghi theory. The solution of the Terzaghi equation gives pore pressure as a function of position (z/H) and time factor.

$$u = \sum_{m=0}^{\infty} \frac{2u_0}{M}\left(\sin \frac{Mz}{H}\right) e^{-M^2T} \quad (15.28)$$

where $M = \pi(2m + 1)/2$

Thus, the hydraulic gradient is

$$i = \frac{\partial}{\partial z}\left(\frac{u}{\gamma_w}\right) = \frac{2u_0}{\gamma_w H} \sum_{m=0}^{\infty} \cos\left(\frac{Mz}{H}\right) e^{-M^2T} \quad (15.29)$$

If a parameter p is defined by

$$p = 2 \sum_{m=0}^{\infty} \cos\left(\frac{Mz}{H}\right) e^{-M^2T} \quad (15.30)$$

equation (15.29) becomes

$$i = \frac{u_0}{\gamma_w H} p \quad (15.31)$$

The real gradient for any layer thickness or loading intensity can be obtained by using the actual values of u_0 and H and the appropriate value of p from Fig. 15.7.

For low values of $u_0/\gamma_w H$ such as would exist in the field, for example, for $u_0 = 0.5$ kg/cm², $H = 5$ m, then $u_0/\gamma_w H = 1$, the field gradients are low throughout most of the layer thickness during the entire consolidation process. On the other hand, for a laboratory sample of 1 cm thickness and the same stress increase, $u_0/\gamma_w H$ is 500, and gradients are very large. A gradient-dependent hydraulic conductivity in such a case could be the cause of substantial discrepancies between laboratory-measured and field values of coefficient of consolidation. Constant rate of strain or constant gradient consolidation testing of such soils would be preferable to the use of load increments, as done in conventional tests, because the lower gradients would minimize particle migration effects.

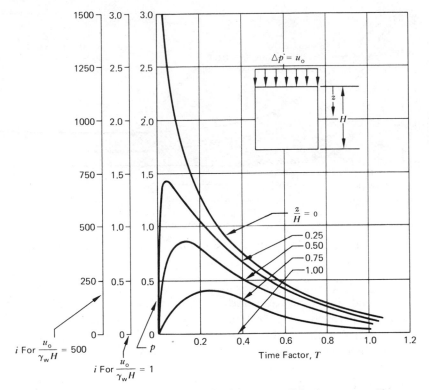

Fig. 15.7 Hydraulic gradients during consolidation according to the Terzaghi theory.

Anisotropy

The development of anisotropic permeability as a result of fabric anisotropy is discussed in Section 12.4. Because of the preferred orientations of elongated or platy particles, ratios of up to 7 in the hydraulic conductivity values in orthogonal directions have been measured, with a typical average of about 2.

In addition to this, gross anisotropy may exist in layered deposits, as for example in the case of a clay deposit with closely spaced, horizontal sand seams, or in a fill compacted in layers. The importance of a high permeability in a horizontal direction depends on the distance to a drainage surface and the type of flow induced. Groundwater flow will clearly be affected, as will the rate of consolidation in a sand drain installation. On the other hand, lateral drainage beneath a newly loaded area may not be greatly influenced by a high ratio of horizontal to vertical hydraulic conductivity if the width of loaded area is large compared to the thickness of drainage layer.

Until recently, varved clays had been thought to have ratios r_k between the horizontal and vertical hydraulic conductivities of 10 to more than 100. Values of r_k of 10 ± 5 for Connecticut Valley varved

clay were determined in the laboratory (Ladd and Wissa, 1970). Similar values were obtained[*] for the varved clay in the New Jersey meadows.

Extensive study of New Liskeard, Ontario, varved soil has been made in both the laboratory (Chan and Kenney, 1973) and in the field (Kenney and Chan, 1973). Both investigations yielded values of r_k less than 5, with smaller values predominating, as shown by Fig. 15.8. Coarse-grained lenses in the coarse layers and preferred particle orientation in the fine layers appear adequate to account for r_k values of 1.5 to 2.0, with the varved macrofabric responsible for the remainder.

For clays deposited at great distances from the points of sediment inflow r_k is likely to be low (Kenney and Chan, 1973). Such clays are considered to be composed of "distal" varves. Depositional conditions would be uniform, sand layers unlikely, and the variability of constituent soils small. On the other hand, "proximal" varves deposited short distances from points of sediment inflow would be less regular and sand layers more likely leading to higher values of r_k.

---

[*] Ladd, C. C. (1974), personal communication.

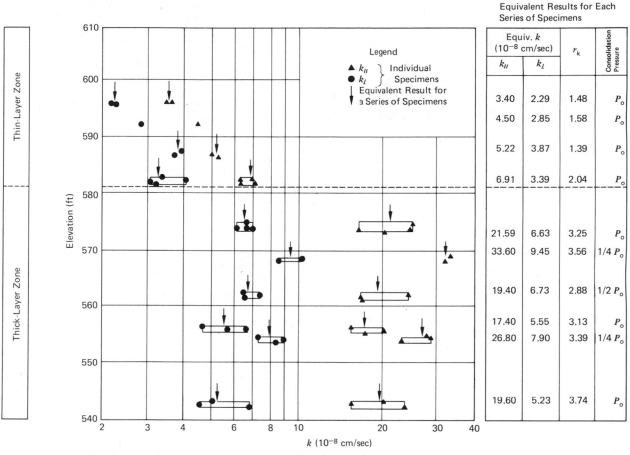

Fig. 15.8 Hydraulic conductivity parallel and perpendicular to varves in New Liskeard varved soil (Chan and Kenney, 1973).

15.4 ELECTROKINETIC PHENOMENA

Coupling between electrical and hydraulic flows and gradients can be responsible for four *electrokinetic phenomena* in systems such as the soil–water–electrolyte system where there are charged particles balanced by mobile counter charges. Each involves relative movements of electricity, charged surfaces, and liquid phases, as shown schematically in Fig. 15.9.

Electro-osmosis

If an electric potential is applied across a wet soil mass, cations are attracted to the cathode and anions to the anode. There is an excess of cations in the system to neutralize the net negative charge on the soil particles. As these cations migrate to the cathode, they drag water with them, causing water movement towards the cathode, as shown in Fig. 15.9a. The anions also drag water with them as they migrate to the anode. This flow is usually much less than the flow to

the cathode; therefore, a net flow to the cathode results.

Streaming potential

When water is caused to flow through a soil under a hydraulic head difference, Fig. 15.9b, double layer charges are displaced in the direction of flow. The result is an electrical potential difference proportional to the hydraulic flow rate, called the streaming potential, between the opposite ends of the soil mass. Streaming potentials of several tens of millivolts have been measured in clays using a high internal impedance volt meter.

Electrophoresis

If a direct current is applied across a colloidal suspension, charged particles are attracted electrostatically to one of the electrodes and repelled from the other. Negatively charged clay particles move towards the anode as shown in Fig. 15.9c. This is called elec-

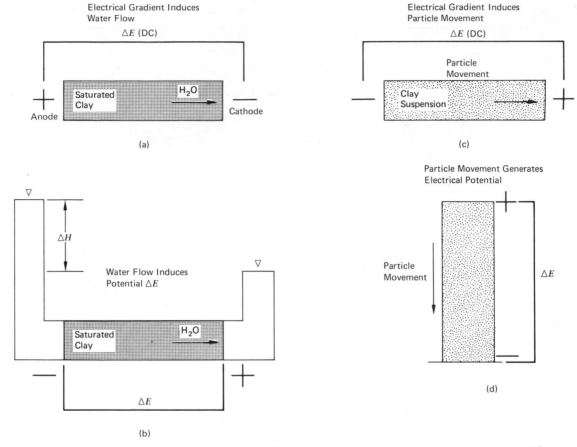

Fig. 15.9 Electrokinetic phenomena. (*a*) Electro-osmosis. (*b*) Streaming potential. (*c*) Electro-phoresis. (*d*) Migration or sedimentation potential.

trophoresis. Electrophoresis involves discrete particle transport through water; electro-osmosis involves water transport through a continuous soil particle network.

Migration or sedimentation potential

The movement of charged particles, for example, clay, relative to a solution, for example, during gravity settling, leads to generation of a potential as indicated in Fig. 15.9*d*. This is caused by the viscous drag of the water that retards the movement of the diffuse layer cations relative to that of the particles.

Of the four electrokinetic effects, electro-osmosis has been given the most attention in geotechnical engineering, because of its practical value for transport of water in fine-grained soils.

15.5 THEORIES FOR ELECTRO-OSMOSIS

Electro-osmosis is a coupled flow, and as such it can be described by equation (15.6). Using the terminol-

ogy of equation (15.7) and for the condition of no hydraulic, chemical, or thermal gradients

$$J_H = L_{HE}X_E \qquad (15.32)$$

In practice, this relationship is often given as

$$q = vA = k_e i_e A \qquad (15.33)$$

where q is the flow rate, v is the flow velocity, k_e is the coefficient of electro-osmotic permeability, i_e is the electrical potential gradient, and A is the total cross-sectional area normal to the direction of flow.

The coefficient of electro-osmotic permeability is a soil property that indicates the hydraulic flow velocity under a unit electrical gradient and is thus analogous to the hydraulic conductivity, which is the hydraulic flow velocity under a unit hydraulic gradient. Measurement of k_e can be made by determination of the flow rate of water through a sample of known length and cross section under a known electrical gradient. Alternatively, null indicating systems may be used

Table 15.3   Coefficient of Electro-Osmotic Permeability

No.	Material	Water Content (%)	k_e in 10^{-5} (cm^2/sec-V)	Approximate k_h (cm/sec)
1.	London clay	52.3	5.8	10^{-8}
2.	Boston blue clay	50.8	5.1	10^{-8}
3.	Kaolin	67.7	5.7	10^{-7}
4.	Clayey silt	31.7	5.0	10^{-6}
5.	Rock flour	27.2	4.5	10^{-7}
6.	Na-Montmorillonite	170	2.0	10^{-9}
7.	Na-Montmorillonite	2000	12.0	10^{-8}
8.	Mica powder	49.7	6.9	10^{-5}
9.	Fine sand	26.0	4.1	10^{-4}
10.	Quartz powder	23.5	4.3	10^{-4}
11.	Ås quick clay	31.0	2.0–2.5	2.0×10^{-8}
12.	Bootlegger Cove clay	30.0	2.4–5.0	2.0×10^{-8}
13.	Silty clay, West Branch Dam	32.0	3.0–6.0	1.2×10^{-8}–6.5×10^{-8}
14.	Clayey silt, Little Pic River, Ontario	26.0	1.5	2×10^{-5}

[a] k_e and water content data for Nos. 1 to 10 from Casagrande (1952). k_h estimated by author. No. 11 from Bjerrum et al. (1957). No. 12 from Long and George (1967). No. 13 from Fetzer (1967). No. 14 from Casagrande et al. (1961).

(Arnold, 1973), or it may be deduced from a streaming potential measurement (Section 15.6). From experience it is known that k_e is generally in the range of 1×10^{-5} to 10×10^{-5} cm^2/sec-V (cm/sec per V/cm) and that it is relatively independent of soil type but sensitive to pore water electrolyte concentration. Some typical values in fresh water permeant are listed in Table 15.3.

Several theories have been advanced to explain electro-osmosis and to provide a basis for quantitative prediction of flow rates.

Helmholtz–Smoluchowski theory

One of the earliest and still widely used theoretical descriptions of electrokinetic phenomena is based on a model introduced by Helmholtz (1879) and later refined by Smoluchowski (1914). A liquid-filled capillary is treated as an electrical condenser with charges of one sign on or near the surface of the wall and countercharges concentrated in a layer in the liquid a small distance from the wall, as shown in Fig. 15.10.* The mobile shell of counterions is assumed to drag water through the capillary by plug flow. There is a high velocity gradient between the two plates of the condenser as shown.

The rate of water flow is controlled by the balance between the electrical force causing movement and

*A derivation using a Poisson–Boltzmann distribution of counter ions adjacent to the wall leads to the same result.

friction between the liquid and the wall. The velocity gradient between the wall and center of positive charge is v/δ; thus, the drag force is $\eta \, dv/dx = \eta \, v/\delta$, where η is the viscosity. The force from the electric field is $\sigma \, \Delta E/\Delta L$, where σ is the surface charge density and $\Delta E/\Delta L$ is the electrical potential gradient. At equilibrium

$$\eta \frac{v}{\delta} = \sigma \frac{\Delta E}{\Delta L} \qquad (15.34)$$

or

$$\sigma\delta = \eta \frac{v}{\Delta E} \Delta L \qquad (15.35)$$

From electrostatics, the potential across a condenser $\mathcal{L}$ is given by

$$\mathcal{L} = \frac{4\pi\sigma\delta}{D} \qquad (15.36)$$

where D is the dielectric constant. Substitution for $\sigma\delta$ in (15.35) gives

$$v = \left(\frac{\mathcal{L}D}{4\pi\eta}\right)\frac{\Delta E}{\Delta L} \qquad (15.37)$$

The potential $\mathcal{L}$ is often termed the "zeta potential" in colloidal systems. It is not the same as the surface potential ψ_0 of the double layer, which is discussed in Chapter 7, although factors that give high values of ψ_0 also give high values of $\mathcal{L}$. A common interpreta-

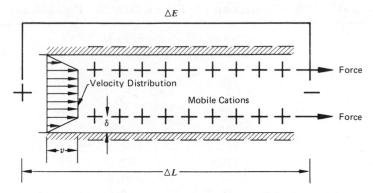

Fig. 15.10   Helmholtz–Smoluchowski model for electrokinetic phenomena.

tion is that the actual slip plane in electrokinetic processes is located some small, unknown distance from the surface of particles; thus, $\mathcal{L}$ should be less than ψ_0. Values of $\mathcal{L}$ in the range of 0 to -50 mV are typical for clays, with the lowest values associated with high pore water salt concentrations.

For a single capillary of area a the flow rate is

$$q_a = va = \frac{\mathcal{L}D}{4\pi\eta} \frac{\Delta E}{\Delta L} a \qquad (15.38)$$

and for a bundle of N capillaries within total cross section A normal to the flow direction,

$$q_A = Nq_a = \frac{\mathcal{L}D}{4\pi\eta} \frac{\Delta E}{\Delta L} Na \qquad (15.39)$$

If the porosity is n, then the cross-sectional area of voids is nA, which must equal Na. Thus,

$$q_A = \left(\frac{\mathcal{L}D}{4\pi\eta}\right) n \frac{\Delta E}{\Delta L} A \qquad (15.40)$$

By analogy with equation (15.33), $i_e = \Delta E/\Delta L$, and

$$k_e = \frac{\mathcal{L}D}{4\pi\eta} n \qquad (15.41)$$

It should be noted that k_e, according to the Helmholtz–Smoluchowski theory, should be independent of pore size. This contrasts with the hydraulic conductivity, k_h, which, as shown in Section 15.3, varies as the square of some effective size. Relative independence of k_e with pore (or particle) size is borne out by the values in Table 15.3, and it is because of this that electro-osmosis is useful in fine-grained soils.

This latter point is well illustrated by the following example. Consider a fine sand and a clay with values

of hydraulic conductivity k_h of 1×10^{-3} cm/sec and 1×10^{-8} cm/sec, respectively. Both have a k_e of 5×10^{-5} cm²/sec-V. For equal flow rates,

$$k_h i_h = k_e i_e$$

or

$$i_h = \frac{k_e}{k_h} i_e \qquad (15.42)$$

If an electrical potential gradient of 0.2 V/cm is used, then, for the fine sand

$$i_h = \frac{5 \times 10^{-5}}{1 \times 10^{-3}} \times 0.2 = 0.01$$

and for the clay

$$i_h = \frac{5 \times 10^{-5}}{1 \times 10^{-8}} \times 0.2 = 1000$$

Thus, for the sand a hydraulic gradient of only 0.01 can move water as effectively as an electrical gradient of 0.2 V/cm, but for the clay, a hydraulic gradient of 1000 would be needed to offset the electro-osmotic flow.

It does not follow, however, that electro-osmosis will always necessarily be an efficient means to move water in clays, since the above example says nothing about the power expenditure needed to develop the desired electrical gradient in the soil or about energy losses at electrode-fluid interfaces.

Schmid theory

The Helmholtz–Smoluchowski theory, by its assumption of a negligible extension of the counterion layer into the pore, is essentially a large pore theory. It also fails to account for an excess of ions over those

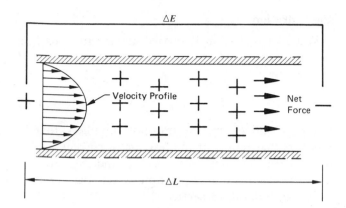

Fig. 15.11 Schmid model for electro-osmosis.

needed to balance exactly the surface charge. Schmid (1950, 1951) proposed a kinetic model that overcomes the first of these two difficulties and, thus, can be considered a small-pore theory.

The counterions are assumed uniformly distributed throughout the entire liquid volume. The electrical force acts uniformly over the entire pore cross section and gives the velocity profile shown by Fig. 15.11. From equation (15.17), the Poisieulle equation for flow rate in a single capillary is

$$q = \frac{\pi r^4}{8\eta} \gamma_w i_h \qquad (15.43)$$

The hydraulic force per unit length causing flow can be isolated as

$$F_H = \pi r^2 \gamma_w i_h \qquad (15.44)$$

so

$$q = \frac{r^2}{8\eta} (F_H) \qquad (15.45)$$

The electrical force per unit length F_E is equal to the charge times the potential; that is,

$$F_E = A_0 F_0 \pi r^2 \frac{\Delta E}{\Delta L} \qquad (15.46)$$

where A_0 is concentration of wall charges in ionic equivalents per unit volume of pore fluid, and F_0 is the Faraday constant. Replacement of F_H by F_E in equation (15.45) gives

$$q_a = \frac{\pi r^4}{8\eta} A_0 F_0 \frac{\Delta E}{\Delta L} = \frac{F_0 A_0}{8\eta} r^2 i_e a \qquad (15.47)$$

and for a total cross section of N capillaries and area A

$$q_A = \frac{A_0 F_0 r^2}{8\eta} n \, i_e A \qquad (15.48)$$

Equation (15.48) shows that k_e should vary as r^2; whereas, the Helmholtz–Smoluchowski theory leads to k_e independent of pore size as previously noted. Of the two theories, the Helmholtz gives the better result in soils. Perhaps this is because most clays have a cluster or aggregate structure (see Chapters 11 and 12), with flow controlled by large pores.

Spiegler friction model

A completely different concept for electrokinetic processes takes into account the interactions of the mobile components (water and ions) on each other and the frictional interactions of these components with the pore walls (Spiegler, 1958). No particular structure is specified for the porous medium. The assumptions used include:

1. Exclusion of co-ions; that is, the medium behaves as a perfect permselective membrane, admitting ions of only one sign.*
2. Complete dissociation of pore fluid ions.

On this basis, the following equation for electro-osmotic transport of water across an ion exchange membrane can be derived

$$\Omega = (W - H) = \frac{C_3}{C_1 + C_3(X_{34}/X_{13})} \qquad (15.49)$$

in which Ω is the true electro-osmotic flow (moles/faraday), W is the measured water transport (moles/faraday), H is the ion hydration (moles/faraday), C_3 is the concentration of free water in membrane (moles/cm^3), C_1 is the concentration of mobile counter ions in membrane (moles/cm^3), X_{34} is the friction coefficient between water and solid wall, and X_{13} is the friction coefficient between cation and water.

Concentrations C_3 and C_1 are hypothetical and probably are less than values measured by chemical analysis, because some ions may be immobile within the membrane. Evaluation of X_{13} and X_{34} requires independent measurements of diffusion coefficients, conductance, transference numbers, and water transport. Thus, equation (15.49) is limited in its usefulness as a predictive equation. Its real value lies in providing a relatively simple physical picture for a complex process.

---

* Ions of the opposite sign as the charged surface are termed counterions. Ions of the same sign are termed co-ions.

From equation (15.49),

$$\Omega = (W - H) = \frac{1}{(C_1/C_3 + X_{34}/X_{13})} \qquad (15.50)$$

At high water contents and for large pores

$$\frac{X_{34}}{X_{13}} \to 0$$

as the importance of X_{34} becomes negligible. Then

$$\Omega_{X_{34} \to 0} = \frac{C_3}{C_1} \qquad (15.51)$$

Thus, a high water to cation ratio implies a high rate of electro-osmotic flow. At low water contents and for small pores, X_{34} will not be zero, thus reducing the flow. An increase in C_1 leads to a decrease in flow per faraday of current passed, because there will be less water per ion. Any increase in X_{13} increases flow due to greater frictional drag in the water by the ions.

Ion hydration

Water of hydration is carried along with ions in a direct current field. The ion hydration transport H is given by

$$H = t_+ N_+ - t_- N_- \qquad (15.52)$$

where t_+ and t_- are the transport numbers, that is, numbers that represent the fraction of current carried by a particular ionic species. N_+ and N_- are the number of moles of hydration water per mole of cation and anion, respectively.

Summary

The four theories are compared in Table 15.4 in terms of their premises, consequences, and limitations. None of the theories appears to have sufficient generality to account for the full range of electrokinetic behavior exhibited by clay–water–electrolyte systems. Concepts from these theories are nonetheless useful

Table 15.4   Summary of Premises and Consequences of Existing Theories for Electro-Osmosis

	Premises	Consequences
1. Helmholtz–Smoluchowski $$k_e = \frac{\mathcal{L}D}{4\pi\eta} n$$	*a.* Counterions concentrated in double layer at solid–liquid interface *b.* DL thickness negligible with respect to pore diameter *c.* Water migrates at same velocity as ions	*a.* Applicable to system with large pores ($d > 1000$ Å) and dilute solutions (concentration < 0.1% by wt) *b.* Independent of pore size—depends on porosity (direct proportion)
2. Schmid theory $$k_e = \frac{A_0 F_0 r^2}{8\eta} n$$ $$A_0 = f(n)$$	*a.* Counterions uniformly distributed in pores *b.* Electrical force $\propto$ charge density and uniform across pore	*a.* $d < 1000$ Å *b.* Unrealistic in high salinity systems *c.* Sensitive function of average pore size *d.* Varies with power of porosity
3. Spiegler friction model $$\Omega = (W - H) = \frac{C_3}{C_1 + C_3(X_{34}/X_{13})}$$	*a.* Function of concentration of water and mobile counterions in a membrane and two friction coefficients *b.* Permselective membrane *c.* Ions completely dissociated	*a.* Increasing water content or decreasing concentration of counter ion leads to increase in water transport
4. Ion hydration $$H = t_+ N_+ - t_- N_-$$	*a.* Water transported as primary water of hydration of migrating ions	*a.* Greater the dissimilarity in $+$ and $-$ transport numbers the greater the water transport

for development of an understanding of the physics of electro-osmosis and for the analysis of some problems.

15.6 PREDICTION OF ELECTRO-OSMOSIS EFFICIENCY

The efficiency and economics of electro-osmosis for use in soils depends on the water transported per unit charge passed, for example, cubic meters per hour per ampere. This quantity may vary over several orders of magnitude depending on such factors as soil type, water content, and electrolyte concentration. A rational basis for examining the influences of these factors can be developed in the following way, without recourse to a specific kinetic model (Gray, 1966, Gray and Mitchell, 1967).

Electro-osmosis occurs when the frictional drag between the ions of one sign and the surrounding water molecules exceeds that caused by ions of the opposite sign.

In fine-grained soils, the adsorbed cations (counterions) provide an excess of cations over anions in the system, and there is a net transfer of water in the direction of the counterion movement. The rate of water movement depends on the applied electric field, the flow resistance of the soil, and the frictional drag of the ions on the water molecules. Thus, of fundamental importance are both the cation–anion distribution, and the water–ion distribution.

Because of the electronegativity of the clay particles, the concentration of cations in the double layer is higher and that of the anions is lower than that in the free external solution, leading to distributions similar to those shown in Fig. 7.1 for a single double layer and Fig. 7.6 for interacting double layers. The cation–anion distribution can be described by means of the double layer theory. Alternatively, the Donnan theory can be used to give the equilibrium distribution (Donnan, 1924). The greater the difference between the concentrations of cations and anions, the greater the net drag on the water in the direction of the cathode during electro-osmosis.

The basis for the Donnan theory is that at equilibrium the potentials of the internal and external solutions are equal and that electroneutrality is required in both phases. It may be shown (Gray, 1966; Gray and Mitchell, 1967) that the ratio of cations to anions in the internal phase R for the case of a symmetrical ($z^+ = z^-$) electrolyte is given by

$$R = \frac{C^+}{C^-} = \frac{1 + (1 + y^2)^{1/2}}{-1 + (1 + y^2)^{1/2}} \quad (15.53)$$

where

$$y = \frac{2C_0 \gamma \pm}{A_0 \bar{\gamma} \pm} \quad (15.54)$$

C_0 is the concentration in external solution, γ is the mean molar activity coefficient in external solution, $\bar{\gamma}$ is the mean molar activity coefficient in double layer, and A_0 is the surface charge density per unit pore volume. The parameter A_0 is related to the cation exchange capacity by

$$A_0 = \frac{(cec)\, \rho_w}{w} \times 100 \quad (15.55)$$

where cec is in meq/g, ρ_w is the density of water (g/cm³), and w is the water content in percent. The higher R is, the greater the electro-osmotic transport is, all other factors being equal.

Equation (15.53) shows that exclusion of anions is favored by a high exchange capacity (active clay), a low water content, and low salinity in the external solution. In inactive clays at high water content, the concentration of anions in the double layer builds up rapidly as the salinity of the external solution increases.

High electro-osmotic efficiency also requires that the water to cation ratio be high, as shown by Spiegler's friction model; that is,

$$W \propto \frac{C_{H_2O}}{C^+}$$

In a high exchange capacity, low water content system there is less water per cation than in a low exchange capacity, high water content system. This effect combined with the cation–anion ratio considerations developed above leads to the water transport–water content–soil type–electrolyte concentration prediction shown schematically in Fig. 15.12. Tests on sodium kaolinite (inactive clay) and sodium illite (more active clay) gave the results shown in Fig. 15.13, which agree with the predictions of Fig. 15.12.

The results of electro-osmosis measurements on a number of materials are summarized in Fig. 15.14, which shows water flow as a function of water content. This figure may be used as a guide for prediction of electro-osmotic flow rates in other soils as well. The flow rates shown are for open systems, with solution admitted at the anode at the same time it is ex-

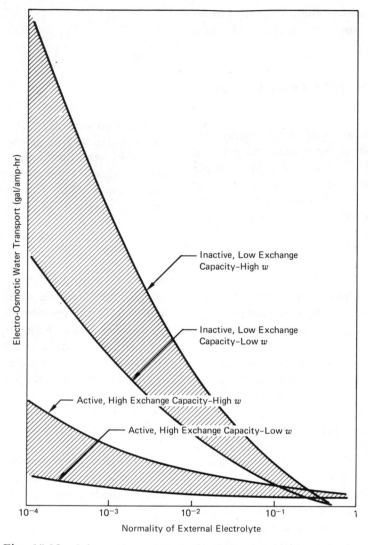

Fig. 15.12   Schematic prediction of electro-osmosis in various clays according to the Donnan concept.

tracted from the cathode. Electrochemical effects (Section 15.8) and water content changes were minimized for the tests used to obtain the data shown. Thus, the values can be interpreted as upper bounds on the flow rates.

To estimate the efficiency of electro-osmosis in any given case, values of the water content, electrolyte concentration of the pore water, and type of clay are required. The former is readily measured, the electrolyte concentration is easily determined using a conductivity cell, and the latter can be estimated on the basis of plasticity and grain size data if mineralogical data are not available. Based on these considerations, electro-osmotic flow rates of 0.03 to 0.06 gal/hr/amp are predicted using Fig. 15.14 for soils 11, 13, and 14 in Table 15.3. Electrical treatment for stabilization

purposes was effective in the field for these soils. For soil 12, a flow rate of only 0.008 to 0.012 gal/hr/amp was predicted, electro-osmosis was not effective in the field.

Saxen's law—predictions of electro-osmosis based on streaming potential

Streaming potentials can be measured directly during a measurement of hydraulic conductivity by using a high ampedence voltmeter and reversible electrodes. An equivalence between streaming potential and electro-osmosis may be derived as follows. Expansion of equation (15.6) for coupled hydraulic and current flows gives

$$q_\mathrm{h} = L_\mathrm{HH}\Delta P + L_\mathrm{HE}\Delta E \qquad (15.56)$$

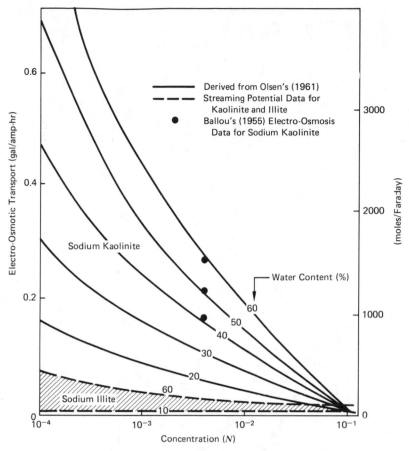

Fig. 15.13 Electro-osmotic water transport vs. concentration of external electrolyte solution for homoionic kaolinite and illite at various water contents.

where q_h is the hydraulic flow rate, ΔP is the pressure drop, and ΔE is the electrical potential drop, and

$$I = L_{EH}\Delta P + L_{EE}\Delta E \qquad (15.57)$$

where I is the electrical current. In a hydraulic conductivity measurement, there is no electrical current flow, so $I = 0$ and ΔE is the streaming potential; equation (15.57) then becomes $\Delta E/\Delta P = - L_{EH}/L_{EE}$. In electro-osmosis, $\Delta P = 0$, so (15.56) is $q_h = L_{HE}\Delta E$ and (15.57) becomes $I = L_{EE}\Delta E$ so $q_h/I = L_{HE}/L_{EE}$. By Onsager's reciprocity theorem

$$L_{EH} = L_{HE} \qquad (15.58)$$

so

$$\left(\frac{q_h}{I}\right)_{\Delta P=0} = -\left(\frac{\Delta E}{\Delta P}\right)_{I=0} \qquad (15.59)$$

This equivalence between streaming potential and electro-osmosis was first shown experimentally by Saxen (1892) and is known as Saxen's law. It has been verified subsequently for a number of clay–water–electrolyte systems by Olsen (1961) and Gray (1966), among others.

Analysis of the units in equation (15.59) shows that the electro-osmotic flow rate in gallons per hour per ampere is equal to 0.0094 times the streaming potential in millivolts per atmosphere.

Energy requirements

The preceding analysis leads to a prediction of amount of water transferred per unit charge passed, that is, gallons or cubic meters per hour per ampere or moles per Faraday. If this quantity is denoted by k_i, then

$$q_h = k_i I \qquad (15.60)$$

Unlike k_e, k_i varies over a wide range, as may be seen in Fig. 15.14. The power P consumed is given by

$$P = \Delta EI = \frac{\Delta E q_h}{k_i}\ W \qquad (15.61)$$

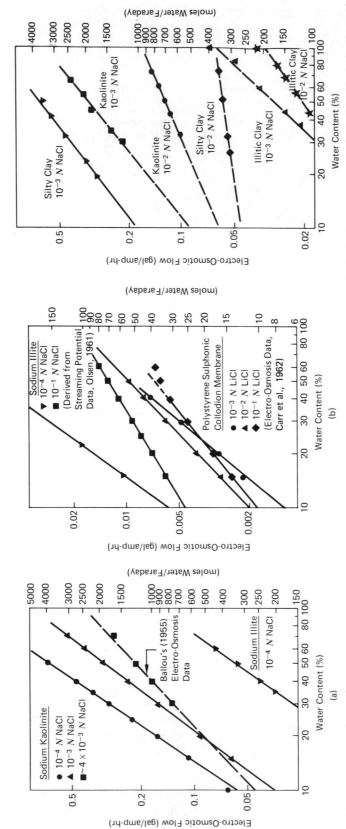

Fig. 15.14 Electro-osmotic water transport as a function of water content, soil type, and electrolyte concentration. (*a*) Homoionic kaolinite and illite. (*b*) Illitic clay and collodian membrane. (*c*) Silty clay, illitic clay, and kaolinite.

for ΔE in volts and I in amperes. The power consumption per unit volume of flow is

$$\frac{P}{q_h} = \frac{\Delta E}{k_i} \times 10^{-3} \text{ kwh} \qquad (15.62)$$

Relationship between k_e and k_i

The electro-osmotic flow rate is given by

$$q_h = K_1 k_i I = K_2 k_e \frac{\Delta E}{\Delta L} A \qquad (15.63)$$

where K_1 and K_2 are unit constants. Thus,

$$k_i = \left(\frac{K_2}{K_1}\right)\left(\frac{A\Delta E}{\Delta L I}\right) k_e \qquad (15.64)$$

Because $\Delta E/I$ is resistance and $\Delta L/\text{resistance} \times A$ is specific conductivity, σ (mhos/cm), equation (15.64) becomes

$$k_i = \frac{K_2}{K_1}\left(\frac{k_e}{\sigma}\right) = 0.94 \left(\frac{k_e}{\sigma}\right) \text{ gal/hr-amp} \qquad (15.65)$$

for k_e in cm²/sec-V. As k_e varies within relatively narrow limits, equation (15.65) shows that the electro-osmotic efficiency measured by k_i is a sensitive function of the conductivity of the soil. For soils 11, 13, and 14 in Table 15.3, σ was in the range of 200 to 300 μmho/cm. For soil 12, in which electro-osmosis was not effective, σ was 2500 μmho/cm.

15.7 CONSOLIDATION BY ELECTRO-OSMOSIS

If, in a compressible soil, electro-osmosis draws water to a cathode, where it is removed from the system, and no water is permitted to enter at the anode, then consolidation of the soil between the electrodes occurs in an amount equal to the volume of water removed. Since water movement away from the anode causes consolidation of the soil in the vicinity of the anode, the effective stress must increase. And because the total stress in the vicinity of the anode remains unchanged, the pore water pressure must decrease. Water drains at the cathode; there is no consolidation; and, therefore, the total, effective, and pore water pressures at the cathode are unchanged. As a result, there develops a hydraulic gradient that tends to cause water flow from cathode to anode. Consolidation continues until the electro-osmotic driving force that moves water toward the cathode is exactly balanced by this hydraulic force, which moves water in the opposite direction.

The usefulness of consolidation by electro-osmosis as a means for soil stabilization has been established by a number of successful field applications (Casagrande, 1959; Bjerrum, Moum, and Eide, 1967). Two questions of importance in any application are how much consolidation will there be? and how long will it take? Answers to these questions can be obtained using the coupled flow equations in place of Darcy's law in classical consolidation theory.

Assumptions

Certain idealizing assumptions are required.

1. Homogeneous and saturated soil.
2. The physical and physico-chemical properties of the soil are uniform and constant with time.*
3. No soil particles are moved by electrophoresis.
4. The velocity of water flow by electro-osmosis is directly proportional to the voltage gradient.
5. All the applied voltage is effective in moving water.†
6. The electric field remains constant with time.
7. The coupling of hydraulically- and electrically-induced flows can be formulated by equation (15.6).
8. There are no electrochemical reactions.

Governing equations

Equation (15.6) written for electro-osmosis is

$$J_H = L_{HH}X_H + L_{HE}X_E \qquad (15.66)$$

For one-dimensional flow between plate electrodes (Fig. 15.15 a), equation (15.66) becomes

$$q_h = -\frac{k_h}{\gamma_w}\frac{\partial u}{\partial x} - k_e \frac{\partial V}{\partial x} \qquad (15.67)$$

and for radial flow for the conditions shown in Fig. 15.15b

$$q_h = -\frac{k_h}{\gamma_w}\left(\frac{1}{r}\frac{\partial u}{\partial r} + \frac{\partial^2 u}{\partial r^2}\right) - k_e\left(\frac{1}{r}\frac{\partial V}{\partial r} + \frac{\partial^2 V}{\partial r^2}\right) \qquad (15.68)$$

---

* Flow of water away from anodes towards cathodes causes a nonuniform decrease in water content along the line between electrodes. This causes changes in hydraulic conductivity, electro-osmotic permeability, compressibility, and electrical conductivity with time and position. To account for these effects analytically would greatly complicate the solution, because the problem becomes highly non-linear. Similar problems arise in classical consolidation theory, but it is known that the simple linear theory as developed by Terzaghi is adequate for most cases.

† In some cases some of the electrical energy may be consumed by generation of heat or by electrolysis. In such cases an effective voltage should be used (Esrig and Henkel, 1968).

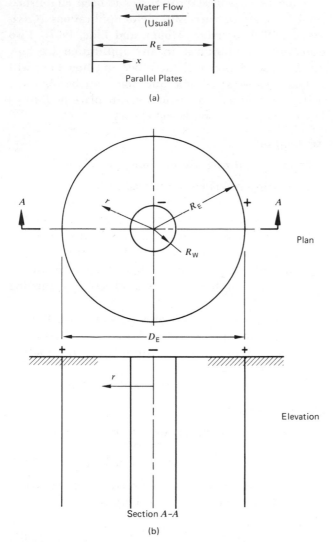

Fig. 15.15 Geometries of electrode systems for one-dimensional and radial flow. (*a*) One-dimensional flow. (*b*) Radial flow.

Introduction of these relationships between flow and gradients in place of Darcy's law in the derivation of the diffusion equation governing consolidation leads to

$$\frac{k_h}{\gamma_w} \frac{\partial^2 u}{\partial x^2} + k_e \frac{\partial^2 V}{\partial x^2} = m_v \frac{\partial u}{\partial t} \quad (15.69a)$$

and

$$\frac{\partial^2 u}{\partial x^2} + \frac{k_e}{k_h} \gamma_w \frac{\partial^2 V}{\partial x^2} = \frac{1}{c_v} \frac{\partial u}{\partial t} \quad (15.69b)$$

where m_v is the compressibility and c_v is the coefficient of consolidation. For the case of radial flow

$$\frac{\partial^2 u}{\partial r^2} + \frac{k_e}{k_h} \gamma_w \frac{\partial^2 V}{\partial r^2} + \frac{1}{r} \left(\frac{\partial u}{\partial r} + \frac{k_e}{k_h} \gamma_w \frac{\partial V}{\partial r} \right) = \frac{1}{c_v} \frac{\partial u}{\partial t} \quad (15.70)$$

Both V and u are functions of position, as shown in Fig. 15.16. V is assumed constant with time; whereas, u varies.

Amount of consolidation

Consolidation by electro-osmosis continues until outflow of water at the cathode ceases. This happens when the hydraulic gradient that develops in response to the differing amounts of consolidation between anode and cathode produce a counterflow $(k_h/\gamma_w)/(\partial u/\partial x)$ that exactly balances the electro-osmotic flow $k_e(\partial V/\partial x)$. At equilibrium there is no flow and q_h in equation (15.67) becomes zero. For this case,

$$\frac{k_h}{\gamma_w} \frac{\partial u}{\partial x} = -k_e \frac{\partial V}{\partial x} \quad (15.71)$$

or

$$du = -\frac{k_e}{k_h} \gamma_w \, dV \quad (15.72)$$

The solution of this equation is

$$u = -\frac{k_e}{k_h} \gamma_w V + C$$

At the cathode, $V = 0$ and $u = 0$; therefore, $C = 0$ and the pore pressure at equilibrium is given by

$$u = -\frac{k_e}{k_h} \gamma_w V \quad (15.73)$$

Equation (15.73) indicates that electro-osmotic consolidation continues at a point until a negative pore pressure, relative to the initial value, has been developed that is dependent on the ratio k_e/k_h, and on the voltage at the point. For conditions of constant total stress, development of this negative pore pressure means an equal increase in the effective stress. It is this increase in effective stress that is responsible for consolidation. For the one-dimensional case consolidation by electro-osmosis is analogous to the loading shown in Fig. 15.17.

For a given voltage, the magnitude of effective stress increase that may be developed depends on k_e/k_h. As k_e varies within narrow limits for different

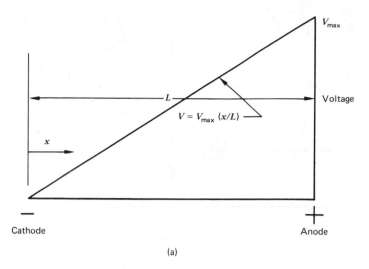

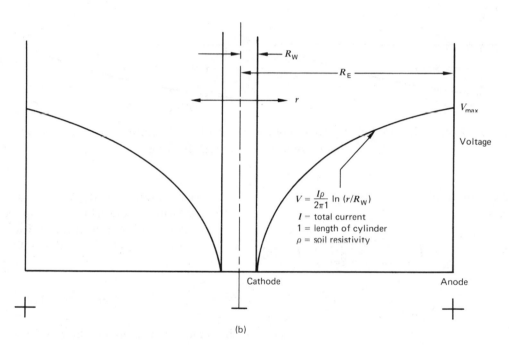

Fig. 15.16 Variation of voltage with distance during electro-osmosis. (*a*) One-dimensional flow. (*b*) Radial flow.

soils, effectiveness of consolidation by electro-osmosis increases with decrease in k_h. Thus, the potential for consolidation by electro-osmosis increases as soil grain size decreases, because the finer-grained the soil, the lower k_h is.

The amount of consolidation in any case depends on the soil compressibility as well as on the change in effective stress. For a linear soil, the coefficient of compressibility a_v is

$$a_v = -\frac{de}{dp} = \frac{de}{du} \qquad (15.74)$$

or

$$de = a_v du \qquad (15.75)$$

It follows, therefore, that electro-osmosis will be of little value in an overconsolidated clay unless the effective stress increases are sufficient to bring the material back into the virgin compression range.

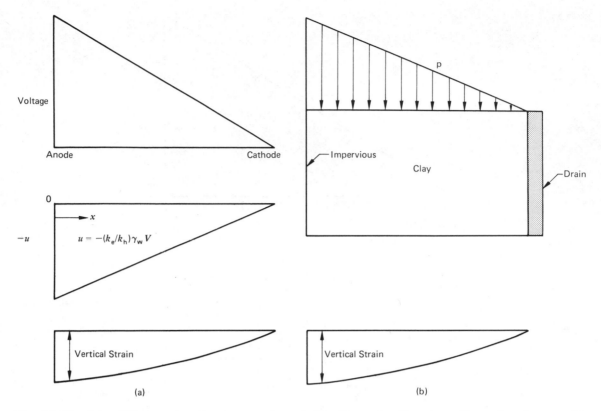

Fig. 15.17   Consolidation by electro-osmosis and by direct loading, one-dimensional case. (a) Electro-osmosis. (b) Direct loading.

For the case of radial consolidation, equation (15.68) becomes

$$\frac{1}{r} \frac{\partial u}{\partial r} + \frac{\partial^2 u}{\partial r^2} = -\frac{k_e}{k_h} \gamma_w \left(\frac{1}{r} \frac{\partial V}{\partial r} + \frac{\partial^2 V}{\partial r^2} \right) \quad (15.76)$$

Define

$$\xi = u + \frac{k_e}{k_h} \gamma_w V \quad (15.77)$$

Then equation (15.76) becomes

$$\frac{1}{r} \frac{\partial \xi}{\partial r} + \frac{\partial^2 \xi}{\partial r^2} = 0 \quad (15.78)$$

A first integration gives

$$\frac{\partial \xi}{\partial r} = \frac{A}{r} \quad (15.79)$$

and a second time yields

$$\xi = A \ln r + B \quad (15.80)$$

The boundary conditions are

1. At the cathode: $r = R_W$, $u = 0$, $V = 0$, so $\xi = 0$.

2. At the anode: $r = R_E$, $V = V_m$, $\partial \xi / \partial r = 0$ (no flow condition. Condition (2) applied to equation (15.79) gives $A = 0$, and condition (1) applied to (15.80) yields $B = 0$. Thus, $\xi = 0$, and from (15.77)

$$u = -\frac{k_e}{k_h} \gamma_w V \quad (15.73)$$

the same equation as for the one-dimensional case.

Rate of consolidation

Solutions for equations (15.69) and (15.70) have been obtained using open-cathode and closed-anode boundary conditions (Esrig, 1968, 1971). For the one-dimensional case the pore pressure is

$$u = \left(\frac{k_e}{k_h} \right) \gamma_w V(x) + \frac{2 k_e \gamma_w V_m}{k_h \pi^2} \quad (15.81)$$

$$\sum_{n=0}^{\infty} \frac{(-1)^n}{(n + \frac{1}{2})^2} \sin \left[\frac{(n + \frac{1}{2}) \pi x}{L} \right] \exp \left[-(n + \frac{1}{2})^2 \pi^2 T_v \right]$$

where $V(x)$ is the voltage at x, V_m is maximum voltage, and T_v is the time factor defined by

$$T_{\mathrm{v}} = \frac{c_{\mathrm{v}}t}{L^2} \qquad (15.82)$$

and

$$c_{\mathrm{v}} = \frac{k_{\mathrm{h}}}{m_{\mathrm{v}}\gamma_{\mathrm{w}}} \qquad (15.83)$$

The average degree of consolidation U as a function of time is

$$U = 1 - \frac{4}{\pi3}\sum_{n=0}^{\infty}\frac{(-1)^n}{(n+1/2)^3}\exp\left[-(n+\tfrac{1}{2})^2\pi^2T_{\mathrm{v}}\right]$$

$$(15.84)$$

Solutions for equations (15.81) and (15.84) are shown in Figs. 15.18 and 15.19. They are applied in just the same way as the theoretical solution for classical consolidation theory.

A numerical solution to equation (15.70) was obtained by Esrig (1968, 1971) with the results shown in Fig. 15.20. For the case of two pipe electrodes, a more realistic field condition than the radial geometry of Fig. 15.15b, Fig. 15.20 cannot be expected to apply exactly. Along a straight line between two pipe electrodes, however, the flow pattern is approximately the same as for the radial case for a considerable distance from each electrode (Esrig, 1968).

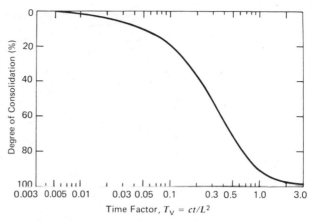

Fig. 15.19 Average degree of consolidation vs. dimensionless time for electro-osmosis between parallel plates.

A further solution is given in Fig. 15.21, which shows the rate of pore pressure buildup at the cathode for no drainage at the electrode. This condition may be of interest in connection with pile driving or pulling, reduction of negative skin friction, or recovery of buried objects. An analysis of phreatic surface conditions in relation to consolidation by electro-osmosis can be made numerically using the finite element method (Lewis and Humpheson, 1973).

One of the most important points to be noted from these solutions is that the rate of consolidation depends completely on the coefficient of consolidation, which varies directly as k_{h} but is completely independent of k_{e}. Thus, low values of k_{h}, as is the case in highly plastic clays, mean long consolidation times.

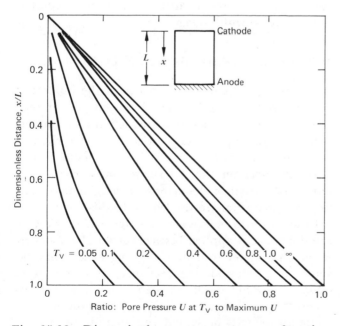

Fig. 15.18 Dimensionless pore pressure as a function of dimensionless time and distance for consolidation between parallel plates by electro-osmosis.

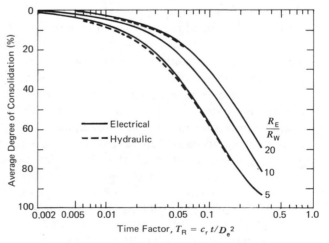

Fig. 15.20 Average degree of consolidiation as a function of dimensionless time for radial consolidation by electro-osmosis (Esrig, 1968).

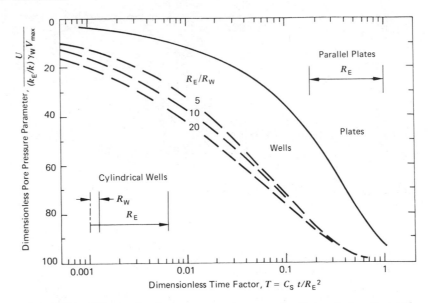

Fig. 15.21   Dimensionless pore pressure at the face of the electrode as a function of dimensionless time for the case of a closed cathode. A swelling condition (Esrig and Henkel, 1968).

The advantage of electro-osmosis in fine-grained soils as a consequence of a high value of k_e/k_h leading to a high increase in effective stress is off-set by the fact that as k_h decreases the time required increases. Optimum should be when k_e/k_h is high enough to permit development of a high pore water tension for reasonable electrode spacing (2 to 3 m) and voltage (50 to 150 V DC), but when k_h is high enough to enable consolidation in a reasonable time. Such conditions would appear best satisfied in silts, clayey silts, and silty clays, as most successful field applications have been in these types of soils.

In the application of the solutions for consolidation by electro-osmosis, it is necessary to consider the directions of drainage and compression so that appropriate interpretations may be made for c_v. The solutions presented herein represent geometries that are well suited for water movement in a horizontal direction, but compression in a vertical direction.

A number of laboratory investigations have established the general validity of the theory for cases where soil properties did not change radically during treatment and electrochemical effects were small (Mise, 1961; Nichols and Herbst, 1967; Esrig and Henkel, 1968; Esrig, 1968, 1971; Wan, 1970; Sundaram and Balasubramaniam, 1973). Although there have been a number of successful applications of electro-osmosis in the field, data permitting assessment of the theory are limited.

Figure 15.22 shows strengths, water contents, and plasticity values after treatment as a function of position between electrodes. Initial values are shown for comparison. The variations in strength and water content after treatment with distance from the anode are consistent qualitatively with the patterns to be expected based on the predicted variation of effective stress increase with voltage.

15.8 ELECTROCHEMICAL EFFECTS

The measured strength increases in the quick clay at Ås (Fig. 15.22) were some 80 percent greater than can be accounted for solely by reduction in water content. Also, the liquid and plastic limits were changed as a result of treatment. Consolidation alone should have no effect on the Atterberg limits; changes in mineralogy, particle characteristics, and/or pore solution characteristics are needed to do this.

In addition to movement of water when a dc voltage difference is applied between metal electrodes inserted into a wet soil, the following effects may develop: ion diffusion, ion exchange, development of osmotic and pH gradients, dessication due to heat generation at electrodes, mineral decomposition, precipitation of salts or secondary minerals, electrolysis, hydrolysis, oxidation, reduction, physical and chemical adsorption, and fabric changes. Because of these effects, continuous changes in soil properties that are

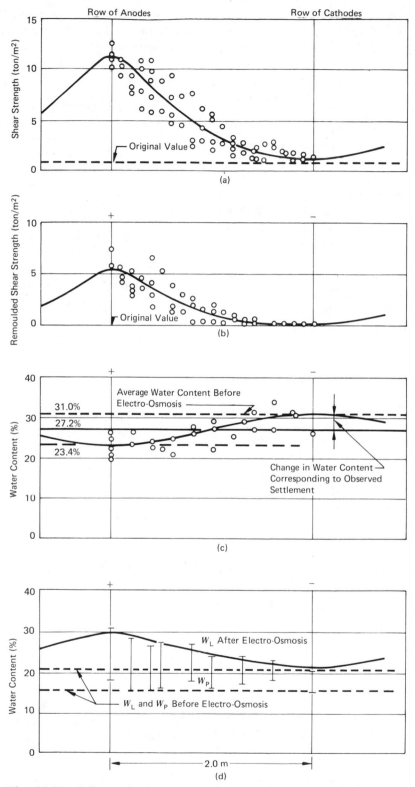

Fig. 15.22 Effect of electro-osmosis treatment on properties of quick clay at Ås (From Bjerrum, Moum, and Eide, 1967). (a) Undrained shear strength. (b) Remolded shear strength. (c) Water content. (d) Atterberg limits.

not readily accountable in terms of the simplified theory previously presented must be expected. Some of these effects may be beneficial in terms of an electrochemical hardening of the soil which results in permanent changes in the strength and plasticity characteristics. Others, such as heating and gas generation, may impair the efficiency of electro-osmosis.

A simplified mechanism for some of the changes that may occur is as follows. At the cathode hydrogen gas is evolved. Cations in solution are driven to the cathode where they combine with the $(OH)^-$ that is left behind to form hydroxides. The pH may rise to values as high as 12 in the vicinity of the cathode (Esrig and Gemenhardt, 1967; Esrig, 1971). Some alumina and silica may go into solution in the high pH environment.

Oxygen gas is evolved at the anode by hydrolysis, and anions in solution react with the freed H^+ to form acids. Chlorine may also form in a highly saline environment. Some of the exchangeable cations on the clay may be replaced by H^+. Because hydrogen clays are generally unstable and high acidity and oxidation cause rapid deterioration of the anodes, the clay will soon alter to the aluminum or iron form depending on the anode material. As a result, the soil is usually strengthened. If reactions at the anode are such that the gas causes cavitation and the dessication leads to crack formation, then two detrimental effects may occur: the negative pore pressure is limited to less than 1 atm, and the electrical resistance near the anode becomes very high leading to a loss in efficiency.

In some cases, gibbsite is formed if aluminum electrodes are used. Hydroxy-aluminum interlayering has developed in montmorillonites; however, interlayering is negligible in illites at low pH. Titkov et al. (1961), Chilingar and Rieke (1967), and Gray and Schlocker (1969) consider electrochemical hardening effects in detail.

15.9   CHEMICO-OSMOTIC EFFECTS IN SOILS

Because of the charged surfaces of the clay minerals and the consequent adsorption of balancing cations, consolidated clay layers behave as semipermeable membranes. A perfect semipermeable membrane prevents the passage of some components in solution while allowing unrestricted flow of others. A "leaky" semipermeable membrane retards or restricts the flow of some constituents but does not prevent flow completely. Fine-grained soils are leaky semipermeable membranes; the flow of salts is restricted relative to

that of water. The more active the clay and the lower the porosity, the more important the effect.

At least three practical consequences of the membrane properties of clays are of concern in geotechnical problems:

1. Chemico-osmotic flow of water through a clay layer as a result of solutions of different concentration on opposite sides of the layer. This flow is considered in Section 15.2 and illustrated in Fig. 15.1. Although comparatively little is known about the magnitude of k_c, it appears much more sensitive to soil type than is k_e.
2. Chemico-osmotic consolidation may occur if solutions of high concentration are placed in contact with an active, highly compressible clay at high water content containing little salt in the pores. The membrane properties of the clay restrict the inflow of salt. The high salt concentration in the external water reduces the free energy of the water relative to that in the pores, so water flows from the pores to establish equilibrium. Figure 15.23 is an example showing consolidation of bentonite. The consolidation observed was comparable to that for a load increment from 1 to 2 kg/cm^2.
3. As a result of ion exclusion, when an electrolyte solution is forced through a clay layer, the effluent may be less concentrated than the original solution. The magnitude of this effect, termed "salt filtering," "reverse osmosis," or "ultrafiltration," increases with increasing exchange capacity, decreasing porosity, and decreasing input solution concentration (Hanshaw and Coplen, 1973).

Volume changes due to chemico-osmotic consolidation are small except in the case of montmorillonite at high water content. Over long time periods chemico-osmotic flow of water through a clay layer and ultrafiltration may be important in problems in geology, groundwater quality, and waste disposal. Ultrafiltration may be important in the formation of saline groundwater and geothermal brines (Hanshaw and Coplen, 1973). Osmotic pressures may be used experimentally to determine the moisture tension–water content characteristics of soils (Waldron and Manbeian, 1970).

Theory of transient flow under combined chemical and hydraulic gradients

Since clay layers are leaky rather than perfect semipermeable membranes, salts, wastes, or other mate-

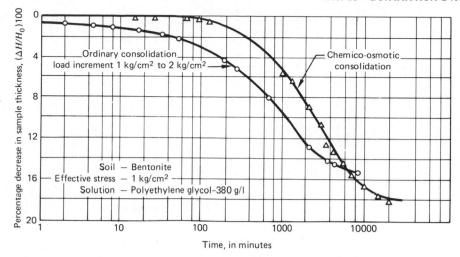

Fig. 15.23   Chemico-osmotic consolidation of bentonite.

rials in solution on one side will ultimately penetrate and pass through to the other in response to chemical and hydraulic gradients acting across the layer. Quantitative analysis of the transient flows subsequent to a change in solution concentration on one or both sides of a clay layer can be done in a manner similar to the analysis of electro-osmotic consolidation. The steps are to develop equations describing simultaneous flow of salt and water, derive equations of continuity for combined movement of salt and water, and combine the resulting equations to yield two diffusion equations: one for salt flow and the other for water flow. The problem is more complex than for electro-osmotic consolidation, because in chemico-osmotic diffusion the chemical gradient at any point changes continuously with time. The theory is presented in abbreviated form here.* An isotropic, homogeneous, saturated clay layer is considered under isothermal conditions. It is assumed that no ion exchange or adsorption reactions occur during diffusion.

From the postulates of irreversible thermodynamics equations for the flow rate of water $\bar{J}_w$ and for the flow rate of salt relative to water $\bar{J}_D$ can be derived (Katchalsky and Curran, 1967; Letey and Kemper, 1969).

$$\bar{J}_w = L_{11} V_w \text{ grad } (-u) + L_{12}\left(\frac{RT}{C_s}\right) \text{grad } (-C_s)$$
$$(15.85)$$

$$\bar{J}_D = L_{21} V_w \text{ grad } (-u) + L_{22}\left(\frac{RT}{C_s}\right) \text{grad } (-C_s)$$
$$(15.86)$$

* Detailed development of the theory is given by Greenberg (1971), Greenberg, Mitchell, and Witherspoon (1973), and Mitchell, Greenberg, and Witherspoon (1973).

where L_{ij} represents phenomenologic coefficients, V_w is the partial molar volume of water (L^3/mole), u is the hydrostatic pressure ($M/L/t^2$), R is the gas constant (ML^2/t^2mole/°K), T is the absolute temperature (°K), and C_s is the molar salt concentration (moles/L^3). Both flows are in units of mole/L^2/t and the "forces" (V_w grad $(-u)$ and (RT/C_s) grad $(-C_s)$) are in units of $ML/$mole$/t^2$. The L_{ij} are in units of mole^2 t/ML^3. According to Onsager's reciprocity theorem, (equation 15.14)

$$L_{12} = L_{21} \qquad (15.87)$$

The flow of salt relative to the fixed soil layer $\bar{J}_s$ is of more significance than that relative to the water $\bar{J}_D$. They are related by

$$\bar{J}_s = \bar{J}_D + \left(\frac{C_s}{C_w}\right)\bar{J}_w \qquad (15.88)$$

where $C_w = 1/V_w = $ molar concentration of water.

By the principle of conservation of mass, the rate of increase of mass density at any point must equal the rate at which matter flows to the point. For water in the pores,

$$\bar{\nabla}\bar{J}_w = -\frac{\partial N_w}{\partial t} \qquad (15.89)$$

and for salt

$$\bar{\nabla}\bar{J}_s = -\frac{\partial N_s}{\partial t} \qquad (15.90)$$

where N_w and N_s are the number of moles of water and salt, respectively, per unit volume of sediment. Combination of the flow equations (15.85) and (15.88) with the continuity equations (15.89) and (15.90) leads, after replacement of $\bar{J}_D$ in (15.88) by (15.86), to

$$-\frac{\partial N_w}{\partial t} = \bar{\nabla}\left[L_{11}V_w\,\text{grad}\,(-u) + L_{12}\frac{RT}{C_s}\,\text{grad}\,(-C_s)\right]$$

(15.91)

$$-\frac{\partial N_s}{\partial t} = \bar{\nabla}\left[\left(L_{21}V_w + \frac{C_s}{C_w}L_{11}V_w\right)\text{grad}\,(-u)\right.$$
$$\left. + \left(L_{22}\frac{RT}{C_s} + L_{12}\frac{RT}{C_w}\right)\text{grad}\,(-C_s)\right]$$

(15.92)

These diffusion equations are similar to those derived by Kirkwood (1954). They are not, however, in a form suitable for direct application. The following relationships may be noted:

$L_{11} = \dfrac{k_h}{V_w^2\gamma_w}$ (from equation (15.85) when grad $(-C_s) = 0$, by analogy with Darcy's law)

$L_{12} = \dfrac{C_s}{RTV_w}k_{hc}$ (from equation (15.85))   (15.93)

$L_{22} = \dfrac{C_s}{RT}D$ (from equation (15.86) when grad $(-u) = 0$ by analogy with Fick's law of diffusion)

$L_{21} = \dfrac{x}{V_w\gamma_w}k_{ch}$ (from equation (15.92) and assuming that L_{21} is a linear function of concentration)

in which γ_w is the specific weight of water $(M/t^2/L^2)$; D is the diffusion coefficient of the salt (L^2/t); k_{hc} and k_{ch} are coupling coefficients $(L^5/\text{mole}/t$ and mole$/t/L^2)$; $x = C_s/C_{sm}$ where C_{sm} is the maximum value of C_s; and k_h is the hydraulic conductivity (L/t).

If the compression behavior of the soil is linear, then

$$\Delta e = a_v\Delta u = (1 + e)\,m_v\Delta u \qquad (15.94)$$

where a_v and m_v are the coefficient of compressibility and the compressibility, respectively. It is assumed that the molar density of the pore water, D, and m_v are constants, then the one-dimensional forms of the diffusion equations (15.91) and (15.92) become, where z is the coordinate direction

$$\frac{1}{c_v}\frac{\partial u}{\partial t} = \frac{\partial^2 u}{\partial z^2} + \frac{k_{hc}\gamma_w}{k_h}\frac{\partial^2 C_s}{\partial z^2} \qquad (15.95)$$

and

$$n\frac{\partial C_s}{\partial t} = \frac{K_{ch}}{\gamma_w}\frac{\partial}{\partial z}\left(x\frac{\partial u}{\partial z}\right) + D'\frac{\partial^2 C_s}{\partial z^2} - m_vC_s\frac{\partial u}{\partial t}$$

(15.96)

where porosity $n = e/(1 + e)$

$$K_{ch} = k_{ch} + C_{sm}k_h$$
$$D' = D + C_sk_{hc}$$
$$c_v = \frac{k_h}{(m_v\gamma_w)} = \text{coefficient of consolidation}$$

In the absence of coupling $k_{hc} = k_{ch} = 0$, equation (15.95) reduces to

$$\frac{\partial u}{\partial t} = c_v\frac{\partial^2 u}{\partial z^2}$$

that is, Terzaghi's equation for one-dimensional consolidation. For no coupling and no hydrostatic pressure gradient, equation (15.96) becomes

$$\frac{\partial C_s}{\partial t} = \frac{D}{n}\frac{\partial^2 C_s}{\partial z^2}$$

which is the salt diffusion equation derived from Fick's law, $\bar{J}_D = D\,\text{grad}\,(-C_s)$. The porosity term accounts for the fact the D is the diffusion rate of salt per unit cross-sectional area of solution. For flow through soil, the area normal to the flow direction is proportional to the porosity.

Interpretation of the coupling factors. Three types of coupling are indicated by equations (15.95) and (15.96). Chemico-osmotic coupling, or the movement of water due to salt concentration differences, is given by the second term on the right-hand side of (15.95). Its effect depends on the value of k_{hc}/k_c.

The first term on the right-hand side of (15.96) refers to salt flow under a hydraulic gradient. It is simply the result of the flow of salt in solution and is termed drag coupling. The third term on the right in equation (15.96) represents salt movement due to void ratio changes, or void ratio coupling.

Unfortunately, few data are available concerning the value of the coupling coefficients. From the consolidation data in Fig. 15.23, Greenberg (1971) computed* $k_{hc} = -8.9 \times 10^{-3}$ cm⁵/mole/sec and $k_{ch} = -0.22 \times 10^{-10}$ mole/sec/cm² for bentonite, and $k_{hc} = -0.6 \times 10^{-6}$ cm⁵/mole/sec and $k_{ch} = -0.7 \times 10^{-17}$ mole/sec/cm² for kaolinite using Olsen's (1972) data for an overburden pressure of 10 kg/cm². Lim-

---

* $k_{hc} = (\Delta\sigma'/\Delta C_s)\,(k_h/_w)$ where $\Delta\sigma'$ is the equivalent effective stress increase due to chemico-osmosis, and $k_{ch} = (\gamma_wC_{sm}/RT)\,k_{hc}$.

ited experimental evidence is available to show agreement between predicted and measured rates of chemico-osmotic consolidation (Mitchell et al., 1973).

The time variation of pore pressure and salt concentration following change in concentration at the boundaries of a clay layer exhibiting coupling effects illustrates application of the theory. If equations (15.95) and (15.96) are put in dimensionless form, then time is given by time factor $T = c_v t / H^2$ and position is given by z/H, where H is the drainage path length. The initial conditions are

$$u = 0; \quad 0 < z < 2H; \quad t = 0$$
$$C_s = C_{so}; \quad 0 < z < 2H; \quad t = 0$$

throughout the layer, and boundary conditions are

$$u = 0; \quad z = 0, 2H; \quad t = t$$
$$C_s = C_{sm}; \quad z = 0, 2H; \quad t = t$$

in which $C_{so} < C_{sm}$.

Figures 15.24 and 15.25 show the results of an analysis of the behavior of a relatively compressible and impermeable clayey soil layer having the properties indicated. The degree of pore pressure equalization (Fig. 15.24) is the average chemico-osmotic pore pressure drop across the layer at a value of T divided by the maximum average pressure drop during the process. If the compression and swelling coefficients were equal, constant, and independent of void ratio, then this curve would indicate consolidation until about $T = 2$ followed by swell until $T = 2000$, when the original thickness would be regained. Since the coefficient of compressibility is usually larger than the

swelling coefficient, however, there would be a residual consolidation in the steady state. The isochrones in Fig. 15.25 show that from $T = 0$ to $T = 2$ very little salt diffuses into the clay during the period that pore water flows out. The water outflow is driven by the osmotic gradient caused by the membrane properties of the clay and the difference in concentration between the pore water and the external solution.

The salt inflow, represented as a proportion of the maximum salt content, increases with time until the internal concentration is in equilibirum with that in the external solution. As the salt concentration increases in the soil the chemico-osmotic consolidation pressure drops. It is this effect that causes the pore pressure equalization curve to reverse and rise back to its original value (Fig. 15.24).

From the results of numerical analyses of several cases, the following characteristics have been observed:

1. Maximum consolidation seems always to occur at a value of $T \approx 2.0$.
2. The amount of consolidation increases with the magnitude of the boundary salt content increase and compressibility.
3. The time to final equilibrium increases as void ratio decreases and compressibility increases.

15.10 HEAT FLOW THROUGH SOIL

Heat transfer through soils is important in problems of insulation, underground storage, underground transmission of electric power, thermal pollution, foundation and pipeline construction in perma-

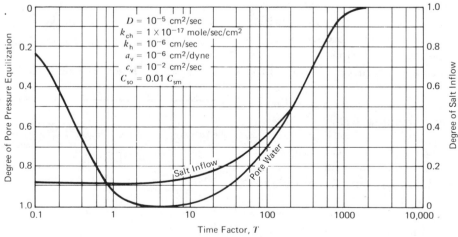

Fig. 15.24 Salt inflow and pore pressure equalization during chemico-osmotic consolidation.

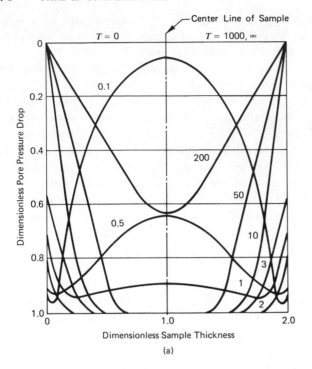
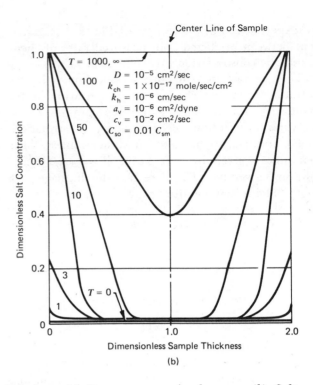

Fig. 15.25   Isochrones during chemico-osmotic consolidation. (a) Pore pressure isochrones. (b) Salt concentration isochrones.

frost, frost action in soils, and construction of frozen soil barriers for use as temporary walls and cutoffs. In principle, the analysis of heat flow is straightforward, with the relationships derived from Fourier's law, as listed in Table 15.1, serving as a basis for computation. In practice, however, the problem may be complex for the following reasons:

1. Difficulties in definition of the initial and boundary conditions.
2. Difficulties in quantification of the thermal loadings and their variation with time and position.
3. Uncertainties in the thermal properties of the soil and their variation with time.
4. Nonhomogeneous and anisotropic thermal properties.
5. Coupled heat flows, most importantly those related to moving ground water.

A detailed consideration of all these factors is beyond the scope of this book. Attention here is confined to three topics: thermal properties of soil, analysis of the depth of frost penetration, which is illustrative of a practical problem that can be studied using the one-dimensional transient heat flow equation, and frost action in soils, which illustrates a phenomenon of great practical importance that can

be understood only through consideration of interactions of the physical and physico-chemical properties of the soil.

Thermal properties of soil

The thermal properties needed for the analysis of thermal problems in soils are the thermal conductivity k_t, the volumetric heat C or heat capacity, and the latent heat of fusion L.

Thermal conductivity. The thermal conductivity is the rate of heat flow under a unit thermal gradient. Some values of thermal conductivity for different materials are listed in Table 15.5. The thermal con-

Table 15.5   Thermal Conductivity of Several Materials

Material	k_t BTU/hr.ft.°F	W/m.°K
Air	0.014	0.00017
Shale	0.9	0.011
Granite	1.6	0.019
Water	0.35	0.0042
Ice	1.30	0.016
Copper	225	2.70
Soil	0.2 to 2.0	0.0024 to 0.024

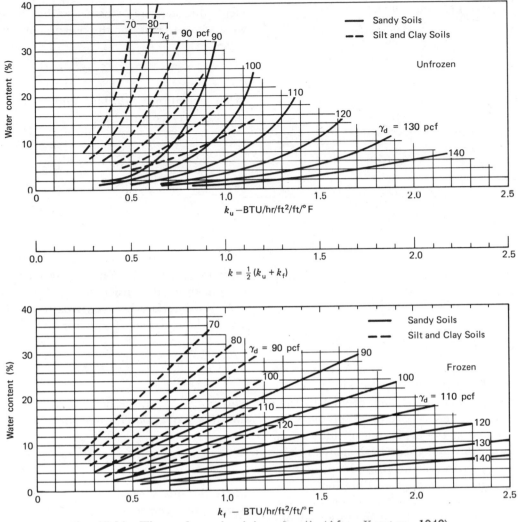

Fig. 15.26 Thermal conductivity of soil (After Kersten, 1949).

ductivity of soil is of the order of 1.0 BTU/hr/ft/°F (0.012 W/m/°K), although it may vary significantly from this value depending on soil type, density, and water content. Some values for unfrozen and frozen soils are summarized in Fig. 15.26. Theories are also available for computation of thermal conductivity from other soil data (American Institute of Electrical Engineers, 1960).

Volumetric heat. Values of volumetric heat C, that is, the heat content per unit volume, of water, ice, and minerals, are given in Table 15.6. The volumetric heat can be estimated using

$$C_u = \gamma_d \left(C_m + \frac{C_w w}{100} \right) \qquad (15.97)$$

and

$$C_f = \gamma_d \left(C_m + \frac{C_i w}{100} \right) \qquad (15.98)$$

for unfrozen and frozen soil, respectively, where γ_d is the dry density, w is the water content in percent, and C_m, C_w, and C_i are the heat capacities of mineral, water, and ice, respectively.

Table 15.6 Heat Capacity Values

Material	BTU/lb-°F	joule/kg-°K
Water	1.0	4186
Ice	0.5	2093
Rock and soil minerals	0.17	710

Latent heat. The latent heat L accompanying the phase change of water to ice is 143.4 BTU/lb (333×10^3 joule/kg). Thus, for freezing or thawing soil, the latent heat L_s is

$$L_s = L \left(\frac{w}{100} \right) \gamma_d \qquad (15.99)$$

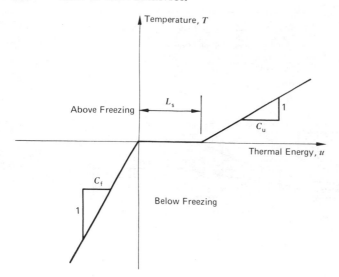

Fig. 15.27   Thermal energy as a function of temperature for a wet soil.

Depth of frost penetration

Accurate prediction of the depth of frost penetration into unfrozen ground during the winter, or the depth of thaw in permanently frozen ground (permafrost) during the summer, is a complex problem for all the reasons stated at the beginning of this section. Both field observations and theoretical analyses show that the dominant factors controlling depth of frost penetration are the intensity and duration of the freezing period. Variations in soil type, density, and water content may account for penetration depth variations by a factor of 2 (Brown, 1964), however.

Theoretical solutions to the problem (Berggren, 1943; Aldrich, 1956; Brown, 1964) have been based on a mathematical analysis developed by Neumann in about 1860.

The relationship between thermal energy u and temperature T for a soil mass at constant water content is shown in Fig. 15.27. In the absence of freezing or thawing,

$$\frac{\partial u}{\partial T} = C \qquad (15.100)$$

The Fourier equation for heat flow is

$$q_t = -k_t \frac{\partial T}{\partial z} \qquad (15.101)$$

In the absence of freezing or thawing, thermal continuity and conservation of thermal energy require that the rate of change in thermal energy of an element plus the rate of heat transfer into the element equal zero; that is, for the one-dimensional case,

$$\frac{\partial u}{\partial t} + \frac{\partial q}{\partial z} = 0 \qquad (15.102)$$

From equations (15.100) and (15.101), equation (15.102) may be written

$$C \frac{\partial T}{\partial t} = k_t \frac{\partial^2 T}{\partial z^2} \qquad (15.103)$$

or

$$\frac{\partial T}{\partial t} = a \frac{\partial^2 T}{\partial z^2} \qquad (15.104)$$

where $a = k_t/C$ is the thermal diffusivity (L^2/t). Equation (15.104) is immediately recognizable as the one-dimensional transient heat flow equation.

At the interface between frozen and unfrozen soil, $z = Z$, the equation of heat continuity is

$$L_s \frac{dZ}{dt} = q_u - q_f \qquad (15.105)$$

where $(q_u - q_f)$ is the net rate of heat away from the interface. Equation (15.105) can be written

$$L_s \frac{dZ}{dt} = k_f \frac{T_f}{z} - k_u \frac{Tu}{z} \qquad (15.106)$$

where the subscripts u and f pertain to unfrozen and frozen soil, respectively. Simultaneous solution of equations (15.104) and (15.106) yields the depth of frost penetration.

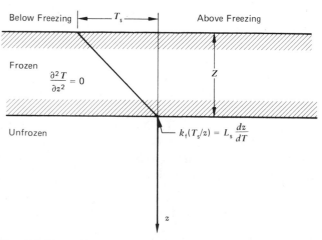

Fig. 15.28   Thermal conditions assumed for the Stefan equation.

Stefan formula. The simplest solution is to assume that the latent heat is the only heat to be removed during freezing. Thermal energy stored as volumetric heat is neglected. This condition is shown by Fig. 15.28. For this case, equation (15.104) does not exist, and (15.106) becomes

$$L_s \frac{dZ}{dt} = k_f \frac{T_s}{Z} \qquad (15.107)$$

where T_s is the surface temperature. The solution of this equation is

$$Z = \left(\frac{2k_f \int T_s dt}{L_s} \right)^{1/2} \qquad (15.108)$$

The integral $\int T_s dt$ is a measure of the intensity of freezing. It is usually expressed by the *freezing index F*, which has units of degrees × time. In practice, F is usually given in degree-days (Fahrenheit). It is shown in relation to the annual temperature cycle in Fig. 15.29. Freezing index values are obtained from meteorological data. Methods for determination of freezing index values are given by Linell et al. (1963), Straub and Wegmann (1965), McCormick (1971) and others. Maps are available showing the distribution of mean freezing index values for different areas.

Modified Berggren formula. The Stefan formula always overpredicts the depth of freezing, because the volumetric heats of frozen and unfrozen soil are neglected. Simultaneous solution for equations (15.104) and (15.106) based on the thermal conditions shown in Fig. 15.30 has been made assuming that the soil has a uniform initial temperature T_o above freezing and that the surface temperature drops suddenly to T_s below freezing (Aldrich, 1956). The solution is

$$Z = \lambda \left(\frac{2kT_s t}{L_s} \right)^{1/2} \qquad (15.109)$$

where k can be taken as an average thermal conductivity for frozen and unfrozen soil. λ is a dimensionless correction coefficient that depends on two parameters as shown in Fig. 15.31. The thermal ratio α is given by

$$\alpha = \frac{T_0}{T_s} \qquad (15.110)$$

and the fusion parameter is

$$\mu = \frac{C}{L_s} T_s \qquad (15.111)$$

In application, the quantity $T_s t$ in (15.109) is replaced by the freezing index, and T_s in (15.111) is given by F/t, where t is the duration of the freezing period.

The coefficient λ corrects the Stefan equation for neglect of volumetric heat. For soils with high water

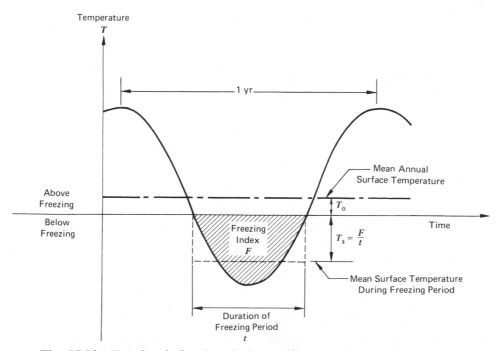

Fig. 15.29   Freezing index in relation to the annual temperature cycle.

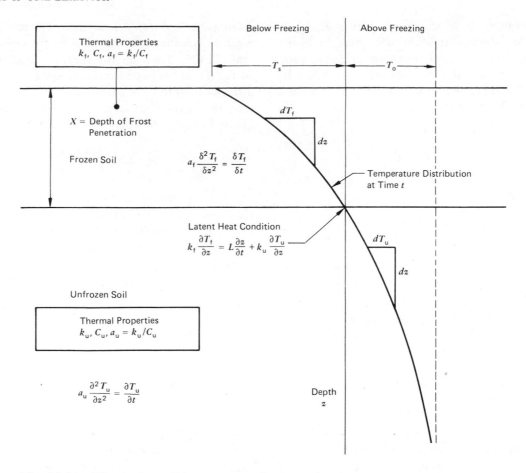

Fig. 15.30   Thermal conditions assumed in the derivation of the modified Berggren formula.

content C is small relative to L_s; therefore, μ is small and the Stefan formula is reasonable. For northern climates where T_o is not much above the freezing point, α is small, and λ is greater than 0.9, and the Stefan formula is satisfactory. The correction becomes important, however, in more temperate climates and in relatively dry or well-drained soils.

Figure 15.32 shows a comparison between theoretical penetration depths for several soil types and a design curve proposed by the Corps of Engineers. This comparison was developed by Brown (1964) using the modified Berggren equations and thermal properties given by Kersten (1949) for the soils indicated.

In practice, consideration must be given to the effect of different types of surface cover on the ground surface temperature. Air temperature and ground temperature are not likely to be the same, and the effects of thermal radiation may be significant. Observed depths of frost penetration in an area may be misleading if estimates for a proposed pavement or other structure are needed, because of differences in surface characteristics and because the pavement or foundation base will be at a different water content and density than the surrounding soil.

The solutions do not account for flow of water into or out of the soil during the freezing period. The former condition may be particularly important when soils susceptible to frost heave are encountered.

Frost heaving

The freezing of some soils is accompanied by the formation of ice layers or "lenses" that may range from a millimeter to several centimeters in thickness. These lenses are essentially pure ice, free from large numbers of contained soil particles. The ground surface in some cases may "heave" by as much as several tens of centimeters, and the overall volume increase can be many times the 9 percent that occurs when water freezes. Heaving pressures of many atmospheres

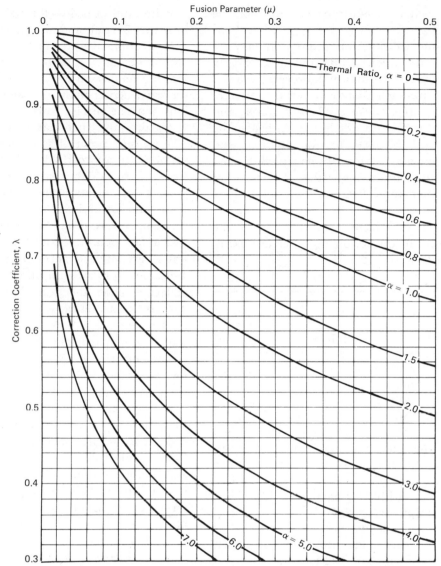

Fig. 15.31 Correction coefficient for use in the modified Berggren formula (From Aldrich, 1956).

are common, and the freezing of frost susceptible soils beneath pavements and foundations can lead to major problems, if not due to uneven uplift during freezing, then to loss of support when the lenses thaw, leaving large water-filled voids in the soil. Ordinarily ice lenses are oriented normal to the direction of cold front movement and become thicker and more widely separated with depth, as shown schematically in Fig. 15.33.

For a given soil and fixed groundwater level, the rate of heave, which may be as high as several millimeters per day, depends on the rate of freezing in a complex manner. The rate of heave increases with rate of frost penetration, but the ratio of heave rate to penetration rate decreases. If the cooling rate is too high, then the soil freezes before water can migrate to an ice lense, and heave becomes that due only to the volume expansion of water during phase change.

There are three necessary conditions for ice lense formation and frost heave:

1. A frost susceptible soil.
2. Freezing temperature.
3. A supply of water.

It follows, therefore, that frost heaving is important in areas with winter temperatures below freezing and in soils where the water table, a perched water table, or a pocket of water is not too distant from the freezing front.

Frost susceptible soils. Almost any soil may be made to heave if the freezing rate and water supply are properly controlled. In nature, however, the usual rates of freezing are such that only certain soils are highly frost susceptible. Clean sands and gravels and fat intact clays generally do not heave. Although the only completely reliable way to evaluate frost susceptibility is by some type of performance test during freezing, soils that contain more than 3 percent of their particles finer than 0.02 mm are potentially frost susceptible.

Frost susceptible soils have been classified by the Corps of Engineers in the following order of increasing frost susceptibility:

Group (increasing susceptibility)	Soil Types
F1	Gravelly soils with 3 to 20 percent finer than 0.02 mm
F2	Sands with 3 to 15 percent finer than 0.02 mm
F3	a. Gravelly soils with more than 20 percent finer than 0.02 mm and sands, except fine silt sands with more than 15 percent finer than 0.02 mm
	b. Clays with PI greater than 12 percent, except varved clays
F4	a. Silts and sandy silts
	b. Fine silty sands with more than 15 percent finer than 0.02 mm
	c. Lean clays with PI less than 12 percent
	d. Varved clays

Mechanism of frost heave. The formation of ice lenses is a complex phenomenon that involves interrelationships between the phase change of water to ice, transport of water to the lens, and general unsteady heat flow in the freezing soil (Taber, 1929, 1930; Beskow, 1935; Martin, 1959; Penner, 1959, 1963d; Kaplar, 1970; Martin and Wissa, 1972).

Four stages may be considered in the ice lens formation cycle:

1. Nucleation of ice.
2. Growth of the ice lens.
3. Termination of ice growth.
4. Heat and water flow between the end of stage (3) and the start of stage (1) again.

Water and heat flow through all four stages, which in reality overlap and are interdependent.

Nucleation of water to form ice is required at the outset. The nucleation temperature T_n is less than the freezing temperature T_o. In soils, T_o is less than the normal freezing point of water, because of dissolved ions, particle surface force effects, and negative pore water pressures that may exist in the freezing zone. The freezing point of soil water decreases as particle surfaces are approached and may be several degrees lower in the double layer than in the center of a pore. Thus, in fine-grained soils there is an unfrozen water film on particle surfaces that may persist until the temperature drops well below 0°C.

The face of an ice front has a thin film of adsorbed water. Freezing advances by incorporation of water molecules from the film into the ice, while new water molecules enter the thin film to maintain its thickness. It is energetically easier to bring water to the ice from adjacent pores than to freeze the adsorbed water on the particles or to propogate the ice through a pore constriction (Martin, 1959). The driving force for water transport to the ice from below is an equivalent hydrostatic pressure gradient that can be generated by freezing point depression, removal of water from the soil at the ice front that creates a higher effective stress in the vicinity of the ice than away from it, interfacial tension at the ice–water interface, and osmotic pressure because of the high concentration of ions in the water adjacent to the ice front. Ice formation continues until the water tension in the pores supplying water becomes great enough to cause cavitation or decreased water flow to the freezing front, or increased heat flow from the ground causes ice formation ahead of the existing lense and initiation of a new lense.

The processes of freezing and ice lens formation proceed in the following way with time. If a homogeneous soil at uniform water content and temperature T_o above freezing is subjected to a surface temperature T_s below freezing, the variation of temperature with depth at some time t_1 will be as shown in Fig. 15.34. The rate of heat flow at any point is given by $-k \, (dT/dz)$. If (dT/dz) at point A is greater than at point B, then the temperature of the element will drop. When water goes to ice, it gives up its latent heat, which flows both up and down and may slow or stop changes in the value of dT/dz for some time period, thus halting the rate of advance of the freezing front into the soil.

Heave results from the formation of a lens at A, with water supplied according to the mechanisms out-

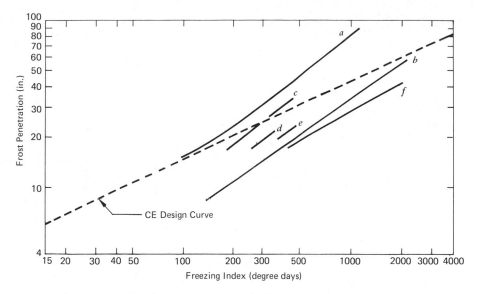

Fig. 15.32 Predicted frost penetration depths compared with the Corps of Engineers' design curve (Brown, 1964). Curve *a*—Sandy soil: dry density 140 lb/ft³ saturated, moisture content 7 percent. *b*—Silt, clay: dry density 80 lb/ft³ unsaturated, moisture content 10 percent, and sandy soil: dry density 90 lb/ft³ unsaturated, moisture content 2 percent. *c*—Sandy soil: dry density 140 lb/ft³ unsaturated, moisture content 2 percent. *d*—Silt, clay: dry density 120 lb/ft³ moisture content 10 to 20 percent (saturated). *e*—Silt, clay: dry density 80 lb/ft³ saturated, moisture content 30 percent. *f*—Pure ice over still water.

lined above. The energy needed to lift the overlying material is available because ice forms under conditions of supercooling at a temperature $T^x < T_{FP}$, where T_{FP} is the freezing temperature (Martin, 1959). The available energy is

$$\Delta F = \frac{L(T_{FP} - T^x)}{T_{FP}} \qquad (15.112)$$

and 1°C of supercooling is sufficient to lift 12.5 kg a distance of 1 cm. Alternatively, the energy for heave

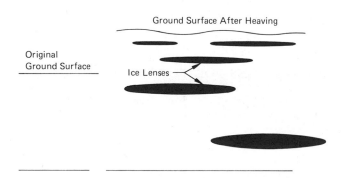

Fig. 15.33 Characteristics of ice lenses and frost heaving.

may originate from the thin water films at the ice surface (Kaplar, 1970).

As long as water can flow to a growing ice lense fast enough, the volumetric heat and latent heat can produce a temporary steady-state condition so that $(dT/dz)_A = (dT/dz)_B$. For example, a silt can supply water at a rate adequate for a heave rate of 1 mm/hr. After some time, however, the ability of the soil to supply water will drop, because the water supply in the region ahead of the ice front will become depleted, and the hydraulic conductivity of the soil will drop owing to increased tension in the pore water. Figure 15.35 illustrates this latter point. Hydraulic conductivity data are shown for a silty sand, a silt, and a clay, all compacted using Modified AASHO effort, at a water content about 3 percent wet of optimum.

For the silty sand, a small negative pore water pressure is sufficient to cause water to drain from the pores, followed by a sharp reduction in hydraulic conductivity. Because the clay can withstand large negative pore pressures without loss of saturation, the hydraulic conductivity is little affected by increasing values of $-\Delta u$. The small decrease observed can be ascribed to consolidation needed to carry the $+\Delta \sigma'$

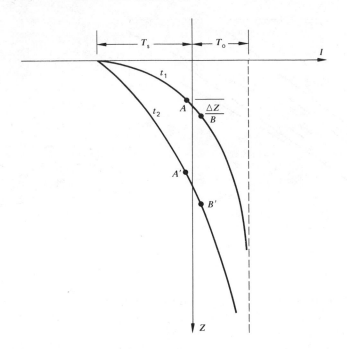

Fig. 15.34 Temperature vs. depth relationships in a freezing soil.

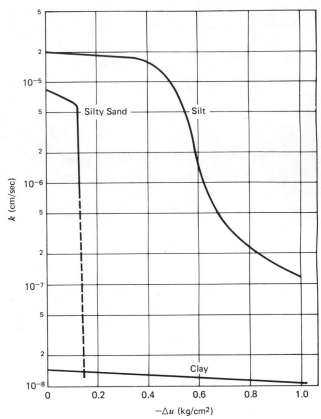

Fig. 15.35 Hydraulic conductivity as a function of negative pore water pressure (Martin and Wissa, 1972).

necessary to balance the $-\Delta u$. For the silt, water drainage is observed when $-\Delta u$ reaches about 0.4 kg/cm², but a continuous water phase remains to substantially greater values.

In a sand, the volume of water in a pore is large, and the latent heat raises the freezing temperature to the normal freezing point. Hence, there is no super-cooling and no heave. Any negative pressure development at the ice front causes the hydraulic conductivity to drop so that water cannot be supplied to form ice lenses. Thus, sands freeze homogeneously with depth. In a clay, the hydraulic conductivity is so low that water cannot be supplied fast enough to maintain the temporary steady-state condition needed for ice lens growth. Heave can develop in clays only if the freezing rate is greatly slowed below that in nature. Silts and silty soils have the optimum combinations of pore size, hydraulic conductivity, and freezing point depression so that normal freezing rates in the field can result in large heave.

Since the freezing temperature penetrates ahead of a completed ice lense, a new lens can only form after the temperature drops to the nucleation temperature. The nucleation temperature for a subsequent lense may be less that previously because of reduced satura-tion and consolidation (caused by previous upward water flow) that reduce the distance water can be from a particle surface. The temperature drop must proceed to a depth where there is sufficient water available after nucleation to supply a growing lense. The thicker the overlying lense, the greater the dis-tance, thus accounting for the increased spacings be-tween lenses with depth. The greater the depth, the smaller the thermal gradient, as may be seen in Fig. 15.34 where $(dT/dz)_A > (dT/dz)_{A'}$, where A' is on the temperature distribution curve for a later time t_2. Be-cause of this, the rate of heat extraction is slowed, and the temporary steady-state condition for lense growth can be maintained for a larger period of time, thus enabling formation of a thicker lens.

SUGGESTIONS FOR FURTHER STUDY

Aldrich, H. P. Jr. (1956), "Frost Penetration Below Highway and Airfield Pavements," *Highway Research Board Bulletin*, No. 135.

Bjerrum, L., Moum, J., and Eide, O. (1967), "Application of Electro-osmosis on a Foundation Problem in Nor-wegian Quick Clay," *Geotechnique*, Vol. 17, pp. 214–235.

Casagrande, L. (1952). "Electro–osmotic Stabilization of Soils," *Journal Boston Society of Civil Engineers*, Vol.

XXXIX, No. 1, reprinted in *Contributions to Soil Mechanics*, 1941–1953.

de Groote, S. R. (1952), *Thermodynamics of Irreversible Processes*, North Holland, Amsterdam.

Esrig, M. I. (1968), "Pore Pressures, Consolidation, and Electrokinetics," *Journal of the Soil Mechanics and Foundations Division, A.S.C.E.*, Vol. 94, No. SM4, pp. 899–921.

Gray, D. H., and Mitchell, J. K. (1967), "Fundamental Aspects of Electro-Osmosis in Soils," *Journal of the Soil Mechanics and Foundations Division, A.S.C.E.*, Vol. 93, No. SM6, pp. 209–236; Closure Discussion: Vol. 95, No. SM3, pp. 875–879, 1969.

Katchalsky, A., and Curren, P. R. (1967), *Nonequilibrium Thermodynamics in Biophysics*, Harvard Univ. Press, Cambridge.

Martin, R. T. (1959), "Rhythmic Ice Banding in Soil," *Highway Research Board Bulletin*, No. 218, pp. 11–23.

Mitchell, J. K., Greenberg, J. A., and Witherspoon, P. A. (1973), "Chemico-Osmotic Effects in Fine-Grained Soils," *Journal of the Soil Mechanics and Foundations Division, A.S.C.E.*, Vol. 99, No. SM4, pp. 307–322.

Olsen, H. W. (1969), "Simultaneous Fluxes of Liquid and Charge in Saturated Kaolinite," *Soil Science Society of America Proceedings*, Vol. 33, No. 3, pp. 338–344.

List of Symbols

a	area		F	free energy
a	crystallographic axis direction or distance		F	freezing index
a	thermal diffusion		F, F_0	faraday constant = 96,500 coulombs
a_v	coefficient of compressibility		ΔF	free energy of activation
A	activity		g	acceleration due to gravity
A	area		G_s	specific gravity of soil solids
A	creep rate parameter		h	head or head loss
A	van der Waal's constant		h	Planck's constant = 6.624×10^{-27} erg sec
A_0	concentration of charges on pore wall		H	ion hydration
A_c	solid contact area		H	total head
A_s	specific surface area per unit weight of solids		H	partial molar heat content
$\overline{A}$	pore pressure parameter = $\Delta u/\Delta(\sigma_1 - \sigma_3)$		H	X-ray diffraction peak height
$\overset{\circ}{A}$	Angstrom unit = 1×10^{-10} m		i	gradient
b	crystallographic axis direction or distance		i_c	chemical gradient
B	parameter in rate process equation = $X(kT/h)$		i_e	electrical gradient
c	cohesion		i_h	hydraulic gradient
c	crystallographic axis direction or distance		i_t	thermal gradient
c	concentration		I	electrical current
c	undrained shear strength		I	intensity of X ray or light
c	velocity of light		J_D	chemical flow rate
c_v	coefficient of consolidation		$\overline{J}_D$	flow rate of salt relative to water
c_w	concentration of water		$\overline{J}_i$	flux
C	clay content		$\overline{J}_s$	flow rate of salt relative to fixed soil layer
C	compressibility		J_w	flow rate of water
C	electrical capacitance		k	Boltzmann's constant = 1.38045×10^{-16} erg/degree
C	short range repulsive force between contacting particles		k	hydraulic conductivity, hydraulic permeability
			k	pore shape factor
C	volumetric heat		k	thermal conductivity
C_c	compression index		k	selectivity coefficient
C_s	shape coefficient		k_c	osmotic conductivity
C_s	swelling index		k_e	electro-osmotic conductivity
d	diameter		k_h	hydraulic conductivity
d	distance		k_t	thermal conductivity
D	diameter of particle		k_0	pore shape factor
D	dielectric constant		K	absolute permeability
D	diffusion coefficient		K	double layer parameter = $(8\pi n_0 e^2 v^2/DRT)^{1/2}$
D	deviator stress		K_0	coefficient of lateral earth pressure at rest
D	stress level = D/D_{max}		K_c	principal stress ratio
e	electronic charge = 4.8029×10^{-10} esu		l	length
	$= 1.60206 \times 10^{-19}$ coulomb		L	latent heat of fusion
e	void ratio		L_{ij}	coupling coefficient or conductivity coefficient
E	experimental activation energy		m	slope of relationship between log creep strain rate and log time
E	potential energy			
$E(\beta)$	distribution function for interparticle contact plane normals		m_s	compressibility of soil solids
			m_v	compressibility
f	shear force		m_w	compressibility of water

n	concentration, ions per unit volume		W	fluid volume
n	porosity		W	water transport
N	Avogadro's number $= 6.0232 \times 10^{23}$ mole^{-1}		W	X-ray diffraction peak width
N	coordination number		x	distance
N	moles of hydration water per mole of cation		X	friction coefficient
N	normal load		X_i	driving force
N	number of particles per cluster in a cluster structure		y	potential function $= ve\psi/kT$
N_s	moles of water per unit volume of sediment		z	potential function $= ve\psi_0/kT$
N_w	moles of salt per unit volume of sediment		Z	elevation or elevation head
p	consolidation pressure		Z	number of molecules per second striking a surface
p_c	preconsolidation pressure		α	angle between b and c crystallographic axes
p_0	present overburden effective stress		α	geometrical packing parameter
p	pressure		α	the slope of the relationship between logarithm of creep rate and creep stress
q	hydraulic flow rate			
q_h	hydraulic flow rate		α	thermal ratio
q_t	heat flow rate		α_s	thermal expansion coefficient of soil solids
Q	charge		α_{st}	thermal expansion coefficient of soil structure
Q	quantity of heat		α_w	thermal expansion coefficient of water
r_k	ratio of horizontal to vertical hydraulic conductivities		β	angle between a and c crystallographic axes
			β	birefringence ratio
R	electrical resistance		β	geometrical packing parameter
R	gas constant $= 1.98726$ cal/degree-mole		β	$8\pi F^2/1000\, DRT$
	8.31470 joules/degree-mole		γ	activity coefficient
	82.0597 cm^3 atm/degree-mole		γ	angle between a and b crystallographic axes
R_H	hydraulic radius		γ	unit weight
s	slope of stress relaxation curve		Γ	surface charge density
S	swell		δ	clay plate thickness measured between centers of surface layer atoms
S	number of flow units per unit area			
S	specific surface area per unit volume of solids		δ	displacement distance
S	fraction of molecules striking a surface that stick to it		δ	deformation parameter in Hertz theory
			δ	solid fraction of a contact area
S	saturation		ϵ	dielectric constant
S_0	specific surface per unit volume of soil particles		ϵ	electronic charge $= 4.8029 \times 10^{-10}$ esu
S	partial molar entropy		ϵ	strain
S_x, S_y, S_z	projected areas of interparticle contact surfaces		$\dot{\epsilon}$	strain rate
t	time		ζ	zeta potential
t	transport member		η	viscosity
T	shear force		θ	liquid to solid contact angle
T	temperature		θ	orientation angle
T	tortuosity factor		λ	wave length of x-ray
T_v	time factor		λ	separation distance between successive positions in a structure
u	midplane potential function			
u	pore water pressure		λ	correction coefficient for frost depth prediction equation
u	thermal energy		μ	coefficient of friction
$\overline{U}$	average degree of consolidation		μ	chemical potential
v	settling velocity		μ	fusion parameter
v	flow velocity		μ	dipole moment
v	valence		μ	viscosity
V	voltage		ν	activation frequency
V	volume		ξ	distance function $= kx$, double layer theory
V_A	attractive energy		π	osmotic pressure
V_m	total volume of soil mass		ρ	charge density
V_R	repulsive energy		σ	area occupied per adsorbed molecule on a surface
V_s	volume of solids		σ	double layer charge
V_w	volume of water		σ	normal stress
V_w	partial molar volume of water		σ	effective normal stress
V_{DR}	volume of water drained		σ	electrical conductivity
w	water content		σ_y	yield strength

τ	surface tension		ϕ	friction angle
τ	shear strength		ψ	electrical potential
τ	shear stress		ψ	intrinsic friction angle
τ	swelling pressure or matrix suction		ψ	total potential of soil water
τ_c	contaminant film strength		ω	angular velocity
τ_m	shear strength of solid material in yielded zone		Ω	true electro-osmotic flow

Bibliography

Abd-El-Aziz, M., and Taylor, S. A. (1964), "Simultaneous Flow of Water and Salt Through Unsaturated Porous Media: I Rate Equations," *Soil Science Society of America Proceedings*, Vol. 29, No. 2, pp. 141–143.

Abdel-Hady, M., and Herrin, M. (1966), "Characteristics of Soil Asphalt as a Rate Process," *Journal of the Highway Division*, A.S.C.E., Vol. 92, No. HW1, pp. 49–69.

Adams, J. I. (1965), "The Engineering Behavior of a Canadian Muskeg," *Proceedings of the Sixth Internation Conference on Soil Mechanics and Foundation Engineering*, Vol. 1, pp. 3–7.

Ahmed, S., Lovell, C. W. Jr., and Diamond, S. (1974), "Pore Sizes and Strength of Compacted Clay," *Journal of the Geotechnical Engineering Division*, A.S.C.E. Vol. 100, No. GT4, pp. 407–425.

Aitchison, G. D., Russam, K., and Richards, B. G. (1965), "Engineering Concepts of Moisture Equilibrium and Moisture Changes in Soils," *Moisture Equilibrium and Moisture Changes in Soils Beneath Covered Areas*, pp. 7–21, Butterworths, Sydney, Australia.

Aitchison, G. D., and Wood, C. C. (1965), "Some Interactions of Compaction, Permeability, and Post-Construction Deflocculation Affecting the Probability of Piping Failure in Small Earth Dams," *Proceedings of the Sixth International Conference of Soil Mechanics and Foundation Engineering, Montreal*, Vol. II, p. 442.

Aldrich, H. P., Jr. (1956), "Frost Penetration Below Highway and Airfield Pavements," *Highway Research Board Bulletin*, No. 135.

Allison, L. E., and Moodie, C. D. (1965), "Carbonate," *Methods of Soil Analysis*, Agronomy, No. 9, Chapter 91, American Society of Agronomy, Madison, Wisc.

American Institute of Electrical Engineers (1960), "Soil Thermal Characteristics in Relation to Underground Power Cables," *A.I.E.E. Transactions*, Vol. 79, Part III, pp. 792–856.

American Society for Testing and Materials (1970), *Special Procedures for Testing Soil and Rock for Engineering Purposes*, Special Technical Publication 479, 5th ed.

Amontons (1699), "De la Resistance Causee dans les Machines," *Academie Royale des Sciences, Paris*, pp. 206–227.

Andersland, O. B., and Akili, W. (1967), "Stress Effect on Creep Rates of Frozen Clay Soil," *Geotechnique*, Vol. 17, No. 1, pp. 27–39.

Andersland, O. B., and Douglas, A. G. (1970), "Soil Deformation Rates and Activation Energies," *Geotechnique*, Vol. 20, No. 1, pp. 1–16.

Anderson, D. M., and Hoekstra, P. (1965), "Crystallization of Clay Adsorbed Water," *Science*, Vol. 149, pp. 318–319.

Anderson, D. M., and Low, P. F. (1958), "The Density of Water Adsorbed by Li , Na-, and K- Bentonite," *Soil Science Society of America Proceeding*, Vol. 22, pp. 99–103.

Arnold, M. (1973), "Laboratory Determination of the Coefficient of Electro-osmotic Permeability of a Soil," *Geotechnique*, Vol. 23, No. 4, pp. 581–588.

Arthur, J. F., and Menzies, B. K. (1972), "Inherent Anisotropy in a Sand," *Geotechnique*, Vol. 22, No. 1, pp. 115–128.

Arulanandan, K., and Mitchell, J. K. (1968), "Low Frequency Dielectric Dispersion of Clay–Water–Electrolyte Systems," *Clays and Clay Minerals*, Vol. 16, No. 5, pp. 337–351.

Arulanandan, K., Sargunam, A., Loganathan, P., and Krone, R. B. (1973), "Application of Chemical and Electrical Parameters to Prediction of Erodability," *Highway Research Board Special Report* 135, Soil Erosion: Causes and Mechanisms, Prevention and Control, pp. 42–51.

Arulanandan, K., and Smith, S. S. (1973), "Electrical Dispersion in Relation to Soil Structure," *Journal of the Soil Mechanics and Foundation Division,* A.S.C.E., Vol. 99, No. SM 12, pp. 1113–1133.

Arulanandan, K., Smith, S. S., and Spiegler, K. S. (1973), "Soil Structure Evaluation by the Use of Radio Frequency Electrical Dispersion," *Proceedings of the International Symposium on Soil Structure, Gothenburg, Sweden,* pp. 29–49.

Aylmore, L. A. G., and Quirk, J. P. (1960), "Domain or Turbostratic Structure of Clays," *Nature,* Vol. 187, p. 1046.

Aylmore, L. A. G., and Quirk, J. P. (1962), "The Structural Status of Clay Systems," *Clays and Clay Minerals, Proceedings of the Ninth National Conference,* pp. 104–130.

Babcock, K. L. (1963), "Theory of the Chemical Properties of Soil Colloidal Systems at Equilibrium," *Hilgardia,* Vol. 34, No. 11.

Bailey, S. W., Brindley, G. W., Johns, W. D., Martin, R. T., and Ross, M. (1971), "Summary of National and International Recommendations on Clay Mineral Nomenclature" and "Report of Nomenclature Committee," *Clays and Clay Minerals,* Vol. 19, pp. 129–133.

Balasubramonian, B., and Morgenstern, N. (1972), "Discussion of Moum et al. (1971)," *Geotechnique,* Vol. 22, pp. 542–544.

Ballou, E. V. (1955) "Electro-osmotic Flow in Homoionic Kaolinite," *Journal of Colloid Science,* Vol. 10, pp. 450–460.

Banin, A., and Low, P. F. (1971), "Simultaneous Transport of Water and Salt Through Clays: 2. Steady-State Distribution of Pressure and Applicability of Irreversible Thermodynamics," *Soil Science,* Vol. 112, No. 2, pp. 69–88.

Barber, E. S. (1958), "Effect of Water Movement on Soil," Highway Research Board Special Report 40, pp. 212–225.

Barden, L. (1973), "Macro- and Microstructure of Soils," *Appendix to the Proceedings of the International Symposium on Soil Structure, Gothenburg, Sweden,* pp. 21–26.

Barden, L., and Berry, P. L. (1965), "Consolidation of Normally Consolidated Clay," *Journal of the Soil Mechanics and Foundations Division,* A.S.C.E., Vol. 91, No. SM5, pp. 15–36.

Barden, L., McGown, A., and Collins, K. (1973),

"Collapse Mechanism in Partly Saturated Soils," *Engineering Geology,* Vol. 7, No. 1, pp. 49–60.

Barden, L., and Sides, G. (1971), "Sample Disturbance in the Investigation of Clay Structure," *Geotechnique,* Vol. 21, No. 3, pp. 211–222.

Barshad, I. (1965), "Thermal Analysis Techniques for Mineral Identification and Mineralogical Composition," *Methods of Soil Analysis,* Agronomy, No. 9, Chapter 50, American Society of Agronomy, Madison, Wisc.

Bates, T. F., Hildebrand, F. A., and Swineford, A. (1950), "Morphology and Structure of Endellite and Halloysite," *American Mineralogist,* Vol. 35, p. 463.

Bawa, K. S. (1957), "Laterite Soils and Their Engineering Characteristics," *Journal of Soil, Mechanics and Foundation Division,* A.S.C.E., Vol. 82, No. SM4, pp. 1428-1–1428-15.

Berggren, W. P. (1943), "Prediction of Temperature Distribution in Frozen Soils," *Transactions of the American Geophysical* Union, Part 3, pp. 71–77.

Bernal, J. D. (1960), "The Structure of Liquids," *Scientific American,* Vol. 203, pp. 124–134.

Bernal, J. D. (1964), "The Structure of Liquids," The Bakerian Lecture, 1962, *Proceedings Royal Society, London,* A230, pp. 299–322.

Beskow, G. (1935), "Soil Freezing and Frost Heaving with Special Application to Roads and Railroads," (Translated by J. O. Osterberg), The Technological Institute, Northwestern University, Evanston, Ill.

Birrell, K. S. (1952), "Physical Properties of New Zealand Volcanic Ash Soils," *Conference on Shear Testing of Soils, Melbourne,* pp. 30–34.

Bischoff, J. L., Greer, R. E., and Luistro, A. O. (1970), "Composition of Interstitial Waters of Marine Sediments: Temperature of Squeezing Effect," *Science,* Vol. 167, pp. 1245–1246.

Bishop, A. W. (1954), Correspondence on "Shear Characteristics of a Saturated Silt, Measured in Triaxial Compression," *Geotechnique,* Vol. 4, pp. 43–45.

Bishop, A. W. (1960), "The Principle of Effective Stress," *Norwegian Geotechnical Institute,* Pub. No. 32, pp. 1–5.

Bishop, A. W. (1966), "The Strength of Soils as Engineering Materials," *Geotechnique,* Vol. 16, pp. 91–128.

Bishop, A. W., Green, G. E., Garga, V. K., Andresen, A., and Brown, J. D. (1971), "A New Ring Shear Apparatus and its Application to the Measurement of

Residual Strength," *Geotechnique,* Vol. 21, No. 4, pp. 273–328.

Bjerrum, L. (1954), "Geotechnical Properties of Norwegian Marine Clays," *Geotechnique,* Vol. 4, pp. 49–69.

Bjerrum, L. (1967), "Engineering Geology of Norwegian Normally Consolidated Clays as Related to Settlements of Buildings," *Geotechnique,* Vol. 17, pp. 81–118.

Bjerrum, L. (1972), "Embankments on Soft Ground," *Proceedings of the A.S.C.E. Specialty Conference on Performance of Earth and Earth-Supported Structures,* Vol. II, pp. 1–54.

Bjerrum, L. (1973), "Problems of Soil Mechanics and Construction on Soft Clays and Structurally Unstable Soils," General Report, Session 4, *Proceedings of the Eighth International Conference on Soil Mechanics and Foundation Engineering,* Vol. 3, pp. 111–159.

Bjerrum, L., Moum, J., and Eide, O. (1967), "Application of Electro-osmosis on a Foundation Problem in Norwegian Quick Clay," *Geotechnique,* Vol. 17, pp. 214–235.

Bjerrum, L., and Rosenqvist, I. Th. (1956), "Some Experiments with Artificially Sedimented Clays," *Geotechnique,* Vol. 6, pp. 124–136.

Bjerrum, L., and Wu, T. H. (1960), "Fundamental Shear Strength Properties of the Lilla Edit Clay," *Geotechnique,* Vol. X, No. 3, pp. 101–109.

Black, C. A., Ed. (1965), *Methods of Soil Analysis,* in Two Parts, Agronomy, No. 9, American Society of Agronomy, Madison, Wisc.

Black, W., DeJongh, J. G. V., Overbeek, J. Th., and Sparnaay, M. J. (1960), *Transactions of the Faraday Society,* Vol. 56, p. 1597.

Blackmore, A. V., and Miller, R. D. (1962), "Tactoid Size and Osmotic Swelling in Calcium Montmorillonite," *Soil Science Society of America, Proceedings,* Vol. 25, pp. 169–173.

Blaser, H. D., and Arulanandan, K. (1973), "Expansion of Soils Containing Sodium Sulfate," *Proceedings of the Third International Conference on Expansive Soils,* Vol. I, pp. 13–16.

Blaser, H. D., and Scherer, O. J. (1969), "Expansion of Soils Containing Sodium Sulfate Caused by Drop in Ambient Temperature," *Highway Research Board Special Report* 103, pp. 150–160.

Bochko, R. (1973), "Types of Microtextural Elements and Microporosity in Clays," *Proceedings of the International Symposium on Soil Structure, Gothenburg, Sweden,* pp. 97–101.

Bolt, G. H. (1955*a*), "Analysis of the Validity of Gouy-Chapman Theory of the Electric Double Layer," *Journal of Colloid Science,* Vol. 10, p. 206.

Bolt, G. H. (1955*b*), "Ion Adsorption by Clays," *Soil Science,* Vol. 79, pp. 267–276.

Bolt, G. H. (1956), "Physico-Chemical Analysis of the Compressibility of Pure Clays," *Geotechnique,* Vol. 6, No. 2, pp. 86–93.

Bolt, G. H., and Miller, R. D. (1958), "Calculation of Total and Component Potentials for Water in Soil," *Transactions of the American Geophysical Union,* Vol. 39, No. 5.

Bolt, G. H., and Page, A. L. (1965), "Ion-Exchange Equations Based on Double-Layer Theory," *Soil Science,* Vol. 99, No. 6, pp. 357–361.

Bolt, G. H., and Peech, M. (1953), "The Application of the Gouy Theory to Soil-Water Systems," *Soil Science Society Proceedings,* pp. 210–213.

Borodkina, M. M., and Osipov, V. I. (1973), "Automatic X-ray Analysis of Clay Microfabrics," *Proceedings of the International Symposium on Soil Structure, Gothenburg, Sweden,* pp. 15–20.

Bowden, F. P., and Tabor, D. (1950), *The Friction and Lubrication of Solids,* Part I, Oxford University Press, London.

Bowden, F. P., and Tabor, D. (1964), *The Friction and Lubrication of Solids,* Part II, Oxford University Press, London.

Bower, C. A., and Wilcox, L. V. (1965), "Soluble Salts," *Methods of Soil Analysis,* Agronomy, No. 9, Chapter 62, American Society of Agronomy, Madison, Wisc.

Boyes, R. G. H. (1972), "Uses of Bentonite in Civil Engineering," *Proceedings of the Instn. Civil Engineers,* Vol. 52, pp. 25–37.

Brekke, T. L., and Selmer-Olsen, R. (1965), "Stability Problems in Underground Construction Caused by Montmorillonite-Carrying Joints and Faults," *Engineering Geology,* Vol. 1, No. 1, pp. 3–19.

Brewer, R. (1964), *Fabric and Mineral Analysis of Soils,* Wiley, New York.

Brinch-Hansen, J., and Gibson, R. E. (1949), "Undrained Shear Strengths of Anisotropically Consolidated Clays," *Geotechnique,* Vol. 1, No. 3, pp. 189–204.

Brindley, G. W., and MacEwan, D. (1953), "Structural Aspects of the Mineralogy of Clays and Related Silicates," in *Ceramics*, A Symposium, British Ceramic Society, pp. 15–19.

Broadbent, F. E. (1965), "Organic Matter," *Methods of Soil Analysis*, Agronomy, No. 9, Chapter 92, American Society of Agronomy, Madison, Wisc.

Bromwell, L. G. (1965), "Adsorption and Friction Behavior of Minerals in Vacuum," Phase Report 2, Research in Earth Physics, Contract Report 3-101, Department of Civil Engineering, M.I.T., Cambridge, Mass.

Bromwell, L. G. (1966), "The Friction of Quartz in High Vacuum," Phase Report 7, Research in Earth Physics, Contract Report No. 3-101, Department of Civil Engineering, M.I.T., Cambridge, Mass.

Brooker, E. W., and Ireland, H. O. (1965), "Earth Pressure at Rest Related to Stress History," *Canadian Geotechnical Journal*, Vol. 2, No. 1, pp. 1–15.

Brown, G., Ed. (1961), *X-ray Identification and Crystal Structure of the Clay Minerals*, 2nd ed., Mineralogical Society of London, London.

Brown, W. G. (1964), "Difficulties Associated with Predicting Depth of Freeze or Thaw," *Canadian Geotechnical Journal*, Vol. 1, No. 4.

Buck, A. D. (1973), "Quantitative Mineralogical Analysis by X-ray Diffraction," *Tenth Meeting of the Clay Minerals Society, Twenty-second Annual Clay Minerals Conference, Banff, Alberta*, October, 1973.

Burst, J. F. (1969), "Diagenesis of Gulf Coast Clayey Sediments and its Possible Relation to Petroleum Migration," *American Association of Petroleum Geologists Bulletin*, Vol. 53, No. 1, pp. 73–93.

Cady, J. G. (1965), "Petrographic Microscope Techniques," *Methods of Soil Analysis*, Agronomy No. 9, Chapter 46, American Society of Agronomy, Madison, Wisc.

Campanella, R. G., and Gupta, R. C. (1969), "Effect of Structure on Shearing Resistance of a Sensitive Clay," *Soil Mechanics Series*, No. 6, University of British Columbia, Vancouver.

Campanella, R. G., and Mitchell, J. K. (1968), "Influence of Temperature Variations on Soil Behavior," *Journal of the Soil Mechanics and Foundations Division*, A.S.C.E., Vol. 94, No. SM3, pp. 709–734.

Campanella, R. G., and Vaid, Y. P. (1972), "Creep Rupture of a Saturated Natural Clay," *Proceedings of the Sixth International Congress on Rheology, Soil Mechanics Series*, No. 16, University of British Columbia, Vancouver.

Campanella, R. G., and Vaid, Y. P. (1974), "Triaxial and Plane Strain Creep Rupture of an Undisturbed Clay," *Canadian Geotechnical Journal*, Vol. 11, No. 1.

Campbell, W. E. (1969), "Solid Lubricants," *Boundary Lubrication: An Appriasal of World Literature*, A.S.M.E., pp. 197–227.

Caquot, A. (1934), *Equilibre des Massifs a Frottement Interne. Stabilite des Terres Pulverulents en Coherentes*. Gauthier-Villars, Paris.

Carlslaw, H. S., and Jaeger, J. C. (1957), *Conduction of Heat in Solids*, Clarendon Press, Oxford.

Carman, P. C. (1956), *Flow of Gases Through Porous Media*, Academic, New York.

Carr, C. W., et al. (1962), "Electro-osmosis in Ion Exchange Membranes," *Journal of the Electrochemical Society*, Vol. 109, pp. 251–255.

Carrier, W. D., III, Mitchell, J. K., and Mahmood, A. (1973), "The Nature of Lunar Soil," *Journal of Soil Mechanics and Foundations Division*, A.S.C.E., Vol. 99, No. SM10, pp. 813–832.

Carroll, D. (1970), "Clay Minerals: A Guide to their X-ray Identification," Geological Society of America, Special Paper 126.

Cary, J. W., Kohl, R. A., and Taylor, S. A. (1964), "Water Adsorption by Dry Soil and its Thermodynamic Functions," *Soil Science Society of America Proceedings*, Vol. 28, pp. 309–314.

Casagrande, A. (1932), "Research on the Atterberg Limits of Soils," *Public Roads*, October.

Casagrande, A. (1932), "The Structure of Clay and its Importance in Foundation Engineering," *Contributions to Soil Mechanics, Boston Society of Civil Engineers, 1925–1940*, pp. 72–112.

Casagrande, A. (1936), "The Determination of the Preconsolidation Load and its Practical Significance," *Proceedings of the First International Conference on Soil Mechanics and Foundation Engineering*, p. 60.

Casagrande, A. (1948), "Classification and Identification of Soils," *Transactions*, A.S.C.E., Vol. 113, pp. 901–991.

Casagrande, A., and Wilson, S. (1951), "Effect of Rate of Loading on Strength of Clays and Shales at Constant Water Content," *Geotechnique*, Vol. II, No. 3.

Casagrande, L. (1952), "Electro–osmotic Stabilization

of Soils," *Journal of the Boston Society of Civil Engineers,* Vol. XXXIX, No. 1, reprinted in *Contributions to Soil Mechanics, 1941–1953.*

Casagrande, L. (1959), "A Review of Past and Current Work on Electro-osmotic Stabilization of Soils," *Harvard Soil Mechanics Series,* No. 45, Harvard University.

Casagrande, L., et al., (1961), "Electro-osmotic Stabilization of a High Slope in Loose Saturated Silt," *Proceedings of the Fifth International Conference on Soil Mechanics and Foundation Engineering,* Vol. II, pp. 555–558.

Casimir, H. B. G., and Polder, D. (1984), "The Influence of Retardation of the London-van der Waals Forces," *Physical Review,* Vol. 73, p. 360.

Chan, H. R., and Kenney, T. C. (1973), "Laboratory Investigation of Permeability Ratio of New Liskeard Varved Soil," *Canadian Geotechnical Journal,* Vol. 10, pp. 453–472.

Chandler, R. J. (1972), "Lias Clay Weathering Processes and Their Effect on Shear Strength," *Geotechnique,* Vol. 22, No. 3, pp. 403–431.

Chapman, D. L. (1913), "A Contribution to the Theory of Electrocapillarity," *Philosophical Magazine,* Vol. 25, No. 6, pp. 475–481.

Chapman, H. D. (1965), "Cation-Exchange Capacity," *Methods of Soil Analysis,* Agronomy, No. 9, Chapter 57, American Society of Agronomy, Madison, Wisc.

Chattopadhyay, P. K. (1972), "Residual Shear Strength of some Pure Clay Minerals," Ph.D. Thesis, The University of Alberta, Edmonton, Canada.

Chilingar, G. V., and Rieke, H. H. (1967), Discussion, *Journal of the Soil Mechanics and Foundations Division,* A.S.C.E., Vol, 93, No. SM6, p. 391.

Christensen, R. W., and Wu, T. H. (1964), "Analyses of Clay Deformation as a Rate Process," *Journal of the Soil Mechanics and Foundations Division* A.S.C.E., Vol. 90, No. 6, pp. 125–157.

Clarke, F. W, (1920), "Data of Geochemistry," *U.S. Geological Survey Bulletin 695,* Washington, D.C.

Clevenger, W. A. (1958), "Experiences with Loess as a Foundation Material," *Transactions,* A.S.C.E., Vol. 123, pp. 151–180.

Collins, K., and McGown, A. (1974), "The Form and Function of Microfabric Features in a Variety of Natural Soils," *Geotechnique,* Vol. 24, No. 2.

Collis-George, N., and Bozeman, J. M. (1970), "A Double Layer Theory for Mixed Ion Systems as Applied to the Moisture Content of Clays Under Restraint," *Australian Journal of Soil Research,* Vol. 8, No. 3, pp. 239–258.

Cornell University (1951), "Final Report, Soil Solidification Research," Ithaca, New York.

Croney, D., and Coleman, J. D. (1954), "Soil Structure in Relation to Soil Suction (pF)," *Journal of Soil Science,* Vol. 5, p. 75.

Croney, D., Coleman, J. D., and Bridge, P. M. (1952), "The Suction of Moisture Held in Soils and other Porous Materials," Technical Paper Road Research Board, No. 24, H.M.S.O., London.

Cullity, B. D. (1956), *Elements of X-Ray Diffraction,* Addison-Wesley, Reading, Mass.

Dantu, P. (1961), "Etude mechanique d'un milieu pulverulent forme de spheres egales de capacite maxima," *Proceedings of the Fifth International Conference on Soil Mechanics and Foundation Engineering,* Vol. 1, pp. 61–66.

Darcy, A. (1856), "Les Fontaines Publiques de la Ville de Dijon," Dalmont, Paris.

Davidtz, J. C., and Low, P. F. (1970), "Relation Between Crystal Lattice Configuration and Swelling of Montmorillonites," *Clays and Clay Minerals,* Vol. 18, No. 6, p. 325.

Day, P. R. (1965), "Particle Fractionation and Particle Size Analysis," *Methods of Soil Analysis,* Agronomy No. 9, Chapter 43, Part I, American Society of Agronomy, Madison, Wisc.

Day, P. R. (1955), "Effect of Shear on Water Tension in Saturated Clay," Annual Reports I and II, Western Regional Research Project W-30, 1954–1955, University of California, Berkeley.

Degens, E. T. (1965), *Geochemistry of Sediments,* Prentice-Hall, Englewood Cliffs, N.J.

de Groote, S. R. (1952), *Thermodynamics of Irreversible Processes,* North Holland, Amsterdam.

Deresiewicz, H. (1958), "Mechanics of Granular Matter," *Advances in Applied Mechanics,* Vol. V, pp. 233–306.

Derjaguin, B. V. (1960), "The Force Between Molecules," *Scientific American,* Vol. 203, No. 1, p. 47, July.

Derjaguin, B. V., and Krylov, N. A. (1944), "Anomalies Observed in the Flow of Liquids Through Hard Fine-Porous Filters," *Proceedings, Conference of Viscosity of Liquids and Colloid Solutions,* Vol. 2, pp. 52–53, USSR Academy of Science Press, Moscow.

DeWit, C. T., and Arens, P. L. (1950), "Moisture Content and the Density of Some Clay Minerals and Some Remarks on the Hydration Pattern of Clay," *Transactions*, Soil Science, Vol. 2, p. 59–62.

Diamond, S. (1970), "Pore Size Distributions in Clays," *Clays and Clay Minerals*, Vol. 18, pp. 7–23.

Diamond, S. (1971), "Microstructure and Pore Structure of Impact Compacted Clays," *Clays and Clay Minerals*, Vol. 19, No. 4, p. 239.

Diamond, S., and Kinter, E. B. (1956), "Surface Area of Clay Minerals as Derived from Measurements of Glycerol Retention," *Clays and Clay Minerals*, NAS-NRC Publ. 556, pp. 334–347.

Dickey, J. W. (1966), "Frictional Characteristics of Quartz," S. B. Thesis, M.I.T., Cambridge, Mass.

Donnan, F. G. (1924), "The Theory of Membrane Equilibrium," *Chemical Reviews*, Vol. 1, pp. 73–90.

Dudley, J. H. (1970), "Review of Collapsing Soils," *Journal of Soil Mechanics and Foundations Division*, A.S.C.E., Vol. 96, No. SM3, pp. 925–947.

Duncan, J. M., and Seed, H. B. (1966), "Anisotropy and Stress Reorientation in Clay," *Journal of the Soil Mechanics and Foundations Division*, A.S.C.E., Vol. 92, No. SM5, pp. 21–50.

Dzyaloshinskii, I. E., Lifshitz, E. M., and Pitaevskii, L. P. (1961), "General Theory of van der Waals Forces," *Soviet Physics, Uspekhi*, Vol. 73, Nos. 3–4, pp. 381–422.

Eden, W. J., and Crawford, C. B. (1957), "Geotechnical Properties of Leda Clay in the Ottawa Area," *Proceedings of the Fourth International Conference on Soil Mechanics and Foundation Engineering*, Vol. 1, pp. 22–27.

Eisenberg, D., and Kauzman, W. (1969), *The Structure and Properties of Water*, Oxford University Press, New York, 296 pp.

El Rayah, H.M.E. and Rowell, D.C. (1973), "The Influence of Iron and Aluminum Hydroxides on the Swelling of Na-Montmorillonite and the Permeability of a Na-soil," *Journal of Soil Science*, Vol. 24, No. 1, pp. 137–144.

El Swaify, S. A., and Henderson, D. W. (1967), "Water Retention by Osmotic Swelling of Certain Colloidal Clays with Varying Ionic Composition," *Journ. of Soil Science*, Vol. 18, No. 2, pp. 223–232.

Esrig, M. I. (1968), "Pore Pressures, Consolidation, and Electrokinetics," *Journal of the Soil Mechanics and Foundations Division*, A.S.C.E., Vol. 94, No. SM4, pp. 899–921.

Esrig, M. I. (1971), "Electrokinetics in Soil Mechanics and Foundation Engineering," *Transactions of the New York Academy of Sciences*, Series II, Vol. 33, No. 2, pp. 234–245.

Esrig, M. I., and Gemeinhardt, J. P., Jr. (1967), "Electrokinetic Stabilization of an Illitic Clay," *Journal of the Soil Mechanics and Foundations Division*, A.S.C.E., Vol. 93, No. SM3, pp. 109–128.

Esrig, M. I., and Henkel, D. J. (1968), "The Use of Electrokinetics in the Raising of Submerged Partially Buried Metallic Objects," *Soil Engineering Series Research Report*, No. 7, Cornell University, Ithaca, N.Y.

Eyring, H. (1936), "Viscosity, Plasticity, and Diffusion as Examples of Absolute Reaction Rates," *Journal of Chemical Physics*, Vol. 4, No. 4, pp. 283–291.

Fanning, F. A., and Pilson, M. E. Q. (1971), "Interstitial Silica and pH in Marine Sediments: Some Effects of Sampling Procedures," *Science*, Vol. 173, pp. 1228–1231.

Fetzer, C. A., (1967), "Electro-osmotic Stabilization of West Branch Dam," *Journal of the Soil Mechanics and Foundations Division*, A.S.C.E., Vol. 93, No. SM4, pp. 85–106.

Fink, D. H., Nakayama, F. S., and McNeal, B. L. (1971), "Demixing of Exchangeable Cations in Free Swelling Bentonite Clay," *Soil Science Society of America, Proceedings*, Vol. 35, pp. 552–555.

Finnie, I. and Heller, W. (1959), *Creep of Engineering Materials*, McGraw-Hill, New York.

Florin, V. A. (1951), "Consolidation of Earth Media and Seepage Under Conditions of Variable Porosity and Consideration of the Influence of Bound Water," *Izvestia Academy of Science, USSR, Section of Technical Sciences*, No. 11, pp. 1625–1649.

Foster, M. D. (1953), "Geochemical Studies of Clay Minerals: II—Relation Between Ionic Substitution, and Swelling in Montmorillonites," *American Mineralogist*, Vol. 38, 994–1006.

Foster, M. D. (1955), "The Relation Between Composition and Swelling in Clays," *Clays and Clay Minerals*, Vol. 3, pp. 205–220.

Foster, R. H. (1973), "Analysis of Soil Microstructure," *Proceedings of the International Symposium of Soil Structure, Gotheburg, Sweden*, pp. 5–13.

Frank, H. S. (1958), "Covalency in the Hydrogen

Bond and the Properties of Water and Ice," *Proceedings of the Royal Society London*, A247, pp. 481–492.

Frank, H. S., and Wen, W. Y. (1957), Structure Aspects of Ion-Solvent Interaction in Aqueous Solutions: A Suggested Picture of Water Structure," *Faraday Society Discussions*, No. 24, pp. 133–140.

Franklin, A. F., Orozco, L. F., and Semrau, R. (1973), "Compaction of Slightly Organic Soils," *Journal of Soil Mechanics and Foundations Division*, A.S.C.E., Vol. 99, No. SM7, pp. 541–557.

Garlanger, J. E. (1972), "The Consolidation of Soils Exhibiting Creep under Constant Effective Stress," *Geotechnique*, Vol. 22, No. 1, pp. 71–78.

Garrels, R. M. (1951), *A Textbook of Geology*, Harper, New York, 511 pp.

Geuze, E. C. W. A., and Tan, T. K. (1953), "The Mechanical Behaviour of Clays," *Proceedings of the Second International Congress of Rheology, Oxford*, p. 247.

Gibbs, H. J., and Bara, J. P. (1967), "Stability Problems of Collapsing Soils," *Journal Soil Mechanics and Foundations Division*, A.S.C.E., Vol. 93, No. SM4, pp. 577–594.

Gibson, R. E. (1953), "Experimental Determination of the True Cohesion and True Angle of Internal Friction in Clays," *Proceedings of the Third International Conference on Soil Mechanics and Foundation Engineering*, Vol. I, pp. 126–130.

Gillott, J. E. (1968), *Clay in Engineering Geology*, Elsevier, New York.

Gillott, J. E. (1969), "Study of the Fabric of Fine-Grained Sediments with the Scanning Electron Microscope," *Journal of Sedimentary Petrology*, Vol. 39, No. 1, pp. 90–105.

Gillott, J. E. (1970), "Fabric of Leda Clay Investigated by Optical, Electron-Optical, and X-ray Diffraction Methods," *Engineering Geology*, Vol. 4, No. 2, pp. 133–153.

Glaeser, R., and Mering, J. (1954), "Isothermes d'hydration des Montmorillonites bi-ioniques (Na, Ca)," *Clay Minerals Bulletin*, Vol. 2, pp. 188–193.

Glasstone, S., Laidler, K. and Eyring, H. (1941), *The Theory of Rate Processes*, McGraw-Hill, New York.

Goldschmidt, V. M. (1926), Undersokelser ved Lersedimenter, Nordisk fordbrugsforskning, Kongress 3, Kobenhavn, pp. 434–445.

Goldstein, M., and Ter-Stepanian, G. (1957), "The Long-Term Strength of Clays and Depth Creep of Slopes," *Proceedings of the Fourth International Conference Soil Mechanics and Foundation Engineering*, A.S.C.E., Vol. II, pp. 311–314.

Gouy, G. (1910), "Sur la constitution de la charge electrique a la surface d'un electrolyte," *Anniue Physique (Paris)*, Serie 4, Vol. 9, pp. 457–468.

Gradwell, M. and Birrell, K. S. (1954, "Physical Properties of Certain Volcanic Soils (from New Zealand)," New Zealand *Journal of Science and Technology*, Vol. 36B, pp. 108–122.

Graham, J., Walker, G. F., and West, G. W. (1964), "Nuclear Magnetic Resonance Study of Interlayer Water in Hydrated Layer Silicates," *Journal of Chemical Physics*, Vol. 40, No. 2, pp. 540–550.

Gray, D. H. (1966), "Coupled Flow Phenomena in Clay–Water Systems," Ph.D. Thesis, University of California, Berkeley.

Gray, D. H. (1969), "Prevention of Moisture Rise in Capillary Systems by Electrical Short Circuiting," *Nature*, Vol. 223, No. 5204, pp. 371–374.

Gray, D. H., and Mitchell, J. K. (1967), "Fundamental Aspects of Electro-Osmosis in Soils," *Journal of the Soil Mechanics and Foundations Division*, A.S.C.E., Vol. 93, No. SM 6, pp. 209–236, Closure Discussion: Vol. 95, No. SM3, pp. 875–879, 1969.

Gray, D. H., and Schlocker, J. G. (1969), "Electrochemical Alteration of Clay Soils," *Clays and Clay Minerals*, Vol. 17, pp. 309–322.

Greenberg, J. A. (1971), "Diffusional Flow of Salt and Water in Soils," Ph.D. Thesis, University of California, Berkeley.

Greenberg, J. A., Mitchell, J. K., and Witherspoon, P. A. (1973), "Coupled Salt and Water Flows in a Groundwater Basin," *Journal of Geophysical Research*, Vol. 78, No. 27, pp. 6341–6353.

Griffin, J. J., Windom, H., and Goldberg, E. D. (1968), "The Distribution of Clay Minerals in the World Ocean," *Deep Sea Research*, Vol. 15, pp. 433–459.

Grim, R. E. (1962), *Applied Clay Mineralogy*, McGraw-Hill, New York.

Grim, R. E. (1968), *Clay Mineralogy*, 2nd edition, McGraw-Hill, New York.

Gupta, R. P., and Swartzendruber, D. (1962), "Flow-Associated Reduction in the Hydraulic Conductivity of Quartz Sand," *Soil Science Society of America, Proceedings*, pp. 6–10.

Hafiz, M. S. (1950), "Strength Characteristics of

Sands and Gravels in Direct Shear," Ph.D. Thesis, University of London.

Hansbo, S. (1960), "Consolidation of Clay with Special Reference to the Influence of Vertical Sand Drains," *Proceedings,* 18, Swedish Geotechnical Institute, Stockholm.

Hansbo, S. (1973), "Influence of Mobile Particles in Soft Clay on Permeability," *Proceedings of the International Symposium on Soil Structure, Gothenburg, Sweden,* pp. 132–135.

Hanshaw, B. B., and Coplen, T. B. (1973), "Ultrafiltration by a Compacted Clay Membrane—II. Sodium Ion Exclusion at Various Ionic Strengths," *Geochimca et Cosmochimica Acta,* Vol. 37, pp. 2311–2327.

Hardcastle, J. H., and Mitchell, J. K. (1974), "Electrolyte Concentration-Permeability Relationships in Sodium Illite-Silt Mixtures," *Clays and Clay Minerals,* Vol. 22, No. 2, pp. 143–154.

Hardy, Sir William Bate (1936), *Collected Scientific Papers of,* Cambridge University Press, 900 pp.

Helmholtz, H. (1879), *Wiedemanns Annalen d. Physik,* Vol. 7, p. 137.

Henkel, D. J., and Sowa, V. A. (1963), "Discussion," *Laboratory Shear Testing of Soils,* A.S.T.M. Spec. Tech. Pub. No. 361

Heselton, L. R. (1969), "The Continental Shelf," *Center for Naval Analysis, Research Contribution* 106, Arlington, Va. (AD 686 703).

Hill, T. L. (1950), "Statistical Mechanics of Adsorption, IX: Adsorption Thermodynamics and Solution Thermodynamics," *Journal of Chemical Physics,* Vol. 18, pp. 246–256.

Hirst, T. J., and Mitchell, J. K. (1968), "Compositional and Environmental Influences on the Stress-Strain-Time Behavior of Soils," Report No. TE 68-4, University of California, Berkeley.

Holtz, W. G., and Gibbs, H. J. (1956), "Engineering Properties of Expansive Clays," *Transactions,* A.S.C.E., Vol. 121, pp. 641–677.

Holzer, T. L., Hoeg, K., and Arulanandan, K. (1973), "Excess Pore Pressures During Undrained Clay Creep," *Canadian Geotechnical Journal,* Vol. 10, No. 1, pp. 12–24.

Horn, H. M., and Deere, D. U. (1962), "Frictional Characteristics of Minerals," *Geotechnique,* Vol. 12, No. 4, pp. 319–335.

Horne, M. R. (1965), "The Behavior of an Assembly of Rotund, Rigid, Cohesionless, Particles," *Proceedings of the Royal Society, London,* Part I, Vol. 286, pp. 62–78, Part II, Vol. 286, pp. 79–97, Part III, Vol. 310, pp. 21–34.

Hörz, F., Carrier, W. D. III, Young, J. W., Duke, C. M., Nagle, J. S., and Fryxell, R. (1973), "Apollo 16 Special Samples," Part B of Sec. 7, of Apollo 16 Preliminary Science Report, NASA SP-315.

Houston, W. N. (1967), "Formation Mechanisms and Property Interrelationships in Sensitive Clays," Ph.D. Thesis, University of California, Berkeley.

Houston, W. N., and Mitchell, J. K. (1969), "Property Interrelationships in Sensitive Clay," *Journal of Soil Mechanics and Foundations Division,* A.S.C.E., Vol. 95, No. SM4, pp. 1037–1062.

Hvorslev, M. J. (1937), "Uber die Festigheit Eigenschaffen Gestorter Bindiger Boden," *Ingerniorvidenshabeliege Skriften,* no. 45, Danmarks Naturvidenshabelige Samfund, Kobenhaven, 159 pp.

Hvorslev, M. J. (1960), "Physical Components of the Shear Strength of Saturated Clays," *Proceedings of the A.S.C.E. Research Conference on the Shear Strength of Cohesive Soils,* Boulder, Colo., pp. 169–273.

Ingles, O. G. (1962), "Bonding Forces in Soils, Part 3: A Theory of Tensile Strength for Stabilised and Naturally Coherent Soils," *Proceedings of the First Conference of the Australian Road Research Board,* Vol. I. pp. 1025–1047.

Ingles, O. G. (1968), "Soil Chemistry Relevant to the Engineering Behavior of Soils," Chapter 1, *Soil Mechanics, Selected Topics* (I. K. Lee, Ed.), Elsevier, New York.

Ingles, O. G., and Aitchison, G. D. (1969), "Soil-Water Disequilibrium as a Cause of Subsidence in Natural Soils and Earth Embankments," *Proceedings of the International Symposium on Land Subsidence (Tokyo),* AIHS Pub., No. 89, Vol. 2, p. 342.

Ingles, O. G. (1972), "Discussion," *Proceedings of the Specialty Conference on Performance of Earth and Earth-Supported Structures,* A.S.C.E., Vol. III, pp. 111–125.

Jackson, M. L., and Sherman, G. D. (1953), "Chemical Weathering of Minerals in Soils," *Advances in Agronomy,* Vol. 5, pp. 219–318, Academic, New York.

Janbu, N. (1963), "Soil Compressibility as Determined by Oedometer and Triaxial Tests," *European Conference on Soil Mechanism and Foundation Engineering, Wiesbaden, Germany,* Vol. 1, pp. 19–25.

Jenny, H. (1941), *Factors of Soil Formation*, McGraw-Hill, New York, 281 pp.

Kallstenius, T., and Bergau, W. (1961), "Research on Texture of Granular Masses," *Proceedings of the Fifth International Conference on Soil Mechanics and Foundations Engineering*, Vol. 1, pp. 165–170.

Kaplar, C. W. (1970), "Phenomenon and Mechanism of Frost Heaving," *Highway Research Record*, No. 304, pp. 1–13.

Karlsson, R. (1961), "Suggested Improvements in the Liquid Limit Test, with Reference to Flow Properties of Remolded Clays," *Proceedings of the Fifth International Conference on Soil Mechanics and Foundation Engineering*, Vol. 1, pp. 171–184.

Karlsson, R. (1963), "On Cohesive Soils and Their Flow Properties," *Swedish Geotechnical Institute Report*, No. 5, Stockholm.

Katchalsky, A., and Curran, P. R. (1967), *Nonequilibrium Thermodynamics in Biophysics*, Harvard University Press, Cambridge.

Kazi, A., and Moum, J. (1973), "Effect of Leaching on the Fabric of Normally Consolidated Marine Clays," *Proceedings of the International Symposium on Soil Structure, Gothenburg, Sweden*, pp. 137–152.

Keller, G. H. (1969), "Engineering Properties of Sea-Floor Deposits," *Journal of Soil Mechanics and Foundations Division*, A.S.C.E., Vol. 95, No. SM6, pp. 1379–1392.

Keller, W. D. (1957), *The Principles of Chemical Weathering*, Revised Edition, Lucas, Columbia, Mo., 111 pp.

Keller, W. D. (1964), "Processes of Origin and Alteration of Clay Minerals," in *Soil Clay Mineralogy* (C. I. Rich and G. W. Kunze, Eds.), pp. 3–76, University of North Carolina Press, Chapel Hill.

Keller, W. D. (1964), "The Origin of High Alumina Clay Minerals—A Review," Clays and Clay Minerals, *Proceedings of the Twelfth National Conference* Monograph No. 19, Earth Sciencies Series, Pergamon, pp. 129–151.

Kellogg, C. E. (1941), *The Soils That Support Us*, Macmillan, New York.

Kenney, T. C. (1959), "Discussion" *Journal of the Soil Mechanics and Foundations Division*, A.S.C.E., Vol. 85, No. SM3, pp. 67–79.

Kenney, T. C. (1967), "The Influence of Mineralogical Composition on the Residual Strength of Natural Soils," *Proceedings of the Oslo Geotechnical Conference on the Shear Strength Properties of Natural Soils and Rocks*, Vol. I, pp. 123–129.

Kenney, T. C., and Chan, H. T. (1972), "Use of Radiographs in a Geological and Geotechnical Investigation of Varved Soil," *Canadian Geotechnical Journal*, Vol. 9, pp. 195–205.

Kenney, T. C., and Chan, H. T. (1973), "Field Investigation of Permeability Ratio of New Liskeard Varved Soil," *Canadian Geotechnical Journal*, Vol. 10, pp. 473–488.

Kenney, T. C., Moum, J., and Berre, T. (1967), "An Experimental Study of Bonds in a Natural Clay," *Proceedings of the Geotechnical Conference, Oslo, on Shear Strength of Natural Soils and Rocks*, Vol. 1, pp. 65–69.

Kersten, M. S. (1949), Final Report, Laboratory Research for the Determination of the Thermal Properties of Soils, Engineering Experiment Station, University of Minnesota.

Kidder, G., and Reed, L. W. (1972), "Swelling Characteristics of Hydroxy-Aluminum Interlayered Clays," *Clays and Clay Minerals*, Vol. 20, pp. 13–20.

Kimoto, S. and Russ, J. C. (1969), "The Characteristics and Applications of the Scanning Microscope," *Materials Research and Standards*, January, 1969, pp. 8–16.

King, F. H. (1898), "Principles and Conditions of the Movement of Groundwater," 19th Annual Report, U.S. Geological Survey, Part 2, pp. 59–294.

Kirkpatrick, W. M., and Rennie, I. A. (1973), "Clay Structure in Laboratory Prepared Samples," *Proceedings of the International Symposium of Soil Structure, Gothenburg, Sweden*, pp. 103–111.

Kirkwood, J. G. (1954), "Transport of Ions Through Biological Membranes from the Standpoint of Irreversible Thermodynamics," *Ion Transport Across Membranes* (H. Clarke, Ed.), Academic, New York.

Kittrick, J. A. (1965), "Electron Microscope Techniques," Chapter 47 and "Electron Diffraction Techniques for Mineral Identification," Chapter 48, *Methods of Soil Analysis*, Agronomy No. 9, American Society of Agronomy, Madison, Wisc.

Klug, H. P., and Alexander, L. E. (1954), *X-Ray Diffraction Procedures*, Wiley, New York.

Kolaian, J. H., and Low, P. F. (1960), "Thermodynamic Properties of Water in Suspensions of Montmorillonite," *Clays and Clay Minerals*, Vol. 9, pp. 71–84.

Kolb, C. R., and Shockley, W. G. (1957), "Mississippi Valley Geology—Its Engineering Significance," *Journal of Soil Mechanics and Foundations Division,* A.S.C.E., Vol. 83, No. SM3, pp. 1–14.

Komamura, F., and Huang, R. J. (1974), "A New Rheological Model for Soil Behavior," *Journal of the Geotechnical Engineering Division,* A.S.C.E., Vol. 100, No. GT7, pp. 807–824.

Kondner, R. L., and Vendrell, J. R. Jr., (1964), "Consolidation Coefficients: Cohesive Soil Mixtures, *Journal of Soil Mechanics and Foundations Division,* A.S.C.E., Vol. 90, No. SM5, pp. 31–42.

Kononova, M. M. (1961), *Soil Organic Matter* (Translated from the Russian by T. Z. Nowakowski and G. A. Greenwood, Pergamon, London.

Kozeny, J. (1927), "Ueber kapillare Leitung des Wassers im Boden," *Wien, Akad. Wiss.,* Vol. 136, Pt. 2a, p. 271.

Krinitzky, E. L., and Turnbull, W. J. (1967), "Loess Deposits of Mississippi," GSA Special Paper No. 94, New York.

Krinsley, D. H., and Smalley, I. J. (1973), "Shape and Nature of Small Sedimentary Quartz Particles," *Science,* Vol. 180, pp. 1277–1279, 22 June 1973.

Krumbein, W. C., and Sloss, L. L. (1951), *Stratigraphy and Sedimentation,* Freeman, San Francisco.

Kruyt, H. R. (1052), *Colloid Science, Vol. I—Irreversible Systems,* Elsevier, Amsterdam, New York, London.

Kuenen, Ph.H. (1959), "Sand—Its Origin, Transportation, Abrasion and Accumulation," Alex, L. du Toit Memorial Lectures No. 6, The Geological Society of South Africa, Annexure to vol. LXII, 33 pp.

Kunze, G. W. (1965), "Pretreatment for Mineralogical Analysis," *Methods of Soil Analysis,* Agronomy No. 9, Chapter 44, American Society of Agronomy, Madison, Wisc.

Lacerda, W. A., and Houston, W. N. (1973), "Stress Relaxation in Soils," *Proceedings of the Eighth International Conference Soil Mechanics and Foundation Engineering,* 1/34, Moscow, pp. 221–227.

Ladd, C. C. (1961), "Physico-Chemical Analysis of the Shear Strength of Saturated Clays," Sc.D. Thesis, M.I.T.

Ladd, C. C., and Martin, R. T. (1967), "The Effects of Pore Fluid on the Undrained Strength of Kaolinite," M.I.T. Civil Engineering Research Report R67-15.

Ladd, C. C., and Wissa, A. E. Z. (1970), "Geology and Engineering Properties of Connecticut Valley Varved Clays with Special Reference to Embankment Construction," Department of Civil Engineering, M.I.T. Res. Rep. R70-56, Soils Publ. 264.

Ladd, R. S. (1974), "Specimen Preparation and Liquefaction of Sand," *Journal of the Geotechnical Division,* A.S.C.E., Vol. 100, No. GT10, pp. 1180–1184.

Lafeber, D., (1965), "The Graphical Representation of Planar Pore Patterns in Soils," *Australian Journal of Soil Research,* Vol. 3, pp. 143–164.

Lafeber, D. (1966), "Soil Structural Concepts," *Engineering Geology,* Vol. 1, No. 4, pp. 261–290.

Lafeber, D. (1967), "The Optical Determination of Spatial (Three Dimensional) Orientation of Platy Clay Minerals in Soil Thin-Sections," Geoderma, Vol. 1, No. 3/4, pp. 359–369.

Lafeber, D. (1969), "Discussion of Morgenstern and Tchelenko (1967b)," *Geotechnique,* Vol. 18, No. 3, pp. 379–382.

Lafeber, D., and Kurbanovic (1965), "Photographic Reproductions of Soil Fabric Patterns," *Nature,* Vol. 208, No. 5010, pp. 609–610.

Lafeber, D. and Willoughby, D. R. (1971), "Fabric Symmetry and Mechanical Anisotropy in Natural Soils," *Proceedings of the Australia-New Zealand Conference on Geomechanics, Melbourne, 1971,* Vol. 1, pp. 165–174.

Lahav, N., and Bresler, E. (1973), "Exchangeable Cation-Structural Parameter Relationships in Montmorillonite," *Clays and Clay Minerals,* Vol. 21, pp. 249–255.

Lambe, T. W. (1949), "How Dry is a Dry Soil?," *Proceedings of the Highway Research Board,* Vol. 29, pp. 491–496.

Lambe, T. W. (1952), "Differential Thermal Analysis," *Proceedings of the Highway Research Board,* Vol. 21, pp. 620–641.

Lambe, T. W. (1953), "The Structure of Inorganic Soil," A.S.C.E. Separate No. 315, October.

Lambe, T. W. (1954a), "The Permeability of Fine-Grained Soils," A.S.T.M. Special Publication 163, pp. 56–67.

Lambe, T. W. (1954b), "The Improvement of Soil Properties with Dispersants," *Journal of the Boston Society of Civil Engineers,* April 1954.

Lambe, T. W. (1958), "The Structure of Compacted

Clay," *Journal of the Soil Mechanics and Foundations Division*, A.S.C.E., Vol. 84, No. SM2, p. 34.

Lambe, T. W. (1960), "A Mechanistic Picture of Shear Strength in Clay," *Proceedings of the A.S.C.E. Research Conference on the Shear Strength of Cohesive Soils*, p. 437.

Lambe, T. W., and Martin, R. T. (1953–1957), "Composition and Engineering Properties of Soil," Highway Research Board Proceedings, I-1953, II-1954, III-1955, IV-1956, V-1957.

Lambe, T. W., and Whitman, R. V. (1969), *Soil Mechanics*, Wiley, New York, 553 pp.

Langston, J. E., and Lee, G. B. (1965), "Preparation of Thin Sections for Moist Organic Soil Materials," *Soil Science Society of America, Proceedings*, pp. 221–223.

Larsen, E. J. (1964), "The New Classification," *Soil Conservation*, Vol. 30, No. 5.

Lee, I. K. (1966), "Stress-Dilatancy Performance of Feldspar," *Journal of Soil Mechanics and Foundations Division*, A.S.C.E., Vol. 92, No. SM2.

Lee, K. L. (1975), "Formation of Adhesion Bonds at High Pressure," University of California at Los Angeles, Engineering Report 7586.

Lee, K. L., and Farhoomand, I. (1967), "Compressibility and Crushing of Granular Soil in Anisotropic Compression," *Canadian Geotechnical Journal*, Vol. IV, No. 1, pp. 68–86.

Lee, K. L., and Seed, H. B. (1967a), "Cyclic Stress Conditions Causing Liquefaction of Sand," *Journal Soil Mechanics and Foundations Division*, A.S.C.E., Vol. 93, No. SM1, pp. 47–70.

Lee, K. L., and Seed, H. B. (1967b), "Dynamic Strength of Anisotropically Consolidated Sand," *Journal of Soil Mechanics and Foundations Division*, A.S.C.E., Vol. 93, No. SM5, pp. 169–190.

Leonards, G. A., and Ramiah, B. K. (1960), "Time Effects in the Consolidation of Clays," ASTM Special Technical Publication 254, pp. 116–130.

Letey, J., and Kemper, W. D. (1969), "Movement of Water and Salt Through a Clay–Water System: Experimental Verification of Onsager Reciprocal Relation," *Soil Science Society of America, Proceedings*, Vol. 33, pp. 25–29.

Lewis, R. W., and Humpheson, C. (1973), "Numerical Analysis of Electro-Osmotic Flow in Soils," *Journal of the Soil Mechanics and Foundations Division*, A.S.C.E., Vol. 99, No. SM8, pp. 603–616.

Lifshitz, E. M. (1955), *Zhur. Eksptl, i Teoret.* (J.E.T.P.) Fig. 29, p. 94.

Linell, K. A., et al. (1963), "Corps of Engineers Pavement Design in Areas of Seasonal Frost," *Highway Research Record*, No. 33, pp. 76–128.

Lo, K. Y. (1969a), "The Pore Pressure-Strain Relationships of Normally Consolidated Undisturbed Clays. Part I. Theoretical Considerations," *Canadian Geotechnical Journal*, Vol. 7, pp. 383–394.

Lo, K. Y. (1969b), "The Pore Pressure-Strain Relationship of Normally Consolidated Undisturbed Clays. Part II. Experimental Investigation and Practical Applications," *Canadian Geotechnical Journal*, Vol. 7, pp. 395–412.

London, F. (1937), "The General Theory of Molecular Forces," *Transactions of the Faraday Society*, Vol. 33, No. 8.

Long, E., and George W. (1967), "Turnagain Slide Stabilization, Anchorage, Alaska," *Journal of the Soil Mechanics and Foundations Division*, A.S.C.E., Vol. 93, No. SM4, pp. 611–627.

Low, P. F. (1961), "Physical Chemistry of Clay-Water Interaction," *Advances in Agronomy*, Vol. 13, pp. 269–327, Academic, New York.

Low, P. F. (1968), "Mineralogical Data Requirements in Soil Physical Investigations," *Minerology in Soil Science and Engineering*, Soil Science Society of America, Spec. Publ. No. 3, pp. 1–34.

Low, P. F., and White, J. L. (1970), "Hydrogen Bonding and Polywater in Clay–Water Systems," *Clays and Clay Minerals*, Vol. 18, No. 1.

Lutton, R. J. (1969), "Fracture and Failure Mechanics in Loess and Applications to Rock Mechanics," Research Report S-69-1, U.S. Army Engineer Waterways Experiment Station, Vicksburg, Miss.

Lutz, J. F., and Kemper, W. D. (1959), "Intrinsic Permeability of Clay as Affected by Clay–Water Interaction," *Soil Science*, Vol. 88, pp. 83–90.

Mackenzie, R. C. (1958), "Density of Water Sorbed on Montmorillonite," *Nature*, Vol. 181, p. 334.

Mahmood, A. (1973), "Fabric-Mechanical Property Relationships in Fine Granular Soils," Ph.D. Dissertation, University of California, Berkeley.

Mahmood, A. and Mitchell, J. K. (1974), "Fabric-Property Relationships in Fine Granular Materials," *Clays and Clay Minerals*, Vol. 22, No. 5/6, pp. 397–408.

Marsal, R. H. (1973), "Mechanical Properties of

Rockfill," *Embankment Dam Engineering,* Casagrande Volume, pp. 109–208, Wiley, New York.

Marshall, C. E. (1964), *The Physical Chemistry and Mineralogy of Soils,* Vol. 1—Soil Materials, Wiley, New York.

Martin, R. T. (1955), "Ethylene Glycol Retention by Clays," *Proceedings of the Soil Science Society of America,* Vol. 19, pp. 160–164.

Martin, R. T. (1959), "Rhythmic Ice Banding in Soil," *Highway Research Board Bulletin,* 218, pp. 11–23.

Martin, R. T. (1960), "Adsorbed Water on Clay: A Review," *Clays and Clay Minerals,* Vol. 9, pp. 28–70.

Martin, R. T. (1966), "Quantitative Fabric of Wet Kaolinite," *Fourteenth National Clay Conference,* pp. 271–287.

Martin, R. T., and Wissa, A. E. Z. (1972), "Mechanism of Frost Action," Department of Civil Engineering, M.I.T., Cambridge, Mass.

Matalucci, R. V., Abdel-Hady, M., and Shelton, J. W. (1970*a*), "Influence of Grain Orientation on Direct Shear Strength of a Loessial Soil," *Engineering Geology,* Vol. 4, pp. 121–132.

Matalucci, R. V., Abdel-Hady, M., and Shelton, J. W. (1970*b*), "Influence of Microstructure of Loess on Triaxial Shear Strength," *Engineering Geology,* Vol. 4, pp. 341–351.

Matalucci, R. V., Shelton, J. W., and Abdel-Hady, M. (1969), "Grain Orientation in Vicksburg Loess," *Journal of Sedimentary Petrology,* Vol. 39, No. 3, pp. 969–979.

Matsuo, S. (1957), "A Study of the Effect of Cation Exchange on the Stability of Slopes," *Proceedings of the Fourth International Conference on Soil Mechanics and Foundations Engineering,* Vol. II, pp. 330–333.

Matuo, S-I., and Kamon, M. (1973), "Microscopic Research on the Consolidated Samples of Clayey Soils," *Proceedings of the International Symposium of Soil Structure, Gothenburg,* Sweden, pp. 194–203.

McCormick, G. (1971), "Estimation of Design Freezing Indices," *Transportation Engineering Journal,* A.S.C.E., Vol. 97, No. TE3, pp. 401–409.

McCrone, W. C. and Delly, J. G. (1973), *The Particle Atlas, Edition Two, Volume I, Principles and Techniques,* Ann Arbor Science, Ann Arbor, 296 pp.

McEwan, D. M. C. (1953), *Clay Minerals Bulletin,* No. 2, p. 73.

McGown, A. (1973), "The Nature of the Matrix in Glacial Ablation Tills," *Proceedings of the International Symposium on Soil Structure, Gothenburg, Sweden,* pp. 87–96.

McKyes, E., and Yong, R. N. (1971), "Three Techniques for Fabric Viewing as Applied to Shear Distortion of a Clay," *Clays and Clay Minerals,* Vol. 19, pp. 289–293.

McNeal, B. L. (1970), "Prediction of Interlayer Swelling of Clays in Mixed-Salt Solutions," *Soil Science America, Proceedings,* Vol. 34, pp. 201–206.

McNeal, B. L., Norwell, W. A., and Coleman, N. T. (1966), "Effect of Solution Composition on the Swelling of Extracted Soil Clays," *Soil Science Society America, Proceedings,* Vol. 30, pp. 313–317.

Meade, R. H. (1961), "X-ray Diffractometer Methods for Measuring Preferred Orientation in Clays," U.S. Geological Survey Paper 424B, pp. B273–B276.

Meade, R. H. (1964), "Removal of Water and Rearrangement of Particles During the Compaction of Clayey Sediments-Review," U.S. Geological Survey Professional Paper 497-B, U.S. Government Printing Office.

Mehra, O. P., and Jackson, M. L. (1960), "Iron Oxide Removal from Soils and Clays by a Dithionite-Citrate System with Sodium Bicarbonate Buffer," *Clays and Clay Minerals,* Vol. 7, pp. 317–327.

Mesri, G. and Olson, R. E. (1970), "Shear Strength of Montmorillonite," *Geotechnique,* Vol. 20, No. 3, pp. 261–270.

Mesri, G. and Olson, R. E. (1971), "Consolidation Characteristics of Montmorillonite," *Geotechnique,* Vol. 21, No. 4, pp. 341–352.

Michaels, A. S., and Lin, C. S. (1954), "The Permeability of Kaolinite," *Industrial Engineering Chemistry,* Vol. 46, pp. 1239–1246.

Mielenz, R. C., and King, M. E. (1955), "Physical–Chemical Properties and Engineering Performance of Clays," *California Division of Mines Bulletin* 169, pp. 196–254.

Millar, C. E., Turk, F. M., and Foth, H. D. (1965), *Fundamentals of Soil Science,* 4th ed., Wiley, New York.

Miller, R. H., and Low, P. F. (1963), "Threshold Gradient for Water Flow in Clay Systems," *Soil Science Society of America, Proceedings,* Vol. 27, No. 6, pp. 605–609.

Miller, R. H., Overman, A. R., and Peverly, J. H.

(1969), "The Absence of Threshold Gradients in Clay-Water Systems," *Soil Science of America, Proceedings,* Vol. 33, No. 2, pp. 183–187.

Mise, T. (1961), "Electro-Osmotic Dewatering of Soil and Distribution of Pore Water Pressure," *Proceedings of the Fifth International Conference Soil Mechanics and Foundation Engineering,* Vol. 1, pp. 255–258.

Mitchell, J. K. (1956), "The Fabric of Natural Clays and its Relation to Engineering Properties," *Proceedings of the Highway Research Board,* Vol. 35, pp. 693–713.

Mitchell, J. K. (1960), "Fundamental Aspects of Thixotropy in Soils," *Journal of the Soil Mechanics and Foundations Division, A.S.C.E.,* Vol. 86, No. SM3, pp. 19–52.

Mitchell, J K. (1962), "Components of Pore Water Pressure and Their Engineering Significance," *Proceeding of the Ninth National Clay Conference, Clays and Clay Minerals,* pp. 162–184.

Mitchell, J. K. (1964), "Shearing Resistance of Soils as a Rate Process," *Journal of the Soil Mechanics and Foundations Division, A.S.C.E.,* Vol. 90, No. SM1, pp. 29–61.

Mitchell, J. K., and Arulanandan, K. (1968), "Electrical Dispersion in Relation to Soil Structure," *Journal of the Soil Mechanics and Foundations Division, A.S.C.E.,* Vol. 94, No. SM2, pp. 447–471.

Mitchell, J. K., Campanella, R. G., and Singh, A. (1968), "Soil Creep as a Rate Process," *Journal of the Soil Mechanics and Foundations Division, A.S.C.E.,* Vol. 94, No. SM1, pp. 231–253.

Mitchell, J. K., Greenberg, J. A., and Witherspoon, P. A. (1973), "Chemico-Osmotic Effects in Fine-Grained Soils," *Journal of the Soil Mechanics and Foundations Division, A.S.C.E.,* Vol. 99, No. SM4, pp. 307–322.

Mitchell, J. K., Hooper, D. R., and Campanella, R. G. (1965), "Permeability of Compacted Clay," *Journal of the Soil Mechanics and Foundations Division, A.S.C.E.,* Vol. 91, No. SM4, pp. 41–65.

Mitchell, J. K, and McConnell, J. R. (1965), "Some Characteristics of the Elastic and Plastic Deformation of Clay on Initial Loading," *Proceedings of the Sixth International Conference on Soil Mechanics and Foundation Engineering,* Vol. 1, pp. 313–317.

Mitchell, J. K., Singh, A., and Campanella, R. G. (1969), "Bonding, Effective Stresses, and Strength of Soils," *Journal of the Soil Mechanics and Foundations Division, A.S.C.E.,* Vol. 95, No. SM5, pp. 1219–1246.

Mitchell, J. K., and Woodward, R. J. (1973), "Clay Chemistry and Slope Stability," *Journal of the Soil Mechanics and Foundations Division, A.S.C.E.,* Vol. 99, No. SM10.

Mitchell, J. K., and Younger, J. S. (1967), "Abnormalities in Hydraulic Flow Through Fine-Grained Soils," *A.S.T.M. Special Technical Publication 417,* pp. 106–141.

Moffatt, W. G., Pearsall, G. W. and Wulff, J. (1965), *The Structure and Properties of Materials,* Vol. 1, Structure, Wiley, New York.

Mooney, R. W., Keenan, A. G., and Wood, L. A. (1951), "Adsorption of Water Vapor on Montmorillonite, II," *Journal of the American Chemical Society,* Vol. 74, p. 1371.

Moore, C. A. (1968), "Quantitative Analysis of Naturally Occurring Multicomponent Mineral Systems by X-Ray Diffraction," *Clays and Clay Minerals,* Vol. 16, pp. 325–336.

Moore, C. A. (1971), "Effect of Mica on K_0 Compressibility of Two Soils," *Journal of the Soil Mechanics and Foundations Division, A.S.C.E.,* Vol. 97, No. SM9, pp. 1275–1291.

Moore, C. A., and Mitchell, J. K. (1974), "Electromagnetic Forces and Soil Strength, *Geotechnique,* Vol. 24, No. 4, pp. 627–640.

Morgenstern, N. R., and Tchalenko, J. S. (1967a), "The Optical Determination of Preferred Orientation in Clays and its Application to the Study of Microstructure in Consolidated Kaolin," *Proceedings of the Royal Society, London,* A300, I: pp. 218–234, II: pp. 235–250.

Morgenstern, N. R., and Tchalenko, J. S. (1967b), "Microscopic Structures in Kaolin Subjected to Direct Shear," *Geotechnique,* Vol. 17, No. 4, pp. 309–328.

Morgenstern, N. R., and Tchalenko, J. S. (1967c), "Microstructural Observations on Shear Zones from Slips in Natural Clays," *Proceedings of the Geotechnical Conference, Oslo,* Vol. 1, pp. 147–152.

Mortland, M. M., and Kemper, W. D. (1965), "Specific Surface," *Methods of Soil Analysis,* Agronomy No. 9, Chapter 42, American Society of Agronomy, Madison, Wisc.

Moum, J., Löken, T., and Torrance, J. K. (1971), "A Geotechnical Investigation of the Sensitivity of a Normally Consolidated Clay from Drammen, Norway," *Geotechnique,* Vol. 21, No. 4, pp. 329–340.

Moum, J., and Rosenqvist, I. Th. (1957), "On the Weathering of Young Marine Clay," *Proceedings of the Fourth International Conference on Soil Mechanics and Foundation Engineering*, London, Vol. 1, pp. 77–79.

Moum, J., and Rosenqvist, I. Th. (1961), "The Mechanical Properties of Montmorillonitic and Illitic Clays Related to the Electrolytes of the Pore Water," *Proceedings of the Fifth International Conference on Soil Mechanics and Foundation Engineering*, Vol. 1, pp. 263–267.

Müller, G. (1967), *Methods in Sedimentary Petrology*, Section on degree of roundness according to Russel-Taylor-Pettijohn (after Schneiderholm) pp. 100–101, Hafner, New York.

Murayama, S. (1969), "Effect of Temperature on Elasticity of Clays," Highway Research Board Special Report 103, pp. 194–203.

Murayama, S., and Shibata, T. (1958), "On the Rheological Characteristics of Clays," Part I, Bulletin No. 26, Disaster Prevention Research Institute, Kyoto, Japan.

Murayama, S., and Shibata, T. (1961), "Rheological Properties of Clays," *Proceedings of the Fifth International Conference on Soil Mechanics and Foundation Engineering*, Vol. 1, pp. 269–273.

Murayama, S., and Shibata, T. (1964), "Flow and Stress Relaxation of Clays (Theoretical Studies on the Rheological Properties of Clay—Part I)," *Rheology and Soil Mechanics Symposium of the International Union of Theoretical and Applied Mechanics*, *Grenoble*.

Nayak, N. V., and Christensen, R. W. (1971), "Swelling Characteristics of Compacted Expansive Soils," *Clays and Clay Minerals*, Vol. 19, pp. 251–261.

Nicholls, R. L., and Herbst, R. L., Jr. (1967), "Consolidation Under Electrical Pressure Gradients," *Journal of the Soil Mechanics and Foundations Division*, A.S.C.E., Vol. 93, No. SM5, pp. 139–151.

Noble, C. A., and Demirel, T. (1969), "Effect of Temperature on the Strength Behavior of Cohesive Soil," Highway Research Board Special Report 103, pp. 204–219.

Noorany, I. and Gizienski, S. F. (1970), "Engineering Properties of Submarine Soils: State-of-the-Art Review," *Journal of the Soil Mechanics and Foundations Division*, A.S.C.E., Vol. 96, No. SM5, pp. 1735–1762.

Norman, L. E. J. (1958), "A Comparison of Values of Liquid Limit Determined with Apparatus with Bases of Different Hardness," *Geotechnique*, Vol. 8, No. 1, pp. 70–91.

Norris, G. (1975), "The Effect of Particle Size and the Natural Variation in Particle Shape and Surface Roughness on the Stress–Strain and Strength Behavior of Uniform Quartz Sands," Ph.D. Thesis, University of California, Berkeley (in progress).

Norrish, K. (1954), "The Swelling of Montmorillonite," *Faraday Society, London, Discussions*, No. 18, pp. 120–134.

Oakes, D. T. (1960), "Solids Concentration Effects in Bentonite Drilling Fluids," *Clays and Clay Minerals*, Vol. 8, pp. 252–273.

O'Brien, N. R. (1971), "Fabric of Kaolinite and Illite Floccules," *Clays and Clay Minerals*, Vol. 19, pp. 353–359.

Oda, M. (1972a), "Initial Fabrics and Their Relations to Mechanical Properties of Granular Material," *Soils and Foundations*, Vol. 12, No. 1, pp. 17–37.

Oda, M. (1972b), "The Mechanism of Fabric Changes During Compressional Deformation of Sand," *Soils and Foundations*, Vol. 12, No. 2, pp. 1–18.

Oda, M. (1972c), "Deformation Mechanism of Sand in Triaxial Compression Tests," *Soils and Foundations*, Vol. 12, No. 4, pp. 45–63.

Odell, R. T., Thornburn, T. H., and McKenzie, L. J. (1960), "Relationships of Atterberg Limits to Some other Properties of Illinois Soils," *Proceedings of the Soil Science Society of America*, Vol. 24, No. 4, pp. 297–300.

Olsen, H. W. (1961), "Hydraulic Flow Through Saturated Clays," Sc.D. Thesis, M.I.T., Cambridge, Mass.

Olsen, H. W. (1962), "Hydraulic Flow Through Saturated Clays," *Proceedings of the Ninth National Conference on Clays and Clay Minerals*, pp. 131–161.

Olsen, H. W. (1965), "Deviations from Darcy's Law in Saturated Clay," *Soil Science Society of America, Proceedings*, pp. 135–140.

Olsen, H. W. (1969), "Simultaneous Fluxes of Liquid and Charge in Saturated Kaolinite," *Soil Science Society of America, Proceedings*, Vol. 33, No. 3, pp. 338–344.

Olsen, H. W. (1972), "Liquid Movement Through Kaolinite under Hydraulic, Electric and Osmotic Gradients," *The American Association of Petroleum Geologists Bulletin*, Vol. 56, No. 10, pp. 2022–2028.

Olson, R. E. (1974), "Shearing Strength of Kaolinite, Illite and Montmorillonite," *Journal of the Geotechnical Division,* A.S.C.E., Vol. 100, No. GT11, pp. 1215–1229.

Olson, R. E., and Mitronovas, F. (1962), "Shear Strength and Consolidation Characteristics of Calcium and Magnesium Illite," *Clays and Clay Minerals,* Vol. 11, pp. 185–209.

Olson, R. E., and Mesri, G. (1970), "Mechanisms Controlling the Compressibility of Clay," *Journal of the Soil Mechanics and Foundations Division,* A.S.C.E., Vol. 96, No. SM6, pp. 1863–1878.

Olson, R. V. (1965), "Iron," *Methods of Soil Analysis,* Agronomy No. 9, Chapter 65, American Society of Agronomy, Madison, Wisc.

Onsager, L. (1931), "Reciprocal Relations in Irreversible Processes," *Physical Review,* Vol. 37, pp. 405–426 and Vol. 38, pp. 2265–2279.

Osipov, J. B., and Sokolov, B. A. (1972), "Quantitative Characteristics of Clay Fabrics Using the Method of Magnetic Anisotropy," *Bulletin of the International Association of Engineering Geology,* No. 5, pp. 23–38.

Oster, J. D., and Low, P. F. (1964), "Heat Capacities of Clay and Water Mixtures," *Soil Science Society of America, Proceedings,* Vol. 28, pp. 605–609.

Osterman, J. (1964), "Studies on the Properties and Formation of Quick Clays," *Clays and Clay Minerals,* Monograph No. 19, Earth Science Series, pp. 87–108.

Paduana, J. A. (1965), "The Effect of Type and Amount of Clay on the Strength and Creep Characteristics of Clay-Sand Mixtures," Ph.D. Thesis, University of California, Berkeley.

Pauling, L. (1960), *The Nature of the Chemical Bond,* Cornell University Press, Ithaca, N.Y.

Peech, M. (1965), "Hydrogen-Ion Activity," *Methods of Soil Analysis,* Agronomy No. 9, Chapter 60, American Society of Agronomy, Madison, Wisc.

Penner, E. (1959), "The Mechanism of Frost Heaving in Soils," Highway Research Board Bulletin 225, pp. 1–13.

Penner, E. (1963a), "Anisotropic Thermal Conduction in Clay Sediments," *Proceedings of the International Clay Conference,* Vol. 1, pp. 365–376.

Penner, E. (1963b), "Sensitivity in Leda Clay," *Nature,* Vol. 197, No. 4865, pp. 347–348.

Penner, E. (1963c), "Anisotropic Thermal Conduction in Clay Sediments," *Proceedings of the International Clay Conference,* Vol. 1, pp. 365–376.

Penner, E. (1963d), " The Nature of Frost Heaving in Soils," *Proceedings of the International Conference of Permafrost,* NAS-NRC publication 1287.

Penner, E. (1964), "Studies of Sensitivity and Electro-Kinetic Potential in Leda Clay," *Nature,* Vol. 204, No. 4960, pp. 808–809.

Penner, E. (1965), "A Study of Sensitivity in Leda Clay," *Canadian Journal of Earth Sciences,* Vol. 2, No. 5, pp. 425–441.

Pettijohn, F. J. (1949), *Sedimentary Rocks,* Harper, New York.

Pickett, A. G., and Lemcoe, M. M. (1959), "An Investigation of Sear Strength of the Clay–Water System by Radio-Frequency Spectroscopy," *Journal of Geophysical Research,* Vol. 64, pp. 1579–1586.

Plum, R. L., and Esrig, M. I. (1969), "Some Temperature Effects on Soil Compressibility and Pore Water Pressure," HRB Special Report 103, pp. 231–242.

Polivka, M., and Best, C. (1960), "Investigation of the Problem of Creep in Concrete by Dorn's Method," University of California, Berkeley, Department of Civil Engineering.

Pople, J. A. (1951), "Molecular Association in Liquids II. A Theory of the Structure of Water," *Proceedings of the Royal Society,* London, A205, pp. 163–178.

Pratt, P. F. (1965), "Potassium," Methods of Soil Analysis, Agronomy No. 9, Chapter 71, American Society of Agronomy, Madison, Wisc.

Procter, D. C., and Barton, R. R. (1974), "Measurements of the Angle of Interparticle Friction," *Geotechnique,* Vol. 24, No. 4, pp. 581–604.

Pusch, R. (1966a), "Investigation of Clay Microstructure by Using Ultra-Thin Sections," *Swedish Geotechnical Institute Preliminary Report,* No. 15, Stockholm.

Pusch, R. (1966b), "Quick-Clay Microstructure," *Engineering Geology,* Vol. 1, No. 6, p. 433.

Pusch, R. (1970), "Microstructural Changes in Soft Quick Clay at Failure," *Canadian Geotechnical Journal,* Vol. 7, No. 1, pp. 1–7.

Pusch, R. (1973a), "Influence of Salinity and Organic Matter on the Formation of Clay Microstructure," *Proceedings of the International Symposium on Soil Structure, Gothenburg, Sweden,* pp. 161–173.

Pusch, R. (1973*b*), "Physico-Chemical Processes which Affect Soil Structure and Vice Versa," *Appendix to the Proceedings of the International Symposium on Soil Structure, Gothenburg, Sweden,* pp. 27–35.

Pusch, R. (1973*c*), "Structural Variations in Boulder Clay," *Proceedings of the International Symposium on Soil Structure, Gothenburg, Sweden,* pp. 113–121.

Pusch, R., and Arnold, M. (1969), "The Sensitivity of Artificially Sedimented, Organic-Free Illitic Clay," *Engineering Geology,* Vol. 3, No. 2, pp. 135–148.

Quigley, R. M., and Thompson, C. D. (1966). "The Fabric of Anisotropically Consolidated Sensitive Marine Clay," *Canadian Geotechnical Journal,* Vol. III, No. 2, pp. 61–73.

Radforth, N. W., and MacFarlane, I. C. (1957), "Correlation of Paleo-botanical and Engineering Studies of Muskeg (Peat) in Canada," *Proceedings of the Fourth International Conference on Soil Mechanics and Foundation Engineering,* Vol. 1, pp. 93–97.

Ranganatham, B. V., and Satyanarayana, B. (1965), "A Rational Method of Predicting Swelling Potential for Compacted Expansive Clays," *Proceedings of the Sixth International Conference Soil Mechanics and Foundation Engineering,* Vol. 1, pp. 92–96.

Ravina, I., and Low P. F. (1972), "Relation Between Swelling, Water Properties, and *b*-Dimension in Montmorillonite-Water Systems," *Clays and Clay Minerals,* Vol. 20, pp. 109–123.

Ravina, I., and Zaslavsky, D. (1972), "The Water Pressure in the Electrical Double Layer," *Israel Journal of Chemistry,* Vol. 10, pp. 707–714.

Ree, R. and Eyring, H. (1958), "The Relaxation Theory of Transport Phenomena," *Rheology* (F. R. Eirich, Ed.), Vol. II, Chapt. 3, Academic, New York.

Reiche, P. (1945), "A Survey of Weathering Processes and Products," *University of New Mexico, Publication in Geology,* No. 1, Albuquerque, N.M.

Reinheimer, G. (1971), *Mikrobiologic der Gerwasser,* VEB Gustav Fischer Verlag, Jena.

Rich, C. I. (1968), "Hydroxy Interlayers in Expansible Layer Silicates," *Clays and Clay Minerals,* Vol. 16, pp. 15–30.

Richards, B. G. (1967), "A Review of Methods for the Determination of the Moisture Flow Properties of Unsaturated Soils," Tech. Memo. No. 5, C.S.I.R.O., Melbourne.

Richards, B. G. (1969), "Psychrometric Techniques for Measuring Soil Water Potential," Tech. Rept. No. 9, C.S.I.R.O., Melbourne

Richards, B. G. (1974), "Behavior of Unsaturated Soils," Chapter 4 of *Soil Mechanics-New Horizons* (I. K. Lee, Ed.), Elsevier, New York.

Richards, L. A. Ed. (1954), *Diagnosis and Improvement of Saline and Alkali Soils,* Agricultural Handbook, No. 60, U.S. Government Printing Office, Washington, D.C., 160 pp.

Rieke, H. H., III, and Chilingarian, G. V. (1974), *Compaction of Argillaceous Sediments,* Elsevier, New York, 424 pp.

Ripple, C. D., and Day, P. R. (1966), "Suction Responses Due to Shear of Dilute Montmorillonite-Water Pastes," *Clays and Clay Minerals,* Vol. 14, pp. 307–316.

Roscoe, K. H., Schofield, A. N., and Wroth, C. P. (1958), "On the Yielding of Soils," *Geotechnique,* Vol. 8, pp. 22–53.

Rosenqvist, I. Th. (1946), "Om Leirers Kvikkaktighet," *Meddelelsen fra Vegdirektoren,* No. 3, pp. 29–36.

Rosenqvist, I. Th. (1953), "Considerations on the Sensitivity of Norwegian Quick Clays," *Geotechnique,* Vol. III, No. 5, pp. 195–200.

Rosenqvist, I. Th. (1955), "Investigations in the Clay-Electrolyte-Water System," Norwegian Geotechnical Institute, Publication No. 9, Oslo, 125 p.

Rosenqvist, I. Th. (1959), "Physico-Chemical Properties of Soils: Soil Water Systems," *Journal of the Soil Mechanics and Foundations Division,* A.S.C.E., Vol. 85, No. SM2, Part 1, pp. 31–53.

Ross, C. S., and Hendricks, S. B. (1945), "Minerals of the Montmorillonite Group," U.S. Geological Survey Professional Paper 205B.

Rowe, P. W. (1962), "The Stress-Dilatancy Relation for Static Equilibrium of an Assembly of Particles in Contact," *Proceedings of the Royal Society,* Vol. A269, pp. 500–527.

Rowe, P. W. (1972), "The Relevance of Soil Fabric to Site Investigation Practice," *Geotechnique,* Vol. 22, No. 2, pp. 195–300.

Rowe, P. W. (1973), "Stress-Strain Relationships for Particulate Materials at Equilibrium," *Proceedings of A.S.C.E. Specialty Conference on Performance of Earth and Earth-Supported Structures,* Vol. 3, pp. 327–357.

Roza, S. A., and Kotov, A. L. (1958), "Experimental Studies of the Creep of Soil Skeletons," *Zapiski Leningradskogo ordenor Lerrinai Trudovovo Krasmorro Znamemo gornono instituta um G. A. Plakhanova,* Vol. 34, No. 2, pp. 203–213.

Samuels, S. G. (1950), "The Effect of Base Exchange on the Engineering Properties of Soils," Building Research Station, Note C176, Great Britain.

Sangrey, D. (1970), "Discussion of Houston and Mitchell (1969), *"Journal of the Soil Mechanics and Foundations Division,* A.S.C.E., Vol. 96, No. SM3, pp. 1067–1070.

Saxen, U. (1892), in "Electrokinetic Potentials," *Textbook of Physical Chemistry* (S. Glasstone, Ed.) Van Nostrand, New York, Chapter XIV, p. 1225.

Schiffman, R. L. (1959), "The Use of Visco-Elastic Stress-Strain Laws in Soil Testing," ASTM Special Technical Pub. No. 254, pp. 131–155.

Schmertmann, J. H. (1969), "Swelling Sensitivity," *Geotechnique,* Vol. 19, No. 4, pp. 530–533.

Schmertmann, J. H., and Osterberg, J. O. (1960), "An Experimental Study of the Development of Cohesion and Friction with Axial Strain in Saturated Cohesive Soils," *Proceedings of the A.S.C.E. Research Conference on the Shear Strength of Cohesive Soils,* pp. 643–694.

Schmertmann, J. H. (1963), "Generalizing and Measuring Hoorslev's Effective Components of Shear Resistance," *Laboratory Shear Testing of Soils,* A.S.T.M. Spec. Tech. Pub. No. 361, pp. 147–162.

Schmid, G. (1950, 1911), "Zur Elektrochemie Feinporiger Kapillarsystems," *Zhurnal fur Elektrochemie,* Vol. 54, p. 425 Vol. 55, p. 684.

Schmidt, N. O. (1965), "A Study of the Isolation of Organic Matter as a Variable Affecting Engineering Properties of Soils," Ph.D. Thesis, University of Illinois, Urbana.

Schmidt, J. D., and Westmann, R. A. (1973), "Consolidation of Porous Media with Non-Darcy Flow," *Journal of the Engineering Mechanics Division,* A.S.C.E., Vol. 99, No. EM6, pp. 1201–1216.

Schweitzer, J, and Jennings, B. R. (1971), "The Association of Montmorillonite Studied by Light Scattering in Electric Fields," *Journal of Colloid Interface Science,* Vol. 37, pp. 443–457.

Scott, R. F. (1964), *Principles of Soil Mechanics,* Addison-Wesley, Reading, Mass.

Seed, H. B., and Chan, C. K. (1957), "Thixotropic Characteristics of Compacted Clays," *Journal of the Soil Mechanics and Foundations Division,* A.S.C.E., Vol. 83, No. SM4.

Seed, H. B., and Chan, C. K. (1959), "Structure and Strength Characteristics of Compacted Clays," *Journal of the Soil Mechanics and Foundations Division,* A.S.C.E., Vol. 85, No. SM5, pp. 87–128.

Seed, H B, and Lee, K. L. (1966), "Liquefaction of Saturated Sands During Cyclic Loading," *Journal Soil Mechanics and Foundations Division,* A.S.C.E., Vol. 92, No. SM6, pp. 105–134.

Seed, H B, Mitchell, J. K., and Chan, C. K. (1962), "Swell and Swell Pressure Characteristics of Compacted Clays," *Highway Research Board Bulletin* 313, pp. 12–39.

Seed, H. B., and Peacock, W. H. (1971), "Test Procedures for Measuring Soil Liquefaction Characteristics," *Journal of Soil Mechanics and Foundations Division,* A.S.C.E., Vol. 97, No. SM8, pp. 1099–1119.

Seed, H. B., Woodward, R. J., and Lundgren, R. (1962), "Prediction of Swelling Potential for Compacted Clays," *Journal Soil Mechanics and Foundations Division,* A.S.C.E., Vol. 90, No. SM4, pp. 107–131.

Seed, H. B., Woodward, R. J., and Lundgren, R. (1964a), "Clay Mineralogical Aspects of the Atterberg Limits," *Journal of Soil Mechanics and Foundations Division,* A.S.C.E., Vol. 90, No. SM4, pp. 107–131.

Seed, H. B., Woodward, R. J., and Lundgren, R. (1964b), "Fundamental Aspects of the Atterberg Limits," *Journal of Soil Mechanics and Foundations Division,* A.S.C.E., Vol. 90, No. SM6, pp. 75–105.

Selmer-Olsen, R. (1964), "Almendelig Geologi og Ingeniorgeologi," TAPIR, Trondheim.

Shainberg, I., Bresler, E., and Klausner, Y. (1971), "Studies on Na/Ca Montmorillonite Systems I. The Swelling Pressure," *Soil Science,* Vol. III, No. 4, pp. 214–219.

Shannon and Wilson, Inc. (1964), "Report on Anchorage Area Soil Studies, Alaska," to U. S. Army Engineer District, Anchorage, Alaska, Seattle, 300 p.

Sherard, J. L., Decker, R. S., and Ryker, N. L. (1972), "Piping in Earth Dams of Dispersive Clay," *Proceedings of the ASCE Speciality Conference on the Performance of Earth and Earth-Supported Structures,* Purdue University, pp. 589–626.

Sherard, J. L., Dunnigan, L. P., Decker, R. S., and Steele, E. F. (1976a), "Pinhole Test for Identifying

Dispersive Soils," *Journal of the Geotechnical Division*, ASCE, Vol. 102, No. GT 1, pp. 69–85.

Sherard, J. L., Dunnigan, L. P., and Decker, R. S. (1976b), "Identification and Nature of Dispersive Soil," *Journal of the Geotechnical Engineering Division*, ASCE, Vol. 102, No. GT 4, pp. 287–301.

Sherif, M. A., and Burrous, C. M. (1969), "Temperature Effects on the Unconfined Shear Strength of Saturated, Cohesive Soil," *Highway Research Board Special Report* 103, pp. 267–272.

Singh, A., and Mitchell, J. K. (1968), "General Stress-Strain-Time Function for Soils," *Journal of the Soil Mechanics and Foundations Division, A.S.C.E.*, Vol. 94, No. SM1, pp. 21–46.

Singh, A., and Mitchell, J. K. (1969), "Creep Potential and Creep Rupture of Soils," *Proceedings of the Seventh International Conference on Soil Mechanics and Foundation Engineering*, Vol. I, pp. 379–384.

Skempton, A. W. (1953), "The Colloidal Activity of Clay," *Proceedings of the Third International Conference on Soil Mechanics and Foundation Engineering*, Vol. I, pp. 57–61.

Skempton, A. W. (1960a), "Significance of Terzaghi's Concept of Effective Stress," in *From Theory to Practice in Soil Mechanics*, Wiley, New York, pp. 43–53.

Skempton, A. W. (1960b), "Effective Stress in Soils, Concrete, and Rocks," *Proceedings of the Conference on Pore Pressure and Suction in Soils*, Butterworths, London, pp. 4–16.

Skempton, A. W. (1964), "Long Term Stability of Clay Slopes," *Geotechnique*, Vol. 14, p. 77.

Skempton, A. W., and Henkel, D. J. (1953), "The Post-Glacial Clays of the Thames Estuary at Tilburg and Shellhaven," *Proceedings of the Third International Conference on Soil Mechanics and Foundation Engineering*, Vol. 1, p. 302.

Skempton, A. W., and Northey, R. D. (1952), "The Sensitivity of Clays," *Geotechnique*, Vol. 3, No. 1, pp. 30–53.

Sloane, R. C., and Kell, R. R. (1956), "The Fabric of Mechanically Compacted Kaolin," *Proceedings of the Fourteenth National Clay Conference, Clays and Minerals*, pp. 289–296.

Smalley, I. J., Cabrera, J. G., and Hammond, C. (1973), "Particle Nature in Sensitive Soils and Its Relation to Soil Structure and Geotechnical Properties," *Proceedings of the Symposium on Soil Structure, Gothenburg, Sweden*, pp. 184–193.

Smart, P. (1971), "Structure of a Red Clay Soil from Nyeri, Kenya," *Quarterly Journal of Engineering Geology*, Vol. 6, pp. 129–139.

Smart, P. (1973), "Statistics of Soil Structure in Electron Microscopy," *Proceedings of the International Symposium on Soil Structure, Gothenburg, Sweden*, pp. 69–76.

Smith, W. O., Foote, P. D., and Busany, P. F. (1929), "Packing of Homogeneous Spheres," *Physical Review*, Vol. 34, pp. 1271–1274.

Smoluckowski, M. (1914), *Handbuch der Elektrizitat und Magnetismus* (L. Graetz, Ed.), Vol. 2, J. A. Barth, Leipzig.

Smothers, W. J., and Chiang, Y. (1958), *Differential Thermal Analysis*, Chemical Publishing, New York.

Soderblom, R. (1966), "Chemical Aspects of Quick Clay Formation," *Engineering Geology*, 1, pp. 415–431.

Soderblom, R. (1969), "Salt in Swedish Clays and its Importance for Quick Clay Formation," *Swedish Geotechnical Proceedings*, No. 22, 63 pp.

Soil Survey Staff (1974), "Soil Taxonomy—A Basic System of Soil Classification for Making and Interpreting Soil Surveys," *Soil Conservation Service, U.S.D.A. Handbook* No. 436, U.S. Government Printing Office.

Sopp, O. I. (1964), "X-ray Radiography and Soil Mechanics: Localization of Shear Planes in Soil Samples," *Nature*, Vol. 202, p. 832.

Soveri, U. (1950), "Differential Thermal Analyses of Some Quaternary Clays of Fennoscandia," *Suomalaisen Tideakaternian Toimituksia, Annaks Academiae Scientiarium Fennicae. Se. A, III Geologica-Geographica 23*, 103 p.

Spiegler, K. S. (1958), "Transport Processes in Ionic Membranes," *Transactions of the Faraday Society*, Vol. 54, pp. 1408–1428.

Sridharan, A., Altschaeffl, A. G., and Diamond, S. (1971), "Pore Size Distribution Studies," *Journal of the Soil Mechanics and Foundations Division, A.S.C.E.*, Vol. 97, No. SM5, pp. 771–787.

Stern, O. (1924), "Zur Theorie der Elektrolytischen Doppelschriht," *Zietschrift Electrochem*, Vol. 30, pp. 508–516.

Stevenson, F. J. (1965), "Gross Chemical Fractionation of Organic Matter," Chapter 94, *Methods of Soil Analysis*, Agronomy No. 9, American Society of Agronomy, Madison, Wisc.

Straub, A. L., and Wegmann, F. J. (1965), "The Determination of Freezing Index Values," *Highway Research Record,* No. 68, pp. 17–30.

Sundaram, P. N., and Balasubramaniam, (1973), "Consolidation of a Bombay Marine Clay under Combined Surcharge and Electrical Gradient," *Journal of the Indian Geotechnical Society,* Vol. 3, Pt. 1, pp. 43–55.

Swartzendruber, D. (1962a), "Modification of Darcy's Law for the Flow of Water in Soils," *Soil Science,* Vol. 93, pp. 27–29.

Swartzendruber, D. (1962b), "Non-Darcy Behavior and Flow Behavior in Liquid-Saturated Porous Media," *Journal of Geophysical Research,* Vol. 67, No. 13, pp. 5205–5213.

Swartzendruber, D. (1963), "Non-Darcy Behavior and Flow of Water in Unsaturated Soils," *Soil Science of America, Proceedings,* pp. 491–495.

Swedish State Railways (1922), Geotechnical Commission 1914-1922, Final Report, Stockholm, Stat. Jarnvagar Medd., No. 2.

Taber, S. (1929), "Frost Heaving," *Journal Geology,* Vol. 37, pp. 428–461.

Taber, S. (1930), "The Mechanisms of Frost Heaving," *The Journal of Geology,* Vol. 38, pp. 303–317.

Taylor, D. W. (1948), *Fundamentals of Soil Mechanics,* Wiley, New York.

Tchalenko, J. S. (1968a), "The Microstructure of London Clay," *Quarterly Journal of Engineering Geology,* Vol. 1, No. 3, pp. 155–168.

Tchalenko, J. S. (1968b), "The Evolution of Kinkbands and the Development of Compression Textures in Sheared Clays," *Tectonophysics,* Vol. 6, pp. 159–174.

Terzaghi, Ch. (1920), "New Facts About Surface Friction," *The Physical Review,* N.S. Vol. XVI, No. 1, pp. 54–61; reprinted in *From Theory to Practice in Soil Mechanics,* pp. 165–172, Wiley, New York, 1961.

Terzaghi, K. (1925a), *Erdbaumechanik auf Bodenphysikalischer Grundlage,* Deuticke, Vienna, 399 pp.

Terzaghi, K. (1925b), "Simplified Soil Tests for Subgrades and Their Physical Significance," *Public Roads,* October.

Terzaghi, K. (1931), "The Influence of Elasticity and Permeability on the Swelling of Two-Phase Systems," in *Colloid Chemistry* (J. Alexander, Ed.), Vol. III, Chemical Catalog Co., New York. pp. 65–88.

Terzaghi, K. (1936), "The Shearing Resistance of Saturated Soils," *Proceedings of the First International Conference on Soil Mechanics,* Vol. 1, pp. 54–56.

Terzaghi, K. (1941), "Undisturbed Clay Samples and Undisturbed Clays," *Journal of the Boston Society of Civil Engineers,* Vol. 28, No. 3, pp. 211–231. Also in *Contributions to Soil Mechanics 1941–1953,* Boston, 1953, pp. 45–65.

Terzaghi, K. (1944), "Ends and Means in Soil Mechanics," *Engineering Journal of Canada,* Vol. 27, p. 608.

Thorp, J., and Smith, G. D. (1948), "Higher Categories of Soil Classification: Order, Suborder, and Great Soil Groups," *Soil Science,* Vol. 67, No. 2, pp. 118–126.

Titkov, N. I., et al. (1961), "Electrochemical Induration of Weak Rocks," Consultants Bureau, New York (Authorized translation from the Russian).

Torrance, J. K. (1974), "A Laboratory Investigation of the Effect of Leaching on the Compressibility and Shear Strength of Norwegian Marine Clays," *Geotechnique,* Vol. 24, No. 2, pp. 155–173.

Tovey, N. K. (1971), "A Selection of Scanning Electron Micrographs of Clays," CUED/C-SOILS/TR5a (1971), University of Cambridge, Department of Engineering.

Tovey, N. K. (1973a), "Techniques of Observation and Methods of Quantification," *Appendix to the Proceedings of the International Symposium on Soil Structure, Gothenburg, Sweden,* pp. 1–18.

Tovey, N. K. (1973b), "Quantitative Analysis of Electron Micrographs of Soil Structure," *Proceedings of the International Symposium of Soil Structure, Gothenburg, Sweden,* pp. 50–57.

Tovey, N. K., Frydman, S., and Wong, K. Y. (1973), "A Study of a Swelling Clay in the Scanning Electron Microscope," *Proceedings of the Third International Conference on Expansive Soils, Haifa, Israel.*

Tovey, N. K., and Wong, K. Y. (1973), "The Preparation of Soils and Other Geological Materials for the S.E.M.," *Proceedings of the International Symposium on Soil Structure, Gothenburg, Sweden,* pp. 59–67.

Trollope, D. H. (1960), "The Fabric of Clays in Relation to Shear Strength," *Proc. 3rd Australia-New Zealand Conference on Soil Mechanics and Foundation Eng.,* pp. 197–202.

Trollope, D. H. (1964), "On the Shear Strength of

Soils," *Proceedings of the Conference on Soil Stabilization, C.S.I.R.O., Melbourne, Australia.*

Trollope, D. H., Rosengren, K. J., and Brown, E. T. (1965), "The Mechanics of Brown Coal," *Geotechnique,* Vol. 15, No. 4, pp. 363–386.

Tschebotarioff, G. P., and Welch, J. D. (1948), "Lateral Earth Pressures and Friction Between Soil Minerals," *Proceedings of the Second International Conference Soil Mechanics and Foundation Engineering,* Vol. 7, p. 135.

United States Department of Agriculture (1951, 1962), Handbook 18, Supplement issued September 1962.

van Olphen, H. (1963), *An Introduction to Clay Colloid Chemistry,* Wiley Interscience, New York, 301 pp.

Verwey, E. J. W., and Overbeek, J. Th. G. (1948), *Theory of the Stability of Lyophobic Colloids,* Elsevier, Amsterdam, New York, London.

Vialov, S., and Skibitsky, A. (1957), "Rheological Processes in Frozen Soils and Dense Clays," *Proceedings of the Fourth International Conference Soil Mechanics and Foundation Engineering,* Vol. I, pp. 120–124.

von Engelhardt, W., and Tunn, W. L. M. (1955), "The Flow of Fluids Through Sandstones," (Translated by P. A. Witherspoon from Heidelberger Beitsage Zur Mineralogie und Petrographie), Vol. 2, pp. 12–25 (Illinois State Geologic Survey, Circular 194).

Waldron, L. H., and Manbeian, T. (1970), "Moisture Characteristics by Osmosis," *Soil Science,* October.

Wan, D. T. Y. (1970), "Consolidation of Soils by Electro-Osmosis," Ph.D. Dissertation, University of California, Berkeley.

Warkentin, B. P. (1961), "Interpretation of the Upper Plastic Limit of Clays," *Nature,* Vol. 190, pp. 287–288.

Warkentin, B. P., Bolt, G. H., and Miller, R. D. (1957), "Swelling Pressure of Montmorillonite," *Soil Science Society of America Proceedings,* Vol. 21, No. 5.

Warkentin, B. P., and Bozozuk, M. (1961), "Shrinking and Swelling Properties of Two Canadian Clays," *Proceedings of the Fifth International Conference Soil Mechanics and Foundation Engineering,* Vol. 1, pp. 851–855.

Warshaw, C. M., and Roy, R. (1961), "Classification and a Scheme for the Identification of the Layer Silicates," *Geological Society of America, Bulletin,* Vol. 72, pp. 1455–1492.

Weaver, C. E., and Pollard, L. D. (1973), *The Chemistry of Clay Minerals,* Developments in Sedimentology 15, Elsevier, Amsterdam, 213 pp.

Weymouth, J. E., and Williamson, W. O. (1953), "The Effects of Extrusion and Some Other Processes on the Micro-Structure of Clay," *American Journal of Science,* Vol. 251, p. 89.

White, W. A. (1955), "Water Sorption Properties of Homoionic Clay Minerals," Ph.D. Thesis, University of Illinois, Urbana.

Whittig, L. D. (1965), "X-Ray Diffraction Techniques for Mineral Identification and Mineralogical Composition," *Methods of Soil Analysis,* Agronomy No. 9, Chapter 49, American Society of Agronomy, Madison, Wisc.

Wilson, S. D. (1973), "Deformation of Earth and Rockfill Dams," *Embankment Dam Engineering,* Casagrande Volume, Wiley, New York, pp. 365–427.

Winterkorn, H. F., and Tschebotarioff, G. P. (1947), "Sensitivity of Clay to Remolding and its Possible Causes," *Proceedings of the Highway Research Board,* Vol. 27.

Wong, K. S., and Duncan, J. M. (1974), "Hyperbolic Stress–Strain Parameters for Nonlinear Finite Element Analyses of Stresses and Movements in Soil Masses," Report TE-73-4, Department of Civil Engineering, University of California, Berkeley.

Wu, T. H. (1960), "Geotechnical Properties of Glacial Lake Clays," *Transactions,* A.S.C.E., Vol. 125, p. 994.

Wu, T. H. (1964), "A Nuclear Magnetic Resonance Study of Water in Clay," *Journal of Geophysical Research,* Vol. 69, pp. 1083–1091.

Wu, T. H., Resindez, D., and Neukirchner, R. J. (1966), "Analysis of Consolidation by Rate Process Theory," *Journal of the Soil Mechanics and Foundations Division,* A.S.C.E., Vol. 92, No. SM6, pp. 229–248.

Yong, R. N., and Sheeran, D. E. (1973), "Fabric Unit Interaction and Soil Behavior," *Proceedings of the International Symposium on Soil Structure,* Gothenburg, Sweden, pp. 176–183.

Yong, R. N., and Warkentin, B. P. (1966), *Introduction to Soil Behavior,* Macmillan, New York.

Yoshinaka, R., and Kazama, H. (1973), "Microstructure of Compacted Kaolin Clay," *Soils and Foundations,* Vol. 13, No. 2, pp. 19–34.

Youd, T. L. (1973), "Liquefaction, Flow, and Associated Ground Failure," *Geological Survey Circular* 688, United States Department of the Interior, Washington, D.C., 12 pp.

Younger, J. S., and Lim, C. I. (1969), "An Investigation into the Flow Behavior through Compacted Saturated Fine-grained Soils with Regard to Fines Content and Over a Range of Applied Hydraulic Gradients," *IAHR International Symposium on the Fundamentals of Transport Phenomena in Flow Through Porous Media, Haifa, Israel.*

Author Index

Subject Index